T0134975

Environmental Footprints and Eco-design of Products and Processes

Series Editor

Subramanian Senthilkannan Muthu, Head of Sustainability - SgT Group and API, Hong Kong, Kowloon, Hong Kong

Indexed by Scopus

This series aims to broadly cover all the aspects related to environmental assessment of products, development of environmental and ecological indicators and eco-design of various products and processes. Below are the areas fall under the aims and scope of this series, but not limited to: Environmental Life Cycle Assessment; Social Life Cycle Assessment; Organizational and Product Carbon Footprints; Ecological, Energy and Water Footprints; Life cycle costing; Environmental and sustainable indicators; Environmental impact assessment methods and tools; Eco-design (sustainable design) aspects and tools; Biodegradation studies; Recycling; Solid waste management; Environmental and social audits; Green Purchasing and tools; Product environmental footprints; Environmental management standards and regulations; Eco-labels; Green Claims and green washing; Assessment of sustainability aspects.

Elena G. Popkova
Editor

Smart Green Innovations in Industry 4.0 for Climate Change Risk Management

 Springer

Editor
Elena G. Popkova
RUDN University
Moscow, Russia

ISSN 2345-7651 ISSN 2345-766X (electronic)
Environmental Footprints and Eco-design of Products and Processes
ISBN 978-3-031-28459-5 ISBN 978-3-031-28457-1 (eBook)
https://doi.org/10.1007/978-3-031-28457-1

This Springer imprint is published by the registered company Springer Nature Switzerland AG
The registered company address is: Gewerbestrasse 11, 6330 Cham, Switzerland

Introduction: Climate-Smart Economy and Business as a Vector for the Sustainable Development of Industry 4.0 in the Decade of Action

The environmental footprint of Industry 4.0 deserves special attention in the Decade of Action because, for the first time in the history of industrial revolutions, it is defined not so much by society and the economy as by technology. On the one hand, as a combination of high-tech market segments and the most advanced digital economic practices, Industry 4.0 opens up new opportunities for the development of green technologies. This brings the fight against climate change to the forefront, gaining intelligent decision support, green monitoring, and control automation in the economy and business of Industry 4.0.

The climate-smart economy and business is a promising vector for the sustainable development of Industry 4.0 in the Decade of Action. Climate risks have been considerably exacerbated in recent years by rapid economic growth, largely driven by industry and, therefore, characterized by high resource and energy intensity combined with large amounts of production and consumption waste. New opportunities for managing these risks are emerging in a climate-smart economy and business that relies on advanced Industry 4.0 technologies, including artificial intelligence (AI), big data, the Internet of Things, and many others.

There are four major trends in the development of the climate-smart economy and business in the Decade of Action. The first trend is the emergence and active development of a new institution—climate-responsible entrepreneurship. Its particularity is the integration of SDG 8, SDG 12, and SDG 13 into the corporate culture, development strategy, and corporate reporting on environmental responsibility and sustainable development.

The considered institution is formed based on the overall high level of climatic responsibility of today's society. The institution is supported by climate-responsive demand for Industry 4.0 products in progressive sustainable communities. Created green jobs allow employees to be environmentally responsible by participating in the creation and implementation of climate-smart innovations in support of the sustainable development of Industry 4.0.

The second trend is intensive creation, diffusion in the economy, and active use of climate-smart green innovations in all sectors of the digital economy. In each industry, this process has its own specifics and relies on its own application solutions. The development of clean energy to support the implementation of SDG 7 and climate-resilient agriculture to support the implementation of SDG 2 are especially important among the implemented practices.

The considered process is characteristic of high-tech segments of industry markets, including financial (FinTech), healthcare (MedTech), education (EdTech), energy (EnergyTech), agricultural (AgtoTech), and other markets. In these segments, the most competitive are climate-responsible business entities that practice smart responsible innovation in support of the sustainable development of Industry 4.0.

The third trend lies in managing climate risks through green innovation in smart regions and cities worldwide. Understanding regional specifics is essential for public management of the climate-smart economy and business based on sustainable communities to support the implementation of SDG 11, SDG 14, and SDG 15. There is now a wealth of international experience in climate risk management based on smart green innovations in the regions.

The fourth trend is the popularization of ESG management of climate risks and green finance in support of the fight against climate change. ESG management is a progressive management practice that balances and systemically achieves the priorities of social progress (improving the quality of life), accelerating economic growth and protecting the environment. This is achieved by relying on the climate-smart technologies of Industry 4.0.

Green finance, particularly ESG finance, involves environmental responsibility, including climate responsibility, in investment decisions and financing innovation in Industry 4.0. Safety for the environment is an additional criterion of efficiency, which allows us to shift the focus from the economy to nature in interpreting the rationality of decisions. With the COVID-19 pandemic and crisis, green investments have gained increased relevance because they support the implementation of SDG 3.

A shortcoming of the existing literature is that the described trends are studied in isolation. This causes a gap in the literature related to the vague outlines of the environmental footprints of Industry 4.0 and the eco-design of products and processes that rely on climate-smart innovation to support sustainable development. This book fills an identified gap in the literature through a comprehensive study of smart green innovations in Industry 4.0 for climate change risk management in the unity of the four trends noted.

The book aims to explore the theoretical and methodological basis, international experience, and prospects for developing a climate-smart economy and business as a vector of the sustainable development of Industry 4.0 in the Decade of Action. The research goal is consistently achieved in four parts of the book. The first part looks at climate-responsible entrepreneurship in support of the sustainable development of Industry 4.0. Particular attention is paid to eco-design and quality certification in implementing smart green innovations from the perspective of combating climate change.

The second part focuses on the systematization of best practices of climate-smart green innovations by sector of the digital economy. Issues of state and corporate governance are being worked through, considering the industry specifics of Industry 4.0. The third part highlights the experience of climate risk management based on smart green innovations in regions and countries. Particular attention is paid to the best practices of the European Union (EU) and the Eurasian Economic Union (EAEU). The final fourth part focuses on ESG climate risk management and green finance in support of combating climate change.

The book is original in that it reveals an understudied and new—actualized in the Decade of Action—climate aspect of smart green innovations and rethinks the environmental footprints of Industry 4.0 from a climate risk and risk management perspective for the first time. The theoretical significance of the book lies in the formation of an innovative concept of climate change risk management, in which the economy, society, nature, and technology are presented and interact effectively. The book's contribution to the literature lies in forming a systemic vision of the environmental footprints of Industry 4.0 in the unity of energy efficiency, green finance, ESG management, and responsible climate-smart innovation in various sectors of the digital economy.

The book's applicability is that it offers a systemic approach to the practical implementation of SDGs 7–9 and 11–15. Through this, the book builds a comprehensive understanding of combating climate change in the economy and business under Industry 4.0. A separate chapter of the book is devoted to the COVID-19 pandemic and the risks of epidemics and pandemics, considered in the context of the climate agenda of the Decade of Action. The management implications of the book are that it offers promising application solutions for improving the eco-design of products and processes in Industry 4.0, as well as solutions for increasing the effectiveness of climate change risk management through smart green innovations in Industry 4.0.

The primary target audience for this book is scientists studying the development of the green economy and combating climate change. In the book, they will find an innovative and systemic vision of smart green innovations in Industry 4.0 for climate change risk management. The book will also be of interest to practicing experts. Government and corporate governance actors will find applied solutions and comprehensive author's recommendations to improve the management of the climate-smart economy and business in support of the development of Industry 4.0 in the Decade of Action.

Moscow, Russia   Elena G. Popkova
                                                              elenapopkova@yahoo.com

Contents

Experiences of Climate Risk Management Based on Smart Green
Innovations in Regions and Countries

Climate-Responsible Entrepreneurship in Support of the Sustainable Development of Industry 4.0

Eco-Design and Quality Certification in Implementing Smart Green Innovations for Maximizing Their Contribution to Combating Climate Change

Elena G. Popkova

Abstract The paper aims to identify prospects for improving the regulation of smart green innovations to maximize their contribution to combating climate change. The research is based on a mathematical apparatus and Numbeo and WIPO statistics for 2022. The author applies the regression analysis method to determine the climate index's dependence on factors such as environmental quality certification in the economy, the freedom of international trade, diversification, and market size. A scenario analysis of climate change up to 2024 is carried out. As a result, the author concludes that environmental certification is a much more effective mechanism for combating climate change than ensuring market freedom. The prospects for better regulation of smart green innovations to maximize their contribution to the fight against climate change are related to the development of environmental certification. It is advisable to reinforce environmental certification with the eco-design of products. On this basis, the author proposes a new approach to regulating the practice of creating and implementing smart green innovations based on eco-design and environmental certification of product quality in the economy. The new approach to developing eco-design of climate-resistant products recommends combining the latest achievements of science with creativity and art. This will create the most vibrant, informative, and attractive eco-design.

Keywords Eco-design · Quality certification · Smart green innovations · Climate-resistant products · Fighting climate change

JEL Classification L15 · O13 · O31 · O32 · O38 · Q01 · Q52 · Q54 · Q56

E. G. Popkova (✉)
RUDN University, Moscow, Russia
e-mail: elenapopkova@yahoo.com

1 Introduction

Smart green innovations are applied solutions to environmental issues using smart technology. The most pressing environmental issues include combating climate change, which is enshrined as a global priority of the Decade of Action in Sustainable Development Goal 13 (SDG 13). The difficulty in developing smart green innovations is the contradictory nature of their essence. They combine the features of smart innovations, the meaning of which is to realize the possibilities of digital technology and strengthen the digital competitiveness of business entities, and the features of green innovation, the meaning of which is to show environmental responsibility to protect the environment [13].

In this regard, smart green innovations have a commercial essence but carry an important non-commercial mission. The problem is that their inconsistency is a barrier to unlocking the potential of combating climate change through smart green innovation. The current approach to regulating the practice of creating and implementing smart green innovations relies on market self-governance. In this case, the regulatory mechanism is that as social progress increases, the level of environmental awareness in society increases, and consumers have an increasing demand for climate-resilient products made with smart green innovations, showing an increased willingness to finance their creation and implementation.

On the one hand, the apparent advantage of the current approach is the low cost of government regulation, reducing the overall cost of smart green innovations and ensuring their mass availability. On the other hand, there is also the serious disadvantage that consumers are far from being able to verify whether smart green innovations have actually been used and whether products are actually climate-resistant [2, 5, 12].

This leads to high risks of the so-called mistakes of the first kind when the business formally approaches the implementation of corporate environmental responsibility. In this case, conventional products are passed off as climate-resistant and sold on the market at an inflated price. If consumers become aware of this, their confidence and demand for climate-resistant products will decrease accordingly. There are also the so-called mistakes of the second kind, where imperfect information flows and insufficient marketing support make consumers unaware of the products' contribution to combating climate change and place low demand on them, thereby holding back the development of smart green innovations.

This research seeks to address the problem and aims to identify prospects for improving the regulation of smart green innovations to maximize their contribution to combating climate change.

2 Literature Review

The theoretical basis of the research is the concept of combating climate change. According to this concept, smart green innovations play an important role in combating climate change by producing and selling climate-resilient products [10].

Best practices in combating climate change through smart green innovations are reflected in the works of Aldieri et al. [1] and Kabir et al. [7].

Nevertheless, the available literature, for example, Qiu [11] and Jiang and Zheng [6], also noted the limitation of the established approach to regulating the practice of creating and implementing smart green innovations. This limitation is due to the fact that this approach does not guarantee full-scale information support for the subjects of market relations on determining whether a product is climate-resistant or not. Uncertainty about the prospects for smart green innovations and the potential for increasing their contribution to combating climate change is a gap in the literature.

This raises the research question (RQ) of how to best unlock the potential of smart green innovations to combat climate change. Based on the works of Kumar et al. [8] and Zhansagimova et al. [14], which point out the benefits of environmental certification for improving the overall quality of products, this research hypothesizes that eco-design and quality certification in smart green innovations will maximize their contribution to combating climate change. To find an answer to the RQ posed, the author models the impact of environmental quality certification on climate change outcomes, drawing on international best practices.

3 Materials and Method

This research relies on a mathematical apparatus to obtain the most accurate and reliable results. Using the method of regression analysis, the author determines the dependence of the climate index (calculated by Numbeo and designated as ClInd) on factors such as environmental quality certification in the economy (designated as EnvSetr) and international trade freedom, diversification, and market size (designated as MarFree) calculated by WIPO. The research model is as follows:

$$ClInd = a + b_{EnvSetr} * EnvSetr + b_{MarFree} * MarFree \tag{1}$$

The reliability of the results of economic and mathematical modeling is tested with the correlation coefficient. The author also conducts Fisher's F-test and Student's t-test. The research hypothesis is recognized as proven if the contribution of environmental certification to climate change outcomes is higher than the contribution to these outcomes of market freedom, that is, if $b_{EnvSetr} > b_{MarFree}$.

To determine the prospects for increasing the contribution of environmental certification to combat climate change, the author conducts a scenario analysis using the least-squares method. The author also makes an automatic forecast of changes in the values of the indicator of environmental quality certification of products in the economy up to 2024. A histogram of the normal distribution of this prediction is created. Climate change scenarios are defined, for which various predicted values of the factor variable are substituted in model (1).

Statistics are taken from the dataset "Corporate Social Responsibility, Sustainability, ESG, and Combating Climate Change: Simulation Modeling and Neural

Table 1 Results of regression analysis

Regression statistics						
Multiple R	0.6421					
R-squared	0.4123					
Normalized R-squared	0.3388					
Standard error	11.0466					
Observations	10					
Variance analysis						
	df	SS	MS	F	Significance of F	
Regression	1	684.8959	684.8959	5.6126	0.0453	
Balance	8	976.2262	122.0283			
Total	9	1661.1221				
Parameters of the regression model						
	Coefficients	Standard error	t-statistics	P-value	Lower 95%	Upper 95%
Constant	75.0289	5.6406	13.3015	0.0000	62.0216	88.0362
EnvSert	11.2819	4.7621	2.3691	0.0453	0.3004	22.2633

Source Calculated and compiled by the author

Network Analysis in Regions of the World 2022" [3]. The sample includes the top 10 countries leading in the ranking of sustainable development and combating climate change based on corporate social and environmental responsibility in the countries of the world in 2022 [4] (Table 1).

As can be seen from Fig. 1, the sample includes countries from different geographic regions of the world and with different levels and rates of socio-economic development, which ensured the representativeness of the sample.

4 Results

The dependence of the climate index on such factors as environmental quality certification in the economy, the freedom of international trade, diversification, and the market size was determined based on data from Fig. 1 using regression analysis. The following econometric model is obtained:

$$ClInd = 64.9804 + 8.4713 * EnvSetr + 0.2523 * MarFree \qquad (2)$$

The multiple correlation coefficient in model (2) is 0.6638. This means that a 66.38% change in the climate change index is explained by a change in the factor variables. The significance of F was 0.1309. That is, the equation corresponds to

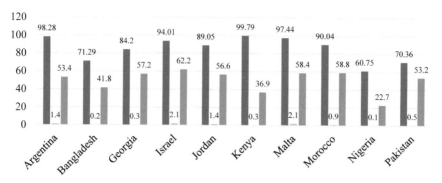

Fig. 1 The empirical basis for the study, 2022. *Source* Compiled by the author based on materials from the Institute of Scientific Communications [3]

a significance level of 0.15, at which the tabular F is 2.5183. The observed F is 2.7572, which is higher than the tabular F. Consequently, Fisher's F-test is passed; the equation is reliable at a given significance level.

This suggests that the contribution of environmental certification to climate change outcomes ($b_{EnvSetr} = 8.4713$) is many times greater than that of market freedom ($b_{MarFree} = 0.2523$), that is, $b_{EnvSetr} > b_{MarFree}$. This proves the hypothesis and allows us to exclude the factor variable from further research. Refined (cleared of the influence of the insignificant variable) results of the regression analysis of the dependence of climate index on such factors as environmental certification of product quality in the economy are shown in Table 1.

Based on the results from Table 1, the following econometric model is obtained:

$$ClInd = 75.0289 + 11.2819 * EnvSetr \qquad (3)$$

The multiple correlation coefficient in model (3) is 0.6421. This means that the 64.21% change in the climate change index is due to environmental certification. The significance of F is 0.0453. That is, the equation corresponds to a significance level of 0.05, at which the tabular F is 5.3177. The observed F is 5.6126, which is higher than the tabular F. The tabular value of the Student's t-test at a given significance level is 2.2621. The observed t-statistic for the factor variable is 2.3691, exceeding the tabular t-value. Consequently, Fisher's F-test and Student's t-test are passed; the equation is reliable at a given significance level.

According to the arithmetic mean (0.93) and standard deviation (0.77) of the indicator of ecological certification of product quality in the economy, 100 random numbers are automatically generated from Fig. 1, which formed the forecast of

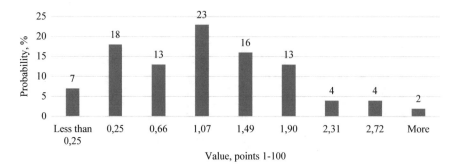

Fig. 2 Forecast of changes in the values of the indicator of environmental certification of product quality in the economy up to 2024. *Source* Calculated and compiled by the author

changes in the values of this indicator up to 2024. A histogram of the normal distribution of these predicted values is shown in Fig. 2.

The histogram in Fig. 2 shows that the probability of values below the level of 2022 (pessimistic scenario) is low and amounts to 38% (7 + 18 + 13). Highly probable (with probability 23 + 16 + 13 = 52%) is a significant increase (from 87.10 points to 96.46 points) in the climate change index due to the development of environmental certification (from 1.07 points to 1.90 points)—this is a realistic scenario. It is also quite probable (with a probability of 4 + 4 + 2 = 10%) to achieve the maximum possible (100 points) climate change index through the development of environmental certification (0.31 points and above)—this is the optimistic scenario.

5   Discussion

The contribution of this research to the literature lies in the development of scientific provisions of the concept of combating climate change through clarification of the tools of this struggle. The results show that environmental certification is a much more effective mechanism for combating climate change than ensuring market freedom. On this basis, the author proposes a new approach to regulating the practice of creating and implementing smart green innovations based on eco-design and environmental certification of product quality in the economy. A comparative analysis of the existing and new approaches to regulating the practice of creating and implementing smart green innovations is conducted in Table 2.

According to Table 2, in contrast to Hart and Chong [2], Jiang et al. [5], and Roulston et al. [12], the proposed new approach proposes government regulation rather than market self-regulation as an instrument for regulating the practice of creating and implementing smart green innovations. In contrast to Aldieri et al. [1] and Kabir et al. [7], the author recommends relying not on the environmental responsibility of society but on a new regulatory mechanism—eco-design and environmental quality certification. In contrast to Qiu [11] and Jiang and Zheng [6], the new approach

Table 2 Comparative analysis of the existing and new approaches to regulating the practice of creating and implementing smart green innovations

Criteria for comparing approaches	Existing approach		Proposed new approach
	Characteristics of the approach	Sources indicating these characteristics	
Regulatory tool	Market self-governance	Hart and Chong [2], Jiang et al. [5], and Roulston et al. [12]	Government regulation
Regulatory mechanism	Environmental responsibility of society	Aldieri et al. [1] and Kabir et al. [7]	Eco-design and environmental quality certification
Guarantee of climatic stability of products	It is absent, which reduces the demand for smart green innovations	Qiu [11] and Jiang and Zheng [6]	It is provided, which increases the demand for smart green innovations

Source Developed by the author

provides a guarantee for climate-resistant products, which increases the demand for smart green innovations.

6 Conclusion

The main conclusion of this research is that the prospects for improving the regulation of smart green innovations to maximize their contribution to the fight against climate change are related to the development of environmental certification. It is advisable to reinforce environmental certification with the eco-design of products. However, it should be noted that science is related to art [9]. When developing eco-design for climate-resistant products, it is advisable to combine the latest science with creativity and art. This will create the most vibrant, informative, and attractive eco-design.

References

1. Aldieri L, Bruno B, Lorente DB, Vinci PC (2022) Environmental innovation, climate change and knowledge diffusion process: how can spillovers play a role in the goal of sustainable economic performance? Resour Policy 79:103021. https://doi.org/10.1016/j.resourpol.2022.103021
2. Hart DM, Chong H (2022) Climate innovation policy from Glasgow to Pittsburgh. Nat Energ 7(9):776–778. https://doi.org/10.1038/s41560-022-01113-7
3. Institute of Scientific Communications (2022a) Dataset "Corporate social responsibility, sustainability, ESG, and combating climate change: simulation modeling and neural network analysis in regions of the world 2022." Retrieved from https://datasets-isc.ru/data2/korporativnaya-sotsialnaya-otvetstvennost. Accessed 17 Nov. 2022

4. Institute of Scientific Communications (2022b) Sustainability and climate change rankings based on corporate social and environmental responsibility in the countries of the world in 2022. Retrieved from https://datasets-isc.ru/data2/korporativnaya-sotsialnaya-otvetstvennost/rejting-ustojchivogo-razvitiya-2022. Accessed 17 Nov. 2022

5. Jiang D, Shu H, Fan Y, Dong Y, Li H (2022) Influence of renewable energy and natural resources on climate change: the role of green innovation in China. Front Environ Sci 10:966656. https://doi.org/10.3389/fenvs.2022.966656

6. Jiang Y, Zheng W (2021) Coupling mechanism of green building industry innovation ecosystem based on blockchain smart city. J Clean Prod 307:126766. https://doi.org/10.1016/j.jclepro.2021.126766

7. Kabir KH, Sarker S, Uddin MN, Leggette HR, Schneider UA, Darr D et al (2022) Furthering climate-smart farming with the introduction of floating agriculture in Bangladeshi wetlands: successes and limitations of an innovation transfer. J Environ Manage 323:116258. https://doi.org/10.1016/j.jenvman.2022.116258

8. Kumar D, Shamim M, Arya SK, Siddiqui MW, Srivastava D, Sindhu S (2021) Valorization of by-products from food processing through sustainable green approaches. In: Mor RS, Panghal A, Kumar V (eds) Challenges and opportunities of circular economy in agri-food sector. Springer, Singapore, pp 191–226. https://doi.org/10.1007/978-981-16-3791-9_11

9. Popkova EG (2022) Art. Retrieved from https://elenapopkova.art. Accessed 17 Nov. 2022

10. Popkova EG, Shi X (2022) Economics of climate change: global trends, country specifics and digital perspectives of climate action. Frontiers Environ Econ 1:935368. https://doi.org/10.3389/frevc.2022.935368

11. Qiu L (2022) Does the construction of smart cities promote urban green innovation? Evidence from China. Appl Econ Lett. https://doi.org/10.1080/13504851.2022.2103497

12. Roulston M, Kaplan T, Day B, Kaivanto K (2022) Prediction-market innovations can improve climate-risk forecasts. Nat Clim Chang 12(10):879–880. https://doi.org/10.1038/s41558-022-01467-6

13. Turginbayeva A, Shaikh AA (2022) How price sensitivity influences green consumer purchase intention? In: Vrontis D, Weber Y, Tsoukatos E (eds) Sustainable business concepts and practices: 15th annual conference of the EuroMed academy of business. EuroMed Press, Palermo, Italy, pp 1388–1390

14. Zhansagimova AE, Nurekenova ES, Bulakbay ZM, Beloussova EV, Kerimkhulle SY (2022) Development of rural tourism based on green technologies in Kazakhstan. In: Popkova EG, Sergi BS (eds) Sustainable agriculture. Springer, Singapore, pp 17–26. https://doi.org/10.1007/978-981-19-1125-5_3

Tourism, Environment, and Sustainability

Filippo Grasso and Daniele Schilirò

Abstract Environment and sustainability are key factors for sustainable development and mitigating climate change risks. This chapter discusses the relevance of sustainability in tourism and the consequent concept of sustainable tourism, emphasizing the significant role of natural resources and the environment. It examines the importance of territorial branding as a strategic tool for communicating a tourist destination in economic terms and in terms of its environment and ability to offer forms of sustainable tourism. A focus is also given to the relationship between territorial branding, the villages, and roots tourism in the Italian case. The chapter argues that tourism creates challenges in terms of environmental sustainability, as in the case of overtourism. Thus, it requires adequate governance that protects all stakeholders involved. Therefore, the role of governance is key in supporting sustainable tourism. Additionally, the chapter suggests that innovation and digital technologies become important tools for achieving a tourist destination's sustainability goal.

Keywords Tourism · Environment · Sustainability · Sustainable tourism · Territorial branding · Roots tourism

JEL Classification Q56 · Z32 · Z38

1 Introduction

Tourism has grown continuously over the past few decades. The world tourism trends reveal an economic sector in constant growth, despite some temporary slowdowns due to various crises (such as the COVID-19 pandemic). Tourism plays a fundamental role in the development of many countries, above all due to the strong impact it produces

F. Grasso (✉) · D. Schilirò
University of Messina, Messina, Italy
e-mail: filippo.grasso@unime.it

D. Schilirò
e-mail: daniele.schiliro@unime.it

© The Author(s), under exclusive license to Springer Nature Switzerland AG 2023
E. G. Popkova (ed.), *Smart Green Innovations in Industry 4.0 for Climate Change Risk Management*, Environmental Footprints and Eco-design of Products and Processes, https://doi.org/10.1007/978-3-031-28457-1_2

on the territories, their economy, and the environment. Countries are interested in supporting tourism development precisely because of the sector's positive socio-economic effects. These mainly refer to the multiple job opportunities, the increase in income levels, local development, the diversification of the economy, the positive effects on the balance of payments, and the increase in public revenues. Tourism is now an industry, the travel industry, and it is a mature economic sector. Tourism represents about 10% of global employment and 10% of the global gross domestic product (GDP).

However, tourism creates challenges in terms of environmental sustainability. Therefore, it requires adequate governance that protects all stakeholders involved.

In the recent years of the new millennium, the paradigm of sustainable tourism development was definitively acquired at a social and environmental level. This must constitute the reference paradigm of tourism policies and governance of tourist destinations.

This chapter discusses the relevance of sustainability in tourism and the consequent concept of sustainable tourism. It examines the importance of territorial branding as a strategic tool for communicating a tourist destination in economic terms and in terms of its environment and ability to offer forms of sustainable tourism. A focus is also given to the relationship between territorial branding, the villages, and roots tourism in the Italian case. Conclusions end the chapter.

2 Materials and Methodology

This chapter represents a theoretical and conceptual study concerning sustainable tourism. The authors analyze the relationship between tourism, the environment, and the issue of sustainability. This contribution focuses on the characteristics of sustainable tourism and provides a perspective on the relationship between tourism and sustainability. Our analysis, while largely theoretical, consists of materials, comparisons, and forecasts based on reliable secondary data sources from international and national tourism organizations. The analysis shows that natural resources, environment, governance, and institutions are key factors for sustainable tourism.

3 Tourism and Sustainability

Sustainability considers how humans might live in harmony with the natural world, protecting it from damage and destruction. More specifically, sustainability means meeting our own needs without compromising the ability of future generations to meet their own needs. Since its introduction [25], the concept of sustainability has been related to the notion of development. Schilirò [18] argues that today's view of sustainability is to keep the environment on a human scale, employ low-carbon technologies, obtain environmentally friendly products, limit the adverse effects of climate change, and implement an approach of permanent recycling. Sustainability requires eco-compatible innovations, a population policy, and adequate institutions. The sustainability paradigm applies to all human activities, including tourism.

Concerns over climate and environmental change have reinforced the need for a more sustainable approach to tourism. Given the centrality of sustainability in tourism, policies and practices must be adequate to implement a balanced system between tourism and sustainability that creates a green economy and post-carbon tourism [11].

In their *Making Tourism More Sustainable—A Guide for Policy Makers*, the United Nations Environment Programme (UNEP) and the World Tourism Organization (WTO) define sustainable tourism as "tourism that takes full account of its present and future economic, social and environmental effects to meet the needs of visitors, industry, the environment, and host communities." Therefore, sustainability principles apply to tourism development's environmental, economic, and socio-cultural aspects.

A common requirement of the various forms of sustainable tourism is its ability to satisfy the needs of tourists and destinations, providing future opportunities for development on an environmental, social, economic, and cultural level.

The UNEP and WTO [23] also identify three essential characteristics of sustainable tourism:

1. Environmental resources must be protected.
2. Local communities must benefit from this type of tourism in terms of income and quality of life.
3. Visitors must have a quality experience.

Following the sustainable tourism paradigm, the UNEP and WTO [23] vision include the concept of a sustainable tourism destination. First, a sustainable destination has an unlimited time horizon and can remain viable for an unlimited time. Second, a sustainable destination respects the environment; that is, it does not alter the environment, be it natural, social, or artistic. Third, a sustainable destination respects other activities and does not hinder or inhibit the development of other social and economic activities. Furthermore, it meets the needs of tourists and current host areas. It protects and enhances opportunities for the future.

Additionally, tourism can contribute directly and indirectly to achieving the 17 Sustainable Development Goals (SDGs) by 2030, as established in the United Nations 2030 Agenda and highlighted in the UNWTO Tourism For SDGs platform. In particular, regarding the goal of decent work and economic growth, tourism is an important economic sector worldwide. Tourism offers decent work opportunities to young people and women. The policies that favor diversification significantly improve its socio-economic impact on the territory and poverty alleviation. Furthermore, to make cities and human settlements inclusive, safe, long-lasting, and sustainable, tourism can incentivize urban infrastructure development and accessibility. It contributes to the enhancement and protection of the cultural and natural heritage of a destination. Investments aimed at the energy efficiency of infrastructures and sustainable mobility should result in smarter and greener cities for residents and tourists.

Moreover, regarding the goal of ensuring sustainable production and consumption models, the tourism sector can and must adopt sustainable consumption and production methods, accelerating the transition towards sustainability. Tools to monitor the impacts of sustainable development for tourism, including energy, water, waste, biodiversity, and job creation, will result in better economic, social, and environmental outcomes. However, above all, tourism must fundamentally contribute to the fight against climate change and take measures to combat climate change and its consequences. Tourism stakeholders should lead in the global response to climate change. By reducing the carbon footprint of transport and hospitality, tourism can substantially reduce carbon emissions and help address one of the most pressing challenges of our time.

Furthermore, sustainable tourism is a phenomenon that has enriched itself with many declensions, including ecotourism, which focuses on protecting natural and rural areas and the educational function regarding the ecological dimension. Geotourism is defined as a form of tourism that looks at the geographical character of the visited place, its environment, heritage, aesthetics, culture, and the well-being of its inhabitants, which reinforces, in addition to the natural character, its history, cultural roots, and traditions. Ecotourism also includes roots tourism. In community tourism, sustainable development is applied to improve residents' quality of life by preserving the natural and anthropogenic landscape, providing a high-quality experience to tourists, and optimizing local economic benefits. Finally, cultural tourism focuses on knowledge and protecting cultural, historical, and natural heritage.

A requirement common to the different declensions of sustainable tourism is its ability to satisfy the needs of tourists and destinations, providing opportunities for future development of the environmental, social, economic, and cultural areas.

In summary, a sustainable destination can be defined as an ecosystem in which all stakeholders are involved in the definition and management of their social, environmental, cultural, and economic priorities, planning a long-term growth strategy centered on a unique positioning of the destination, realizing it through integrated networks of value without compromising its environment and the equilibrium of its natural resources.

However, despite the interest and the attention received, many studies show that tourism is unsustainable globally. For example, Sharpley [21] believes that sustainable tourism (i.e., environmentally sound tourism development) is essential; however, he argues that sustainable development through tourism is unachievable. According to his critique, there is a divide between tourism theory or policy and practice. The trajectories of tourism on a global scale contrast starkly with the policies and principles of the sustainable tourism development agenda. Scott [20] warns that the challenges of climate change and the consequences of delayed action and overall poor industry readiness in tourism should be of particular interest to tourism stakeholders. He hopes for an accelerated collective response that requires a broad commitment to three tasks: (1) better communication and knowledge mobilization, (2) greater research capacity, and (3) strategic policy and planning commitment. Unfortunately, although many salient knowledge gaps have been identified in tourism and climate change, tourism policymakers seem disconnected from the substantive scientific literature available; therefore, practices in tourism management and tourism flow management remain disengaged from the concept of sustainability. Shu-Yuan et al. [22] highlight the challenges and obstacles to tourism sustainability, such as high energy use, extensive water consumption, and habitat destruction. They also discuss the key interdisciplinary elements in sustainable tourism, including green energy, green transport, green buildings, green infrastructure, green agriculture, and smart technologies. They propose some implementation strategies to overcome the challenges and barriers to achieving sustainable tourism. They also face the aspects of regulation, institution, finance, technology, and culture, together with a system of key performance indicators.

Institutional efficiency is a critical aspect of sustainable tourism practice, which has often proved weak, except in some countries. Therefore, sustainable tourism is not achieving the desired results, mainly due to institutional challenges. In this regard, the best example is the environmental issue, which has long been advocated to reduce greenhouse gas emissions to mitigate the climate crisis but has always been evaded.

Regarding sustainable tourism, each tourist destination has its threshold regarding the ability to absorb tourists and their needs in its environment. The "load capacity" represents this threshold. An increase in tourist activity can lead to environmental and social damage with the consequent decline in activities. This threshold is not the same for all tourist areas; some can withstand a large number of activities, while others are more fragile ecologically and can be damaged even by low-level flows [24]. Institutions have a key role in determining the "load capacity" and the protection of the environment since institutions matter in the successful development policy, as North [14] highlighted. This is also true in the case of the sustainable tourism policy.

Overtourism concerns the "load capacity" and the pursuit of sustainable tourism. Overtourism has a devastating impact on the natural and cultural heritage of many attractive places at a world level. Several global trends are contributing to the emergence and continuation of overtourism. They are still immune to mitigation and resolutions [2]. Overtourism also damages heritage and nature in general as a cause of pollution environment (waste, harmful emissions), leading to the degradation of entire ecosystems. When tourism creates overcrowded destinations, they become unsustainable; it is necessary to prevent this from happening. One way out is to increase the supply of destinations, diversifying the offer to serve the differentiated needs of the tourists. To avoid the phenomenon of overtourism, it is also necessary to favor de-seasonalization to make tourism more sustainable. Providing new infrastructures to serve the ever-increasing numbers is another possibility but on condition of respect for the environment and "load capacity."

The European Union has also promoted sustainable economic development practices in tourism. Sustainability is a necessary ingredient of the competitiveness of a territory in the long term, becoming an essential reference of the Lisbon "Europe 2020" strategy. Tourism must foster prosperity in Europe, respond to social concerns, territorial cohesion, the protection and enhancement of natural and cultural heritage, and, finally, strengthen the ability of the territories to resist the impact of climate change.

Nowadays, a characteristic of world tourism is that tourism involves broader and more differentiated sections of the population than ever. Moreover, the growing competition in the tourism market implies that producers must differentiate their products, transforming them into experiences that involve the consumer. Thus, mass tourism is increasingly losing ground, while highly personalized forms of tourism are affirming, which can give tourists a unique experience that can increase most of the time dedicated to the holiday, which makes them feel good physically and relationally and enriches them culturally [8].

For example, tourism linked to cultural heritage can entirely fall within the paradigm of sustainable development. It represents one of Europe's largest tourism market segments. It is considered one of the most formidable drivers capable of developing the destinations and territories involved.

Furthermore, tourism requires the conscious use of technology, which is developing with great speed and has disruptive characteristics. New technologies, especially digital ones (artificial intelligence, blockchain, virtual reality, extended reality, and the metaverse), offer enormous potential in directing tourists and favoring destination territories [19]. They can also act as a tool for sustainable tourism and become a winning weapon for the territories if used with awareness and adequately explained to all who intend to use them. Therefore, developing new technologies and applications is important. Nevertheless, it must be focused on the possible creative use of ICTs and other digital technologies that allows destinations to represent themselves through the cultural, historical, architectural, and environmental heritage by defining the models and tools necessary to make this possible. After all, the world of information and communication has profoundly changed for about a decade, precisely due to the development of digital technologies, while the COVID-19 pandemic, which has hit the tourism sector hard, has accelerated these changes.

Consequently, faced with a world connected globally by the network, the constant and real-time circulation of information, and the need to meet sustainability requirements, the tourism ecosystem we face for the next few years has profoundly changed. For example, new ways of working, such as smart working and virtual meetings, will reduce the flow of business travel and the congress tourism industry in and between countries, bringing eventual benefits to the environment.

In conclusion, we can say that an approach to sustainable tourism allows a tourist destination to become sustainable. This implies that all stakeholders involved (public at various levels, companies, public institutions, associations, and citizens) are aware of the mutual benefits, making themselves available to give up part of their advantage to achieve the environmental, social, and economic equilibrium of the system. Technology and innovation become essential tools for achieving a destination's sustainability goal.

Additionally, another exciting trend of tourism is that it has become an important potential source of income for large cities of art but also for many destinations such as small municipalities, small villages of hilly and mountainous inland areas, and areas near the sea. This is observed in several European countries, such as Portugal, Spain, France, Greece, and Italy. Competition in the tourism sector appears to be increasingly fierce internationally than within the tourist resorts of a country.

The survival of these small villages and their territories is linked to their ability to effectively and efficiently develop the tourist offer and be able to communicate to potential tourists an environment that is still "intact" from an environmental point of view.

Territorial branding becomes a strategic tool for communicating a tourist destination, its uniqueness, its environment, and the ability to offer forms of sustainable tourism.

4 Territorial Branding: A Tool for Promoting Sustainable Tourism

The governance of sustainable tourism involves all tools to strengthen the image and identity of tourist destinations, also improving their communication. Territorial branding can be one of the important means to convey a tourist destination as a place that respects the environment and is framed in the paradigm of sustainable tourism.

When we talk about brands in the tourism industry, we must first consider the tourism supply chain and the set of stakeholders involved in the supply chain. Therefore, it is necessary to identify the actors involved and define their respective roles. It is also required to analyze the offer systems and the promotion activity to which a correct information activity must be coupled.

In general, a product is what is sold, while a brand, which represents an essential intangible asset of the company, is the perceived image of the product sold. On the other hand, branding is the strategy to create this image. Brand differentiation is the opportunity to distinguish one product from the competition and succeed in the market.

A successful branding strategy is one in which customers must be convinced that there are significant differences between existing brands within a product category.

Entrepreneurs must create a powerful, unique value proposition to include in their brand strategy to implement differentiation.

In the case of tourism, we are talking about territorial branding. The goal of a tourism branding strategy related to the territory is to create an emotional connection between the inhabitants of a destination and their guests. As Pollice and Spagnuolo [15] argue, territorial branding should be interpreted not as a simple marketing action that supports the competitive affirmation of a local territory in the national or international market but rather as a strategy that attributes relevance to the link with the territorial identity. For Pollice and Spagnuolo [15], it is precisely the link between branding and territorial identity that constitutes the key to identifying how branding can take on a propulsive role in the development processes of the territory, contributing to the concept of sustainable tourism and strengthening its competitiveness.

Globalization affecting the tourism industry tends to exacerbate the competition between tourist destinations and territories. However, it becomes essential for the territories to build their recognizability, their relationship with the environment, and their landscape characteristics. Such elements increase the territories' attractiveness towards the flows that transit on global digital networks and platforms and concern people, capital, projects, and innovations.

In particular, if we look at the brand strategy of a small tourist destination, such as the villages, it must constitute a project where all stakeholders are involved through feedback processes, where attention to the environment and proposal of sustainable tourism are key factors, rather than proposing a project on the tourist destination (the villages) detached from the needs of the territory and imposed from above (top-down).

4.1 Territorial Branding: An Identity Issue

Territorial branding of a tourist destination is the process of discovery, creation, development, and realization of ideas and concepts to redefine the identity, the distinctive features, and the genius loci of a place and, consequently, to build its overall sense.

Territorial branding defines the identity in time and space of the tourist destination. In this regard, it is necessary to refer to two fundamental concepts:

- The ecosystem. This term refers to all natural aspects of a given place: the botanical, geological, biological, or naturalistic characteristics in the strict sense.
- The territory. When we talk about the territory, we are talking about everything generated by human action within a place by interacting with the ecosystem: architecture, art, archeology, agriculture, craft, economic production, etc.

The landscape, which reveals the environment and its natural resources, is the spatial representation of the two concepts (ecosystem and territory) where nature and the work of people are intertwined and which give substance to the identity of the places. It is the perceivable whole and is a cohesive whole that is difficult to separate internally. For this reason, it must be protected, enhanced, and promoted with targeted branding policies.

As Avena states [1], a territorial branding strategy goes beyond the classic concept of tourism, which has the characteristics of immateriality and a natural indisposition in the standardization of services. Territorial branding is about the narrative of places composed of ecosystems and territory, landscapes that are by their nature complex, rich in history and stories, characterized by different aspects and elements, where, however, minimum common denominators can be identified.

The goal of territorial branding is precisely to systematize these common factors. With territorial branding, we go to the destination's heart. The image of the landscape can become the primary identity element and the logo of the tourist destination, where the deeper levels of meaning are transformed into the distinctive features of its attractiveness towards the outside.

The communication of the image and contents of the territory that aims at its enhancement requires different types of investments, mainly concerning infrastructure and buildings that make up the hardware, and must consider environmental sustainability and landscape protection. The software includes the events and stories linked to the identities of the places. Finally, there is the coordination of the organizational structures and the "virtual" elements: logos (e.g., the landscape image of the territory), symbolic actions, and websites.

Territorial branding assumes that the public connects the brand to an image of a destination they believe in and trust. In fact, for the tourist-consumer, the brand has an identifying function and a fiduciary function, which is why it is important to preserve the landscape of the tourist destination and its image.

Beyond the enhancement of the landscape and its image, storytelling is one of the most interesting strategies for proposing a tourist destination brand, making the public connect the brand to what they want.

Storytelling does not mean just telling a story. It is necessary to merge the narration with the possibility of conveying what the potential visitor wants and, in some way, connect it to the tourist destination, its ecosystem, the characteristics of the territory, and its landscape. More than a message aimed at buying, it is a message that underlines some intrinsic values of the tourist destination that can act as leverage with respect to the visitor's choice.

Additionally, the branding strategy also involves the integration of the brand within marketing programs and activities.

Promoting hiking and short breaks in smaller areas from the point of view of tourist flows, such as the villages and their surrounding territories, means encouraging forms and experiences of slow and sustainable tourism in more direct contact with the natural environment, as well as supported and hoped for in various contributions by Grasso [7, 13], reiterated in Grasso and Sergi [10] and Grasso and Schilirò [9], but also cultural tourism and personal enrichment.

These initiatives are all aimed at psycho-physical well-being, personal growth, and respect for the environment. Therefore, it is important in the name of sustainability to strengthen the relationship between the center and the periphery, enhancing the lesser-known areas such as the villages and rural areas of many countries in Europe and other areas of the world. This is to develop networks and connections from domestic and proximity markets. Also, during the pandemic, the internal and proximity market in many countries proved more attractive and resilient than the international market.

As far as territorial branding policies are concerned, the adoption of new communication tools is important, i.e., digital technologies and the technological innovations connected to them, which represent an essential element for achieving the objectives of promotion and enhancement of experiences and touristic destinations. In particular, Popkova et al. [16] underline the advantages of digital technology development, which are more apparent in developed countries. They highlight the social advantages of creating highly efficient, highly paid, knowledge-intensive, and creative jobs. The ultimate goal is to give a strong innovative push capable of revolutionizing the stakeholders' approach to tourism communication. For example, the use of a cloud platform, as in the case of Italy, makes it possible to improve the operational efficiency of

IT systems, achieve significant cost reductions, improve security and data protection, and speed up the delivery of services. In Italy, ENIT, the Italian National Tourism Agency, has adopted the cloud computing model on its official tourism website. With this strategy, ENIT has identified the migration from the website www.italia.it to the cloud as a fundamental prerequisite for the qualification of services in terms of reliability, security, scalability, reversibility, and data protection. However, digital communication tools must be declined with sustainability, responsibility, inclusion, and diversity. Thus, it is possible to offer the possibility of creating personalized itineraries within a sustainable tourism perspective, where the person and the environment, made up of landscapes, nature, culture, and traditions are at the center. Even villages and rural and peripheral areas can be given the opportunity to develop branding strategies for their destinations, transmitting their identity and positioning.

Avena [1] also highlights the importance of digital channels for implementing branding strategies in the tourism sector. The Travel 2.0 model is the new generation of digital tourism based on social interaction, sharing, and collaboration between various travelers. Avena [1] stresses that the change brought about in this tourism model supported by digital technology is disruptive, above all, from a socio-cultural point of view because it makes the new traveler more competent and aware of their choices. Furthermore, the use of digital channels and websites dedicated to tourism allows those who want to develop a branding strategy to carry out the so-called Content Analysis, i.e., the objective, systemic, and quantitative description of the communication content [1]. This Content Analysis makes it possible to systematically and objectively describe the characteristics of the messages that appear on websites and evaluate their contents referring to the travelers-tourists' experiences in terms of sustainability characteristics. It is, therefore, a further analysis tool to develop adequate territorial branding strategies.

5   Territorial Branding, the Villages, and Roots Tourism: The Italian Case

In deepening the topic of territorial branding and highlighting forms and segments of tourism that can be important for sustainable tourism, a brief digression on the topic of villages and roots tourism is appropriate. As Iavarone [12] reminds us, similar to many other European countries, Italy has a large number of municipalities with less than two thousand inhabitants; of these small municipalities, many are real villages, found, above all, in the regions of Southern Italy and, unfortunately, in the process of depopulation, if not already emptied. Given this reality of small municipalities, particularly in Italy, a brand strategy calibrated to the needs of the villages is very appropriate, aiming, among other things, at the important phenomenon of roots tourism because the connection that unites millions of expatriates with the many villages and towns scattered throughout the Italian national territory is very strong [12, p. 198].

In regions such as, for example, Puglia and Calabria, roots tourism represents over 50% of arrivals with an induced activity that tends to move an overall tourist flow of about 80 million people, with an audience of second and third-generation of Italians who do not know Italy and the territories of their ancestors' origin. As Grasso pointed out [6], roots tourism is important to regenerate places, enhance the cultural heritage of the territories, and re-inhabit villages.

Therefore, roots tourism represents a significant segment of tourism with enormous potential that small municipalities and villages must be able to exploit. The country's system must focus on developing adequate intervention and promotion policies. In this regard, it is right to mention the initiative carried out by the Directorate General for Italians Abroad of the Ministry of Foreign Affairs and International Cooperation, which recognized the potential offered by this tourism segment and, with the collaboration of various public agencies and entities, created the technical coordination table on roots tourism, giving life to a series of projects supported by the Ministry of Foreign Affairs. These are mainly research, training, identification, and promotion projects for tourist destinations and to enhance the needs and expectations of Italians living abroad who are interested in roots tourism.

Additionally, beyond the important initiative launched by the Directorate General for Italians Abroad, Grasso and Schilirò [9] argue that different lines of intervention can be proposed to stimulate roots tourism towards small towns and villages of origin. For example, creating "laboratories of emigration" where tour operators will offer information to potential tourists and digitize valuable material for tourists who want to deepen their knowledge of the places and their families of origin, also identifying monuments, natural places, stories, and culinary traditions.

ENIT [3] also took action to promote regional priorities on projects relating to villages and, therefore, indirectly favor roots tourism through national and international communication campaigns by integrating itineraries and related tourist products in national and international communication campaigns. The theme of roots tourism continues to be central to ENIT's communicative interests. The large catchment area of root tourism of about 80 million people, as mentioned above, has as its main markets of origin, from a geographical point of view, immigrants of Italian origin from Brazil (about 25 million), Argentina (about 20 million), the USA (about 17 million), followed by France, Switzerland, Germany, and Australia. Focusing on these markets can create a process of self-reinforcement of the effectiveness of communication actions, where these renewed sensitivities are taken into account [3, p. 51].

In turn, according to Ferrari and Nicotera [5], the destination marketing strategies to be put in place to encourage the development of roots tourism and, therefore, the recall to Italy of the communities of Italians around the world in the coming years will have to make use of four different areas of intervention, namely:

- National and regional planning;
- Targeted offer;
- Involvement of all stakeholders and creation and strengthening of networks;
- Communication.

Ferrari [4] highlights the strategies for developing tourism in the villages, promoting the destination and, at the same time, favoring the tourist flow linked to roots tourism, always within the scope of the targeted offer. Ferrari argues that events aimed at local emigrants and their descendants, which involve promoting products and elements of local identity, are a handy promotional tool.

The most recent tourism literature supports Ferrari's thesis regarding events. Indeed, event tourism has become highly effective in developing tourist destinations [17]. Event tourism undoubtedly has many positive effects on the territories involved. Paying attention and respect to the environment increases the income possibilities of tourist destinations, even the smallest and marginal ones. It is also effective in making the tourist season longer. Connecting the destination's image with one or more events, which recall the identity and the history of the places, is undoubtedly a valuable and effective strategy for promoting tourism products and branding.

This leads to the thesis, shared by the authors of this contribution that the enhancement of villages through roots tourism is a path of research and enhancement of the identities of places and people. It is consistent with the paradigm of sustainable tourism.

6 Results and Conclusions

This chapter highlighted the importance of sustainability in tourism, discussing the theme of the sustainable tourism paradigm. The result of our analysis is that natural resources and the environment play an increasingly key role in tourism, as claimed by United Nations Environment Programme and World Tourism Organization and shown in several international reports on tourism.

Another result of our analysis is that the governance and institutions supporting the sustainable tourism model have a key role. More specifically, governance for sustainable tourism must involve all stakeholders, making them aware of mutual benefits. Furthermore, innovation and digital technologies become important tools for achieving a destination's sustainability goal. Therefore, a sustainable tourism approach allows a tourist destination to become sustainable.

Indeed, there are challenges to sustainable tourism, like overtourism, excessive load capacity, climate change, and the consequences of delayed action toward sustainability and the protection of the environment.

Within an approach that aims at sustainability in tourism, branding referring to a tourist destination can represent a strategic tool. It is not limited to proposing its image. However, it implies creating an emotional connection between the inhabitants of the destination and their guests, which involves values. Sustainability and the environment can constitute strong and qualifying points of reference.

In this chapter, we argued that the territorial branding of a tourist destination is a process of discovery, creation, and realization of concepts to redefine the distinctive features of a place that concerns the territory and the ecosystem and reflects its most accurate image in the landscape.

Through territorial branding, places are told about places made up of ecosystems and territory, complex landscapes rich in history and stories, with elements of diversity, where it is possible to identify minimum common factors. The brand in tourism has an identifying function for the destination and a fiduciary function for the tourist-consumer.

Our analysis also revealed the importance of the territorial branding strategy with regard to the countless villages scattered throughout many European countries. Such a strategy for these small containers of treasures of history, traditions, art, and nature must constitute an organic project where all stakeholders are involved through feedback processes.

Village tourism and branding strategies should also be linked to root tourism or return tourism, a significant phenomenon in quantity and growing attraction, given the strong bond that unites expatriates with the numerous villages. Moreover, villages are scattered throughout the country, as the case of Italy reveals.

Through their local institutions, many small villages in the southern regions must do their part by committing to developing eco-sustainable tourism and innovative channels for communication, information, and tourism marketing. However, as statistics tend to demonstrate, tourists are increasingly looking for personalized experiences. Thus, branding strategies that use digital technologies must be oriented in this direction.

Finally, although the COVID-19 pandemic has hit the tourism sector hard, it has highlighted two aspects that will characterize tourism in the coming years and necessarily condition territorial branding strategies. The first aspect concerns the renewed importance of domestic tourism, despite increased competition in the international tourism market. The second aspect concerns tourists' interest in sustainable tourism, attentive to quality rather than quantity. Tourism tends to focus more on the essentials, on discovering nature and the environment, on holidays as a life experience, and on the rediscovery of values from a sustainability perspective.

References

1. Avena G (2021) Le opinioni degli utenti rilevate sui siti di attrazione turistica del territorio siciliano valutate con la Content Analysis [The opinions of users collected on tourist attraction sites in Sicily and evaluated with the Content Analysis]. Humanities 10(2):1–19. https://doi.org/10.13129/2240-7715/2021.2.1-19
2. Butler RW, Dodds R (2022) Overcoming overtourism: a review of failure. Tourism Rev 77(1):35–53. https://doi.org/10.1108/TR-04-2021-0215
3. ENIT (2021) Piano Annuale di marketing e promozione 2021 [Annual marketing and promotion plan 2021]. Ente Nazionale del Turismo Italiano, Roma, Italia. Retrieved from https://tsf2016venice.enit.it/wwwenit/images/amministrazionetrasparenteepe/disposizionigenerali/Piano%202021.pdf. Accessed 4 Oct 2022
4. Ferrari S (2021) Turismo delle radici e altri fenomeni di consumo. Futuri filoni di ricerca. Gli eventi come attrattive per i turisti delle radici [Roots tourism and other consumption phenomena. Future lines of research. Events as an attraction for roots tourists]. In: Ferrari S, Nicotera T (eds) Primo rapporto sul turismo delle radici in Italia. Dai flussi migratori ai flussi turistici: strategie

di destination marketing per il richiamo in patria delle comunità italiane nel mondo [First report on roots tourism in Italy. From migratory flows to tourist flows: destination marketing strategies for attracting Italian communities around the world to their homeland]. Egea, Milano, Italy, pp 199–202

5. Ferrari S, Nicotera T (eds) (2021) Primo rapporto sul turismo delle radici in Italia. Dai flussi migratori ai flussi turistici: strategie di destination marketing per il richiamo in patria delle comunità italiane nel mondo [First report on roots tourism in Italy. From migratory flows to tourist flows: destination marketing strategies for attracting Italian communities worldwide to their homeland]. Egea, Milano, Italy

6. Grasso F (2018, April 15) Il turismo di radice: guardare oltre per rigenerare i luoghi [Roots tourism: looking beyond to regenerate places]. Pickline. Retrieved from https://pickline.it/2018/04/11/il-turismo-di-radice-guardare-oltre-per-rigenerare-i-luoghi/. Accessed 28 July 2022

7. Grasso F (2018) Turismo: governare il territorio, gestire le risorse, promuovere la destinazione [Tourism: governing the territory, managing resources, promoting the destination]. Maurfix, Roma, Italia

8. Grasso F, Schilirò D (2021a) Tourism, economic growth and sustainability in the Mediterranean region. In: Grasso F, Sergi BS (eds) Tourism in the mediterranean sea: an Italian perspective. Emerald Publishing Limited, Bingley, UK, pp 129–142. https://doi.org/10.1108/978-1-80043-900-920211011

9. Grasso F, Schilirò D (2021b) Per un turismo sostenibile: il turismo di ritorno. Aspetti socioeconomici e politiche turistico-territoriali [For sustainable tourism: return tourism. Socio-economic aspects and tourism-territorial policies]. Turistica 30(4):27–38

10. Grasso F, Sergi BS (2021) Tourism in the mediterranean sea: an Italian perspective. Emerald Publishing Limited, Bingley, UK. https://doi.org/10.1108/9781800439009

11. Hall CM, Gössling S, Scott D (2017) The Routledge handbook of tourism and sustainability. Routledge, Abingdon-on-Thames

12. Iavarone S (2021) Turismo delle radici e altri fenomeni di consumo. Futuri filoni di ricerca. I borghi in via di spopolamento come possibili destinazioni per la ricerca di radici e identità [Roots tourism and other consumption phenomena. Future lines of research. Depopulated Villages as possible destinations for the search for roots and identity]. In: Ferrari S, Nicotera T (eds) Primo rapporto sul turismo delle radici in Italia. Dai flussi migratori ai flussi turistici: strategie di destination marketing per il richiamo in patria delle comunità italiane nel mondo [First report on roots tourism in Italy. From migratory flows to tourist flows: destination marketing strategies for attracting Italian communities around the world to their homeland]. Egea, Milano, Italy, pp 196–198

13. Lombardo G, D'Andrea P, Grasso F, Arrigo S, Cicero N (2021) Monti Peloritani. Borghi, cammini, pellegrinaggi, itinerari del vino [Peloritani mountains. Villages, paths, pilgrimages, wine itineraries]. Edizioni Edas, Messina, Italy

14. North DC (1994) Institutions matter. Economic History 9411004. Munich University Library, Munich, Germany. Retrieved from https://econwpa.ub.uni-muenchen.de/econ-wp/eh/papers/9411/9411004.pdf. Accessed 6 Sept 2022

15. Pollice F, Spagnuolo F (2009) Branding, identità e competitività [Branding, identity and competitiveness]. Geotema 37:49–56

16. Popkova EG, De Bernardi P, Tyurina YG, Sergi BS (2022) A theory of digital technology advancement to address the grand challenges of sustainable development. Technol Soc 68:101831. https://doi.org/10.1016/j.techsoc.2021.101831

17. Sak M, Eren S, Bayram GE (2022) Role of event tourism in economic development. In: Arora S, Sharma A (eds) Event tourism in Asian countries: challenges and prospects. Apple Academic Press, New York, NY, pp 253–260

18. Schilirò D (2019) Sustainability, innovation, and efficiency: a key relationship. In: Ziolo M, Sergi BS (eds) Financing sustainable development: key challenges and prospects. Palgrave Macmillan, Cham, Switzerland, pp 83–102. https://doi.org/10.1007/978-3-030-16522-2_4

19. Schilirò D (2021) Digital transformation, COVID-19, and the future of work. Int J Bus Manage Econ Res 12(3):1945–1952. Retrieved from http://www.ijbmer.com/docs/volumes/vol12issue3/ijbmer2021120303.pdf. Accessed 4 Oct 2022

20. Scott D (2021) Sustainable tourism and the grand challenge of climate change. Sustainability 13(4):1966. https://doi.org/10.3390/su13041966
21. Sharpley R (2020) Tourism, sustainable development and the theoretical divide: 20 years on. J Sustain Tour 28:1932–1946. https://doi.org/10.1080/09669582.2020.1779732
22. Shu-Yuan P, Gao M, Kim H, Shah KJ, Pei S-L, Chiang P-C (2018) Advances and challenges in sustainable tourism toward a green economy. Sci Total Environ 635:452–469. https://doi.org/10.1016/j.scitotenv.2018.04.134
23. United Nations Environment Programme and World Tourism Organization (2005) Making tourism more sustainable—a guide for policy makers. Retrieved from https://www.unep.org/resources/report/making-tourism-more-sustainable-guide-policy-makers. Accessed 30 Aug 2022
24. Weaver D (2006) Sustainable tourism: theory and practice. Butterworth Heinemann, Oxford, UK
25. World Commission on Environment and Development (1987) Our common future (The Brundtland Report). Oxford University Press, Oxford, UK

The Environmental Component of Sustainable Socio-economic Development

Alexander S. Tulupov⦿, Bogdan E. Kosobutsky, Ivan A. Titkov, and Artem A. Belichko

Abstract The paper shows the contradictions of the existing national security system in Russia, considered a basis for sustainable socio-economic development. The authors propose an innovative approach to building a discrete model and ensuring national security. The authors show the close interdependence of environmental and informational types of security that determines the development of Industry 4.0. Moreover, the authors provide a list of basic types of security in the structure of national security. Using environmental security in the national security model as an example, the authors propose to structure the sources of threats according to the following criteria: kinds of threat, place of origin and formation of threat, and the degree of controllability and manageability of the impact. The variants of the states of each criterion are highlighted. The methodological basis of the conducted work is Russian and foreign publications and normative legal documents on the problems of safety, the creation of Industry 4.0, and sustainable development. The main scientific tools of the research are economic analysis, including environmental and economic analysis, as well as system analysis, including theories of sets and multidimensional information spaces, content analysis, and information modeling. The application of the proposed discrete model of national security will level the existing contradictions and shortcomings and provide a new qualitative level of individual types of national security and its entire complex.

Keywords National security · Industry 4.0 · Sustainable development · Socio-economic system · Green management

JEL Classification O38 · O44 · Q56

A. S. Tulupov (✉) · B. E. Kosobutsky · I. A. Titkov · A. A. Belichko
Market Economy Institute of RAS, Moscow, Russia
e-mail: tul@bk.ru

B. E. Kosobutsky
e-mail: bk.k@bk.ru

E. G. Popkova (ed.), *Smart Green Innovations in Industry 4.0 for Climate Change Risk Management*, Environmental Footprints and Eco-design of Products and Processes, https://doi.org/10.1007/978-3-031-28457-1_3

1 Introduction

Ensuring sustainable social and economic development in the Russian Federation is regulated by a whole complex of normative legal documents, of which one of the basic ones is the National Security Strategy of the Russian Federation [6], in which paragraph 3 states, "this Strategy is based on the inextricable relationship and interdependence of national security of the Russian Federation and the socio-economic development of the country." This relationship is also seen in the official interpretation of national security, which is understood as "the state of protection of national interests of the Russian Federation from external and internal threats, which ensures the implementation of constitutional rights and freedoms of citizens, decent quality and standard of living, civil peace and accord in the country, and protection of the sovereignty of the Russian Federation, its independence and national integrity, and the socio-economic development of the country" [6, p. 5].

The study of the works of Russian and foreign researchers shows that the problem of the relationship between different types of security that form a system of national security has not been sufficiently investigated in the development of the mechanism to ensure national security and, accordingly, sustainable socio-economic development. The goals pursued and the mechanisms used to achieve them for different types of security are not balanced, do not consider the complex interaction of the components of the national security system, and often contradict each other. The authors showed this circumstance in their previous research [11].

The basic law [7], as well as the National Security Strategy of the Russian Federation [6], does not describe the structure of national security with clearly structured components (by type of security provided, level of consideration, type of danger, regulation tool, etc.). In our opinion, this makes it possible to monitor the state and develop approaches to ensure national and individual types of security at the proper level. Moreover, this leads to duplication and contradiction of functions provided by different regulators in ensuring sustainable socio-economic development.

It is important to note that advanced global development trends also dictate the need for digitalization, the transition to Industry 4.0 [10], and the consideration of the environmental factor in developing approaches to socio-economic regulation and national security [9, 12]. It is also necessary to note that the list of national interests and priorities of the Russian Federation, outlined in the National Security Strategy of the Russian Federation [6], indicates the need for "environmental protection, conservation of natural resources, environmental management, and adaptation to climate change," as well as "environmental security and environmental management." Decree "On the national development goals of the Russian Federation until 2030" [5] and Decree "On the national goals and strategic objectives of the development of the Russian Federation until 2024" [4] pay much attention to the environmental components in the development of the national economy.

Simultaneously, contradictions in the approaches to the provision and assessment of environmental, economic, and national security are clearly manifested in the Strategies [1–3, 6], which, declaring the interrelation of national, economic, and environmental types of security, are not interconnected. For example, the Strategy of Environmental Security of the Russian Federation until 2025 [2] assesses environmental security by 18 indicators (section V, paragraph 28), and the Strategy of Economic Security of the Russian Federation until 2030 [3] assesses economic security by 40 indicators (section IV, paragraph 27). In turn, the National Security Strategy of the Russian Federation [1] assesses the unifying type of security, the national security, only by ten indicators, and these indicators are not integral, interconnecting different types of security. The National Security Strategy of the Russian Federation [1] took one indicator (out of 18) from the Strategy of Environmental Security of the Russian Federation until 2025 [2] ("the share of the territory of the Russian Federation that does not comply with environmental standards...") and three indicators (out of 40) from the Strategy of Economic Security of the Russian Federation until 2030 [3] ("gross domestic product per capita," "inflation rate," and "decile ratio." The priority and validity of the inclusion of these indicators in the National Security Strategy of the Russian Federation [1] raise many questions. In the new version of the National Security Strategy of the Russian Federation [6], the section with the list of indicators of the national security state is absent. There are also many questions about the content of official documents regulating national security and its components— what in developed countries is considered a driver of sustainable development and economic growth is presented as a threat to the national economy in Russia. This research discusses the identified contradictions for these types of security in detail [8].

Thus, there is an urgent need to streamline and systematize the components of national security from the perspective of a unified integrated approach of consideration, with an urgent task to consider the influence of the environmental factor as a major component of sustainable socio-economic functioning of the national economy.

2 Methodology

The main research methods are economic analysis, including environmental and economic analysis, and system analysis, including theories of sets and multidimensional information spaces, content analysis, and information modeling.

The research methodology is based on fundamental scientific works of Russian and foreign authors on the problems of national, environmental, economic, information, and other types of security, the creation of Industry 4.0, and ensuring sustainable development.

The validity of the research conclusions is confirmed by a systemic approach to the problems addressed and the use of methods and techniques that are adequate to the research objectives.

The following types of analysis have been applied: economic, ecological and economic, and system analysis, including theories of sets and multidimensional information spaces, content analysis, and information modeling.

3 Results

To streamline the structure and develop consistent approaches to ensuring national security within the framework of the Russian Science Foundation Project No. 22-28-01458 "Ensuring National Security Based on the Theoretical and Methodological Framework of Sustainable Development" (https://rscf.ru/project/22-28-01458/), the authors propose a discrete model of national security, which can be represented in the matrix-morphological form in the first approximation (Table 1).

In the multidimensional information-analytical national security space (NSS), which is a geometric representation of the presented morphological matrix, the environmental security sector has interconnections with all components of the NSS. Thus, ensuring environmental security is necessary for the entire typology of identified objects of protection: from the individual person to all living and non-living nature and the country as a whole. Simultaneously, ensuring environmental safety for designated objects of protection is carried out through legislative, administrative, informational, and economic instruments of regulation.

Legislative tools include all normative legal documentation: The Constitution of the Russian Federation, Codes, Federal laws, and the entire relevant hierarchy, from federal and branch to regional and local government levels.

Administrative safety documents include licensing, certification, examination, audit, environmental impact association, etc.

Information tools include monitoring, mapping, cadastre maintenance, and various information systems, including geographic information systems (GIS).

Table 1 The structure of national security (matrix-morphological model)

Evaluation parameters	Criterion for consideration, i		
	Type of protected object (object), O	Type of security (safety), S	Regulatory tools (regulation tools), R
Option name, m	Person (personality)	Economic	Legislative
	Society	Social	Informational
	Business entity (organization, enterprise)	Industrial	Administrative
	Living nature (fauna and flora)	Food	Economic
	Inanimate nature (including outer space)	Energy	
	State	Environmental	
		Informational	
		Military (defense)	

Source Developed and compiled by the authors

Economic instruments include taxes, grants, subsidies, subventions, transfers, benefits, quotas, loans, payments, duties, fines, insurance, refundable deposits, target financing, etc.

It is important to note the presence of information security in the national security model, including the development of information technology, artificial intelligence, and the automation of business processes. These components define the fourth industrial revolution, Industry 4.0. Simultaneously, in the presented national security model, digitalization and implementation of the principles of Industry 4.0 should be aimed at achieving a state of functioning of national security in which economic development goals do not contradict environmental imperatives.

To ensure environmental security, the national security model proposes structuring the sources of threats according to the following criteria: K—a kind of threat, P—a place of origin and formation of threat, and M—the degree of controllability and manageability of the impact. The state options for each criterion are shown in Table 2.

Table 2 Sources of threats to environmental sustainability in the national security model

Evaluation parameters	Criterion for consideration, i		
Option name, m	Kind of threat, K	Place of origin and formation of threat, P	Degree of controllability and manageability of the impact, M
	Chemical	Within the object (violation of technology, inadequate organizational and managerial impacts, violation of tax law, discipline, inadequate economic impacts)	Uncontrollable and spontaneous (a natural phenomenon, the result of spontaneous human activity, etc.)
	Biological	External, in the home region or country	Managed and targeted (including criminal and military)
	Radiation	External, but in a different region or country	
	Thermal		
	Noise and vibration		
	Fire		
	Explosion		
	Natural phenomena (seismicity, flood, hurricane, etc.)		

Source Developed and compiled by the authors

The structure presented in Table 2 is a three-dimensional discrete space in which any source of threat can be expressed by a point or region with coordinates (K, P, and M). Thus, the threat to people living in the region where the nuclear power plant is located, which operates in violation of the operating parameters of the modern automation system of Industry 4.0 class, can be expressed in the proposed space by the coordinates M1 (non-targeted, natural threat), K3 (radiation), and P2 (external, in the home region).

Thus, the source of threat considered in the above example can be described by a point with coordinates (M1, K3, and P2).

For comprehensive and systemic protection against the above threat, it is necessary to use the tools of the proposed discrete national security model. Simultaneously, in the assessment and regulation of the type of safety of the necessary type of object, it is necessary to select the factors of influence and the detail of each allocated criterion. The scope of consideration and the detail of structuring increase the credibility of the assessment and, accordingly, the assurance of a particular type of security and the entire complex of national security.

4 Conclusions

The paper shows the close relationship between environmental security in the complex of different types of security that form a system of national security. Separate attention is paid to the relationship between environmental and informational types of security, which determines the development of Industry 4.0. The authors proposed a multidimensional information-analytical national security space (NSS) in which the assessment and regulation of individual types of security or national security as a whole are subject to strictly structured and systemic algorithms that exclude duplication and contradictions. The space of national security is not closed. As the knowledge system of individual types and national security evolves, the NSS fills empty cells, and new consideration aspects emerge.

The application of the proposed discrete model of national security will level the existing contradictions and shortcomings and provide a new qualitative level of individual types and national security as a whole.

Acknowledgements The research was funded by the grant of the Russian Science Foundation No. 22-28-01458 "Ensuring National Security Based on the Theoretical and Methodological Framework of Sustainable Development" (https://rscf.ru/project/22-28-01458/).

References

1. Presidential Executive Office (2015) National Security Strategy of the Russian Federation (approved by Presidential Decree of December 31, 2015 No. 683). Moscow, Russia. Retrieved from http://www.kremlin.ru/acts/bank/40391. Accessed 10 Sept 2022
2. Presidential Executive Office (2017a) Strategy of environmental security of the Russian Federation until 2025 (approved by Presidential Decree of April 19, 2017 No. 176). Moscow, Russia. Retrieved from http://www.kremlin.ru/acts/bank/41879. Accessed 10 Sept 2022
3. Presidential Executive Office (2017b) Strategy of economic security of the Russian Federation until 2030 (approved by Presidential Decree of May 13, 2017 No. 208). Moscow, Russia. Retrieved from http://www.kremlin.ru/acts/bank/41921. Accessed 10 Sept 2022
4. Presidential Executive Office (2018) Decree "On the national goals and strategic objectives of the development of the Russian Federation until 2024" (May 7, 2018 No. 204). Moscow, Russia. Retrieved from http://www.kremlin.ru/acts/bank/43027. Accessed 10 Sept 2022
5. Presidential Executive Office (2020) Decree "On the national development goals of the Russian Federation until 2030" (July 21, 2020 No. 474). Moscow, Russia. Retrieved from http://www.kremlin.ru/acts/bank/45726. Accessed 10 Sept 2022
6. Presidential Executive Office (2021) National security strategy of the Russian Federation (approved by Presidential Decree of July 2, 2021 No. 400). Moscow, Russia. Retrieved from http://www.kremlin.ru/acts/bank/47046. Accessed 10 Sept 2022
7. Russian Federation (2010) Federal law "On security" (No. 390-FZ on December 28 2010, as amended by Federal laws on October 5, 2015 No. 285-FZ, on February 6, 2020 No. 6-FZ, and on November 9, 2020 No. 365-FZ). Moscow, Russia
8. Tulupov AS (2018) The imbalance of economic and environmental policy of the state on the example of the Strategy of Economic Security of the Russian Federation. In: Tsvetkova VA, Zoidova KKh (eds) Collection of plenary papers of the VII international forum "Russia in the xxi century: global challenges and prospects of development". MEI RAS, Moscow, Russia, pp 210–216
9. Tulupov AS (2019) Environmental and economic aspects of national security. In: Bobylev SN, Solovyova SV, Khovavavko IYu (eds) Proceedings of the international scientific conference "Sustainable development and new models of the economy". Lomonosov Moscow State University, Moscow, Russia, pp 40–41
10. Tulupov AS (2019) Environmental resources management and the transition to the cyber economy. In: Filippov V, Chursin A, Ragulina J, Popkova E (eds) The cyber economy. Springer, Cham, Switzerland, pp 305–313. https://doi.org/10.1007/978-3-030-31566-5_31
11. Tulupov AS (2022) The system of national security of the Russian Federation: directions for improvement. In: Proceedings of the international scientific and practical conference "Russia in the XXI century in the context of global challenges: modern problems of risk management and security of socio-economic and socio-political systems and natural-technogenic complexes." Moscow, Russia, Presidium of the RAS
12. Tulupov AS, Mudretsov AF, Prokopiev MG (2020) Sustainable green development of Russia. In: Bogoviz A (ed) Complex systems: innovation and sustainability in the digital age. Springer, Cham, Switzerland, pp 135–140. https://doi.org/10.1007/978-3-030-44703-8_15

Capabilities of Business Analysis in Developing Data-Driven Decision Solutions

Elena N. Makarenko⬢, Yulia G. Chernysheva⬢, Irina A. Polyakova⬢,
Irina A. Kislaya⬢, and Tatiana V. Makarenko⬢

Abstract The modern world is characterized by a high rate of change, uncertainty, and instability. To function and develop successfully, organizations need to be able to respond quickly to changing external conditions. To do this, it is necessary to be able to develop solutions that will be of the greatest value for a particular organization and a particular situation, that is, consider the context. A data-driven solution is a new approach to solution development. These solutions allow us to influence the causes of the problem and ensure that the same problem will no longer arise after implementing the solution. Therefore, they ensure the successful development of the business and its competitiveness even in conditions of instability. The business analysis allows us to implement this approach and ensure the development of the best solutions that have significantly more value for business than solutions based on the opinion of management and expert assessments. A business analyst studies the context in detail, mediates between the various interests and expectations of stakeholders and individual parts of organizations, and develops a data-based solution using, among other things, big data, and works according to agile principles. Therefore, it is necessary to explore the possibilities of business analysis further when developing solutions in terms of business problems and in using the capabilities of the organization.

Keywords Business analysis · VUCA-world · Data-driven decision · Value of the solution · Business analytic

JEL Classification M11 · M21

E. N. Makarenko · Y. G. Chernysheva (✉) · I. A. Polyakova · I. A. Kislaya · T. V. Makarenko
Rostov State University of Economics, Rostov-on-Don, Russia
e-mail: julia282001@mail.ru

E. N. Makarenko
e-mail: makarenko.rsue@yandex.ru

1 Introduction

The world has changed almost beyond recognition. We now live in a VUCA world. The VUCA world is a stressful, constantly changing world with many challenges, a world in which huge amounts of information need to be processed quickly and adequately. This world requires new approaches in all areas, from the competencies of specialists to the organization's response to the challenges that it constantly faces. Since the late 1980s, the VUCA concept has caused the need to work in conditions of instability and rapid changes. Since 2020, the BANI world has been called one of the manifestations of the VUCA world. There are different opinions on whether the BANI world is a new world or one of the manifestations of VUCA. Analysts believe that this is one of the manifestations of VUCA because, to overcome it, the same competencies as for VUCA are required. The answers to the VUCA world and the BANI world are vision, understanding, clarity, and speed.

In the VUCA world, there can no longer exist specialists who are "stuck" in place, organizations that reject new products, and businessmen who do not invest in development. It is no longer possible to rely on clear guidelines. One needs to constantly monitor the situation and respond to it promptly, be able to learn new things, and abandon old habits. The profession of a business analyst arose in such a world. This specialist has all the competencies to overcome the difficulties of this world.

It is no longer enough just to use descriptive, predictive, or normative analytics. Contemporary economic development has globally changed the context in which modern business exists. Therefore, new approaches are needed. Business analysis was created by the need of the business to develop and be successful. A business should be able to change along with the external environment in which it is located. Therefore, we need a specialist business analyst who will be able to ensure these changes, find the problem and its true cause, develop a solution based on data, and ensure the transition of the organization from the current state to the desired future state. The business analysis makes it possible to solve any business problem in any organization, regardless of its size and type of activity.

In fact, the business analysis combines all previously developed and separately applied management and analysis technologies. It also uses and helps to implement new flexible practices of working on projects (all the analyst's activities are project work). It includes data analysis and uses digital technologies, data mining capabilities, and analysis in the field of cybersecurity. It can be considered a new analytical concept, the highest level of analytical work, which gives businesses significantly greater opportunities in developing solutions to emerging problems and using business opportunities for its development [3].

Business analysts are often called intermediaries between the various interests and expectations of stakeholders and individual parts of organizations. They guarantee that the value from the solution will be delivered to them. This value is generated by the organization's employees and digital technologies [1]. Simultaneously, even greater value is created if their interests coincide. Establishing these connections has always been one of the main roles of business analysts; in today's organizations, even higher demands are placed on this role because organizations need to connect all stakeholders in business ecosystems to remain competitive in the digital economy. Studies of systems theory have long proved that the greatest value is achieved if all elements of the system do not work separately, each achieving its own high results, considering the needs of each other and the organization. Only in this case can the best results and the greatest value be achieved. The business analyst is precisely the connecting link that will ensure the coordination of all needs of all parts of the organization to achieve its goals. This is their special value and the value of business analysis.

2 Methods

One of the most important tasks for organizations is the ability to respond quickly to changes in the external environment [6]. It is possible to ensure successful business development only if the developed and implemented solutions can influence the causes of the problem and ensure that the same problem will no longer arise after implementing the solution [5].

Businesses have currently come to understand the need to move from making decisions based on the opinion of management to the data-driven decision approach. The business analysis allows us to implement this approach and ensure the development of the best solutions with a significantly greater value for the business, ensuring its sustainable development and competitiveness even in conditions of instability. Working according to agile principles, a business analyst always develops a solution based on data, using, among other things, big data.

Therefore, it is necessary to conduct further research in this area and determine and justify the long-term value of data-driven solutions for business owners and their significance for a wider range of stakeholders, including investors, employees, customers, suppliers, local communities, and governments.

The research methodology should involve the following:

- The study of the essence of the development of a data-based solution;
- The study of the possibilities and value of business analysis when applied in related fields and digital technologies;
- The identification of the reasons for the insufficient dissemination of this practice.

3 Discussion

To understand what business analysis is, what its capabilities are in management and decision-making, what a business analyst does in an organization, and, most importantly, what the value of business analysis is, it is necessary to consider its conceptual model, areas of knowledge, perspectives, and conceptual apparatus.

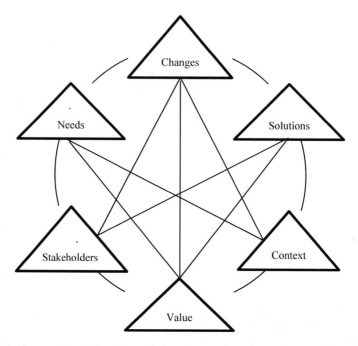

Scheme 1 Conceptual model (main scheme) of business analysis. *Source* Compiled by authors based on BABOK: A guide to the Business Analysis Body of Knowledge (Version 3.0) [4]

The conceptual framework is enclosed in the conceptual model of business analysis (Scheme 1) and includes six basic concepts—stakeholders, needs, context, solutions, value, and changes. The conceptual apparatus also includes key concepts—risk, plan, design, and benefit. Business analysis involves working within the framework of these six concepts. None of these concepts is predominant, and none of them can be excluded from the field of activity of a business analyst.

Areas of knowledge are the areas of competence of a business analyst. They include the ability to work with stakeholders and requirements, develop a data-based solution and its design, evaluate its value, and conduct strategy analysis, that is, conduct business analysis to use business opportunities for its successful development and carry out activities in the field of planning and monitoring business analysis. Strategy analysis and business analysis planning are the highest level of a business analyst's competencies.

Angles can be viewed as a prism through which business analysts look at their tasks in the current context. Perspective allows one to focus on tasks and methods specific to a particular context or initiative. Any initiative can be viewed from one or more perspectives—agile, business intelligence, information technology, business architecture, and business process management. Most initiatives involve one or more perspectives. Simultaneously, the listed angles are not exhaustive. They reflect the most common of them. As the profession develops, their list expands.

The types and tasks of a business analyst are as follows:

- Interaction with stakeholders;
- Identification, generalization, and analysis of information from various sources (inside and outside the organization), while the main source is the data obtained by the business analysts themselves, most often when interacting with an interested party;
- Identification of problems and their causes and development of solutions based on the data-driven decision approach, that is, a data-based solution;
- Needs analysis and working with requirements;
- Development of business process models and their analysis;
- Assessment of the organization's capabilities and limitations for developing a solution and its implementation, that is, making changes;
- Evaluation of the usefulness (value) of the developed solution (analysis of the effectiveness of the solution) for the interested party;
- Development of an organizational change strategy and description of the future and transitional states;
- Risk assessment in the context of various aspects: when working with stakeholders (e.g., the risk of not identifying all stakeholders); when developing a solution (e.g., the risk of forming an incorrect solution due to insufficient information about the problem); when making changes (changes almost always involve abandoning something familiar; there is the risk of problems arising as a result of the lack of complete information about the future new state after the changes).

The main regulatory document in the field of business analysis is the BABOK, which describes what business analysis is [4]. Business analysts use almost all existing analysis and management techniques. Business analysts independently decide which technique will be used when solving a particular problem based on the understanding of what technique can bring the greatest benefit in a particular case. That is, the business analysis methodology seems to adapt to a specific situation and is almost always unique, as well as its results [2].

The results of the research conducted by the International Institute of Business Analysis in the field of describing the opportunities and value of business analysis for an organization, conducted annually, are indicative. Most business analysts work in commercial organizations and large organizations in such leading industries as information technology, banking and finance, manufacturing, and other industries.

The main competencies of business analysts are problem solving, critical thinking, working with stakeholders, and decision-making. The main responsibilities of a business analyst in an organization are direct business analysis, business process development, data analysis, and project management.

The most commonly used techniques are techniques related to working with stakeholders and requirements, modeling techniques, and techniques for presenting and analyzing information.

Business analysts have three main certification levels. The first level assumes knowledge (in the absence of experience) of working with stakeholders and requirements, the second—experience from 2 to 5 years, and the third—experience over 5 years.

In addition to these levels, there are four more certification levels that confirm the competence of business analysts in various related fields and allow them to understand the broad possibilities and value of business analysis for an organization when developing solutions.

Certification in the field of agile business analysis (flexible business analysis) is one of the angles of business analysis based on agile thinking. Agile thinking is the ability to work on a project using flexible methods that involve the ability to quickly respond to a change in the request of an interested party and change the techniques of work, which almost always provides a unique method of analysis, which is important when developing a solution that can meet the needs of interested parties.

A business analyst working with flexible approaches often performs work in the field of product ownership analysis (POA). POA is a set of practices for solving problems related to creating successful and exceptional products and services.

Certification in the field of business data analytics is a combination of the competencies of a business analyst and data scientists, which business analysts also possess. The creation of this certification was a consequence of the digital transformation in the activities of organizations. In 2021, 77% of business analysts who participated in the annual survey from IIBA indicated that their organization has implemented digital technologies or is at the planning stage of their implementation. According to recent estimates (made in 2020), more than 70% of digital transformations have failed, and the main reason for this is the lack of competent analytical interpretation and use of the data obtained. Consequently, there is an increasing need for specialists in business data analytics.

Digital initiatives mostly affect such business functions as basic business processes and operations, customer service, marketing, finance, sales, maintenance, and logistics. There are the following activities in which business analysts are involved in digital initiatives: planning, vision development (envisioning), business case development, analysis, execution, rollout, and testing. About 70% of business analysts surveyed indicated that they are involved in digital initiatives in the organization or are heavily involved.

The use of digital advantages certainly brings additional value to the organization. However, as the experience of implementing IT technologies has shown, there has been some paradox at the stage of their emergence: the implementation of projects using the latest technologies did not bring the desired result and almost did not add value to the organization. This situation was typical until the benefits of the role of a business analyst in IT projects were recognized. Currently, most of these projects are already carried out with the participation of a business analyst. Nowadays, most projects have a business analyst whose tasks are to identify the real needs of the business, understand the root causes, and develop alternative solutions and recommend the one that will bring the expected value to the business.

With the advent of big data processing technologies (big data), wide possibilities of their application, and high potential value, the previous scenario is practically repeated—the use of big data for making data-driven decisions does not bring the desired positive effect without the participation of a specialist (business analyst) who knows how to work with them, interpret them, and integrate them into the solution. The role of a business analyst in digital transformation projects is even more important now than ever. To avoid the trap of choosing a slick solution and find the best solution to achieve organizational goals and objectives in this context, a business analyst will cooperate with technical experts only after he or she understands the needs of the business.

Data scientists (experts, data processing specialists) perform work on predicting the future state based on data, using statistical, mathematical, and programming skills for this. Business analysts are also needed for the efficient use of the results obtained in the development of a solution. They perform the role of a business data analyst and work closely with data scientists to develop and support data-driven solutions, thereby providing business value.

Business data analysis is a practice that applies a specific set of methods, competencies, and procedures to continuously study and research past and current business data to obtain business information that can lead to better decisions. The basis of business decisions and improvements is the evidence obtained from the data. That is, evidence is selected not to support a biased opinion or a point of view, but a point of view is formed based on data. Competencies in the field of business data analysis are not limited only to the ability of an organization to perform analytical activities. They also include capabilities such as innovation, creating a culture of data analysis, and process design.

Business data analytics provides evidence-based identification of problems and their solution. It includes six main actions related to data:

1. Access to data;
2. Study (data research);
3. Aggregation;
4. Data analysis;
5. Interpretation of data;
6. Presentation of results.

Business data analytics focuses on data collection and analysis within the framework of questions: who, what, when, where, what, why, and how. These questions determine the preliminary study of the data. After that, the question arises in the following approximate format: if something happens (or does not happen), then it will happen (or will not happen) as a result of an event (the result of something, etc.) that is different from (something, for example, existing), or will affect (something). That is, business data analytics requires a further business analysis based on the data generated by it to ensure that the analysis of this data will provide valuable information for solving important business situations (problems or opportunities).

Therefore, for the success of data-driven interaction, there must be partnership and collaboration between those who provide business experience (stakeholders and business analysts) and those who have technical skills: data analysts and data experts (data analysts and scientists). These roles interact with each other to ensure a proper understanding of the business context and find the best ways to derive value from the available data. A business analyst will first identify the needs of the business. After that, a business analyst will move on to cooperation with a data scientist and a data analyst to find the best solution to achieve the goals and objectives of the organization. As a result of this collaboration, the business analyst uses the results of the data analysis presented (data analyst) to develop business solutions and implement the final solutions. This approach avoids the mistake of choosing a "trend" solution. In small organizations, these roles can be performed by one specialist.

The 2020 and 2021 surveys showed that about 38% of business analysts are already involved in business data analytics. They note the following most important domains of business data analytics for their role in the field of business analysis:

- Data analysis—90%;
- Interpretation of reporting results—87%;
- Use of analysis results for business decision-making—87%;
- Work with data sources—86%;
- Guidance on business development strategy—84%;
- Identification of research questions—74%;
- Data analytics—67%.

The main reasons for the development of business data analytics are the use of its results in working out a development strategy, improving customer experience, developing and delivering products that customers need, ensuring revenue growth from taking advantage of innovations, obtaining additional competitive advantages, and improving the organization's efficiency.

Thus, the role of a business analyst in digital transformation projects has become extremely relevant. Therefore, the certification in Business Data Analytics (IIBA–CBDA) was opened, which confirms the competence of business analysts in the field of data analysis for successful digital transformation.

A business analyst often plays the role of a business data analyst, working closely with data science specialists to support data-driven solutions and provide business value. This allows us to develop the best solutions for this context, which, in fact, can be called the business card of a business analyst, along with his or her work on agile methods, since this approach ensures that the solution developed by business analysts will be the best in this context.

Cybersecurity issues have become even more acute since 2021. Cyber-attacks occurred every minute, while at least half of them were aimed at small businesses, which most often did not have sufficient capabilities to protect themselves from them. Even basic data protection was at risk. The largest number of cyber-attacks is still aimed at organizations providing financial services compared to organizations of other types of activity (300 times more).

The main reasons for cyber-attacks are insufficiently reliable security measures in IoT (Internet of Things) devices, low awareness of organizations in the field of protection against cyber-attacks, and insufficient attention to the risks associated with digitalization. The results of many ongoing studies in the field of cybersecurity predict a significant number of vacant jobs in this area because most organizations do not have advanced skills in the field of cybersecurity. This indicates the need for cybersecurity specialists, and this need is increasing, along with the growth of vulnerability. One of the tasks of ensuring cybersecurity is to identify the most likely areas of attack and predict possible attack scenarios. Each organization should take security measures at the strategic level. Thus, the expansion of technology in business and life has made cybersecurity analysis one of the most important issues of organizations, governments, and individuals. It has led to the inclusion of cybersecurity as part of a holistic analysis in business analysis.

Business analysis in the context of cybersecurity is aimed at ensuring the security of every aspect of the organization, identifying needs, and providing solutions. In 2021, about 20% of business analysts (and about 18% in 2020) who participated in the survey noted that their activities are related to cybersecurity in the organization and highlighted that the most important skills needed to work in this area are risk management and knowledge of regulatory requirements.

In 2020, IIBA and IEEE (Institute of Electrical and Electronics Engineers, Institute of Electrical and Electronics Engineers) developed and launched an innovative program to support practicing business analysts for whom cybersecurity has become an urgent component of their activities. This program allows one to get the basics of knowledge for understanding cybersecurity from the point of view of business analysis. In addition to this program, a Certificate in Cybersecurity Analysis (IIBA–CCA) was opened, confirming the relevant competencies of business analysts in cybersecurity.

One of the tasks of ensuring cybersecurity on the part of a business analyst is to identify the most likely areas of attack, identify the most significant data requiring priority protection, and predict possible attack scenarios, as well as the development of actions in case a cyberattack still occurs.

If we summarize the competencies of business analysts, we can say that their possible roles in the organization are both in the field of business and in the field of technology.

4   Findings

It is necessary to focus on such a feature of business analysis as the application of the data-driven decision approach. Data-driven decision-making (DDDM)—information-based decisions (or data-driven decisions) has become an alternative to the HiPPO (Highest Paid Person's Opinion) approach—decision-making based on the opinion of management, which is already outdated. Making a decision based on the HiPPO approach is not based on data, charts, reports, trends, figures, or facts. Managers who practically do not use data when making decisions are very self-confident, believing that they do not need data to confirm the correctness of their beliefs.

The priority is their expert opinion, intuition, and experience. It often happens that the situation had changed a long time before, and they did not even find out about it. This can be disastrous for business. The economic situation in the VUCA world is unstable; instability, uncertainty, and risks have only increased in the context of the pandemic. It is necessary to make important decisions carefully, correctly interpreting the information received and studying the context of the situation based on qualitative data. HiPPO's "sluggishness" can undermine the organization's authority in the market.

The data-driven approach significantly increases the opportunities for successful business development and gaining advantages over competitors as solutions. Nowadays, successful tactics and management strategies are based on this approach.

The data-driven approach has replaced the development of a solution based on expert assessments and the opinion and experience of the manager. The problem with the HiPPO approach is that the manager cannot be competent and objective in all matters and know all the features and subtleties of the context.

The data-driven approach assumes that the stage of solution development should include a full understanding of what this decision will affect, who and how it will affect, what will need to be changed in the process of its implementation, and what result can be achieved in the end.

Working with data, including big data, a business analyst will first identify the needs of the business and only then proceed to develop a solution to achieve the goals and objectives of the organization in accordance with its limitations and capabilities. As a result, when developing solutions, it is possible to avoid the mistake of choosing a "trend" solution; each solution is unique because it affects the cause of its occurrence in a particular context.

Despite the high value of the data-driven approach, according to repeatedly conducted research, the number of data-driven companies is not yet large. For example, Melbourne Business School studied how companies in 46 countries use analytics. The results showed that only 6% could be considered leaders in this direction. This is a business in which the analytical strategy is well developed, and all departments are included in it. Top management makes decisions solely based on data.

It should be noted that 49% of companies fall into the category of "Researchers" because they partially use data for decision-making and have not fully developed the infrastructure for a full-fledged data-driven approach.

The rest of the companies are in the group of "Imitators" and "Laggards." They use data only in one specific area or do not develop analytics at all.

The reason for the low level of use of this approach can be called insufficient awareness of managers about the possibilities and advantages of this approach. The difficulty lies in overcoming one's own views and commitment to the HiPPO approach, as well as underestimating the importance of business analysts in developing solutions based on the data-driven approach. Simultaneously, annual studies conducted by IIBA show that decisions made based on data after conducting a business analysis can increase the ROI (return on investment ratio) by three times. Data-driven decisions make it possible to make an organization the best in its field or superior to competitors and have a return on investment of 50% or higher. Additionally, organizations have 2.8 times more failed projects if they do not conduct business analysis and do not develop data-based solutions.

5 Conclusion

The data-driven model is the only way to change the approach to decision-making in an organization, resist preconceived opinions, and make objective decisions. It is not positions and titles that should win but conclusions supported by data. The market and the rules of the game are changing so fast that trends and deviations can be identified in time, and they can be used efficiently only with the help of analytical technologies. Data-based solutions will create a new level of the corporate culture. The developed solutions will be of particular value and ensure that the problem is completely solved (will not arise again).

References

1. Albekov A, Romanova T, Vovchenko N, Epifanova T (2017) Study of factors which facilitate increase of effectiveness of university education. Int J Educ Manag 31(1):12–20. https://doi.org/10.1108/IJEM-02-2016-0037
3. Chernysheva YuG, Shepelenko GI (2018) The new profession of "Business analyst" and the new occupational standards: the case of Russia. Eur Res Stud J 21(SI1):86–94. https://doi.org/10.35808/ersj/1161
2. Chernysheva YG, Shepelenko GI, Gashenko IV, Orobinskaya IV, Zima YS (2017) Business analysis as an important component of ensuring enterprise's economic security. Eur Res Stud J 20(3B):250–259. Retrieved from https://www.um.edu.mt/library/oar/bitstream/123456789/30807/1/Business_Analysis_as_an_Important_Component_of_Ensuring_Enterprises_Economic_Security_2017.pdf. Accded 28 Aug 2022
4. International Institute of Business Analysis (IIBA) (2015) BABOK: a guide to the business analysis body of knowledge (version 3.0). IIBA, Toronto, Canada. Retrieved from https://tinyurl.com/mrx4ptrf. Accded 28 Aug 2022
5. Kotler P, Turner RE (1985) Marketing management: analysis, planning and control. Prentice-Halt, Ontario, Canada
6. Porter ME (1980) Competitive strategy: techniques for analyzing industries and competitors. Free Press, New York, NY (Republished with a new introduction, 1998)

A Smart Approach to In-House Analytics and Business Management 4.0

Victoria A. Bondarenko[ID], **Olesya V. Ivanchenko**[ID], and **Natalia V. Przhedetskaya**[ID]

Abstract This research discusses the use of a smart in-house analytics approach as part of accelerating digital transformation and achieving strategic business management goals. The paper aims to determine the main directions of strategic management in the digital economy and the content of a contemporary business process management system. The authors investigate business intelligence systems as tools for implementing smart business management 4.0 approaches to analyze and make real-time management decisions in an uncertain environment. The authors highlight the reasons for using business intelligence (BI) systems and the challenges that these systems address in terms of implementing a smart approach. The functionality of BI systems and the cycle of data analysis and control decisions in providing a complete and reliable analysis of a company's business processes are explored. Considering the emergence of new areas of technology, the authors discuss the business and technology fundamentals of Business 4.0. The analysis shows the need for management transformation using the agile methodology within the framework of digitalization of business processes and innovation.

Keywords Management · Business 4.0 · Smart approach · Digitalization · Business analytics · Business intelligence · Artificial Intelligence

JEL Classification M10 · M15 · M21 · O14 · O32 · O33

V. A. Bondarenko (✉) · O. V. Ivanchenko · N. V. Przhedetskaya
Rostov State University of Economics, Rostov-on-Don, Russia
e-mail: b14v@yandex.ru

N. V. Przhedetskaya
e-mail: nvpr@bk.ru

1   Introduction

Business management is now focused on coping with the COVID-19 pandemic. In this context, the role of digitalization has become central to the continuation of economic and social activity. Overall, the COVID-19 pandemic has accelerated the digital transformation of business management as businesses move towards distributed employment models and the provision of digital services and products. The COVID-19 pandemic has pushed consumers and businesses to adopt digital services and technologies, accelerating the digital transformation of consumer behavior and business activity over the long term.

Developing IT skills and capabilities is a key priority for businesses to accelerate digital transformation as a means of delivering results. Moreover, the development of appropriate content, applications, and services and the creation of an enabling environment will encourage consumers and businesses to adopt digital services and digital technologies further. Artificial intelligence (AI), the Internet of Things (IoT), cloud computing, distributed registry technology, digital commerce, autonomous mobility, and many other emerging technology areas will shape approaches to organizational management [12, 17].

Developing the skills and capacity to take a smart approach to in-house analytics is a key priority to accelerate digital business transformation as a means of achieving the company's strategic goals [4, 5].

The current development and deployment of IT infrastructure and related services continue the trends toward digital transformation for society in general and business in particular [2, 14].

2   Materials and Method

As part of the study of smart in-house analytics as an organizational solution to accelerate digital transformation, the research considers how necessary it is to achieve the strategic goals of business management 4.0.

Scientific and practical developments of Russian and foreign authors in the field of new technological sectors, digitalization, and advanced methods of organizing business processes and business models have formed an informational and empirical basis for our research. Additionally, the authors analyzed secondary data from open Internet sources. The collected empirical data allow us to get the complete picture of management approaches to introducing innovations and a "smart" approach to in-house analytics in business management 4.0.

Such general scientific methods and techniques as a system and logical-semantic analysis, generalization, synthesis of the received information, expert evaluation, and graphical interpretation of empirical and factual information were used as research tools.

3   Results

Strategic management in the digital economy includes the following areas:

- Strategy and state management of the digital enterprise;
- Development management based on the goal tree and strategic map;
- New products as company growth points and digital developments;
- Using advanced technology to improve productivity and enterprise value;
- Business intelligence systems, artificial intelligence;
- Requirements management in building complex technical systems;
- Digital ecosystem architecture and disruptive technologies;
- Systems engineering (SE) approaches to deliver guaranteed results on schedule;
- Advanced methodology for designing complex technical systems;
- Corporate architecture and the benefits of the 'digital enterprise';
- Digitalization of management—digital twin management and design automation;
- Integrated communications of the digital enterprise and digital marketing.

The goal of smart business management 4.0 is to link all information resources of an enterprise into a single digital environment, plan production resources more intelligently, and make necessary management decisions in time based on up-to-date data.

The company's business process management system includes the following:

1. Business process architecture;
2. Management of business processes by objectives and indicators;
3. An incentive system for managers to improve business processes by KPI;
4. Business process description and analysis practices;
5. The practice of business process optimization and change implementation;
6. Automation of business processes;
7. Standardization of business processes;
8. Control and audit of business processes;
9. Corporate system of personnel training to the methods of process management;
10. Process office [22].

Business intelligence (BI) systems are a tool for implementing a smart approach to business management 4.0. They can quickly process large amounts of data from various sources and visualize it in easy-to-use reports to analyze and make management decisions [6–8, 16, 21].

The reasons for using business intelligence systems are as follows:

- Lack of a single information source;
- Lack of up-to-date information in real-time;
- The necessity to support heterogeneous systems;
- Difficulty in obtaining reporting, lack of full automation of the process [9, 11, 19].

Challenges addressed by BI systems in implementing a smart approach to in-house analytics include the following:

- A huge amount of data arises during the digitalization of an organization's activities. This data is generated by different information systems and has different formats. The BI system brings the data to a single format, collects it in one place, and shows it as a clear infographic.
- Manual generation of reports requires the time and human power of specially trained personnel. The BI system quickly generates reports in an intuitive interface that does not require knowledge of programming languages.
- The origin of certain indicators, as well as their reliability, is difficult to verify. The BI system clearly demonstrates the origin of each indicator.
- The reciprocal influence of some organizational performance indicators on others is not always apparent. Therefore, steps taken to optimize performance may not achieve their objectives. The BI system reveals cause-effect relationships between indicators and takes the speed, depth of analysis, and quality of decisions to a new level.

The cycle of data analysis and control decisions using the BI system is shown in Fig. 1.

The functionality of typical BI systems includes the following:

- Creation of reports based on data from various sources;
- Integrity, reliability, and consistency of information provided by using special procedures for working with data;
- Multivariate analysis in all profiles: one can add new indicators and analytics without changing the structure of the repository;
- Automatic generation of pre-configured reports;
- Self-creation of reports using a visual constructor that does not require knowledge of programming languages;
- Versioning support: operative and historical reports required for making tactical and strategic decisions;

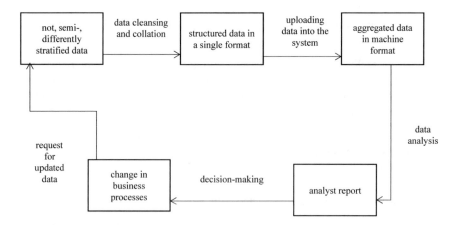

Fig. 1 BI-based data analysis and management decision cycle. *Source* Developed by the authors

- Visualization options: charts, graphs, and tables;
- Access via a web browser from any device, including mobile devices;
- Deployment of the system on dedicated servers and in the cloud;
- Flexible configuration of user roles with various access rights;
- Work in open and closed loops.

Experts highlight the following trends in the development of business intelligence:

1. Active business intelligence. This means the ability to automatically react to various process deviations, automate decision-making processes based on available historical data, and finally bring processes to a reference state. The user is no longer required to monitor dashboards and analyze KPIs constantly. The system reacts to deviations and takes action as required for the business.
2. Self-service analytics. Gartner called self-service or self-service analytics a trend back in 2018 based on the jump in the volume of information, the ever-increasing demand for big data, and the growing interest in Data Science. Self-service is an opportunity to make analytics for decision-making based on facts, not emotions and hypotheses, available to all people, not just business analysts. Users want data to be available anytime and anywhere. In turn, companies want simplified access to data while ensuring consistency and high-quality data and results.
3. Process mining as part of business intelligence. Process mining is process analytics. Tools with this capability extract data from a company's internal systems to provide a complete and reliable analysis of business processes. This allows one to reconstruct the actual flow of the required process and find bottlenecks and deviations from the regulated model [3].

4 Discussion

The advent of technology that enabled devices to connect to the Internet, store and process information in virtual clouds, automate repetitive tasks, and innovate led to the development of Business 4.0, comprising four business aspects and four technological pillars [1, 13, 15, 18, 23].

The business aspects are as follows:

1. Mass personalization—the ability to customize offers to meet the requirements of individual customers;
2. Environmental uncertainty and risk-taking;
3. Exponential growth in data;
4. Formation of ecosystems based on different technology platforms.

The technology pillars are as follows:

1. Agile—an approach that enables high performance in the workplace and adapts the business to customer changes;
2. Business process automation;
3. Innovation;

4. Cloud technologies [20].

A significant proportion of companies are now adopting the agile project implementation methodology and are successfully meeting market demand. Although agile projects are more successful than traditional projects, a certain level of risk remains. However, it can be reduced by implementing appropriate controls to realize value for the business, minimizing the risk of creating an unclaimed product, and improving overall development efficiency [10].

If a company adopts an Agile project delivery methodology, those in charge of corporate governance need to understand how they can maximize the benefit to the organization by developing the necessary resources, freeing up capacity, and gaining confidence in their ability to bring the best product to market while minimizing risk by implementing controls early and regularly.

Agile does not offer perfect solutions in terms of project risks. Risks remain. Nevertheless, by using the agile project delivery methodology, the project team can respond to risk earlier in the development lifecycle through the ability to provide greater visibility into operations and continuous product improvement.

To become an agile organization, it is necessary to synchronize release procedures with strategy, improve technical resources and tools, transform the mindset of employees, and accelerate processes.

Process acceleration can follow the following pattern:

1. Implement short product release cycles with regular demonstration meetings and continuous involvement of the business units;
2. Update the change and defect management procedure to ensure consistency and sequencing of product improvement activities;
3. Ensure continuous "inspect and adapt" training to ensure the practical application of what has been learned;
4. Agree on an approach for implementing the agile project implementation methodology and assess whether the necessary assessment values and agile principles are being adhered to.

Transformation of management includes the following:

- Creating cross-functional, self-organized, and motivated project teams that are entirely focused on the project;
- Increasing the level of employee engagement by giving them greater independence, a specific mission and purpose, and maximum discretion;
- Investing in coaching programs to develop an agile mental mindset and provide continuous training in agile methods and practices;
- Strengthening a culture of open collaboration and cooperation.

Synchronization of release procedures with the strategy includes the following:

1. Prioritization of projects and stories based on their value to the business in line with the overall company strategy;
2. Decentralization of decision-making to reduce delays and ensure that a valuable product is received promptly;

3. Increase of transparency through regular monitoring of velocity and burndown schedules to assess project risks;
4. Minimization of risks through frequent demonstration meetings and encouraging feedback from stakeholders.

The improvement of technical resources and facilities can be carried out according to the following scheme:

- Eliminate factors negatively affecting productivity through proactive, comprehensive automation, version control of the source code, and organizational and technical support;
- Stimulate a process of continuous integration of changes made by agile teams into common code libraries;
- Improve quality through automated code reviews and continuous adaptation;
- Use agile tools for transparent and efficient management, support, collaboration, and reporting.

Developing a better understanding of customers is becoming increasingly strategic because rapidly changing markets, new technologies, and new business models are changing what customers want and how they shop.

5 Conclusion

The COVID-19 pandemic has accelerated the implementation of digital projects in Russian companies. It was important to ramp up production volumes and increase the speed and efficiency of business processes in a short timeframe. Business analytics systems that implement a smart approach make it possible to link all information resources of an enterprise into a single digital environment to make tactical and strategic decisions aimed at improving the efficiency of operations in an uncertain external environment.

In terms of further digitalization of business processes and the development of ICT infrastructure and integrated technologies, there is still considerable scope for capacity development in the area of business intelligence systems and AI. The IoT market is still developing, especially in megacities. Cloud services are still dominated by foreign firms due to a lack of Russian platforms and declining international partnerships. The key challenges that need to be overcome to accelerate the development of BI systems, artificial intelligence, IoT, and cloud technologies are issues related to data sharing and protection and bridging the social, economic, and digital divide.

References

1. Alam M, Khan IR (2020) Business 4.0—a new revolution. In: Information technology for management. KD Publications, Delhi, India, pp 41–57. https://doi.org/10.6084/m9.figshare. 14369636
2. Ananyin VI, Zimin KV, Lugachev MI, Gimranov RD, Skripkin KG (2018) Digital organization: transformation into the new reality. J Bus Inform 2(44):45–54. https://doi.org/10.17323/1998-0663.2018.2.45.54
3. Bochkin A (2022, July 28) 3 trends in the development of Russian business analytics for the coming year. Retrieved from https://www.e-xecutive.ru/management/itforbusiness/1995276-3-trenda-razvitiya-rossiiskoi-biznes-analitiki-na-blizhaishii-god 2022. Accessed 10 Sept 2022
4. Bondarenko V, Guzenko N, Romanishina T, Leventsov V, Gluhov V (2022) Information and communications technology in the development of territories based on designing "smart cities." In: Koucheryavy Y, Balandin S, Andreev S (eds) Internet of things, smart spaces, and next generation networks and systems. Springer, Cham, Switzerland, pp 59–68. https://doi.org/10. 1007/978-3-030-97777-1_6
5. Bondarenko V, Romanishina T, Guzenko N, Mukhanova N, Salkutsan S (2022) Developing smart cities: the risks of using information and communications technology. In: Lecture notes in computer science, 13158 LNCS, pp 71–80. https://doi.org/10.1007/978-3-030-97777-1_7
6. Bruskin SN (2017) Models and tools of predicting analytical research for digital corporation. Vestnik Plekhanov Russ Univ Econ 5:135–139. Retrieved from https://vest.rea.ru/jour/article/view/375/340. Accessed 10 Sept 2022
7. Business Portal "TAdviser" (2022, August 4) AI in analytics: what's beyond BI? Retrieved from https://www.tadviser.ru/index.php/Статья:ИИ_в_аналитике:_что_за_пределами_ BI. Accessed 10 Sept 2022
8. Coria JAG, Castellanos-Garzón JA, Corchado JM (2014) Intelligent business processes composition based on multi-agent systems. Exp Syst Appl 41(4):1189–1205. https://doi.org/10.1016/ j.eswa.2013.08.003
9. Dolganova OI, Deeva EA (2019) Company readiness for digital transformations: problems and diagnosis. J Bus Inform 13(2):59–72. https://doi.org/10.17323/1998-0663.2019.2.59.72
10. Ghezzi A, Cavallo A (2020) Agile business model innovation in digital entrepreneurship: lean startup approaches. J Bus Res 110:519–537. https://doi.org/10.1016/j.jbusres.2018.06.013
11. Grekul VI, Isaev EA, Korovkina NL, Lisienkova T (2019) Developing an approach to ranking innovative IT projects. J Bus Inform 13(2):44–58. https://doi.org/10.17323/1998-0663.2019. 2.43.58
12. Ivanchenko OV, Mirgorodskaya ON, Baraulya EV, Putilina TI (2019) Marketing relations and communication infrastructure development in the banking sector based on big data mining. Int J Econ Bus Adm 7(2):176–184. https://doi.org/10.35808/ijeba/382
13. Kapustin V, Danilov I (2020, August 13) What strategy should I choose for a win-win transformation of the enterprise into the Industry 4.0 format? Retrieved from https://integral-russia.ru/2020/08/13/kakuyu-strategiyu-vybrat-dlya-besproigryshnoj-tra nsformatsii-predpriyatiya-v-format-industrii-4-0/. Accessed 10 Sept 2022
14. Khasanov M, Krasnov F (2019) Transactional digital transformation in scientific organizations. PROneft. Professionally about Oil 1:64–67. Retrieved from https://proneft.elpub.ru/jour/art icle/view/301/301?locale=en_US. Accessed 10 Sept 2022
15. Kleimenova L (2021, May 4) What is Industry 4.0 and what you need to know about it. RBC Trends. Retrieved from https://trends.rbc.ru/trends/industry/5e740c5b9a79470c22 dd13e7. Accessed 10 Sept 2022
16. Krylov SI (2018) Integrated management analysis of innovation: a conceptual and methodological framework. Digest Finan 23(4):466–482. https://doi.org/10.24891/df.23.4.466
17. Mirgorodskaya ON, Ivanchenko OV, Dadayan NA (2020) Using digital signage technologies in retail marketing activities. In: DTMIS'20: Proceedings of the international scientific conference—digital transformation on manufacturing, infrastructure and service, Article 76.

Association for Computing Machinery, New York, NY, pp 1–7. https://doi.org/10.1145/344 6434.3446476

18. Müller JM, Buliga O, Voigt KI (2018) Fortune favors the prepared: how SMEs approach business model innovations in Industry 4.0. Technol Forecast Soc Chang 132:2–17. https://doi. org/10.1016/J.TECHFORE.2017.12.019
19. Nissen V, Tatiana L, Saltan A (2018) The role of IT-management in the digital transformation of Russian companies. Foresight STI Govern 12(3):53–61. https://doi.org/10.17323/2500-2597. 2018.3.53.61
20. Ojala A, Tyrvainen P (2011) Developing cloud business models: a case study on cloud gaming. IEEE Softw 28(4):42–47. https://doi.org/10.1109/MS.2011.51
21. Park S (2017) Understanding digital capital within a user's digital technology ecosystem. In: Park S (ed) Digital capital. Palgrave Macmillan, London, UK, pp 63–82. https://doi.org/10. 1057/978-1-137-59332-0_4
22. Repin V (n.d.) Assessment of maturity of the business process management system. Retrieved from https://www.lobanov-logist.ru/library/all_articles/64181/. Accessed 10 Sept 2022
23. Vaidya S, Ambad P, Bhosle S (2018) Industry 4.0—a glimpse. Procedia Manuf 20:233–238. https://doi.org/10.1016/j.promfg.2018.02.034

Business Management in the Context of the Sustainable Development Paradigm

Olga V. Konina and **Alexander V. Tekin**

Abstract The paper aims to determine the contribution of business in implementing the SDGs and clarify the concept of business management in the context of the sustainable development paradigm. The author applies the regression analysis method to determine the contribution of business to the implementation of the SDGs, which assesses the impact of the ease of doing business index on the sustainable development index. The sample includes the top 10 developed markets and the top 10 dynamic emerging markets, which are the leaders in the UNDP Doing Business 2020 ranking. As a result, the authors identified and measured the significant contribution of businesses to implementing the SDGs. The authors also clarified the concept of business management in the context of the sustainable development paradigm and demonstrated its features in each management area. The contribution of the research to the literature is that the resulting econometric model revealed a pattern of sustainable development with business support, which can be used to predict sustainable development with the most accurate and reliable account of the influence of the factor of doing business. The applied significance for business management is that the refined concept of business management in the context of the sustainable development paradigm can be used in business practice to integrate the SDGs in the field of management.

Keywords Business management · Sustainable development paradigm · Sustainable Development Goals (SDGs) · Pure market competition · Corporate responsibility

JEL Classification G34 · M14 · M21 · Q01

O. V. Konina (✉)
Moscow Pedagogical State University, Moscow, Russia
e-mail: koninaov@mail.ru

A. V. Tekin
Volgograd State Technical University, Volgograd, Russia

© The Author(s), under exclusive license to Springer Nature Switzerland AG 2023
E. G. Popkova (ed.), *Smart Green Innovations in Industry 4.0 for Climate Change Risk Management*, Environmental Footprints and Eco-design of Products and Processes, https://doi.org/10.1007/978-3-031-28457-1_6

1 Introduction

The paradigm of sustainable development is a set of fundamental scientific and practical principles of the Noosphere doctrine [11], the essence of which is reduced to a systemic consideration and management of society, economy, and the environment under their unity, equivalence, and harmony. The key principle of sustainable development is to abandon the consumerism of human and natural resources in favor of their equivalence with economic resources. A quantitative measure of sustainable development is the sustainability index, which reflects the degree to which the SDGs have been achieved [16]. In business management, this principle is implemented in practice through the mechanism of corporate social (and environmental) responsibility.

Although the existing literature notes active support for the SDGs in business and a significant business contribution to the implementation of the SDGs, this contribution has not been quantified and is contradictory from a business management perspective [3]. On the one hand, today's business cannot stay away from the popular worldwide paradigm of sustainable development supported by society and the country [14]. In addition to this, many business people are altruistic and seek to maintain the SDGs not so much under external stimulus but by intrinsic motivation [6].

On the other hand, the sustainable development paradigm does not fit into the model of a pure market economy, as it violates free competition and requires businesses to give up part of their profits in favor of society and the environment [19].

In this regard, the revision of the concept of business management is relevant, aimed at finding a new Pareto optimum, which will balance the interests of profit maximization and its limitations associated with the implementation of the SDGs in business. This study seeks to determine the contribution of businesses in implementing the SDGs and clarify the concept of business management in the context of the sustainable development paradigm.

2 Literature Review

The theoretical basis of this research is formed using the business management concept. The provisions of this concept and the essence of business management in the pure market competition are disclosed in sufficient detail in the works of Corrales-Garay et al. [4], Garina et al. [5], Schrage and Rasche [9], Starinov et al. [15], and Zhuravleva and Grigoryan [20]. A significant impact of the sustainable development paradigm on business management is noted in the works of Al-Baghdadi et al. [1], Arshad et al. [2], Popkova and Sergi [7], Popkova et al. [8], Shabaltina et al. [10], Silva and Nunes [12], Singh et al. [13], and Waiyawuththanapoom et al. [17].

However, existing publications insufficiently elaborate on the essence and characteristics of business management in the context of the sustainable development paradigm, which represents a research gap. Another gap in the literature is related to the uncertainty of the scale of business contributions to the SDGs. The need to fill the gaps is due to the fact that they hinder the reliable assessment and disclosure of business management capacity to support sustainable development. This paper is devoted to filling the identified gaps.

3 Materials and Method

To determine the contribution of business to the implementation of the SDGs, the authors applied the method of regression analysis to assess the impact of the ease of doing business index (according to the World Bank [18]) on the sustainable development index (according to UNDP [16]). The sample includes the top 10 developed markets and the top 10 dynamic emerging markets, which are the leaders in the UNDP [16] Doing Business in 2020 (Table 1).

According to Table 1, in developed markets, ease of doing business averaged a score of 83.53, and sustainability averaged a score of 79.42. In dynamic emerging markets, ease of doing business averaged a score of 77.90, and sustainability averaged 70.02.

4 Results

To determine the dependence of the sustainable development index (y) on the ease of doing business index (x), the authors performed a regression analysis of the data on the entire sample from Table 1. The results of the analysis are shown in Table 2.

Based on the data in Table 2, we can make the following model of pairwise linear regression:

$$y = -7.86 + 1.02x \tag{1}$$

According to model (1), when the business operation is simplified by 1 point, the sustainable development index increases by 1.02 points. The 53.18% change in the sustainable development index is due to a change in the ease of doing business index. The model is reliable at the 0.05 significance level. Based on the model (1), it was found that with the growth of the ease of doing business index from 80.72 points on average in the sample to the maximum possible 100 points (+23.89%), the index of sustainable development will increase from 74.72 points on average in the sample in 2021 to 94.45 points (i.e., by 26.41%).

Consequently, the business makes a meaningful contribution to sustainable development. To determine what makes this contribution, the authors conducted a comparative analysis of business management in pure market competition, in the paradigm of sustainable development, and in the context of various areas of management

Table 1 Statistics on doing business and sustainability in developed and dynamic markets in 2021

Category of countries by level and pace of market development	Place in the ranking of doing business in 2020	Country	Ease of doing business in 2020, score 0–100	Sustainable development index in 2021, score 0–100
Developed markets	1	New Zealand	86.8	79.1
	2	Singapore	86.2	69.9
	4	Denmark	85.3	84.9
	5	Republic of Korea	84.0	78.6
	6	USA	84.0	76.0
	8	UK	83.5	80.0
	9	Norway	82.6	82.0
	10	Sweden	82.0	85.6
	14	Australia	81.2	75.6
	22	Germany	79.7	82.5
Rapidly growing markets	12	Malaysia	81.5	70.9
	16	UAE	80.9	70.2
	21	Thailand	80.1	74.2
	25	Kazakhstan	79.6	71.6
	28	Russia	78.2	73.8
	31	China	77.9	72.1
	33	Turkey	76.8	70.4
	38	Rwanda	76.5	57.6
	49	Belarus	74.3	78.8
	56	Kenya	73.2	60.6

Source Compiled by the authors based on the materials of the UNDP [16] and World Bank [18]

(project management, marketing management, innovation management, financial and investment management, strategic management, risk management, information management, and environmental management).

According to comparative analysis, in the context of the sustainable development paradigm, a serious transformation of the business management concept in terms of its areas is necessary. Let us consider the necessary changes in these areas. Thus, project management implies the priority of efficiency (optimization of the cost–benefit ratio) in conditions of pure market competition. In the context of the sustainable development paradigm, project management implies the priority of sustainability (stability of business development and its support of the SDGs).

Marketing management in conditions of pure market competition implies the promotion of business as commercially attractive. In the context of the sustainable development paradigm, marketing management implies the promotion of business as supporting the SDGs. Innovation management in conditions of pure market

Table 2 Regression statistics of the relationship between the index of sustainable development and the index of ease of doing business

Regression statistics					
Multiple R	0.53176				
R-squared	0.28277				
Normalized R-squared	0.24292				
Standard error	6.3244				
Observations	20				
Variance analysis					
	df	SS	MS	F	Significance of F
Regression	1	283.847	283.847	7.09652	0.01582
Remainder	18	719.965	39.9981		
Total	19	1003.81			
Parameters of the regression equation					
	Coefficients	Standard error	t-statistics	P-value	Lower 95%	Higher 95%
Constant	−7.8578	31.0307	−0.2532	0.80296	−73.051	57.3354
x	1.02308	0.38405	2.66393	0.01582	0.21622	1.82993

Source Calculated and compiled by the authors

competition implies a preference for the most effective innovations. In the context of the sustainable development paradigm, it implies a preference for responsible (sustainable) innovation.

Financial and investment management in conditions of pure market competition implies a preference for commercial investments. In the context of the sustainable development paradigm, it implies a preference for ESG investments (investments in the implementation of SDGs). Strategic management in pure market competition involves the adoption and implementation of a market strategy. In the context of the sustainable development paradigm, strategic management implies the adoption and implementation of a sustainable development strategy.

Risk management in a pure market competition assumes management primarily of economic risks. In the context of the sustainable development paradigm, it assumes system management of social, economic, and environmental risks. In pure market competition, information management involves the publication of financial statements. In the context of the sustainability paradigm, it involves the publication of sustainability reports. Environmental management in a pure market competition involves compliance with environmental legislation. In the context of the sustainable development paradigm, it involves corporate environmental responsibility.

5 Conclusion

Thus, a significant contribution of business to implementing the SDGs has been identified and measured quantitatively. The change in the index of sustainable development by 53.18% is due to changes in the ease of doing business index. In the future, due to the simplification of doing business, the sustainable development index can increase up to 26.41%. The authors also clarified the concept of business management in the context of the sustainable development paradigm and demonstrated its features in each area of management.

The contribution of the research to the literature is that the resulting econometric model revealed a pattern of sustainable development with business support, which can be used to predict sustainable development with the most accurate and reliable account of the influence of the factor of doing business on it. The practical significance of these results is that the clarified role of business management in supporting sustainable development has opened up additional opportunities for improving (increasing the flexibility of) state regulation of sustainable development—through improving the business climate, which was previously seen exclusively as a measure to support business.

The business-management implications are that the refined concept of business management in the context of the sustainable development paradigm can be used in business practice to integrate the SDGs in the field of management. Nevertheless, the results of this research are limited to a generalized study of developed and dynamically developing countries. In further research, it is advisable to study their experiences separately to identify and consider their specificities and develop specific recommendations for integrating the SDGs into business management practices in the countries of each category.

References

1. Al-Baghdadi EN, Alrub AA, Rjoub H (2021) Sustainable business model and corporate performance: the mediating role of sustainable orientation and management accounting control in the United Arab Emirates. Sustainability 13(16):8947. https://doi.org/10.3390/su13168947
2. Arshad NI, Bosua R, Milton S, Mahmood AK, Zainal-Abidin AI, Ariffin MM et al (2022) A sustainable enterprise content management technologies use framework supporting agile business processes. Knowl Manag Res Prac 20(1):123–140.https://doi.org/10.1080/14778238.2021.1973352
3. Cammarano A, Perano M, Michelino F, Del Regno C, Caputo M (2022) SDG-oriented supply chains: business practices for procurement and distribution. Sustainability 14(3):1325. https://doi.org/10.3390/su14031325
4. Corrales-Garay D, Ortiz-de-Urbina-Criado M, Mora-Valentín E-M (2022) Understanding open data business models from innovation and knowledge management perspectives. Bus Process Manag J 28(2):532–554. https://doi.org/10.1108/BPMJ-06-2021-0373
5. Garina EP, Romanovskaya EV, Andryashina NS, Kozlova EP, Efremova AD (2021) Study of restructuring strategies: decentralization of management and enterprise structure. In: Popkova

EG, Ostrovskaya VN (eds) Meta-scientific study of Artificial Intelligence. Information Age Publishing, Charlotte, NC, pp 559–565

6. Pizzi S, Rosati F, Venturelli A (2021) The determinants of business contribution to the 2030 Agenda: introducing the SDG reporting score. Bus Strateg Environ 30(1):404–421. https://doi.org/10.1002/bse.2628

7. Popkova EG, Sergi BS (2021) Dataset modeling of the financial risk management of social entrepreneurship in emerging economies. Risks 9(12):211. https://doi.org/10.3390/risks9120211

8. Popkova E, Bogoviz AV, Sergi BS (2021) Towards digital society management and 'capitalism 4.0' in contemporary Russia. Humanities Soc Sci Commun 8(1):77. https://doi.org/10.1057/s41599-021-00743-8

9. Schrage S, Rasche A (2022) Inter-organizational paradox management: how national business systems affect responses to paradox along a global value chain. Organ Stud 43(4):547–571. https://doi.org/10.1177/0170840621993238

10. Shabaltina LV, Karbekova AB, Milkina E, Pushkarev IY (2021) The social impact of the downturn in business and the new context of sustainable development in the context of the 2020 economic crisis in developing countries. In: Popkova EG, Sergi BS (eds) Modern global economic system: evolutional development vs. revolutionary leap. Springer, Cham, Switzerland, pp 74–82. https://doi.org/10.1007/978-3-030-69415-9_9

11. Shoshitaishvili B (2021) From Anthropocene to noosphere: the great acceleration. Earth's Future 9(2):e2020EF001917. https://doi.org/10.1029/2020EF001917

12. Silva ME, Nunes B (2022) Institutional logic for sustainable purchasing and supply management: concepts, illustrations, and implications for business strategy. Bus Strateg Environ 31(3):1138–1151. https://doi.org/10.1002/bse.2946

13. Singh RK, Kumar Mangla S, Bhatia MS, Luthra S (2022) Integration of green and lean practices for sustainable business management. Bus Strateg Environ 31(1):353–370. https://doi.org/10.1002/bse.2897

14. Sinkovics N, Sinkovics RR, Archie-Acheampong J (2021) The business responsibility matrix: a diagnostic tool to aid the design of better interventions for achieving the SDGs. Multinatl Bus Rev 29(1):1–20. https://doi.org/10.1108/MBR-07-2020-0154

15. Starinov GP, Tseveleva IV, Pershina EY (2020) Foresight system in the structure of criminological forecasting of corruption in the business environment in the multicultural space of Russia. In: Solovev DB, Savaley VV, Bekker AT, Petukhov VI (eds) Proceeding of the international science and technology conference "FarEastCon 2019". Springer, Singapore, pp 379–390. https://doi.org/10.1007/978-981-15-2244-4_35

16. UNDP (2021) Sustainable Development Report 2021. Retrieved from https://www.sdgindex.org/reports/sustainable-development-report-2021/. Accessed 6 May 2022

17. Waiyawuththanapoom P, Thammaboosadee S, Tirastittam P, Jermsittiparsert K, Wongsanguan C, Sirikamonsin P et al (2022) The role of human resource management and supply chain process in sustainable business performance. Uncertain Supply Chain Manag 10(2):517–526.https://doi.org/10.5267/j.uscm.2021.11.011

18. World Bank (2022) Doing Business 2020. Retrieved from https://documents1.worldbank.org/curated/en/688761571934946384/pdf/Doing-Business-2020-Comparing-Business-Regulation-in-190-Economies.pdf. Accessed 6 May 2022

19. Yankovskaya VV, Bogoviz AV, Lobova SV, Trembach KI, Buravova AA (2022) Framework strategy for developing regenerative environmental management based on smart agriculture. In: Popkova EG, Sergi BS (eds) Smart innovation in agriculture. Springer, Singapore, pp 281–286. https://doi.org/10.1007/978-981-16-7633-8_31

20. Zhuravleva NA, Grigoryan MG (2021) The principle of subjectification in the assessment of the management performance of the organization in terms of the development of artificial intelligence. In: Popkova EG, Ostrovskaya VN (eds) Meta-scientific study of Artificial Intelligence. Information Age Publishing, Charlotte, NC, pp 567–573

Study of the Potential of Public–Private Partnership Mechanisms in Projects of Smart Environmental Modernization of Residential Infrastructure

Svetlana B. Globa⑩, **Evgeny P. Vasiljev**⑩, and **Viktoria V. Berezovaya**⑩

Abstract The paper studies the possibilities of public–private partnerships in the field of residential (utility) infrastructure and the development of a promising model for financing public–private partnership projects based on smart technology. High-quality and advanced infrastructure, engineering networks, and communications that serve residential buildings are integral elements in the formation of a favorable living environment for the population and ensuring the compliance of the resources received with the established standards. The authors use methods based on a quantitative and qualitative assessment of the potential for using public–private partnership mechanisms to justify management decisions on the possibility of applying certain models and forms of public–private partnership when concluding agreements with goals of modernization of residential infrastructure using advanced technology. The research object is contracts on the terms of public–private partnership in smart environment projects aimed at the environmental modernization of residential infrastructure. It is shown that this mechanism contributes to the fact that the funds of the state, regional, and municipal budgets can be utilized more cost-effectively due to the replacement of public investment resources with private investments within the framework of public–private partnership. This provides an opportunity to increase the volume of investment in the country's municipal infrastructure through various combinations of public and private funds. The research proposes a mechanism for analyzing the potential of public–private partnership schemes, which allows for choosing the most optimal forms and models for a specific project associated with the development of municipal infrastructure.

Keywords Public–private partnership · Concession · Financial risks · Utility infrastructure · Project potential

JEL Classification Q48 · Q56 · Q57 · Q53 · O33 · L97

S. B. Globa (✉) · E. P. Vasiljev · V. V. Berezovaya
Siberian Federal University, Krasnoyarsk, Russia
e-mail: sgloba@sfu-kras.ru

V. V. Berezovaya
e-mail: VVBerezovaya@sfu-kras.ru

1 Introduction

Problems related to utility infrastructure directly affect the improvement of the national level and the quality of life. Even though considerable attention is given to the development of municipal infrastructure in Russia, the situation remains quite difficult. Particularly, the depreciation rate of communal infrastructure facilities in some subjects of the Russian Federation is still 50–60%.

With the growing digital maturity of cities and regions and the need to increase the level of security, there is an increasing importance of introducing digital technologies in the housing and utility sector, providing automation of work performed and communication with consumers, energy efficiency, and optimization of resource use. Nowadays, automated systems for metering water and energy consumption, technologies for remote transmission of meter readings, and electronic models for handling municipal waste are actively introduced.

The unsatisfactory state of utility networks and underfunding of their care, nonpublic increase in prices for utility infrastructure services, difficulties in reforming it, and the need to improve the efficiency of using public funds for its development necessitates the use of new approaches to solving the problems facing utility infrastructure companies [1–5].

An acute problem associated with communal infrastructure is the lack of objective information about the real situation and the state of networks as a whole and their individual objects. To solve the above problems, significant monetary resources are still needed in addition to the country's will. One of these solutions is the widespread use of public–private partnership mechanisms in the utility infrastructure.

Public–private partnership is already used in projects for the development of utility infrastructure. Analysis of statistical data allows us to conclude that the results of the implementation of such projects in certain subjects of the country are a 47% reduction in accidents in utility and energy infrastructure, as well as a reduction in losses in networks by 18%, which suggests the expediency of their widespread use [6, 7].

2 Methodology

Problems and prospects for the development of residential infrastructure systems are studied in the works of Katanandov and Demin [3], Oksogoev [4], Chanyshev [1], Sviatokha [5], Surnina et al. [7], Babkin and Zdolnikova [6], Allin and Walsh [8], Shvelidze [9], and Nadezhdina et al. [10].

Features and prospects for the use of environmental and smart technologies and the development of Industry 4.0 are studied in the works of Popkova and Sergi [11–13], Nurgissayeva and Tamenova [14].

The development of public–private partnership mechanisms was studied in the works of Buranbaeva et al. [15], Minnibaeva et al. [16], Vyshnivska and Kireitseva

[17], Koguashvili and Chipashvili [18], Maslova [19], Dyeyeva and Khmurova [20], Maslov [21], Savina [22], Jokovic [23], Lakhyzha and Yehorycheva [24], Sheppard and Beck [25], Hrytsenko et al. [26], Sobol [27], and Nikiforov et al. [28].

As a rule, the implementation of public–private partnership projects in communal infrastructure is carried out on a long-term basis. Simultaneously, in such projects, there are risks and certain limitations due to the peculiarity of the industry, which can affect their effectiveness. In this regard, there is a need to justify management decisions for the use of public–private partnership mechanisms, as well as to assess its potential. The capacity assessment allows us to decide on the feasibility of implementing public–private partnership projects [8, 29].

The following forms of public–private partnership have become well-established in Russian practice:

- Life cycle contract;
- Investment contract (agreement);
- Lease agreement with investment obligations;
- Offset contract;
- Special investment contract;
- Energy service contract with signs of public–private partnership;
- Concession agreement;
- Agreement on public–private partnership;
- Projects in other forms that correspond to the characteristics of a public–private partnership.

The objects of the concession agreement are as follows:

(1) Transport and logistics infrastructure;
(2) Energy and heat supply facilities;
(3) Objects of social sphere and communal services;
(4) Objects of the defense sphere;
(5) Agro-industrial facilities;
(6) Objects of the environmental sphere;
(7) Objects of the digital economy.

Article 111.4 of the Federal law "On the contract system in the field of procurement of goods, works, and services to meet state and municipal needs" (April 5, 2013 No. 44-FZ) [30] additionally indicates the possibility of concluding several offset contracts for homogeneous or identical goods or services between customers and investors. However, since the practice of concluding offset contracts is just beginning to take shape, it seems that the mechanism for implementing this opportunity will be specified in the future.

The subject of the offset contract includes the following:

- Obligations of the public party (the customer, who concluded an offset contract according to the results of the tender, undertakes to purchase goods or services from the investor and enter such an investor in the register of sole suppliers);

- Obligations of the investor (general obligations: supply of goods or provision of services; counter investment obligations or master the production of goods or create or reconstruct property for the provision of services).

An offset contract can be concluded in respect of any object. The list of offset objects is not established. The main thing is that it should be the supply of goods or the provision of services. In addition to pharmaceuticals and food products, offset contracts can be concluded in other areas, including housing and utility services, information technology, communication services, transport, etc. [30, 31].

The conclusion of offset contracts in various fields makes it possible to create its own production in the constituent entity of the Russian Federation and can, in a number of areas, ensure a reduction in the dependence on the supply of goods from foreign countries since new or modernized production should ensure the supply of Russian goods to the market.

An important advantage of offset contracts is the possibility of increasing the scale of import substitution and implementing plans for the localization of production of technically complex products. The conclusion of offset contracts was originally designed to create new high-tech industries through private investment, making it possible to reduce dependence on foreign manufacturers and suppliers. So far, this has not happened, but the range of applications for offset contracts is expanding.

The concession form has become the most widespread form of public–private partnership in quantitative terms in the Russian Federation. The second place among the forms of public–private partnership is occupied by a lease agreement with investment obligations. The third place is occupied by an energy service contract with signs of a public–private partnership.

Having studied the main problems, risks, and threats of global challenges of our time, it is necessary to determine what impact they have on strategic management when public–private partnership projects are implemented in the field of regional housing and utility services. Additionally, it is necessary to develop ways to improve efficiency in questions of interaction between authorities and private investors in implementing projects related to regional housing and communal services within the framework of public–private partnership, considering these challenges.

When using public–private partnership mechanisms, it is necessary to assess their potential. The very assessment of the potential for using the public–private partnership mechanism is one of the tasks that is associated with the justification of the management decision on the possibility of using certain models and forms of public–private partnership when concluding an agreement.

3   Results

In general, the price of the potential for the use of public–private partnership mechanisms is carried out based on the analysis of available resources. Simultaneously, there may be two main approaches. The first approach is based on methods for quantifying the potential for using public–private partnership mechanisms, and the second—on methods of qualitative assessment and the potential for using the mechanisms of public–private partnership.

The approach based on a quantitative assessment of the potential for the use of public–private partnership mechanisms can be applied if the necessary statistical information is available.

In the approach based on a quantitative assessment of the potential for using public–private partnership mechanisms, a method for calculating the integral indicator of its potential can be used (Kp). This indicator combines the values of several types of resources used in public–private partnerships. These include, among others, the share of the value of output (provision of services), the share of costs, the share of jobs, and the share of investments in fixed assets used in the framework of public–private partnership activities. A weighted average indicator is calculated, where the weight coefficients reflect the specifics of the targets (obstacles) of developing public–private partnerships. It also considers the model and form. Let us present the final data for calculating the integral indicator of potential (Kp).

1. The indicator of the share of the cost volume of production (provision of services) ($KProd$) has a weighting coefficient equal to 0.4 and is calculated by the following formula:

$$K\,Prod = V\,Fk.prod/V\,tot.prod \qquad (1)$$

where:

VFk.prod—the actual volume of products, works, and services produced (performed) within the framework of public–private partnership;
Vtot.prod—the total volume of products, works, and services produced (performed) within the framework of public–private partnership and outside it.

2. The indicator of the share of costs ($KCosts$) has a weighting factor of 0.25 and is calculated by the following formula:

$$K\,Costs = V\,Fk.costs/V\,total.costs \qquad (2)$$

where:

VFK.costs—the value of costs for industrial products, raw materials, materials and components, works, and services within the framework of public–private partnership;
Vtotal.costs—the total value of costs for the production of works and services produced (performed) within the framework of public–private partnership and outside it.

3. The indicator of the share of jobs ($Kjobs$) has a weighting coefficient equal to 0.20 and is calculated by the following formula:

$$K jobs = N Fk.jobs / N Total.jobs \qquad (3)$$

where:

N *Fk.jobs*—the number of jobs employed within the framework of public–private partnership;
N *Total.jobs*—the total number of jobs employed within the framework of public–private partnership and outside it.

4. The indicator of the share of investment volume (*KInvest*) has a weighting factor of 0.15 and is calculated by the following formula:

$$K Invest = V Fk.invest / Vtot.invest \qquad (4)$$

where:

VFk.invest—the volume of investments in fixed assets within the framework of public–private partnership;
Vtot.invest—the total volume of investments in fixed assets within the framework of public–private partnership and outside it.

The calculation of the integral indicator of the potential of public–private partnership is carried out by writing the specific weight of the corresponding indicator and its individual resource value with their subsequent addition (formula 5):

$$K_p = \sum_{t=1}^{4} W_I \times K_I \qquad (5)$$

where:

W_I the share of the indicator;
K_I the *i-indicator* used in the calculations.

The normative level of the integral indicator of the potential of public–private partnerships may vary depending on the model and form used. The average recommended value of the indicator is 0.4–0.6, which corresponds to the situation when the main goal is to use the economies of scale of production (provision of services) and reduce production costs (provision of services).

A qualitative assessment approach can be applied in the absence of the necessary statistical information on public–private (municipal-private) partnerships.

The approach is the receipt of information from expert specialists (at least 3–5 experts), its analysis, and generalization to choose an assessment of the potential for using public–private partnership mechanisms. It has the following interrelated stages:

(1) In the first stage, specific resources used in public–private partnerships are identified. Experts define them jointly, usually using a resource tree that allows

Table 1 Influence of the integral indicator on the assessment of the potential for the use of public–private partnership mechanisms

The value of the integral indicator	Capacity assessment
Up to 0.05	Very weak potential
0.05–0.1	Weak potential
0.1–0.2	Medium potential
0.2–0.3	Above average potential
0.3–0.4	Strong potential
0.4–0.7	Very strong potential

Source Developed and compiled by the authors

one to decompose resource groups into smaller ones that can be evaluated and described.

(2) In the second stage, the importance of each resource is revealed. Based on the identified resources, each expert compiles its combined table for these resources, where the importance of each resource (from 0 to 1) for public–private partnerships is determined.

(3) In the third stage, an integral arithmetic average is calculated for each resource, which is determined based on the data of all experts, considering the level of their competence and the importance of risk. The expert's level of competence is assessed jointly by all experts on a 10-point scale (from 0 to 1).

(4) The fourth stage determines the impact of each resource and resource group on the final output of the project.

For this purpose, a summary table is compiled, where all risks are grouped according to the resource tree, and the resulting integral indicator is determined for each group, which makes it possible to assess their impact on the final result.

Simultaneously, there is a certain dependence on the integral indicator of influence on assessing the potential for the use of public–private partnership mechanisms. This influence is shown in Table 1.

The analysis of the table in relation to the integral indicator for assessing the potential use of public–private partnership mechanisms allows us to draw general conclusions, which can then be clarified or refuted by various methods.

4 Conclusion

Based on the research, it is necessary to note that several issues remain relevant in the analysis of the potential for the application of concession agreements. Thus, when using a concession, almost the only source of formation of the concessionaire's cash flow is gross revenue, the size of which is equal to the volume of y agreed upon in the concession agreement, which is provided to the concessionaire in accordance with Russian legislation. This requires additional analysis and justification of the decision to use the concession agreement to implement projects for the development of utility infrastructures. As such a justification, we propose a mechanism for analyzing the potential of public–private partnership models, which makes it possible to search for the most optimal forms and models of public–private partnership for a specific project for developing communal infrastructures.

Acknowledgements The study was carried out within the framework of the research grant of the Krasnoyarsk regional fund to support scientific and technological activities on the topic "Development of models of financial support for investments in the communal infrastructure of the region, taking into account the best Russian and world practices and features of the spatial and territorial development of the Krasnoyarsk Territory," No KF-835, agreement on the procedure for targeted financing No. 226 of April 20, 2021.

References

1. Chanyshev IR (2015) Clustering of housing and utility complex as one of directions of its reform. Finan Anal Prob Solutions 8(2):55–64
2. InfraOne Research (2020) Infrastructure of Russia: development index 2020 [Analytical review]. Retrieved from https://infraoneresearch.ru/index_id/2020?index2020. Accessed 25 Sept 2022
3. Katanandov SL, Demin AY (2021) Problems and prospects for the development of the municipal infrastructure management system in the Russian Federation. Manage Consult 6:80–93. https://doi.org/10.22394/1726-1139-2021-6-80-93
4. Oksogoev AN (2013) Communal infrastructure of the region. ESSUTM Bull 1(40):122–126
5. Sviatokha NYu (2013) Spatial-temporal organization of the region housing sphere: the cluster approach. Vestnik Orenburg State Univ 8(157):140–147
6. Babkin IA, Zdolnikova SV (2020) Conceptual model for evaluating the effectiveness of public-private partnership projects in the industrial sector of the economy. Theor Prac Serv Econ Soc Sphere Technol 4(46):11–15
7. Surnina NM, Ilyukhin AA, Ilyukhina SV (2016) Development of social and utility infrastructure of a region: ontological, institutional, informational aspects. J Ural State Univ Econ 5(67):54–65
8. Allin S, Walsh C (2010) Strategic spatial planning in European city-regions: parallel processes or divergent trajectories? NIRSA Working Paper No. 60. National Institute for Regional and Spatial Analysis, Maynooth, Ireland. Retrieved from https://www.maynoothuniversity.ie/sites/default/files/assets/document/WP60_Allin_Walsh_0_0.pdf. Accessed 25 Sept 2022
9. Shvelidze DA (2017) The application of the mechanism of public-private partnership for the development of infrastructure in Russia. In: Proceedings of international conference "process management and scientific developments". Scientific Publishing House "Infinity", Birmingham, UK, pp 7–14
10. Nadezhdina SD, Voronina NI, Pyankova LM (2021) Public-private partnership as a resource for the territorial infrastructure development. Vestnik NSUEM 2:35–45. https://doi.org/10.34020/2073-6495-2021-2-035-045
11. Popkova EG, Sergi BS (2018) Will Industry 4.0 and other innovations impact Russia's development? In: Sergi BS (ed) Exploring the future of Russia's economy and markets: towards sustainable economic development. Emerald Publishing Limited, Bingley, UK, pp 51–68. https://doi.org/10.1108/978-1-78769-397-520181004
12. Popkova EG, Sergi BS (eds) (2019) Digital economy: complexity and variety vs. rationality. Springer, Cham, Switzerland. https://doi.org/10.1007/978-3-030-29586-8
13. Sergi BS (ed) (2019) Tech, smart cities, and regional development in contemporary Russia. Emerald Publishing Limited, Bingley, UK. https://doi.org/10.1108/9781789738810

14. Nurgissayeva AA, Tamenova SS (2022) Public-private partnerships in the field of the "green" economy of the megapolis. Central Asian Econ Rev 3(144):75–87. https://doi.org/10.52821/2789-4401-2022-3-75-87
15. Buranbaeva LZ, Garifullina AF, Zhilina EV (2022) Public-private partnership in the system of regional management. Vestnik BIST (Bashkir Inst Soc Technol) 3(56):48–53. https://doi.org/10.47598/2078-9025-2022-3-56-48-53
16. Minnibaeva AZ, Vaslavskaya IYu, Koshkina IA, Ziyatdinov AF (2021) Improving the mechanisms of public-private partnership. Int J Finan Res 12(2):242–250.https://doi.org/10.5430/ijfr.v12n2p242
17. Vyshnivska B, Kireitseva O (2022) Peculiarities of application of public-private partnership as a mechanism for implementation of innovation activity. Three Seas Econ J 3(1):35–41. https://doi.org/10.30525/2661-5150/2022-1-5
18. Koguashvili P, Chipashvili A (2022) The role of public-private partnerships in land administration. Economics 105(4–5):169–187. https://doi.org/10.36962/ecs105/4-5/2022-169
19. Maslova SV (2021) On the concept of international standards of public-private partnership. Int Law 3:15–25. https://doi.org/10.25136/2644-5514.2021.3.36518
20. Dyeyeva N, Khmurova V (2018) Public-private partnership: stakeholders' interests. Econ Ukraine 9:99–111. https://doi.org/10.15407/economyukr.2018.09.099
21. Maslov VO (2021) Essential characteristics of public-private partnership models. Vestnik Sumy State Univ 1:173–178. https://doi.org/10.21272/1817-9215.2021.1-20
22. Savina AV (2019) On the issue of public and private interest in the institute of public-private partnership. Curr Issues State Law 3(11):346–353. https://doi.org/10.20310/2587-9340-2019-3-11-346-353
23. Jokovic S (2022) Public-private partnerships and concessions in Serbia: manual, 2nd edn. Retrieved from https://www.kobo.com/us/en/ebook/second-updated-edition-public-private-partnerships-and-concessions-in-serbia. Accessed 27 Sept 2022
24. Lakhyzha N, Yehorycheva S (2019) Institutional support of the public-private partnership in the Republic of Poland. Univ Sci Notes 1–2(69–70):145–155. https://doi.org/10.37491/unz.69-70.13
25. Sheppard G, Beck M (2020) Stakeholder engagement and the future of Irish public-private partnerships. Int Rev Adm Sci 88(3):843–861. https://doi.org/10.1177/0020852320971692
26. Hrytsenko L, Boiarko I, Tverezovska O, Polcyn J, Miskiewicz R (2021) Risk-management of public-private partnership innovation projects. Mark Manage Innov 2:155–165. https://doi.org/10.21272/mmi.2021.2-13
27. Sobol TS (2022) Public-private partnership as a tool for increasing the investment attractiveness of a region (by the example of the Moscow region). Bull Udmurt Uni Ser Econ Law 32(5):854–861. https://doi.org/10.35634/2412-9593-2022-32-5-854-861
28. Nikiforov P, Zhavoronok A, Marych M, Bak N, Marusiak N (2022) Conceptual principles of state policy of regulation of development of public-private partnerships: conceptual principles of state policy regulating the development of public-private partnerships. Polit Issues 40(73):417–434. https://doi.org/10.46398/cuestpol.4073.22
29. Ministry of Regional Development of the Russian Federation (2011) Order "On the development of programs for the integrated development of communal infrastructure systems of municipalities" (May 6, 2011 No. 204). Electronic Fund of Legal and Regulatory Documents "Codex", Moscow, Russia. Retrieved from https://docs.cntd.ru/document/902279091. Accessed 27 Sept 2022
30. Russian Federation (2013) Federal Law "On the contract system in the field of procurement of goods, works, services to meet state and municipal needs" (April 5, 2013 No. 44-FZ). Legal Reference System "ConsultantPlus", Moscow, Russia. Retrieved from https://www.consultant.ru/document/cons_doc_LAW_144624/. Accessed 25 Sept 2022
31. Government of the Russian Federation (2013) Decree "On approval of requirements for programs for the integrated development of utility infrastructure systems for settlements, urban districts" (June 14, 2013 No. 502). Legal Reference System "Garant", Moscow, Russia. Retrieved from https://base.garant.ru/70398922/. Accessed 25 Sept 2022

Models of Concession and Public–Private Partnership as an Effective Tool for Achieving the Goals of Sustainable Development of the Region in the Digital Economy

Svetlana B. Globa, **Nina M. Butakova**, **Evgeny P. Vasiljev**, and **Viktoria V. Berezovaya**

Abstract The paper aims to study and select models of the concession agreement in projects to develop communal infrastructure. One of the important aspects of creating such an environment is the creation of favorable conditions for human life, the supply of quality resources, and the provision of work and services to achieve a high-quality environment for a person while at work and at home. The authors use the methods of empirical (observation, comparison, etc.) and theoretical (abstraction, analysis, synthesis, induction, deduction, etc.) research to develop models for the implementation of concession mechanisms and public–private partnership to consistently implement the goals of sustainable development of territories and conclude agreements and contracts on the terms of public–private partnership in projects for the modernization of residential infrastructure, considering the directions of sustainable development of the territory. The authors consider the possibility of using a public–private partnership model, which includes a specialized project financing company for large concession agreements in the communal infrastructure, considering the risks of the public and private partners. A reasonable assessment and management of the effectiveness of concession projects and public–private partnerships will ensure the implementation of such a project, considering the possibility of ensuring the required level of socio-economic efficiency. The proposed models make it possible to reduce the level of risks of concession agreements for communal infrastructure and develop a list of measures to reduce them or transfer them to third parties at all stages of the project life cycle.

S. B. Globa (✉) · N. M. Butakova · E. P. Vasiljev · V. V. Berezovaya
Siberian Federal University, Krasnoyarsk, Russia
e-mail: sgloba@sfu-kras.ru

N. M. Butakova
e-mail: nbutakova@sfu-kras.ru

V. V. Berezovaya
e-mail: VVBerezovaya@sfu-kras.ru

E. G. Popkova (ed.), *Smart Green Innovations in Industry 4.0 for Climate Change Risk Management*, Environmental Footprints and Eco-design of Products and Processes, https://doi.org/10.1007/978-3-031-28457-1_8

Keywords Project finance · Public–private partnership · Concession · Utility infrastructure · Financial society

JEL Classification Q48 · Q56 · Q57 · Q53 · O33 · L97

1 Introduction

Nowadays, we witness the active development and implementation of sustainable development goals and indicators. These processes are changing the requirements of society for the quality of the environment—people are increasingly willing to live in a more environmentally friendly, comfortable, and convenient living environment, an important role in the creation of which is played by the uninterrupted provision of residents of all territories of the region with high-quality public services. For this purpose, it is necessary to ensure the reliable and uninterrupted functioning of systems and objects of communal infrastructure, ensuring their energy efficiency, environmental friendliness, and economic efficiency [1, 2, 4, 5].

High quality and comfort of housing are one of the main criteria for the standard of living of the population, providing basic needs—safety, comfort, quality of food, and leisure (and in today's conditions of the spread of remote work—the quality of labor activity). The quality of public services determines and, in many respects, forms the human environment and affects health and moral well-being.

Problems in the development of communal infrastructure directly affect the following aspects:

- Quality of life of the population, including the state of the equipment used and the compliance of technology with modern requirements in the field of digitalization, ecology, and safety;
- Condition and quality of facilities;
- The workload of fixed assets of enterprises and their moral and physical deterioration;
- The state of networks, the availability of reserves, and other factors that cause an increase in the likelihood of human-made accidents.

2 Methodology

Features and prospects for the use of concession schemes and public–private partnerships in projects for the modernization of infrastructure systems are studied in the works of Calugareanu and Veronica [5], Manukhina [18], Oktaviani et al. [19], Oyebode [20], Panikarova et al. [21], Shiyan [24], Skolubovich and Matveeva [25], Stukalova and Provalenova [26], Vasyutynska [29], Yescombe and Farquharson [30].

Problems and trends in the use of public–private partnership mechanisms are studied in the works of Artyakov and Chursin [1], Avdokushin and Bednyakov [2],

Bacchini et al. [3], Belyakova [4], Chepelenko [6], de Albornoz Portes [7], Dyagileva et al. [9], Khakimova [10], Kociemska [11], Kolodiziev et al. [12], Kravchenko [13], Kurylovich [14], Kuznyetsova and Maslov [15], Kwame Sundaram et al. [16], Lakomy-Zinowik [17], Parlak and Hashi [22], Sergi et al. [23], de Faria Silva, et al. [8], Tireuov et al. [27, 28], Zapivakhin et al. [31].

A concession agreement is concluded between the grantor (the State) and the concessionaire (private investor), under which the concessionaire finances, designs, builds, and operates the object of the concession agreement within a specified period.

Certain rights and obligations of the grantor in the development of communal infrastructure can also be exercised by state and municipal unitary enterprises, which own centralized systems and facilities of hot water supply, cold water supply, waste disposal, heat supply facilities, waste processing facilities, and other facilities on the right of economic management.

For large concession agreements in public utilities infrastructure, taking into account the risks of public and private partners, it is possible to apply the public–private partnership model using a specialized project financing company.

The use of the model of a public–private partnership with the participation in the concession agreement of a specialized project financing company and a syndicate of commercial banks can significantly reduce the level of risks that may be present in the concession agreement on communal infrastructure.

This is due to the fact that a specialized project financing company and a syndicate of commercial banks, before opening financing for a concession agreement, will necessarily conduct a series of comprehensive examinations related to the verification of all necessary procedures that must be carried out by the concessionaire and the grantor when concluding a concession contract. That is, they will assess the presence of risks and their level for the pre-investment stage of the concession agreement. Additionally, they will conduct comprehensive examinations of the existence of risks in the investment and operational stages. This makes it possible to prevent the occurrence of risks in the concession agreement or to develop a list of measures to reduce them or transfer them to third parties at all stages of the life cycle of the concession agreement.

For the implementation of medium-scale concession agreements in the communal infrastructure, including water supply and sanitation, solid municipal waste management, electricity supply, etc., a model of a concession agreement using a specialized financial company is possible, which reduces the risks of the concessionaire and the grantor.

It should be noted that the activities of a specialized company are quite strictly regulated by the Federal law "On the securities market." A specialized financial company can be a joint venture, which is created on a parity basis by the state and private business, as a rule, in the joint-stock organizational and legal form.

The use of the public–private partnership model with the participation of a specialized financial company in the concession agreement makes it possible to reduce the level of risks that may be present in the concession agreement for utility infrastructure. This is due to the fact that a specialized financial company, before issuing its own bonds, must evaluate the objects of investment of funds from issued bonds.

For this purpose, a comprehensive audit of all necessary procedures is carried out, which must be carried out by the concessionaire and the grantor in the concession agreement. This makes it possible to prevent the occurrence of risks in the concession agreement or develop a list of measures to reduce them or transfer them to third parties at all stages of the life cycle of the concession agreement.

3 Results

For the implementation of long-term concession agreements in the communal infrastructure, it is possible to use the model of public–private partnership where the issue of concession bonds issued by the concessionaire itself is used as a financing tool. The model of a concession agreement based on concession bonds is presented in Fig. 1.

In this model of the concession agreement, the issue of concession bonds, the circulation period of which is tied to the term of the concession agreement, is used as the main source of financing. The yield of these bonds considers the yield of the object of the concession agreement for the concessionaire.

The use of concession bonds will reduce the risks of the concession agreement itself due to the fact that when registering the issue of a concession bond loan, the accuracy of the information and the correctness of its interpretation are checked by the registering authority and the organizer of their issue (broker). Simultaneously,

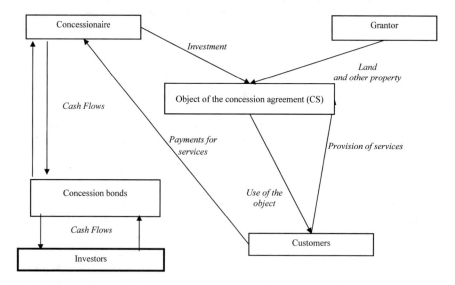

Fig. 1 Scheme of the model of the concession agreement for the development of communal infrastructure on the basis of a specialized financial society. *Source* Developed and compiled by the authors

the organizer of the issue can act as a financial consultant who signs all the emission documents for the concession bond loan and bears joint and several liabilities with the issuers for damage caused by the inaccuracy of the data.

When implementing several concession agreements on communal infrastructure in the region, it is possible to use the public–private partnership model, where there is a management company that performs the functions of an institution for the development of the territory (limited liability company). The model of public–private partnership based on the management company in the region is presented in Fig. 2.

Through this organization, the regional administration, municipal administrations of the region, and private businesses will conduct concession agreements in public infrastructure:

- Subsidies for reimbursement of part of the costs of paying interest to loan recipients;
- Subsidies to reimburse part of the costs of paying lease payments paid to Russian leasing companies for property acquired under leasing agreements;

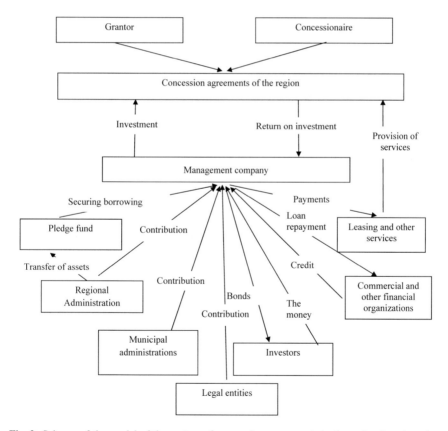

Fig. 2 Scheme of the model of the system of concession agreements in the region based on the management company. *Source* Developed and compiled by the authors

- State guarantees;
- Compensation of part of the costs for the construction of concession agreement facilities;
- Subsidies for compensation of part of the cost of machinery, equipment, and modular facilities;
- Grants;
- Provision of informational, consulting, and organizational assistance to concession agreements in the region, etc.

For the successful use of the management company in the region, it is necessary to create a certain infrastructure consisting of a number of elements. The first element is the expert council, which should conduct the full range of expertise in the audit of draft concession agreements, including institutional, commercial, technical, social, environmental, economic, financial, and other types of audits. The expert council should identify the depth and complexity of the study of the entire range of issues related to the concession agreement, including technical feasibility, effectiveness, and efficiency of the concession agreement.

The second element of the scheme should be the pledge (guarantee) fund of concession agreements of the region. It is formed from the property owned by the region. The pledge (guarantee) fund of concession agreements will be used in the issuance of concession bonds as their collateral.

The third element should be the organizer of the bond issue. As part of bond financing, the organizer can perform various functions, for example, as a financial consultant, underwriter, etc. As is known, a financial consultant helps an enterprise in developing and issuing documents when issuing shares and bonds. A financial consultant signs under the issuing documents he or she develops. In turn, the underwriter carries out the placement of bonds among investors.

The management company transfers the funds received from the placement of securities on a fixed-term, paid, and repayable basis to the concessionaire of the concession agreement. For this purpose, a comprehensive audit of all necessary procedures that must be carried out by the concessionaire and the grantor in the concession agreement will be carried out. This makes it possible to prevent the emergence of risks in the concession agreement or develop a list of measures to reduce them or transfer them to third parties at all stages of the life cycle of the concession agreement.

4   Conclusion

A significant degree of depreciation of public infrastructure facilities, as well as an increase in the cost of building new facilities in this area, has currently led to the fact that projects related to public infrastructure are capital-intensive. They require joint funding from state and non-state sources and funds from extra-budgetary and budgetary sources, which makes it possible to put into practice various mechanisms of public–private partnership and, first of all, various models and forms, including concession ones.

Simultaneously, the sphere of communal infrastructure is significantly regulated. There are certain difficulties with the use of debt financing. The implementation of projects in this area is largely dependent on the policy of the tariff authority, which can differ significantly in different regions.

Acknowledgements The study was carried out within the framework of the research grant of the Krasnoyarsk regional fund to support scientific and technological activities on the topic "Development of models of financial support for investments in the communal infrastructure of the region, taking into account the best Russian and world practices and features of the spatial and territorial development of the Krasnoyarsk Territory," No KF-835, agreement on the procedure for targeted financing No. 226 of April 20, 2021.

References

1. Artyakov VV, Chursin AA (2020) Fundamentals of public-private partnership project management: textbook. Moscow, Russia: Infra-M. https://doi.org/10.12737/1078233
2. Avdokushin YF, Bednyakov AS (2022) Public-private partnership development in Russia and abroad: Issues and solutions. Region Reg Stud Russ Eastern Eur Central Asia 10(2):281–302. https://doi.org/10.1353/reg.2021.0016
3. Bacchini F, Golinelli R, Jona-Lasinio C, Zurlo D (2020) Modeling public and private investment in innovation. Retrieved from https://www.growinpro.eu/wp-content/uploads/2020/03/working_paper_2020_06-1.pdf (Accessed 25 Sept 2022)
4. Belyakova J (2021) Sustainable development mechanisms as a key factor in the development of public-private partnerships. Issues Risk Anal 18(5):72–78. https://doi.org/10.32686/1812-5220-2021-18-5-72-78
5. Calugareanu I, Veronica B (2022) World practice in the evolution of public-private partnership of infrastructure projects. Int J Econ Bus Manag Stud 9(1):1–12. https://doi.org/10.55284/ijebms.v9i1.599
6. Chepelenko A (2021) Public-private partnership as a mechanism for effective management of state property. VUZF Rev, 6(3):133–144. Retrieved from https://www.researchgate.net/publication/354940071_Public-private_partnership_as_a_mechanism_for_effective_management_of_state_property (Accessed 25 Sept 2022)
7. de Albornoz Portes FJC (2017) Alliances: an innovative management model for public and private investments. In: Moya BL, de Gracia MOS, Mazadiego LF (eds) Case study of innovative projects—successful real cases. IntechOpen, London, UK, pp 79–99. https://doi.org/10.5772/intechopen.68228
8. de Faria Silva R, Souza A, Kaczam F, Dalazen LL, da Silva WV, da Veiga CP (2022) Public-private partnerships and value for money. Public Works Manag Policy 27(4):347–370. https://doi.org/10.1177/1087724X221108149
9. Dyagileva O, Liubokhynets L, Zabashtanska T, Shuklina V, Bezuhlyi I (2022) The use of the mechanism of public-private partnership in the investment processes management in the context of digitalization. Cuestiones Politicas 40(72):368–384. https://doi.org/10.46398/cuestpol.4072.20
10. Khakimova AF (2022) Essential conditions of public-private partnership agreement. Juridical Sci Pract 17(4):46–53. https://doi.org/10.25205/2542-0410-2021-17-4-46-53
11. Kociemska H (2010) Public-private partnership project success circumstances. J Mod Account Auditing 6(11):53–58

12. Kolodiziev O, Tyschenko V, Azizova K (2017) Project finance risk management for public-private partnership. Invest Manag Financ Innov 14(4):171–180. https://doi.org/10.21511/imfi. 14(4).2017.14

13. Kravchenko O (2019) Public-private partnership as a mechanism for financing infrastructure modernization. Baltic J Econ Stud 5(1):112–117. https://doi.org/10.30525/2256-0742/2019-5-1-112-117

14. Kurylovich M (2022) Public-private partnership as a legal instrument linking the elements of sustainable development. Vestnik of Polotsk State University. Part D. Econ Legal Sci, 60(5):96–103. https://doi.org/10.52928/2070-1632-2022-60-5-96-103

15. Kuznyetsova A, Maslov V (2021) Methodical principles of analysis of the concept of "organizational and economic mechanism of public-private partnership." Herald Econ 3:137–146. https://doi.org/10.35774/visnyk2021.03.137

16. Kwame Sundaram J, Chowdhury A, Sharma K, Platz D (2016) Public-private partnerships and the 2030 Agenda for Sustainable Development: Fit for purpose? DESA Working Paper No. 148. DESA, New York, NY. Retrieved from https://www.un.org/esa/desa/papers/2016/wp148_2016.pdf (Accessed 11 Oct 2022)

17. Lakomy-Zinowik M (2017) Public-private partnership as an alternative source of financing of public tasks. Ekonomia i Prawo [Econ Law] 16(1):75–86. https://doi.org/10.12775/EiP.201 7.006

18. Manukhina LA (2015) The concession agreement as the most promising mechanism for reforming the housing and communal services industry. Constr Sci Educ, 3:4

19. Oktaviani PP, Muhtar EA, Karlina N (2020) Public-private partnership in water supply of DKI Jakarta. J Public Serv Manag 4(1):109. https://doi.org/10.24198/jmpp.v4i1.27214

20. Oyebode OJ (2022) Public-private partnership: a veritable tool for handling environmental pollution and infrastructural degeneration in Nigeria. In: Ayeni AO, Oladokun O, Orodu OD (eds) Advanced manufacturing in biological, petroleum, and nanotechnology processing. Springer, Cham, Switzerland, pp 131–141. https://doi.org/10.1007/978-3-030-95820-6_11

21. Panikarova SV, Kuklinov ML, Yugov VV (2022) The role of public-private partnership in the development of the Russian regions' infrastructure. Reg Econ: Theory Pract 20(6):1118–1142. https://doi.org/10.24891/re.20.6.1118

22. Parlak B, Hashi AS (2021) Public private partnership in selected countries: a comparative analysis. In: Cobanoglu C, Della Corte V (eds) Advances in global services and retail management. USF M3 Publishing, Tampa, FL, pp 1–12. https://doi.org/10.5038/9781955833035

23. Sergi BS, Popkova EG, Borzenko KV, Przhedetskaya NV (2019) Public–private partnerships as a mechanism of Financing Sustainable Development. In: Ziolo M, Sergi B (eds) Financing sustainable development. Palgrave Macmillan, Cham, Switzerland, pp 313–339. https://doi.org/10.1007/978-3-030-16522-2_13

24. Shiyan AA (2021) On trends in infrastructure development using the public-private partnership mechanism in the G20 countries. Econ Manag 26(11):1271–1277. https://doi.org/10.35854/1998-1627-2020-11-1271-1277

25. Skolubovich AY, Matveeva MV (2018) On the question of public-private partnerships in the field of municipal water supply. Izvestiya vuzov. Investitsii. Stroitelstvo. Nedvizhimost [Proc Univ Invest Constr Real Estate], 8(3):45–53. https://doi.org/10.21285/2227-2917-2018-3-45-53

26. Stukalova IB, Provalenova NQ (2018) Foreign experience of development of the market of housing and communal services on the basis of state-private partnership. Bull NGIEI 1(80):76–88

27. Tireuov K, Mizanbekova S, Pechenaya L (2021) Integration processes—the component of the effective implementation of public-private partnerships. Probl AgriMarket 2:62–69. https://doi.org/10.46666/2021-2.2708-9991.08

28. Tireuov K, Yespolov T, Shatohina N (2022) The experience of public-private partnership in Kazakhstan: modern approaches. Probl AgriMarket 2:85–93. https://doi.org/10.46666/2022-2.2708-9991.09

29. Vasyutynska L (2021) Financing of public infrastructure projects on the basis of public-private partnership: forms and tools. Sci Opin: Econ Manage 3(73):65–68. https://doi.org/10.32836/2521-666X/2021-73-10
30. Yescombe ER, Farquharson E (2018) Public-private partnerships for infrastructure: principles of policy and finance, 2nd edn. Butterworth-Heinemann, Oxford, UK
31. Zapivakhin I, Ilin I, Levina A (2018) Public-private partnership as city project management technology. MATEC Web Conf 170:01037. https://doi.org/10.1051/matecconf/201817001037

Protectionism and Technological Development: The Green Agenda

Igor M. Stepnov⬤, Julia A. Kovalchuk⬤, and Tatiana V. Kolesnikova⬤

Abstract The paper aims to determine the impact of environment-related trade protectionist policies on the technological inequality of countries. This research utilizes the decomposition method with regard to countries participating in the international division of labor. Additionally, the research uses comparative analysis with regard to factors influencing the way different groups of countries set agendas when adopting energy strategies. Moreover, the research applies the generalization and clustering method to instruments of trade protectionism. The implementation of a green protectionism policy determines the development of technological competition during the energy transition. Its implementation contributes to the emergence of new players in the international energy and technology market. Along with benefits for developed countries, there is a high risk of maintaining technological backwardness in developing countries that import energy resources. The research determines the endogenous and exogenous factors influencing the variability of the implementation of green protectionism in developed and developing countries, considering the structure of their foreign trade. Four groups of protectionist measures within the framework of the policy of achieving carbon neutrality are identified: technical instruments regulating environmental standards in the energy sector, tools of fiscal stimulus, fiscal instruments of restrictive nature, and customs and tariff and non-tariff restrictions on foreign trade. Based on the present study, the authors conclude on the countries' benefits and risks of applying the tools of green protectionism depending on their level of development and foreign trade structure.

Keywords Protectionism · Technology · Governance · Public administration · Green protectionism · Environmental protectionism · Technological transition · Green economy · Energy security · Energy transition · Technological competition

JEL Classification O14 · Q43 · Q48

I. M. Stepnov · J. A. Kovalchuk
MGIMO University, Moscow, Russia

T. V. Kolesnikova (✉)
Saint-Petersburg State University of Aerospace Instrumentation, Saint Petersburg, Russia
e-mail: kolesnikova-tv@mail.ru

© The Author(s), under exclusive license to Springer Nature Switzerland AG 2023
E. G. Popkova (ed.), *Smart Green Innovations in Industry 4.0 for Climate Change Risk Management*, Environmental Footprints and Eco-design of Products and Processes, https://doi.org/10.1007/978-3-031-28457-1_9

1 Introduction

The development of the renewable energy industry is announced as a key factor in reducing carbon footprint and anthropogenic pressure on nature and the planet's climate. A number of initiatives have been proposed to combat global warming. The most global initiative is the promise to shift from coal energy by the mid-twentieth century. The brown sources of energy (non-renewable, natural fossil fuels) must be replaced with green sources (renewable, solar, and wind energy, as most often listed). Leaving aside the environmental effect of implementing relevant projects, this research focuses on the economic implications for nations of different development levels and economic structures. In the context of the transition to green energy initiated by developed countries enjoying a high level of technological development, primarily by the member states of the European Union (EU) that possess insufficient natural energy resources to meet the domestic needs of industry and population, it seems necessary to assess the importance of green technological transition for developing countries. Instruments of trade protectionism used by developed countries to support renewable energy sources (RES) pose a threat of technological backwardness in developing economies, which could lead to growing disparities in the living standards of the world's population.

In contrast to classical protectionism, green protectionism entails juxtaposing economic and environmental agendas. This makes society more tolerant of welfare deterioration because, in people's minds, it is associated with environmentally responsible policies and is seen as a necessary measure to improve the environment. According to List, if protectionism was "a payment for learning" [23], then green protectionism is a payment for ecology.

The paper aims to determine the impact of green protectionism policies on the technological inequality of countries.

2 Methodology

The research is based on scholarly papers on the rise of green energy protectionism and the benefits and costs of alternative foreign economic policies.

Understanding the role of ecology in trade and imposing trade restrictions to improve the environment has been discussed by scientists since the early 1990s [4, 36]. This analysis considers the accumulated scientific results of modern researchers:

1. Environmental protectionism is one of the current vectors of increased protectionism [2, 17, 18] in the world under the deglobalization trend.
2. Environmental protectionism is usually disguised and difficult to detect [21] because, contrary to traditional protectionism, trade restrictions are explained not in terms of the expenses of the domestic economy, employment, and living standards in a free trade environment but by the desire to improve the environment and curb global warming.

3. Scientists have found that a society concerned about the environment is more supportive of protectionist policies. This dependence is associated with the fact that people see the causes of environmental degradation in globalization and consider its negative impact to prevail over its positive impact in terms of the price and quality of foreign goods [10].
4. The demand for higher environmental standards in foreign trade is often driven by politicians' desire to meet the demands of voters who view the environmental impact of free trade negatively [9, 16, 19].
5. The impact of environmental protectionism in the transportation sphere on the environment and the economy is not unequivocal [5, 15, 22, 29, 35, 37, 38].
6. In countries where the form of state structure allows sub-national territories (regions, provinces, states, lands, etc.) to set their own rules aligned with national policy, environmental protectionism (as well as other types of protectionism) can be implemented at the local level. The experience of analyzing management decisions taken by local authorities to protect territorial interests through administrative measures [6–8, 21, 26] can be considered at the international level with certain generalizations and assumptions.
7. The implementation of protectionist measures for developing green industries can be seen as a necessary condition for countries implementing a catch-up development strategy [20].

The research identifies four groups of countries with different development levels and foreign trade structures (in terms of energy resources), which helps outline a particular set of factors determining the adoption of state decisions in the field of green energy transition and implementation of protectionist policy. The generalization of protectionist policy tools used for the green energy transition made it possible to form a comprehensive picture of environmental protectionist measures. The identified groups of factors, differences in nations' agenda-setting, and the said tools formed the basis for conclusions on the potential benefits and threats of environmental protectionism.

3   Results

The consensus of the global community on the need to move to low-carbon development is reflected in the Paris Agreement of 2015. The signing of the agreement by 197 countries is binding on countries and establishes obligations to achieve climate goals [24] but provides opportunities to develop their own strategies in accordance with the priorities of socio-economic policy, potential benefits, and threats. In this regard, countries have begun to actively introduce measures to reduce and limit greenhouse gas emissions and include these steps in national energy strategies [32]. Environmental issues are an integral part of trade deals. International treaties often include an environmental clause, forming requirements for partner countries to improve environmental standards for products. Various countries have announced the achievement of their goal on carbon neutrality in the perspective of 25–40 years. More than 100 countries and 400 major cities of the world have promised to eliminate carbon dioxide emissions by 2050 [39]. The combined share of solar and wind energy in electricity production by 2050 could amount to 26–70% [11].

Let us consider the influence of external and internal factors on the adoption of public policies aimed at achieving carbon neutrality in countries depending on their level of development and the structure of their foreign trade.

Let us identify four groups of countries by considering the development level and the structure of foreign trade in energy resources interrelated with the capacity for self-sustainable energy:

- Developed countries, net importers of energy resources;
- Developed countries, net exporters of energy resources;
- Developing countries, net energy importers;
- Developing countries, net energy exporters.

Developed countries have a higher level of technology. Thus, we can talk about their ambition to maintain technological leadership and protect their positions in competition with developing countries. Therefore, the growing external competition from developing countries (e.g., China, India, and others) forces developed countries to design tools to protect their technological superiority.

Developed countries are divided into two groups. The first group is the countries that mainly import energy resources, indicating the scarcity of resources needed to meet domestic demand. The most striking example is the EU. The second group is net exporters that possess enough energy resources to meet domestic energy needs. A case in point is the USA.

Both groups are characterized by the influence of the following two factors:

1. The exogenous factor lies in the intensification of technological competition on the part of rapidly developing states, especially China.
2. The endogenous factor lies in the aspiration of developed societies to satisfy the needs of a higher order as they enjoy a higher living standard than those of developing societies, as illustrated by Maslow's pyramid. These needs apply to influencing the environmental agenda and curbing climate change.

For the first group of countries, the decision-making process is additionally influenced by an endogenous factor related to the ambition of developed countries, which are net importers of energy resources, to ensure energy security by reducing imports of energy products. The addition of this factor to the two factors mentioned above leads to the necessity to develop technologies in the energy and environmental spheres (that are interrelated).

The second group of countries, which do not need energy imports but seek to maintain their technological leadership in the global market, is additionally driven by the desire to ensure a share in the emerging alternative energy market. Therefore, the development of green energy in these countries will further ensure the competitiveness of the relevant industries in the global market and secure a position of technological leadership in the international arena.

Developing countries have a common need to respond to the external challenge of meeting the environmental standards imposed by developed countries through bilateral relations and supranational regulation of international organizations.

The agenda-setting of the third group of countries (developing countries, net energy exporters) is influenced by the following exogenous factors: the threat of declining energy exports and the threat of technological backwardness and sanctions imposed by developed countries in the event of non-compliance with the green energy agenda.

The fourth group of countries (developing countries, net energy importers) is in the least advantageous position as they are forced to purchase energy resources in other countries and do not have a sufficient level of technological development to create a competitive generation of green energy. They are usually forced to follow the trend formed by developed countries.

To sum up, we can conclude that developed countries are the initiators and developers of green standards in the energy field; developing countries accept and follow the new standards. The goal of developed countries is to ensure technological leadership and competitiveness in the market of alternative energy promoted by them. Simultaneously, developing countries, embedded in the green agenda, seek to either develop appropriate technologies that can ensure shares in the new energy market or take advantage of innovations in developed countries and then adapt them for domestic purposes of energy diversification.

China and India occupy a separate place in this structure. The top five countries in solar energy production are China, Japan, the USA, Germany, and India. As for the production of wind energy, the top five are China, the USA, Germany, India, and Spain (according to the 2020 database) [31]. Thus, while the EU is the initiator of the green agenda and environmental protectionism, Chinese and Indian companies may become the biggest beneficiaries of the energy transition due to active technological development.

Let us classify the set of instruments of environmental protectionism based on their nature:

1. Technical tools. The evolution of requirements to the eco-friendliness of industries, vehicles, fuel types, and greenhouse gas emissions reduces pressure on the environment. Simultaneously, these standards affect foreign trade. For example, the change in the eco-standards of automotive fuel in 2011 in the USA led to increased tariffs on imported cars in the amount of $50–$200 per car [21].

2. Tools of fiscal stimulus. Direct subsidization of production in the form of funds for research, development, and implementation of technologies in the production process is essential, especially for a nascent industry in the context of poor access to credit sources and the high cost of borrowed capital [34]. According to McKinsey, to achieve carbon neutrality, Germany would need one trillion euros of investments in fixed assets and five trillion euros in the modernization of infrastructure and housing. Annual investment in the industry is estimated at 240 billion euros, equivalent to 7% of the GDP [25]. Although a large part of the capital is raised through private investors, the burden on the budget is still

very high. Indirect subsidization of production includes the use of special financial instruments, such as solar renewable energy credits (SRECs), to stimulate
renewable energy production. The USA implements this practice at the level
of states [14]. Demand-side subsidies are essential, which was confirmed by the
study of China's experience in developing the green industry [20]. The economic
efficiency of some protectionist measures is questioned by scientists [39].

3. Restrictive fiscal tools. The decision of the European Commission to achieve
    EU carbon neutrality by 2050 [1] provides for an increase in the tax burden for
    companies within the EU that emit greenhouse gases (carbon tax). A carbon tax
    is considered a measure aimed at maintaining and improving the competitiveness
    of European companies abandoning traditional energy sources [12]. This group
    of instruments also includes the sale of greenhouse gas emission quotas. In 2020,
    the global markets of the greenhouse gas emission trading system reached $229
    billion, which is a five-fold increase from 2017. The greenhouse gas emissions
    trading system in Europe covers almost 90% of its volume [40]. Financial insti-
    tutions, including banking and credit institutions, stock exchange, and insurance
    institutions, are involved in the trading of quotas. Between 2019 and 2021, the
    price per emissions allowance increased by 60%, reaching 40 euros in February
    2021 [28]. The impact of fiscal tools of a stimulating and restrictive nature on
    the reduction of the carbon intensity of European countries' GDP is estimated
    as comparable to the 2005–2016 period [32]. This allows us to talk about the
    overall effectiveness of these instruments and their application in the future.

4. Customs and tariff restrictions and non-tariff limitations on foreign trade.
    Concerning foreign trade regulation, close attention of international companies,
    governments, experts, and academia is paid to the possible introduction of a
    carbon border adjustment tax proposed by the EU. Carbon border adjustment tax
    is considered as discriminatory by several countries, including the USA, China,
    the countries producing and exporting traditional energy resources, etc. Addi-
    tionally, the expert and scientific community discusses the compliance of the
    introduction of the tax with the norms of the World Trade Organization. This
    prevents the accelerated imposition of the tax by the EU but does not mean that
    it is unlikely to be adopted in the foreseeable future [12]. Despite the opposition
    from foreign counterparts, the public supporting the EU environmental vector
    raises the probability of introducing the carbon border tax [2].

Let us consider the risk–benefit ratio in implementing environmental protectionism in the context of the green technological transition for the four groups of
countries identified earlier.

Developed countries that import energy resources rely on the emergence of new
powerful industries that ensure energy security, diversify energy sources, generate
additional jobs and tax revenues, and strengthen the economy. Developed and developing countries, which are suppliers of energy resources, have a similar set of benefits
and threats, the difference of which lies in the plane of efforts and the plane of technological capabilities. Decarbonization is projected to lead to a significant increase
in oil and gas consumption by 2040 (by 15% and 50%, respectively) [3], along with

the development of alternative energy. One should expect the diversification of the geographical structure of trading partners due to the restrictions for conventional energy getting to the EU market [13]. As scientists note, the countries that are the EU trading partners are forced to accelerate the development of alternative energy technologies envisaging the threat of a carbon border adjustment tax [12]. The adoption of state strategies related to decarbonization and carbon neutrality provides for the diversification of the investment portfolio of large oil and gas companies with an increasing share of investment in renewable energy [33]. This situation is reflected in companies operating in countries with strict environmental requirements, primarily the EU countries [27, 30]. As for coal-producing and exporting countries, the commitments made to achieve carbon neutrality presuppose the development of alternative energy sources.

For developing countries that import energy resources, the risk of maintaining the technological gap is high.

In the context of the green energy transition, we can predict a redistribution of the energy market with the participation of key players of traditional energy industries and states that are actively developing alternative types of energy.

4 Conclusion

The research shows that implementing environmental protectionism under the green energy transition determines the platform for technological competition. The countries' agenda-setting in identifying carbon neutrality strategies is in line with the desire to maintain or increase their technological competitiveness, along with issues of energy security and the reduction of negative impacts on the environment. The growing green energy trend causes the expansion of the set of instruments of environmental protectionism and the involvement of more countries in the technological competition. Profits and threats to countries lie in the plane of foreign trade, socio-economic development, and technological competition. It is possible to assume that in the struggle to redistribute the energy market, the developed countries that import and export energy resources and developing countries that export energy resources will manifest themselves most strongly. In case their competitive advantages in energy technology are used, they will obtain shares in the emerging energy market. For developing countries that import energy, the risk of technological backwardness is significant. The speed of transformation plays the most important role in the technological transition, as it predetermines the redistribution of the global renewable energy market in favor of those players who will be the fastest in providing it with products of proper quality.

References

1. European Commission (2019) The European green deal. Communication from the commission to the European parliament, the European council, the council, the European economic and social committee and the committee of the regions (11 Dec 2019 COM (2019) 640 final). Brussels, Belgium. Retrieved from https://ec.europa.eu/info/sites/default/files/european-green-deal-communication_en.pdf (Accessed 21 Oct 2022)
2. Afontsev SA (2020) Politics and economics of trade wars. J New Econ Assoc 1(45):193–198. https://doi.org/10.31737/2221-2264-2020-45-1-9
3. Ahmad T, Zhang D (2020) A critical review of comparative global historical energy consumption and future demand: the story told so far. 6:1973–1991. https://doi.org/10.1016/j.egyr.2020.07.020
4. Anderson K, Blackhurst R (1992) The greening of world trade issues. Harvester-Wheatsheaf, London, UK
5. Auffhammer M, Kellogg R (2011) Clearing the air? The effects of gasoline content regulation on air quality. Am Econ Rev 101(6):2687–2722. https://doi.org/10.1257/aer.101.6.2687
6. Bai J, Liu J (2019) The impact of international trade barriers on exports: evidence from a nationwide VAT rebate reform in China. NBER Working Papers 26581. National Bureau of Economic Research, Cambridge, MA. https://doi.org/10.2139/ssrn.4164363
7. Bai J, Li S, Xie D, Zhou H (2021) Environmental protection or environmental protectionism? Evidence from tail pipe emission standards in China. In: AEA papers and proceedings, vol 111, pp 381–385. https://doi.org/10.1257/pandp.20211033
8. Barwick PJ, Cao S, Li S (2021) Local protectionism, market structure, and social welfare: China's automobile market. Am Econ J Econ Pol 13(4):112–151. https://doi.org/10.1257/pol.20180513
9. Bechtel MM, Tosun J (2009) Changing economic openness for policy convergence: when can trade agreements induce convergence of environmental regulation? Int Stud Quart 53(4):931–953. https://doi.org/10.7910/DVN/IMB3J2
10. Bechtel MM, Bernauer T, Meyer R (2012) The green side of protectionism: environmental concerns and three facets of trade policy preferences. Rev Int Polit Econ 19(5):837–866. https://doi.org/10.1080/09692290.2011.611054
11. Bloomberg NEF (2021) New Energy Outlook 2021. Retrieved from https://about.bnef.com/new-energy-outlook/ (Accessed 21 Oct 2022)
12. Bobylev PM, Semeikin AY (2020) Green protectionism in Europe. Energy Policy 10(152):24–33. https://doi.org/10.46920/2409-5516_2020_10152_24
13. Bondar EG (2022) Prospects for the export of energy resources in Russia in modern conditions. Probl Mod Econ 1(81):69–72. Retrieved from http://www.m-economy.ru/art.php?nArtId=7269 (Accessed 21 Oct 2022)
14. Cohen JJ, Elbakidze L, Jackson R (2022) Interstate protectionism: the case of solar renewable energy credits. Am J Agr Econ 104(2):717–738. https://doi.org/10.1111/ajae.12248
15. Davis LW (2008) The effect of driving restrictions on air quality in Mexico City. J Polit Econ 116(1):38–81. https://doi.org/10.1086/529398
16. Drezner DW (2005) Globalization, harmonization, and competition: the different pathways to policy convergence. J Eur Publ Policy 12(5):841–859. https://doi.org/10.1080/135017605001 61472
17. Fajgelbaum PD, Goldberg PK, Kennedy PJ, Khandelwal AK (2020) The return to protectionism. Quart J Econ 135(1):1–55. https://doi.org/10.1093/qje/qjz036
18. Grinberg RS, Komolov OO (2022) Protectionism in Russia: new trends in the context of the import of institutions. Econ Soc Changes: Facts, Trends, Forecast 15(2):44–54. https://doi.org/10.15838/esc.2022.2.80.3
19. Hultberg PT, Edward BB (2003) Cross-country policy harmonization with rent-seeking. Contr Econ Anal Policy 3(2):1–20. https://doi.org/10.2202/1538-0645.1293

20. Landini F, Lema R, Malerba F (2020) Demand-led catch-up: a history-friendly model of late-comer development in the global green economy. Ind Corp Chang 29(5):1297–1318. https://doi.org/10.1093/icc/dtaa038
21. Levinson A (2017) Environmental protectionism: the case of CAFE. Econ Lett 160:20–23. https://doi.org/10.1016/j.econlet.2017.08.019
22. Li S, Xing J, Yang L, Zhang F (2020) Transportation and the environment in developing countries. Ann Rev Resour Econ 12(1):389–409. https://doi.org/10.1146/annurev-resource-103119-104510
23. List F (1841) The national system of political economy. Longmans, Green & Co. Retrieved from https://oll.libertyfund.org/title/lloyd-the-national-system-of-political-economy (Accessed 21 Oct 2022)
24. Maskalenko EV (2022) The role of the northwest federal district in implementing Russia's energy policy. Econ North-West: Probl Dev Prospects 2(69):88–92. https://doi.org/10.52897/2411-4588-2022-2-88-92
25. McKinsey & Company (2021) Net-Zero Deutschland: Chancen und Herausforderungen auf dem Wegzur Klimaneutralität bis 2045. Retrieved from https://www.mckinsey.de/~/media/mckinsey/locations/europe%20and%20middle%20east/deutschland/news/presse/2021/2021-09-10%20net-zero%20deutschland/210910_mckinsey_net-zero%20deutschland.pdf (Accessed 21 Oct 2022)
26. Miravete EJ, Moral MJ, Thurk J (2018) Fuel taxation, emissions policy, and competitive advantage in the diffusion of European diesel automobiles. Rand J Econ 49(3):504–540. https://doi.org/10.1111/1756-2171.12243
27. Pickl MJ (2019) The renewable energy strategies of oil majors—from oil to energy? Energ Strat Rev 26:100370. https://doi.org/10.1016/j.esr.2019.100370
28. Popadko NV, Pankov SV, Popadko AM (2020) Hydrogen energy: stages of development, problems and prospects. Innov Invest 1:293–296
29. Salvo A, Wang Y (2017) Ethanol-blended gasoline policy and ozone pollution in Sao Paulo. J Assoc Environ Resour Econ 4(3):731–794. https://doi.org/10.1086/691996
30. Salygin VI, Lobov DS (2021) Defining major oil and gas companies' development strategies in the era of energy transition. MGIMO Rev Int Relat 14(5):149–166. https://doi.org/10.24833/2071-8160-2021-5-80-149-166
31. Sidorovich V (2020) The Russian RES industry in international comparisons: solar and wind energy. Moscow, Russia: Information and Analytical Center "New Energy." Retrieved from https://renen.ru/wp-content/uploads/2015/09/Russian-RES-industry-international-comparisons-RenEn-Final.pdf (Accessed 21 Oct 2022)
32. Stepanov I (2019) Energy taxes and their contribution to greenhouse gas emissions reduction. HSE Econ J 23(2):290–313. https://doi.org/10.17323/1813-8691-2019-23-2-290-313
33. Stepnov IM, Kovalchuk JA, Gorchakova EA (2019) On assessing the efficiency of intracluster interaction for industrial enterprises. Stud Russ Econ Dev 30:346–354. https://doi.org/10.1134/S107570071903016X
34. Stepnov IM, Kovalchuk JA, Melnik MV, Petrovic T (2022) Public goals and government expenditures: are the solutions of the "modern monetary theory" realistic? Finance: Theory Pract 26(3):6–18. https://doi.org/10.26794/2587-5671-2022-26-3-6-18
35. Viard VB, Fu S (2015) The effect of Beijing's driving restrictions on pollution and economic activity. J Public Econ 125:98–115. https://doi.org/10.1016/j.jpubeco.2015.02.003
36. Whalley J (1991) The interface between environmental and trade policies. Econ J 101(405):180–189. https://doi.org/10.2307/2233810
37. Wolff H (2014) Keep your clunker in the suburb: low-emission zones and adoption of green vehicles. Econ J 124(578):F481–F512. https://doi.org/10.1111/ecoj.12091
38. Zhang W, Lin Lawell C-YC, Umanskaya VI (2017) The effects of license plate-based driving restrictions on air quality: theory and empirical evidence. J Environ Econ Manag 82:181–220. https://doi.org/10.1016/j.jeem.2016.12.002

39. Zharikov MV (2021) The price of decarbonization of the world economy. Econ Taxes Law 14(4):40–47
40. Zubakin VA (2019) State stimulation of transformation of power industry. Strateg Decisions Risk Manag 10(4):320–329. https://doi.org/10.17747/2618-947X-2019-4-320-329

Investment Risks of a Depressed Region in the Context of the Sustainable Development Paradigm

Marina V. Palkinaⓘ, **Nadezhda K. Savelyeva**ⓘ, **Andrey I. Kostennikov**ⓘ, and **Maria V. Ivanova**ⓘ

Abstract The relevance of this research is due to insufficient knowledge of the state of investment risks and mechanisms to reduce them in depressed regions. The research aims to analyze the current state and dynamics of investment risk development in a depressed region. To achieve this goal, the authors used the data of the rating of investment attractiveness of the regions of the Russian agency "Expert RA" for 2011–2020 in depressed regions. State and municipal authorities of depressed regions can use the obtained results for developing measures to reduce the level of investment risk, increase investment inflows, and provide sustainable development of the regional economy.

Keywords Investments · Investment risk · Investment risk factors · Depressed region · Kirov region

JEL Classification R11 · R58 · Z18

1 Introduction

The sustainable development of the economy of any region is connected with investment flows. In turn, the speed and scale of the investment flows into the region depend on the state of the investment climate of the territory. Although many different factors

M. V. Palkina · N. K. Savelyeva (✉)
Vyatka State University, Kirov, Russia
e-mail: nk_savelyeva@vyatsu.ru

M. V. Palkina
e-mail: palmavik@ya.ru

A. I. Kostennikov
Belgorod State National Research University, Belgorod, Russia

M. V. Ivanova
Maykop State Technological University, Maykop, Russia

© The Author(s), under exclusive license to Springer Nature Switzerland AG 2023
E. G. Popkova (ed.), *Smart Green Innovations in Industry 4.0 for Climate Change Risk Management*, Environmental Footprints and Eco-design of Products and Processes, https://doi.org/10.1007/978-3-031-28457-1_10

influence the state of the investment climate in the region, the key one is investment risk. Therefore, the reduction of investment risk is the priority task. Its success depends on the investment flows and the sustainable development of any region. Reducing investment risk is the most acute task for the authorities in depressed regions. Palkina and Kislitsyna [1], Savelyeva and Saidakova [2], Soboleva et al. [3], and others have studied the problems of depressed regions. There are some depressed regions in the Russian Federation, for example, the Oryol Region, the Ulyanovsk Region, the Smolensk Region, the Kirov Region, the Pskov Region; the Volgograd Region, the Ivanovo Region, the Chuvash Republic, the Kurgan Region, and the Altay Territory.

This research aims to analyze the state and development trends of the investment risk of a depressed region and develop proposals for their reduction. For this purpose, the authors had to solve the following tasks:

- To analyze the state and dynamics of the investment risk of one of the depressed regions over a decade;
- To determine directions for reducing the investment risk of the studied depressed region.

2   Methodology

Manifestations and factors of sustainable development of the contemporary economy are considered by such scientists as Andronova et al. [4], Fokina et al. [5], Karanina et al. [6], Patsyuk [7], Sergi et al. [8], Sergi et al. [9], Sozinova and Meteleva [10], Sozinova et al. [11, 12], and Zabaznova et al. [13].

Articles and methodological developments on regional risk management by Gorbunov et al. [14], Kabirova [15], Narolina [16], Sekletsova et al. [17], Snegireva [18], Tkhakushinov [19], Vladimirova [20], Zakharov and Ivanova [21], Trachenko and Dzhioev [22], Shevchenko and Zhabin [23], and Shirinkina [24] describe many approaches to interpreting the concept of investment risk in a region or regional investment risk. A distinctive feature of this risk is that it can occur while implementing investment activities within the boundaries of a particular region. The likelihood of such a risk depends on the state of the investment climate in the region. We used data from the rating of investment attractiveness of Russian regions by the Russian agency "Expert RA" as a basis for the analysis of investment risk. The methodology for compiling the rating of investment attractiveness of Russian regions by the Russian agency "Expert RA" makes it possible to study the structure of investment risk, make a general conclusion about the state of investment risk at the regional level, and conduct an analysis. According to the methodology of the Russian agency "Expert RA," the investment risk of the region consists of the following six private risks (investment risk factors):

1. Economic (general trends in the development of the region's economy);
2. Social (the degree of social tension in the region);

3. Financial (the level of balance between the budget of the region and the finances of enterprises);
4. Ecological (the degree of environmental pollution);
5. Criminal (the state of crime in the region);
6. Managerial (the level of management in the region).

Each investment risk factor is characterized by a group of indicators. The contribution of these components to the overall investment risk rating is assessed according to a survey of experts, potential investors, and representatives of the banking community. The process of data aggregation has several stages. At the stage of calculating the index of each type of investment risks in the region, the methodology uses the "minimax method." At the stage of calculating the integral investment risk index for each region, the weighted average sum of private investment risks (investment risk factors) of the region is calculated. In the next stage, all regions are ranked by the value of the integral investment risk. The region with the lowest integral risk gets first place. According to the methodology of the Russian agency "Expert RA," investment risk, together with investment potential, make up the investment attractiveness of the region and affect the state of the investment climate in the region.

3 Results

The research object was the Kirov Region, one of the depressed regions of the Russian Federation. The data for the calculations were the materials of the Russian agency "Expert RA." The calculations were performed in the Excel software environment.

The analysis of the state and development trends of the investment risk of the Kirov region was carried out in comparison with other depressed regions of the Russian Federation based on the data of the investment attractiveness rating of the regions of the Russian agency "Expert RA" for 2011–2020. According to the results of the analysis, we can make the following conclusions.

The depressed regions of the Russian Federation are characterized by a rather low value of the rank of investment risks (Table 1). Among the depressed regions, the least risky in terms of investment are the Volgograd Region (26th place) and the Altay Territory (37th place). The riskiest is the Oryol Region (64th place) and the Pskov Region (62nd place).

A comparison of the investment risk ranking of the Kirov Region for 2020 with the average median investment risk ranking for all depressed regions has shown (Fig. 1).

The values of three (financial, managerial, and social) of the six components of investment risk in the Kirov Region were below the average median values of these factors calculated for all depressed regions of the Russian Federation (Fig. 1). The largest gap with the average median value of the rank of depressed regions and the lowest value of the rank of the Kirov Region was in terms of social risk.

The Kirov Region had an excess of investment risk in three components: environmental, economic, and criminal risks. The greatest excess of the value of investment

Table 1 The structure of investment risk in depressed regions by its constituent factors for 2020

Indicator	Regions										
	Kirov region	Volgograd region	Ivanovo region	Pskov region	Ulyanovsk region	Oryol region	Chuvash republic	Smolensk region	Kurgan region	Altay territory
Investment risk	39	26	57	62	41	64	55	53	37	27
Social risk	39	37	27	59	54	33	62	52	10	57
Economic risk	54	46	77	50	69	45	44	61	21	41
Financial risk	37	34	70	69	38	64	58	43	28	22
Criminal risk	32	19	36	60	13	17	31	43	11	26
Environmental risk	56	40	34	31	16	23	32	15	4	43
Managerial risk	39	48	51	63	36	58	26	43	24	60

Source Compiled by the authors based on RAEX-Analytics [25]

Fig. 1 The structure of the investment risk ranking and its constituent factors by depressed regions for 2020. *Source* Compiled by the authors based on RAEX-Analytics [25]

risk over the average median value of investment risk was for environmental risk (the excess amounted to 24.5 points).

We have analyzed the data on the ranks of its components for 2011–2020 (Table 2) to form a complete picture of the state and development trends of investment risk and determine adequate measures to reduce it.

Table 2 Dynamics of the investment risk of the Kirov Region by its constituent factors for 2011–2020

Indicator	Years									
	2011	2012	2013	2014	2015	2016	2017	2018	2019	2020
Investment risk	55	58	64	61	40	42	44	41	45	39
Social risk	72	77	63	46	24	23	44	38	38	39
Economic risk	57	59	70	74	63	54	42	52	56	54
Financial risk	67	54	52	50	41	39	33	43	40	37
Criminal risk	3	10	30	42	38	58	54	39	35	32
Environmental risk	54	55	55	55	59	57	43	57	56	56
Managerial risk	20	21	51	49	34	51	48	25	46	39

Source Compiled by the authors based on the Federal State Statistics Service of the Russian Federation [26] and RAEX-Analytics [25, 27–35]

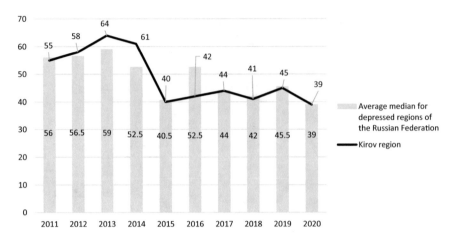

Fig. 2 Dynamics of the investment risk ranking in depressed regions for 2011–2020. *Source* Compiled by the authors based on the Federal State Statistics Service of the Russian Federation [26] and RAEX-Analytics [25, 27–35]

In general, there is a decrease in investment risk in the Kirov Region; the values of its rank are significantly lower than the median values of the investment risk of the depressed regions of the Russian Federation (Fig. 2). It distinguishes the Kirov Region from other depressed regions in terms of attracting investment.

Let us consider the dynamics of the investment risk components of the Kirov Region for 2011–2020. There is a positive trend in rank values for three components of investment risk (social risk, financial risk, and economic risk).

In particular, in 2011, the Kirov Region ranked 72nd in terms of social risk ranking. By 2020, it has improved to 39th place. The difference in ranks was 33 points. Similar dynamics were observed for financial risk. The Kirov Region ranked 67th in 2011 and improved its position to 37th place by 2020; the difference in ranks was 30 points. In terms of economic risk, the Kirov Region experienced significant fluctuations in rank over the analyzed decade: a decrease from 57th place in 2011 to 74th place in 2014, followed by an increase to 54th place by 2020.

Only two components of investment risk—criminal and managerial ones—had negative dynamics of the rank value during the entire analyzed period.

The strongest change in the rank was found in the criminal risk: the Kirov Region moved in the ranking from 3rd place in 2011 to 32nd place in 2020. The influence of criminal risks has increased three times over the decade. The dynamics of the managerial risk ranking can also be characterized as negative. By 2020, the Kirov Region had shifted from 20 to 39th place in this component of investment risk. Moreover, in terms of managerial risk, the Kirov region periodically (in 2013, 2016, and 2019) experienced a significant drop in rank.

According to the environmental component of the investment risk in the Kirov Region, the situation can be characterized as relatively stable: for 20 years, environmental risk fluctuations were within four points. The region managed to improve its rank to 43 only in 2017.

As the Kirov Region has a significant deterioration in the state and dynamics of development in terms of the managerial and criminal components of investment risk, it is advisable to consider the factors that reduce these types of risks.

Managerial risk characterizes the level of management in the region, and its assessment can be carried out based on the following indicators:

- Index of the physical volume of investments in fixed assets;
- Investment in fixed assets per capita;
- Infant mortality rate (number of deaths under the age of one year per 1000 live births).

During the analyzed period, the Kirov Region has a low level of the index of the physical volume of investments in fixed assets. The values of this indicator are significantly lower than all-Russian values. The volume of investments in fixed assets per capita in the Kirov Region for 2018–2020 is also lower (almost 2.5 times) than in Russia. These facts adversely affect the level of managerial risk in the Kirov Region. The situation is more favorable in terms of the infant mortality rate in the Kirov Region: its values for 2018–2020 were below national. Simultaneously, the dynamics of the values of this coefficient in the region are unstable: there are decreases and increases in values. This adversely affects the level of managerial risk as well. Thus, ensuring the investment inflows in fixed assets and reducing infant mortality in the region can be defined as priority areas for the regional authorities to reduce managerial risk.

Criminal risk characterizes the level of crime in the region; its assessment can be carried out based on the basis of the following indicators:

- Total registered crimes per capita;
- The number of economic crimes registered in the reporting period per capita.

In the dynamics for three years in the Kirov region, the total number of registered crimes per capita decreased. Simultaneously, in terms of the number of economic crimes, there has been a growing trend, which adversely affects the level of criminal risk in the region. Thus, the decrease in the number of crimes in the Kirov Region, in general, and the economic focus, in particular, requires close attention from the authorities to reduce the criminal risk of the region.

The emphasis on these areas of reducing the investment risk of the Kirov Region does not negate the need to develop and implement measures to improve its other components (economic, financial, etc.).

4 Conclusion

The research makes it possible to establish the level of investment risk and identify priority areas for its reduction in a depressed region. The level of investment risk in the Kirov Region can be characterized as moderate. Two components are of the greatest importance for reducing investment risk: managerial and criminal risk. The development and implementation of measures to reduce managerial and criminal risk should be included in the priority tasks by the state authorities of the region. It will reduce investment risk, attract additional investment to the depressed region, and create conditions for its sustainable socio-economic development.

Data Availability Data on the dynamics of the investment risk rank of the Kirov region, its components for 2011–2020, and indicators of managerial and criminal risk for 2018–2020, which are described in the next section of the research, are available at https://figshare.com/ with ID https://doi.org/10.6084/m9.figshare.21563796.

References

1. Palkina M, Kislitsyna V, Chernyshev K (2021) Analysis of the relationship of investment and demographic factors in the development of depressed regions. J Urban Reg Anal 13(1):113–124. https://doi.org/10.37043/JURA.2021.13.1.7
2. Savelyeva NK, Saidakova VA (2022) The level of economic security as an indicator of the region depression. In: Bogoviz AV, Popkova EG (eds) Digital technologies and institutions for sustainable development. Springer, Cham, Switzerland, pp 155–160. https://doi.org/10.1007/978-3-031-04289-8_26
3. Soboleva O, Sozinova A, Spengler A, Fokina O, Savelyeva N (2017) Mechanisms of regulation of economic processes in a region. In: Popkova E (ed) Overcoming uncertainty of institutional environment as a tool of global crisis management. Springer, Cham, Switzerland, pp 403–415. https://doi.org/10.1007/978-3-319-60696-5_51
4. Chernova VY, Starostin VS, Degtereva EA, Andronova IV (2019) Study of sector-specific innovation efforts: the case from Russian economy. Entrepreneurship Sustain Issues 7(1):540–552. https://doi.org/10.9770/jesi.2019.7.1(38)
5. Fokina OV, Sozinova AA, Glebova AG, Nikonova NV (2022) Improving the quality of project management at energytech through marketing in support of sustainable and environmental development of energy economics. Front Energy Res 10:943447. https://doi.org/10.3389/fenrg.2022.943447
6. Karanina EV, Sozinova AA, Bunkovsky DV (2022) Quality management in industry 4.0 in the post-COVID-19 period for economic security and sustainable development. Int J Qual Res 16(3):877–890. https://doi.org/10.24874/IJQR16.03-15
7. Patsyuk EV, Krutilin AA, Savelyeva NK, Chernitsova KA (2022) Methodological approach to the polycriteria assessment of agricultural sustainability: digitalization, international experience, problems, and challenges for higher education in Russia. In: Popkova EG, Sergi BS (eds) Sustainable agriculture. Springer, Singapore, pp 43–53. https://doi.org/10.1007/978-981-16-8731-0_5
8. Sergi BS, Popkova EG, Borzenko KV, Przhedetskaya NV (2019) Public-private partnerships as a mechanism of financing sustainable development. In: Ziolo M, Sergi BS (eds) Financing sustainable development: key challenges and prospects. Palgrave Macmillan, Cham, Switzerland, pp 313–339. https://doi.org/10.1007/978-3-030-16522-2_13

9. Sergi BS, Popkova EG, Sozinova AA, Fetisova OV (2019) Modeling Russian industrial, tech, and financial cooperation with the Asia-Pacific region. In: Sergi BS (ed) Tech, smart cities, and regional development in contemporary Russia. Emerald Publishing Limited, Bingley, UK, pp 195–223. https://doi.org/10.1108/978-1-78973-881-020191012
10. Sozinova AA, Meteleva OA (2022) Sites of states with a dynamically developing socio-political structure and economy: analyzing forms and methods of obtaining competitive advantages of transnational (Global) companies. In: Popkova EG, Andronova IV (eds) Current problems of the world economy and international trade. Emerald Publishing Limited, Bingley, UK, pp 233–242. https://doi.org/10.1108/S0190-128120220000042022
11. Sozinova AA, Kosyakova IV, Kuznetsova IG, Stolyarov NO (2021) Corporate social responsibility in the context of the 2020 economic crisis and its contribution to sustainable development. In: Popkova EG, Sergi BS (eds) Modern global economic system: evolutional development vs. revolutionary leap. Springer, Cham, Switzerland, pp 83–90. https://doi.org/10.1007/978-3-030-69415-9_10
12. Sozinova AA, Sofiina EV, Safargaliyev MF, Varlamov AV (2021) Pandemic as a new factor in sustainable economic development in 2020: scientific analytics and management prospects. In: Popkova EG, Sergi BS (eds) Modern global economic system: evolutional development vs. revolutionary leap. Springer, Cham, Switzerland, pp 756–763. https://doi.org/10.1007/978-3-030-69415-9_86
13. Zabaznova TA, Akopova ES, Sozinova AA, Sofiina EV (2022) The benefits of reconstructive agriculture for food security and rural tourism in present and future: innovations and sustainable development. In: Popkova EG, Sergi BS (eds) Sustainable agriculture. Springer, Singapore, pp 207–213. https://doi.org/10.1007/978-981-19-1125-5_24
14. Gorbunov VN, Khanzhov IS, Bainishev SM (2014) Assessing the level of economic development based on the definition of integrated potential and risk. Mod Sci Res Innov 8–2(40). Retrieved from https://web.snauka.ru/issues/2014/08/36660 (Accessed 2 Nov 2022)
15. Kabirova AS (2017) Analysis and management of regional risks. Econ Manag: Probl Solutions 2(6):111–119
16. Narolina YuV (2009) Investment potential and investment risk as the main components of the investment attractiveness of the region. Tambov Univ Rev Ser: Humanit 12(80):137–143
17. Sekletsova OV, Kuznetsova OS, Ponkratova TA (2010) To the question of investment risk analysis in Kemerovo Region. Food Process: Tech Technol 2(17):93–97
18. Snegireva TK (2017) Regional investment processes, risk assessment and ways to solve problems. Manag Issues, 6(49)
19. Tkhakushinov EK (2009) Economic and institutional mechanism for monitoring investment risks in the region. Terra Economicus 7(3–2):242–246
20. Vladimirova IS (2019) Statistical assessment of investment attractiveness of regions of Volga Federal District taking into account risk factors and potential. Managing Sustain Dev 2(21):13–19
21. Zakharov SS, Ivanova EI (2016) Methods for assessing the investment attractiveness of the region. Bulletin of Vladimir State University named after A.G. and N.G. Stoletovs. Ser: Econ Sci 2(8):82–90
22. Trachenko MB, Dzhioev VA (2018) Express analysis of the investment appeal of regions. Finan Credit 24(9):2151–2165. https://doi.org/10.24891/fc.24.9.2151
23. Shevchenko IV, Zhabin VV (2007) Ways to improve methods for assessing the investment climate of regions. Reg Econ: Theory Pract 5(14):47–52
24. Shirinkina EV (2010) Assessment of regional investment risk. Acad Bull 1(11):19–23
25. RAEX-Analytics (2020) Ranking: investment risk of Russian regions in 2020. Retrieved from https://raex-rr.com/pro/regions/investment_appeal/regiona_investment_risk_rating/2020// (Accessed 2 Nov 2022)
26. Federal State Statistics Service of the Russian Federation (Rosstat) (2021) Regions of Russia: socio-economic indicators—2021. Retrieved from https://rosstat.gov.ru/storage/mediabank/Region_Pokaz_2021.pdf (Accessed 2 Nov 2022)

27. RAEX-Analytics (2011) Ranking: investment risk of Russian regions in 2011. Retrieved from https://raex-rr.com/pro/regions/investment_appeal/regiona_investment_risk_rating/2011/ (Accessed 2 Nov 2022)
28. RAEX-Analytics (2012) Ranking: investment risk of Russian regions in 2012. Retrieved from https://raex-rr.com/pro/regions/investment_appeal/regiona_investment_risk_rating/2012/ (Retrieved from 2 Nov 2022)
29. RAEX-Analytics (2013) Ranking: investment risk of Russian regions in 2013. Retrieved from https://raex-rr.com/pro/regions/investment_appeal/regiona_investment_risk_rating/2013/ (Accessed 2 Nov 2022)
30. RAEX-Analytics (2014) Ranking: investment risk of Russian regions in 2014. Retrieved from https://raex-rr.com/pro/regions/investment_appeal/regiona_investment_risk_rating/2014/ (Accessed 2 Nov 2022)
31. RAEX-Analytics (2015) Ranking: investment risk of Russian regions in 2015. Retrieved from https://raex-rr.com/pro/regions/investment_appeal/regiona_investment_risk_rating/2015/ (Accessed 2 Nov 2022)
32. RAEX-Analytics (2016) Ranking: investment risk of Russian regions in 2016. Retrieved from https://raex-rr.com/pro/regions/investment_appeal/regiona_investment_risk_rating/2016/ (Accessed 2 Nov 2022)
33. RAEX-Analytics (2017) Ranking: investment risk of Russian regions in 2017. Retrieved from https://raex-rr.com/pro/regions/investment_appeal/regiona_investment_risk_rating/2017/ (Accessed 2 Nov 2022)
34. RAEX-Analytics (2018) Ranking: investment risk of Russian regions in 2018. Retrieved from https://raex-rr.com/pro/regions/investment_appeal/regiona_investment_risk_rating/2018/ (Accessed 2 Nov 2022)
35. RAEX-Analytics (2019) Ranking: investment risk of Russian regions in 2019. Retrieved from https://raex-rr.com/pro/regions/investment_appeal/regiona_investment_risk_rating/2019/ (Accessed 2 Nov 2022)

Key Aspects of Sustainable Business Model Innovation

Elena S. Ratushnyak◉ **and Vladimir V. Shapovalov**◉

Abstract Sustainability is one of the new competitive advantages on the market. However, companies cannot maintain planetary boundaries in practice when implementing ESG principles in management. Therefore, it is necessary to implement sustainable practices into the operational process as a whole. The research aims to formalize sustainable business model innovation (SBM-I) as a new tool for social and business value integration. The authors used three analytical tools to formulate the main features of the sustainable business model innovation: business model concept, innovation strategies, and sustainable goals. Using the business model as a common tool to visualize a rational way to create and deliver market value helps to create a framework for inclusively implementing sustainability to improve the outcome on the path to sustainability. The analysis of current advances in business model innovation and practical views of BCG consulting company contributed to the connection between theory and practice. It is pointed out that there is an incremental path to SBM-I, which signifies the level of the company's maturity. Consistency paths of developing SBM-I in a methodological way provide the connection between strategy and sustainability, making them mutually reinforcing, giving way to action instead of reporting, and realizing a multilevel approach instead of a company-centric one. SBM-I helps to consider all interests of company stakeholders to develop business in an appropriate business ecosystem. Sustainable business model innovation provides a way applicable for companies regardless of activity field, business size, and access to resources.

Keywords Business ecosystem · Business model innovation · Corporate social responsibility · ESG · Management · Sustainability

JEL Classification M19 · M21 · O12 · O21 · O31

E. S. Ratushnyak (✉) · V. V. Shapovalov
MGIMO University, Moscow, Russia
e-mail: helenarat88@gmail.com

© The Author(s), under exclusive license to Springer Nature Switzerland AG 2023
E. G. Popkova (ed.), *Smart Green Innovations in Industry 4.0 for Climate Change Risk Management*, Environmental Footprints and Eco-design of Products and Processes, https://doi.org/10.1007/978-3-031-28457-1_11

1 Introduction

At the current stage of economic development, sustainable development is a key component of the new competitive advantage of companies. It includes the necessity to address closely the planetary boundaries—the borders within which humanity can continue to develop and thrive [9, 10]. Besides, societal needs constrain companies when their opportunity space is narrowing, restricting their ability to pursue business value creation as usual. Therefore, it appears necessary for management to change the conventional business development approach to the new one.

In this context, a business model accumulates a view of business development and visually demonstrates how the firm creates market value, which contributes to making money. According to current research, as a general guide, the business model includes basic elements organized in a special way, called canvas. Simultaneously, according to BCG, in the last fifty years, the average business model lifespan has fallen from about fifteen years to less than five years. There is a new tool on the way to sustainability within developing a business framework [14].

Concurrently, the business model is one of the company-specific factors that are considered to measure the company's exposure to industry-specific material ESG risks and how well a company is managing these risks [11, 12]. Thus, the appropriate business model can be seen as one that has a strong potential to create environmental and social surpluses (net environmental and societal benefits) and connects them to the drivers of business advantage and value creation [3, 6].

In the BANI world, it is highly important for companies to look for and capture long-lasting competitive advantage. In this context, business model innovation can bring results by providing breakout growth, reinvigorating a lagging core, or defending against industry disruption or decline.

Business model innovation should realize in a special way. It is a simultaneous and mutually supportive reimagination of two key elements of the business: the organization's value proposition to customers and the operating model [3, 8].

Currently, many companies are trying to take different actions to realize business model innovation on the path to sustainability [2, 5, 15, 16]. This can be due to a variety of factors. Nevertheless, the most important of them is the need to embrace and maintain customer loyalty by providing real customer value (value proposition) as the business model's core [4].

Thus, the main questions in this context are as follows:

- What are the features of a sustainable business model?
- Are these features cross-functional?
- Is the model innovative?

To get the appropriate answers, it is necessary to solve different tasks, including the following:

- To analyze the existing approaches to business model innovation to highlight specific features;
- To schematize the elements in a business model that could make it sustainable;

- To match the innovation and sustainable features of the business model;
- To analyze the case of realizing a sustainable business model in the market.

Reimagining the business models of companies for sustainability to optimize social and business value in a new economic ecosystem is essential in developing a new concept of economic ecosystems [1, 7, 13]. The newest business aim is combining sustainable goals and profit.

According to recent research, few companies have attempted to formulate a systematic understanding of the limits of sustainability, vulnerabilities, and their current business models and ecosystem potential, resulting in the risk of reduced competitiveness in the future. This can be interpreted as an omission by managers regarding sustainability and social issues, as they often view them separately from the core business.

Additionally, one of the main problems of developing sustainable business is that ESG metrics have become an end goal instead of measuring action and progress against a strategic plan. Companies are increasingly trying to focus on reporting and compliance per se, as well as on metrics for materiality and sustainability.

This paper could help formalize the common approach to creating system understanding, where sustainable and social aspects are closely connected or even entwined into blocks of business operations to pore management on strategy, action, and advantage on the way to sustainable business.

Understanding the concept of a systemic, sustainable business model is also important in terms of creating a link between business value drivers and progress on sustainability metrics, as there is currently a common problem across firms of little connection between them.

Another challenge in implementing sustainable international business development is that many companies report on ESG metrics at the individual level, sometimes including elements of the supply chain but rarely including the full business ecosystem, industry, or the broader network of stakeholders. Simultaneously, all of them have a role in constraining or enabling advantage and sustainability. Thus, there is a great necessity to consider this aspect when implementing actions in the field of sustainability.

Hence, the concept of sustainable business model innovation contributes to the concept of sustainable business management on a systemic basis; it can act as a framework for implementing sustainability issues in business operations as a whole.

2   Materials and Method

The analytical method was used as the basis of this research. The authors begin by analyzing existing views on business model innovation. The authors used scientific theoretical research and practical materials of a consulting company, in particular BCG, one of the activities of which is consulting in the field of business model development. It allowed the authors to combine theoretical and practical aspects and highlight specific characteristics.

3   Results

Figure 1 shows the two main changes in the business model innovation.

Decisions about changes at the levels of the operating model are made based on the formation of a value proposition by finding the answers to the following questions:

- Where to play along the value chain?
- What cost model is needed to ensure attractive returns?
- What organizational structure and capabilities are essential to success?

For companies seeking to grow through business innovation, there are common critical issues such as:

- What is the extent of the changes?
- What is the acceptable risk level?
- Is the process single or ongoing?

Therefore, as an answer and help to fill these gaps, there are contemporary approaches to business model innovation, each with its own characteristics (Table 1).

In terms of sustainable business management, there is a gradual path to achieving sustainable business model innovation (Fig. 2). This path connects strategy and sustainability on the process base at the enterprise level and at higher levels.

Value Proposition Level

- The choice of target segment;

- product or service offering;

- Revenue model.

Operating Model Level

- Ways of delivering profitability increase;

- Ways of delivering competitive advantage;

- Ways of delivering value creation.

Fig. 1 Changes within business model innovation. *Source* Compiled by the authors

Table 1 Key features of the business model innovation approaches

No.	Approach name	Key indicators signaling the need to implement the approach	Main features of the approach
1	The reinventor approach	– There are fundamental industry challenges (commoditization or new regulation) – The current business development can be characterized as a process of slow aggravation, with no prospects for growth	Reimagination of customer-value proposition and all business processes to deliver the new one = *Transform the core + Provide defense against industry decline or disruption*
2	The adapter approach	For all practical purposes, the current core business cannot hurdle key disruption in the industry even if it is reinvented	There is a possibility of exiting the core business. It is necessary to look closely at adjacent businesses or markets *Expand into noncore + Provide defense against industry decline or disruption*
3	The maverick approach	Zooming potentially more successful core business based on business model innovation	Mavericks could be startups and insurgent established companies. Leveraging the core advantage helps revolutionize the industry and set new standards due to their own individual view of traditional things = *Transform the core + Support breakout growth*
4	The adventurer approach	Promoting new markets using new tools to apply the company's current competitive advantage	Persistently and suppressively expand business by exploring or developing new or adjacent territories = *Expend into noncore + Support breakout growth*

Source Compiled by the authors based on [3]

To be effective, it is necessary to realize SBM-I through several consistent steps. They are as follows:

- The first step is to test the current business model for sustainability against a broader temporal, societal, and spatial context to find out all vulnerability to externalities, its sustainability limits, and its potential to create new environmental and societal value;
- The second step is to explore business model innovations by applying a combination of modular transformations to address limits and leverage potentials;

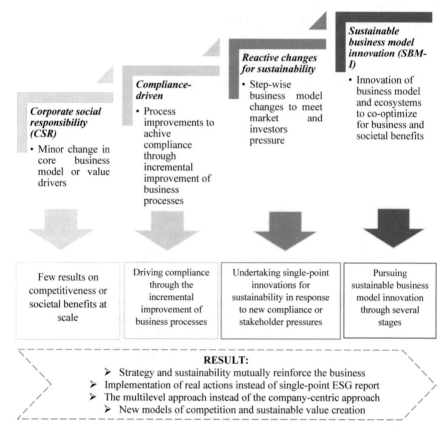

Fig. 2 From CSR to SBM-I: key stages of developing company maturity within business model innovation. *Source* Compiled by the authors based on [16]

- The third step is to connect business model innovations back to the core drivers of business advantage and financial performance to assess how they can deliver value and sustainability.

New models are piloted and tuned to capture advantage in the market and with investors' and stakeholders' interests. It is also essential to understand what changes are needed in the business ecosystem or at the industry level to create the right context for success.

Thus, implementing all steps helps connect strategy and sustainability and reinforce them mutually. However, it provides a way to action instead of reporting and implements a multilevel approach instead of a company-centric one.

To demonstrate the results of realizing SBM-I, it is interesting to analyze the frontrunners' experience of sustainable business model innovation, including Schneider Electric—a digital energy management firm. Initially, this French-based company was a traditional B2B electrical distribution equipment manufacturer on the market.

Due to developing the business model, Schneider Electric became a global leader in digital sustainability solutions. As a result, the business is effective in the context of realizing sustainability in the business ecosystem.

At the heart of the developed sustainable business model is an ecosystem based on the EcoStruxure platform, incorporating various digital technologies such as the Internet of Things. The use of advanced technologies helps manage the data and controls in buildings, infrastructure, data centers, power grids, and industries. The main directions of Schneider Electric's sustainable business model include the following:

- The company showed its customers the way to manage their energy consumption and sustainability footprint, which resulted in saving 90 million metric tons of CO_2 per year collectively;
- The company works on improving environmental and societal impact in other ways, particularly through ground-breaking collaboration with Walmart. This collaboration aims to provide increased access to renewable energy across the retailer's US-based supply chain. The initiative, called the Gigaton Power Purchase Agreement (GPPA) Program, is designed to educate Walmart suppliers about renewable energy purchases and facilitate adoption through aggregate purchase agreements;
- The company works with countries worldwide to develop fast-track healthcare facilities throughout the COVID-19 pandemic. For instance, in China and Turkey, Schneider Electric helped get new hospitals operational. In the UK and Italy, it helped convert exhibition centers into temporary hospitals. In France, India, and Spain, it helped manufacture respirators. On a global scale, it worked on the rapid deployment of solutions to enable intensive care units to manage the challenges of the COVID-19 pandemic better.

4   Conclusion

Having a sustainable business model means having a new competitive advantage in the current business environment by being able to provide sustainability, longevity, and value creation in a changing business, community, and investor environment.

To be a pacesetter in some industries, companies must esteem sustainability as a source of business advantage. It means the necessity to take a new approach that seeks out innovative ways to address environmental and societal needs through the core business models.

There is a path in successive steps to sustainable business development, starting from CSR to SBM-I. SBM-I helps businesses to make sustainability a source of advantage, solving two main issues. The first issue is the possibility of making efforts in the field of developing sustainability initiatives (for instance, reducing greenhouse gas emissions, improving the general quality of life, reducing vulnerability to diseases, or lifting people out of poverty) in a systemic rather than fragmented way, to scale up and give them greater strength. The second issue is the possibility of integrating all efforts into their core business.

Involving different stakeholders in achieving a sustainable goal by embracing different elements of the business model makes all sustainability efforts more effective. Simultaneously, there is no universal way of doing this for all players in the international market. In the current phase, companies can use business model innovation approaches as thought inspiration in developing sustainable business model innovations.

References

1. Auerswald PE, Lokesh D (2017) Economic ecosystems. In: Clark GL, Feldman MP, Gertler MS, Wójcik D (eds) The new oxford handbook of economic geography. Oxford University Press, New York, NY, pp 245–248. Retrieved from https://ssrn.com/abstract=3494495 (Accessed 10 Oct 2022)
2. Biloslavo R, Bagnoli C, Massaro M, Cosentino A (2020) Business model transformation toward sustainability: the impact of legitimation. Manag Decis 58(8):1643–1662. https://doi.org/10.1108/MD-09-2019-1296
3. Boston Consulting Group (BCG) (2022) Business model innovation. Retrieved from https://www.bcg.com/capabilities/innovation-strategy-delivery/business-model-innovation (Accessed 10 Oct 2022)
4. Ferlito R, Faraci R (2022) Business model innovation for sustainability: a new framework. Innov Manag Rev 19(3):222–236. https://doi.org/10.1108/INMR-07-2021-0125. Retrieved from https://doi.org/10.1108/INMR-07-2021-0125/full/pdf?title=business-model-innovation-for-sustainability-a-new-framework (Accessed 18 Oct 2022)
5. Geissdoerfer M, Vladimirova D, Evans S (2018) Sustainable business model innovation: a review. J Clean Prod 198:401–416. https://doi.org/10.1016/j.jclepro.2018.06.240
6. Gerlach R (2019) The tools & business practices to promote sustainable product & business model innovation. Retrieved from https://www.threebility.com/post/the-tools-business-practices-to-promote-sustainable-product-business-model-innovation (Accessed 11 Oct 2022)
7. Guirui Y, Shilong P, Yangjian Z, Lingli L, Jian P, Shuli N (2021) Moving toward a new era of ecosystem science. Geogr Sustain 2(3):151–162. https://doi.org/10.1016/j.geosus.2021.06.004
8. Lindgardt Z, Hendren C (2014) Using business model innovation to reinvent the core: doing something new with something old. Boston Consulting Group. Retrieved from https://www.bcg.com/publications/2014/growth-innovation-using-business-model-innovation-reinvent-core?awsPersonalize=true&awsPersonalizeView=mainNavigation (Accessed 3 Oct 2022)
9. Persson L, Carney Almroth BM, Collins CD, Cornell S, de Wit CA, Diamond ML, Fantke P, Hassellöv M, MacLeod M, Ryberg MW, Søgaard Jørgensen P (2022) Outside the safe operating space of the planetary boundary for novel entities. Environ Sci Technol 56(3):1510–1521. https://doi.org/10.1021/acs.est.1c04158
10. Rockström J, Steffen W, Noone K, Persson Å, Chapin III FS, Lambin E, Lenton TM, Scheffer M, Folke C, Schellnhuber HJ, Nykvist B (2009) Planetary boundaries: exploring the safe operating space for humanity. Ecol Soc 14(2):32. Retrieved from http://www.ecologyandsociety.org/vol14/iss2/art32/ (Accessed 8 Sept 2022)
11. Strategyzer (n.d.) Business models. Retrieved from https://www.strategyzer.com/expertise/business-models (Accessed 25 Aug 2022)
12. Sustainalytics (2022) Sustainalytics' 2022 top-rated ESG companies. Retrieved from https://www.sustainalytics.com/corporate-solutions/esg-solutions/top-rated-companies (Accessed 3 Oct 2022)

13. The Club of Rome (2020) A system change compass: implementing the European green deal in a time of recovery. Retrieved from https://www.clubofrome.org/wp-content/uploads/2020/10/System-Change-Compass-Full-report-FINAL.pdf (Accessed 14 Sept 2022)
14. Threebility (2022) The sustainable business model canvas. Retrieved from https://www.threebility.com/sustainable-business-model-canvas (Accessed 3 Oct 2022)
15. Young D, Gerard M (2021) Four steps to sustainable business model innovation. Boston Consulting Group. Retrieved from https://www.bcg.com/publications/2021/four-strategies-for-sustainable-business-model-innovation (Accessed 3 Sept 2022)
16. Young D, Reeves M (2020) The quest for sustainable business model innovation. Boston Consulting Group. Retrieved from https://www.bcg.com/publications/2020/quest-sustainable-business-model-innovation (Accessed 25 Oct 2022)

Synergetic Approach in Public Administration

Lela V. Chkhutiashvili⊙, **Nana V. Chkhutiashvili**⊙, **Alexander M. Gubin**⊙, and **Galina F. Golubeva**⊙

Abstract The research focuses on the relevant issue of applying the synergetic approach in public administration. The research aims to show the advantages and disadvantages of a synergetic model that is more competitive. The authoritarian management model and the synergetic model are analyzed by comparison methods. As part of synergetics, self-developing systems offer a new understanding of the fundamental foundations of systems theory. Self-regulating systems reflect the object as a process, while self-developing systems characterize the transition from one process to a completely different one. Systems in public administration are now not self-regulating in time and space but also evolving. The synergetic approach to public administration obliges consideration of the complex indirect impact on society of non-state and non-political factors, including economy, ecology, culture, social sphere, geography, and demography. A synergetic approach in public administration is the best solution today because it fits the current realities. However, it has certain drawbacks.

Keywords Interdisciplinary direction · Management model · Synergetic approach · Contemporary trends · Manager · Public administration · Instability of the system

JEL Classification D91 · E01 · F42 · F43 · F64 · Q01 · Q15 · O31 · O32 · O33 · O38 · Q56 · Q57 · O13 · O41 · O43 · O44 · O47

L. V. Chkhutiashvili (✉) · N. V. Chkhutiashvili · A. M. Gubin
Kutafin Moscow State Law University (MSAL), Moscow, Russia
e-mail: lvachhutiashvili@msal.ru

A. M. Gubin
e-mail: amgubin@msal.ru

G. F. Golubeva
Plekhanov Russian University of Economics, Moscow, Russia

1   Introduction

In the post-war period, many countries retained an authoritarian management model, which took over the organization of meeting the basic needs of the population: social security, medicine, education, etc. Authoritarianism implies direct subordination in a hierarchical structure and bureaucratic methods of solving managerial tasks. With this management model, any action is performed through a certain system of instances; other organizations of this system do not participate in solving the problem. Everything is subject to standard rules and procedures established by the system. If these rules are not followed, this may entail consequences that will affect the entire structure. This management model is distinguished by a monopoly on the provision of services. It is quite structured and suitable for any consumer needs, but it cannot function effectively with the new technologies that have come into our lives together with the information revolution, globalization, and the Fourth Industrial Revolution. All public services for residents of a city or country should be provided exclusively by state authorities to eliminate the interference of private entrepreneurs in a particular area.

2   Materials and Method

The solution to the research task was carried out through comparative, logical, and statistical analysis, structure analysis, mutual respect, and the dynamics of the phenomena in the documentation and information areas based on information from books, newspapers, legal acts, and the Internet. The descriptive method, including the techniques of observation, interpretation, comparison, and generalization, is fundamental for the undertaken research.

3   Results

The synergetic approach in public administration is more flexible and more suitable for the current requirements for the work of public authorities. Under the new (synergetic) management model, there is participation management, in which creative groups are created that take an active part in the strategic promotion of the company and solving problems in various sectors.

The synergetic management model is more competitive and provides an opportunity for expansion, unlike the traditional model, in which everything is structured, and any competition is destroyed at the root. Since competition is the main engine of progress, which does not allow the organization to stand still, it forces it to develop and implement new solutions to problems that entail the population's demand.

To fully understand the essence of the problem, it is worth understanding the definition of an interdisciplinary scientific direction. The interdisciplinary focus of scientific research comprehensively solves such important problems as the integration of all stakeholders (organizations and departments). The synergetic approach is the methodological basis for the transition to sustainable development [1]. The interdisciplinary direction of scientific knowledge is unique because an interdisciplinary approach is a universal method that selects methods from all possible fields of knowledge.

Speaking about the interdisciplinary direction of scientific knowledge, such terms as "transdisciplinarity" and "polydisciplinarity" are often used. Transdisciplinarity considers studies that have horizontal connections of reality that go beyond specific disciplines. In turn, polydisciplinarity is considered a subject of study from different sides using the capabilities of various disciplines [1]. Nowadays, interdisciplinarity is a landmark in the development of science as such.

If we talk about synergetics and the synergetic approach, it is a direction of interdisciplinary scientific research. This branch is a direct ideological continuation of the entire interdisciplinary direction of scientific knowledge.

Synergetics is a synthesis of complex self-organizing systems [6]. That is, a group of elements takes part in finding solutions to certain problems. In the case of an interdisciplinary direction of scientific knowledge, many disciplines are used; in the case of a synergetic approach, all links of the system take part.

The term synergetics was derived by Professor Herman Haken of the University of Stuttgart. He noticed that during the transition from disorder to order, similar behavior of elements occurs in all phenomena, which the scientist called a cooperative, synergetic effect [5]. This term comes from the ancient Greek *synergea*—joint, coordinated action.

Synergetics is a generalized science that studies the basic laws of the self-organization of systems. At its core, a synergetic system is a kind of living organism that undergoes the process of evolution; the effect of the unpredictability of the system appears just as in the process of evolution. This unpredictability of the evolution of synergetic systems has been called stochasticity.

Within the framework of synergetics, systems are considered not as self-regulating entities but rather as a result of the interaction of these entities with each other. They are expressed through the category of self-developing systems. The work of V. Stepin [8] reveals the essence of the previously established and rebuilds them, as a result of which the system acquires new integrity. With the emergence of new levels of the organization, new, relatively independent subsystems are formed in the system. Simultaneously, the control unit is being rebuilt, new order parameters are being added, and new types of direct connections and feedback connections are established.

The synergetic approach in public administration is fundamental for contemporary trends in the world. This approach is the most flexible system in which there is no clear hierarchy in the management of the system elements because these elements are a self-organizing system, where the manager only creates this system and maintains its integrity.

The nonlinearity of the synergetic approach should be considered because such an approach in public administration implies the self-organization of the system. This nonlinearity means the indirect development of the system. Nonlinearity consists of three fundamental provisions: multivariability, rhythmicity, wave character, resonance, and the possibility of rapid development of the system.

In a synergetic sense, the goal of public administration is to create conditions for the dynamic development of a society capable of overcoming extreme crisis states. If we consider the synergetic approach in public administration from the side of the manager and not from the side of the system, then we may encounter a misunderstanding of the essence of the manager's role because, at first glance, the system develops without the manager's direct participation. However, this is not the case. In public administration, managers play the role of a parent who, step by step, raise their "child" (organization) and does everything to make this child competitive and, most importantly, independent in the future.

In public administration, an objectively synergetic approach is manifested in developing close interaction between the state, business, and society. A public–private partnership (PPP) is a special system of interaction between functioning on the principles of institutional quality, accessibility, provision, and realization of the common and private interests of the participants. Public authorities and the private sector of the economy participate in various fields of activity based on the consolidation of limited resources and redistribution of risks between the participants of the partnership. In terms of volume, PPP is part of the concept of interaction between businesses and the government. Moreover, the interaction of the government with business does not consist only of partnerships. Cooperation between businesses and the government can be called a partnership if they stimulate the joint growth of productivity and efficiency of the private and public sectors of the economy.

The development of Russian entrepreneurship in the twenty-first century has broad prospects, which largely depend on the state policy of supporting small and medium-sized businesses, the development of legal regulation, and the formation of a favorable business environment. For this purpose, it is important to introduce new production and management technologies and form a qualitatively new system of corporate values of the Russian business community. During the digital transformation of society and the creation of a socially oriented economy, the state and business become strategic partners.

Distinctive features of a new business in Russia are the ability to work in innovative high-tech areas of production, interact with government authorities, and skillfully use administrative resources [7]. The more efficiently the market functions, the fewer economic functions are assumed by the state. Significant state support is required in areas where the business has not acquired stability.

State support for small and medium-sized businesses in the Russian Federation is aimed at increasing the social responsibility of domestic entrepreneurs in the development of society. Russian business should feel like an integral and most important part of the social organism, an integral part of the mechanism of constructive interaction between society, the state, and business.

Nevertheless, the synergetic approach in public administration also has negative sides, for example, the instability of the system itself [2]. Since most of the rights in solving the problems that have arisen are transferred to the participants of the system, the further development of this system may be under great question. Development with such an approach may not go in the direction that the manager sought.

It is worth mentioning the disadvantages of this system for the manager. Although managers play the role of "gray cardinals" who do not take an active part in this system, they still play one of the most important (if not the most important) roles in this system. Since managers monitor the solution of problems during the workflow and must invest much effort in the development of this system, it subsequently becomes exactly what the manager sought from the very beginning.

Therefore, it is fair to put forward the following principles for managing the state system at the time of a crisis situation and overcoming the extreme of the crisis:

- Management impact should be directed to the most profitable alternative for further development of the system;
- Management impact should be carried out at the right time.

Synergetics aims society to perceive the reform process as a movement towards higher and socially significant goals and develops the ability and willingness to use the new in the interests of social progress comprehensively. That is why the national idea is an important factor in the self-organization of society because only the appeal to national symbols of faith and the emergence of high spiritual motives in society will increase the level of moral self-awareness, as a result of which destructive factors in public administration will be organized in such a way that a reliable counterbalance is represented by civil society, which is interested in the success of the reforms carried out by the state [9]. In other words, the synergetic approach in government requires analyzing the impact of internal components, considering the complex impacts on society from other non-political and non-state factors (e.g., family, spiritual, and cultural).

Synergetics perceive the reform as a movement towards socially significant goals and promote the development of readiness to use everything new in the interests of social progress. The national idea becomes an important factor in the self-organization of society, a symbol of changing relations, and the most complex self-organizing system. For Russia, this circumstance largely determines the inefficiency of many traditional methods of public administration, which, for example, work successfully in Western countries [4]. Therefore, in our opinion, the truly important and effective use of the synergetic approach is in the field of public administration in today's Russia.

Thus, as part of synergetics, self-developing systems offer a new understanding of the fundamental foundations of systems theory. The synergetic approach in public administration is a breath of fresh air for the entire system. The classical administration system does not allow for competition with private enterprises that are also engaged in providing services to citizens. Therefore, the synergetic approach to management is the best option.

4 Conclusion

Summing up the above, we can conclude that the public administration system should change along with changes in technology, society, and the environment. The key to the survival of systems is the search for a permanent compromise between authoritarian and democratic governance; the structure of a self-organizing system should ensure a balance between autonomy and automation. For a system to survive, it cannot be centralized or decentralized. Its structure should be regulated by the ratio "centralism—autonomy" to unlock the potential of self-organization and maintain its positive trends, restraining negative ones in response to the challenges of the time [10]. That is why the synergetic approach in public administration is currently the best solution.

The synergetic approach obliges public administration to consider complex indirect impacts on society from non-state and non-political factors, including economy, ecology, culture, social sphere, geography, and demography. Special attention should be paid to developing a state policy that ensures the development of advanced technologies to bring Russia among the leading countries in the field of the digital economy. For this purpose, it is necessary to create good legislative support for the digital economy management system [3]. It fits the current realities, although it has certain drawbacks.

References

1. Aliyev ShM, Nabieva DN (2010) Sustainable development and interdisciplinary direction of scientific research. South Russ: Ecol Dev 5(1):8–11. https://doi.org/10.18470/1992-1098-2010-1-8-11
2. Chkhutiashvili LV (2020) Transition to the digital economy: the essence and factors of improving the efficiency of the economy of the Russian Federation. In: Kosorukov OA, Pechkovskaya VV (eds) Collection of materials of the III International Scientific and practical Conference "Innovative economics and management: Methods and technology". LLC "Green Print.", Moscow, Russia, pp 121–123
3. Chkhutiashvili LV, Chkhutiashvili NV (2020) Economic security of a person and society in the digital economy. In: Sokolov AV, Frolov AA (eds) Opportunities and threats of the digital society: materials of the conference of the All-Russian scientific and practical conference. Yaroslavl, Russia, pp 232–236
4. Gurina MA (2017) Increasing the of competitiveness organizational systems: a paradigm shift in social management. In: Nesterova NN, Koryakina OY, Kidinov AV, Kukina EE, Morozova NS (eds) Strategic initiatives of socio-economic development of economic entities of the region in the conditions of external constraints: materials of the international scientific and practical conference organized jointly with the SEZ administration PPT "Lipetsk". Ritm, Voronezh, Russia, pp. 72–76
5. Kharchenko KV (2008) Sociology of management: from theory to technology. Regional Printing House, Belgorod, Russia
6. Knyazeva EN (2019) Philosophical sciences. Interdisciplinary research strategies: textbook. Yurayt, Moscow, Russia
7. Poletaev VE (2011) Public-private partnership as a form of cooperation between the state and business in modernizing Russia. Almanac Mod Sci Educ 9(52):12–15

8. Stepin V (2009) Self-developing systems and the philosophy of synergetics. Econ Strat 11(7):24–35
9. Volkova OA (2017) Substantiation of the need for a synergetic approach in modern public administration. In: Moiseev AD, Gurina MA (eds) The science and art of management: materials of the IV International Competition of scientific papers. Ritm, Voronezh, Russia, pp 61–64
10. Zyryanov AV (2013) Forms of realization of the government (synergetic campaign). Bull Omsk Univ Ser: Law 4(37):13–16

Advanced Climate-Smart Technology as the Basis for the Activities of Green Entrepreneurship in the Digital Economy Markets

Elena G. Popkova⬥ and Bruno S. Sergi⬥

Abstract This paper analyses the use of ICT in the field of climate management as the basis of the growth of the effectiveness of the environmentally oriented entrepreneurship's functioning. We elaborate on the state and problems of the implementation and adaptation of digital technologies by market members on the example of partner countries, namely Azerbaijan, Turkey and Kazakhstan. The study demonstrates the technological advantages of the Turkish technology sector, which positively influence the digitalization of companies in Azerbaijan and Kazakhstan and contribute to environmental conservation. The scientific contribution of this research is the identification of forms (models) of the implementation of ICT according to the policy and financial capabilities of companies and the description of the pros and cons of their usage. This work is aimed at the assessment of the effect of the implementation of digital technologies on the effectiveness of entrepreneurship activities. This goal implies the following tasks: determining the dynamics of the indicators of digitalization and effectiveness of certain companies of the selected countries; determining the dependence between two groups of indicators; discovering tendencies of international partnership between companies of the selected countries (Turkey, Azerbaijan and Kazakhstan) in the sphere of the digital economy, which led to an increase in effectiveness.

Keywords Advanced climate-smart technologies · Digitalisation · Financial capabilities · Digital economy · Transfer of technologies · Green entrepreneurship

JEL Classification L24 · M21 · O14 · O32 · O33 · Q55

E. G. Popkova (✉)
RUDN University, Moscow, Russia
e-mail: elenapopkova@yahoo.com

B. S. Sergi
Harvard University, Cambridge, USA
e-mail: bsergi@unime.it

University of Messina, Messina, Italy

1 Introduction

The entrepreneurial sector—similarly to the population—quickly and effectively reacts to innovations that appear in the ICT markets. Government institutes, and—in some countries—the research sphere—are often behind in the implementation and application of advanced technologies. The reason is the use of budget finance according to annual programmes. Striving toward international cooperation influences the technological development of business processes and technological provision which is implemented due to flexible management of financing. Certain companies achieve results in the attraction of investments in technological development, successfully implement projects, return their investments, demonstrate positive credit history and develop further projects on the implementation of new tools of the digital economy. Other companies do not have investments and fail to reach the designated indicators of digitalization. The results of the entrepreneurial sector's functioning in the markets of the digital economy become the basis of the economic development of national economies and the indicator of the state of the digital provision of sectors. Close cooperation between countries in the financial, economic, technological, trade and environmental spheres is a result of interaction at the level of governments and the level of business structures and individual enterprises. Modern tendencies of the development of technological partnership in such a form lead to less developed countries starting to work on new directions for digitalization, which facilitates economic growth and affects climate conservation. The study of the issues of implementation of ICT in the entrepreneurial sphere in partner countries is relevant given the necessity to raise the competitiveness of developing countries' sectors by means of technological growth taking into account the needs of greening processes and industries. A special place in the field of digitalization belongs to climate new technologies that can contribute to the environmental achievements of entrepreneurship and their support from international partners.

The purpose of this study is to assess the impact of the introduction of advanced climate-smart technologies on the performance of the state's green entrepreneurship sector. To achieve the presented goal, the following tasks are defined: to establish the dynamics of digitalization indicators (including those related to environmentally smart solutions) and the efficiency of individual enterprises of the countries under study; identify the relationship between the two groups of indicators and establish trends in international partnerships between companies in the countries under study (Turkey, Azerbaijan and Kazakhstan) in the digital economy, which contributed to increased efficiency and improved climate impact.

2 Materials and Method

The issues of the use of advanced technologies in the activities of entrepreneurship and individual enterprises were studied in the works of [1, 6–8, 12]. Materials of these works allowed determining a possible link between ICT and the effectiveness of indicators of production and trade. For an in-depth analysis of the issue of digitalization's impact on the level of effectiveness of the entrepreneurial sector, we conduct a comparative study by the example of three countries: Turkey (industrial economy, characterised by quick growth); Azerbaijan and Kazakhstan (countries of the Caspian region, partners of Turkey).

In this research, statistical analysis is used to collect data on the studied indicators of digital technologies implementation (an indicator of investments in non-material assets) and the indicators of sales revenue in dynamics; the comparative analysis is used to compare the change in the estimated factors; the trend method is used to determine the tendencies of digitalization and partnership at the level of countries.

Let us present the change in estimated indicators (implementation of ICT (capital investments in non-material assets)) and the effectiveness of development (revenue) of the companies of Turkey, Azerbaijan and Kazakhstan. We study national companies that cooperate with companies from one or two selected partner countries.

Table 1 presents the indicators of the activities of Turkish technology company Turkey Netaş Telekomünikasyon A.Ş. in 2019–2021.

The data of the study of the activities of Netaş Telekomünikasyon A.Ş. (Turkey) will be further used for assessing the dependence of the analysed indicators. As for the estimated indicators, we note their growth and the existence of a direct positive connection between investments in digitalization and revenue, investments in the use of ICT and environmental protection results (green investment).

Table 2 presents the indicators of the state and dynamics of the change in the estimated variables in the context of the activities of the Azerbaijani company Bakcell (mobile communications and Internet). In 2020–2021, the company used integrated technological solutions from Netaş Telekomünikasyon A.Ş. (Turkey).

The growth of investments in ICT allowed the Azerbaijani company Bakcell to raise its revenue and export of services to countries with its subsidiaries—national operators of mobile communications and Internet.

Table 3 presents estimated indicators of investments in ICT and the level of sales revenues of the company Kazakhtelecom (Kazakhstan).

We study Kazakhtelecom because it cooperates with the Azerbaijani company AzerTelecom, which, in its turn, works with the Turkish technology company Netaş Telekomünikasyon A.Ş. in the sphere of technology transfer [3]. For more than five years, Kazakhtelecom worked with AzerTelecom in the direction of technology transfer. In 2022, these companies from Kazakhstan and Azerbaijan created a project for the joint laying of an underwater fibre optic cable line in the maritime territory of the Caspian Sea. This partnership led to the technological growth of Kazakhtelecom in 2019 and 2020.

Table 1 Volumes of investments in ICT and the volume of revenue of Netaş Telekomünikasyon A.Ş

Indicator	Value			Change, ± , %	
	2019	2020	2021	2019/2020	2020/2021
1. Investments in the creation (purchase) of digital technologies, USD, including	3,304,573	3,427,914	4,428,364	+3.73	29.19
1.1. Investments in ICT abroad, % of total investments (row 1)	11	13	15	+2	+2
1.1.1. Investments in ICT (technologies of data transfer, cloud technologies and cybersecurity) in Azerbaijan, % of total investments in ICT (row 1)	2	3	5	+1	+2
1.1.2. Investments in ICT in Kazakhstan (mobile and broadband Internet, telecommunication technologies), % of total investments in ICT (row 1)	0.5	1	2	+0.5	+1

(continued)

Table 1 (continued)

Indicator	Value			Change, ± , %	
	2019	2020	2021	2019/2020	2020/2021
1.2. Investments in startups	0	Startup eco-systems in rural areas of the country Delivered by drone 20,000 boxes of seeds for women's alternative work in the field of growing and selling medicinal plants (instead of income from deforestation)	Startup eco-systems in rural areas of the country Delivered by drone 20,000 boxes of seeds for women's alternative work in the field of growing and selling medicinal plants (instead of income from deforestation)	0	0
2. Revenue, USD, including:	71,592,652.27	93,460,577.16	123,830,054.27	+30.54	+32.49
2.1. Export, % of revenue	17	22	24	+5	+2
2.1.1. Export to Azerbaijan, % of revenue from the provision of services on data storing and transfer and cybersecurity	4	6	7	+2	+1
2.1.2. Export to Kazakhstan, % of revenue from the provision of Internet and telecommunication services	2	3	4	+1	+1

(continued)

Table 1 (continued)

Indicator	Value			Change, ± , %	
	2019	2020	2021	2019/2020	2020/2021
3. Other results		Providing an opportunity for 20,000 rural households previously focused on selling products from deforestation, to receive income without harming the environment. The seed delivery was carried out with minimal harm to the environment (drones). Measures to protect the environment at the national level have been implemented with the help of climate-smart technologies	Providing an opportunity for 20,000 rural households previously focused on selling products from deforestation, to receive income without harming the environment. The seed delivery was carried out with minimal harm to the environment (drones). Measures to protect the environment at the national level have been implemented with the help of climate-smart technologies		

Source Constructed by the author based on [9, 10]

Table 2 Indicators of investments in digital technologies and sales revenue of Bakcell

Indicators	Value			Change, ± , %	
	2019	2020	2021	2019/2020	2020/2021
1. Investments in the purchase of ICT, USD, including	2,045,600	3,127,400	4,832,500	+52.8	+54.5
1.1. Investments in ICT abroad (roaming), % of total investments (row 1)	4	5.5	6	+1.5	+0.5
2. Revenue, USD, including:	102,316,520	107,300,500	124,347,600	+4.87	+15.89
2.1. Export, % of revenue from the sale of services	7	8.3	9.2	+1.3	+0.9

Source Constructed by the author based on [5]

Table 3 The volumes of investments in ICT and sales revenues of Kazakhtelecom

Indicator	Value			Change, ± , %	
	2019	2020	2021	2019/2020	2020/2021
1. Investments in the purchase of ICT, thousand tenges, including	221,344,251.8	109,816,529	0	−50.39	−100
1.1. Investments in ICT abroad (roaming), % of total investments (row 1)	2.6	3.3	0	+0.7	−3.3
2. Revenue, USD, including:	914,068,293.75	1,131,920,208.8	1,263,553,641.85	+23.83	+11.63
2.1. Export, % of revenue from sale of services	3	3.5	3.6	+0.5	+0.1

Source Constructed by the author based on [11]

3 Results

A study of the effect of investments in ICT on the growth of revenue from sales of products (services) of the studied companies and an analysis of the character of cooperation at the international level showed the following.

The leader in the digital technologies market of Turkey—Netaş Telekomünikasyon A.Ş. works with enterprises from 21 countries. The strongest cooperation is observed with enterprises from Azerbaijan and Kazakhstan [9]. In the course of three years, there was a growth of revenues from the export of technological products (services), which were received due to investments in the technological development of offices (branches) located in other countries (including Kazakhstan and Azerbaijan).

The data in Table 1 demonstrate the effectiveness of investments in ICT on the whole and in foreign branches (offices) in particular. It is also possible to note the positive influence of investments in digitalization in network structures that are located in Azerbaijan (2–5% of total investments) and Kazakhstan (0.5–2%) on revenues from the activities in these countries (export: 4–7% and 2–4%, accordingly).

Against the background of the growth of revenue in 2018–2019, Netaş Telekomünikasyon A.Ş. faced a problem with the growth of profitability (the same was observed in 2020–2021). This was due to an increase in the cost of activities of offices and subsidiaries located in other countries [6]. The leading telecommunication and production enterprises of Azerbaijan and Kazakhstan are customers of Netaş Telekomünikasyon A.Ş. We should mention Azerbaijani providers: Azercell Telekom provides 80% of the broadband Internet in the country and uses technological solutions to raise network efficiency that were purchased by the terms of technology transfer [2]; Bakcell is the national leader in the sphere of mobile communications and mobile Internet operator; Azertelecom provides the services of cell coverage and implements international projects in laying fibre optic cable lines [4]. There are also enterprises working in pharmaceutical production and companies in the oil industry.

An important indicator of the functioning of Netaş Telekomünikasyon A.Ş. (Turkey) was investing in a startup eco-system in the rural areas of the state. It was determined that in cooperation with the social companies startups E-cording. Netaş Telekomünikasyon A.Ş. implemented a project to preserve the forests of Turkey in the countryside. Namely, the project assumed: promoting the possibility of growing and helping to sell for export medicinal herbs; conducting a program to promote the benefits of participating in a project that provides employment for 20 thousand families during 2020, 2021 (annually); implementation of sending planting material using drones (climate-smart technology that does not affect the climate during the delivery of goods). As a result, the company made a positive impact on climate protection for two years in a row, ensuring that 20,000 households stopped deforestation.

The analysis also demonstrated positive results of the influence of investing in digital technologies on revenues from the sale of services in the sphere of communications and the Internet, provided by the Azerbaijani company Bakcell (Table 2). This company successfully integrated technological solutions of Netaş Telekomünikasyon A.Ş. (Turkey), purchased by the terms of technology transfer. There was a growth in investments in ICT (52.8% in 2019–2020, 54.5% in 2020–2021), which predetermined the expansion of the activities by means of an increase in cell and Internet coverage in Azerbaijan and abroad. This led to an increase in sales revenue (4.87% in 2019–2020, 15.89% in 2020–2021).

A study of the technological development of Kazakhtelecom (Table 3) showed that this company reduced investments in ICT from 2019 to 2020 (by 50.39%). However, the company invested in digitalization, which, in whole, led to the growth of revenue (23.83% in 2019–2020). In 2021, compared to 2020, there was no investment in technologies, which influenced the revenues and export. We can state the negative impact of the reduction of investments on the growth of the financial indicator of activities—sales revenue.

Thus, we determined the characteristics of technological transformations of companies of the partner countries in the spheres of high technologies, economy and trade. The growth of investments in ICT is a factor of the economic growth of market participants at the national and international levels. Companies that cooperate with more technologically developed partners achieve a certain potential, which is required for large-scale projects of the international level. We should also mention a decrease in economic growth due to the termination or reduction of technological improvement of processes and materials.

4 Discussion

A study of the models of advanced technology management of the selected companies from three countries demonstrated their effectiveness in the sphere of activities within the main direction (management of non-material assets and technological solutions in the sphere of data storing and data transfer (cloud services), cybersecurity and telecommunication technologies) and the presence of certain problems in the administrative and adjacent spheres. The latter affects the effectiveness of the formation of the national digital economies, which was shown by the example of the activities of Turkey's leading technology company Netaş Telekomünikasyon A.Ş. This enterprise was not able to effectively control the work of its offices abroad and switched to virtual representation in certain partner countries.

The model of implementation of digital technologies, which is based on the pattern "development-use-transfer" (the example of Netaş Telekomünikasyon A.Ş.) has some pros and cons. The pros include the possibility to increase the market presence due to the growth of demand for products (services) in new markets and among new segments of consumers; an increase in sales revenue, the growth of which will allow, in the long-term, ensuring R&D for new technological solutions. The risks include unprofitability of expenditures for R&D in the sphere of implementation of technology production; risks of information security; risks of digital compatibility (consumers in a range of countries may have different requirements to digital products (services); risks of regulation of the digital economy at the national level (tax issues, issues of regulation of labour laws in the sphere of management of branches, subsidiaries, etc.) [7, 8]. Netaş Telekomünikasyon A.Ş. (Turkey) faced these risks, due to which it had to perform a reorganisation of its presence in certain countries, e.g., in Azerbaijan and Algeria.

The model of using ICT in the form of transfer-integration allows participants to avoid risks at the stage of the creation, implementation and production of technologies. The focus on the application of this model is more acceptable for companies that do not have opportunities for R&D or technology production but support the priority of constant growth.

If a company focuses on the model of using digital technologies within the "transfer-integration" without constant improvement (a passive form of technology transfer), there are risks of losses of market advantages that lead to the reduction of demand and sales revenue. The absence of technological improvement may be caused by the internal policy of companies and by the absence of financial opportunities for capital investments in this sphere. An example of such an experience is the activities of the Kazakh company Kazakhtelecom in 2021, which, due to the consequences of the pandemic, which led to the reduction of the financial opportunities for investing, was not able to reach the results of the previous periods.

Consideration of the above models of implementation of ICT and their influence on the level of the digital economy can allow avoiding problems of losing competitive positions, occupied by enterprises and can help in the case of participating in projects on the development of own ICT in the case of insufficient financial and technological potential. We think that refusal of risks in the sphere of R&D will not allow companies to enter a new level of digitalization and will not facilitate market growth.

The activity of Netaş Telekomünikasyon A.Ş. (Turkey) deserves special attention in the field of environmental conservation. Namely, the use of startup ecosystems based on the use of climate-smart technologies (drones) allowed the company to leave a positive climate footprint in the field of conservation of national forest resources. Although this project is not measured by economic outcome, an environmental outcome was achieved. This allows concluding that this company has effectively used ICT to achieve environmental results related to national interests in the field of environmental conservation. Such an indicator of the functioning of the digital economy is as important as providing a contribution to the growth of national GDP.

5 Conclusion

In this research, we distinguished possible forms of technological growth an environmental conservation (focus on climate-smart technologies). The results of the analysis allowed for the assessment of factual specifics of the enterprises' participation in technological development and the determination of differences that form the above forms (models) of digitalization. The forms of implementation and use of ICT, which are connected with the implementation of R&D, further transfer of technologies and continuous transfer of technologies, could be adapted by enterprises that are aimed at market growth. The passive form of technology transfer without the focus on constant upgrade cannot always protect a company from market influences, which is a sign of its unsuitability to adaptation in the modern business environment, which is changing very quickly.

The formation of new models of implementation and use of digital technologies is a mandatory condition of the modern market of the digital economy. Consequently, the entrepreneurial sector should be ready for a reaction to the emergence of new tendencies for digitalization in each sector of the economy, Creation of own models in this sphere will allow companies to influence the market and avoid increased dependence on the external environment.

References

1. Ayaydin H, Pilatin A, Barut A, Pala F (2019) Do R&D investments influence market performance? A research on Borsa Istanbul (BIST) technology index (XUTEK). Glob J Econ Bus Stud 8(16):64–75
2. Azercell (2022) About us. https://www.azertelecom.az/en/whoweare/. Accessed: 10 Nov 2022
3. Azertag.az (2022) AzerTelecom, Kazakhtelecom sign strategic partnership memorandum on Trans-Caspian project. https://azertag.az/en/xeber/AzerTelecom_Kazakhtelecom_sign_strategic_partnership_memorandum_on_Trans_Caspian_project-2315015. Accessed: 10 Nov 2022
4. Azertelecom.az (2022) About us. https://www.azertelecom.az/en/whoweare/. Accessed: 10 Nov 2022
5. Bakcell (2022) Annual report. https://www.bakcell.com/en/sustainability. Accessed: 10 Nov 2022
6. Ergun I, Ozcan I (2022) Intellectual capital in the period of industry 4.0: a research on the information sector. J Account Inst 66:79–94
7. Koch BA (2020) Liability for emerging digital technologies: an overview. J Eur Tort Law 11(2):115–136
8. Luo Y (2022) A general framework of digitization risks in international business. J Int Bus Stud 53:344–361
9. Netas (2022) Financial reports. https://netas.com.tr/investor-relations/financial-reports/?lang=en. Accessed: 10 Nov 2022
10. Netas (2022) Netaş, the global face of Turkey! https://netas.com.tr/international-markets/?lang=en. Accessed: 10 Nov 2022
11. Telecom.kz (2022) Annual reports. https://telecom.kz/en/pages/11643/171802. Accessed: 10 Nov 2022
12. Turginbayeva A, Domalatov Y (2019) Strategic direction to support public-private partnerships in the innovation sector of Kazakhstan: problems and prospects. Ad Alta—J Interdisc Res 9(1):54–61. Retrieved from http://www.magnanimitas.cz/ADALTA/090107/papers/A_09.pdf. Accessed: 12 Nov 2022

Climate-Responsible Entrepreneurship's Competitiveness Management Features in the Digital Economy Markets

Juliana A. Kitsai, **Alexey V. Tolmachev**, **Karina A. Chernitsova**, and **Nikolay I. Litvinov**

Abstract *Purpose* studying the influence of digital transformations on the parameters of market interaction and competition, as well as searching and generalising the competitiveness management features in the digital economy and socially-responsible entrepreneurship. *Design/methodology/approach* the systems approach allowed for a comprehensive analysis of the interaction between the market subjects in the digital economy and for the determination of the key features of competition in digital markets. The systems approach, supplemented with the methods of economic theory and macro-economics, allowed for a deeper analysis of the specifics of the market interaction between the subjects of the digital economy and the identification of the factors of formation of competitive advantages and the management of competitiveness under the conditions of dynamic development and wide use of digital innovations with the focus on their climate-responsible behaviour. *Findings* we determined the important role of the market parameters in the management of competitiveness in the digital economy and discovered that digital markets are subject to principles similar to traditional markets. Digital markets are characterised by such features as multilateralism, network effects and scale effect, the possibility to use zero prices, dependence on intellectual property, etc. These features define the character of competitive relations in the market and impact the conditions of formation of the mechanism of managing the competitiveness of market participants. Certain tools of traditional management of competitiveness are not effective in the digital economy. The digital economy offers its tools for managing competitiveness, which are based on technical innovations, better integration into the management system,

J. A. Kitsai (✉)
Immanuel Kant Baltic Federal University, Kaliningrad, Russia
e-mail: Juliana_kn666@mail.ru

A. V. Tolmachev
Kuban State Agrarian University, Krasnodar, Russia

K. A. Chernitsova
Plekhanov Russian University of Economics, Moscow, Russia

N. I. Litvinov
Volgograd State University, Volgograd, Russia

adaptation to the conditions of the life cycle and synchronization of the speed of business processes and the speed of innovations implementation. The use of digital innovations allows forming new markets and opportunities, as well as improving value creation chains. *Originality/value* the performed research allowed determining a close interconnection between the character of the market, competitive relations and methods of managing the competitiveness of entrepreneurial structures, which are mutually dependent.

Keywords Management · Competitiveness · Entrepreneurship · Market · Digital economy · Competition · Digital innovations · Digital markets

JEL Classification D41 · F41 · F43 · M13 · O31 · O33 · O39

1 Introduction

The peak of the development of the digital economy, which took place at the beginning of the twenty-first century and accelerated as a result of the COVID-19 pandemic, actualised the issues of studying digital markets from the position of the search for the tools of the management of competitiveness for digital entrepreneurship. A wide set of such methods that is inherent to the analogue economy, on the one hand, does not allow for the constant achievement of the desired result; on the other hand, it is often unsuitable for usage in the digital economy.

It should be noted that innovations that were developed due to digitalisation allowed creating new values and opportunities, raising the level of convenience and accessibility of a large number of goods and services and ensuring diversity and other consumer advantages. The implemented changes also became a driver for development and digital transformation. Thus, the parameters of the market and economic changes, interrelations between market subjects and chains of value creation were also transformed, which was reflected in the mechanisms of competition and functioning of digital markets [2].

Under such conditions, the competitiveness of entrepreneurial structures requires further research of the parameters of market interaction, determination of the impact of digital transformation on market proportions and mechanisms and substantiation of the possibilities of using new conditions for the competitive struggle of digital entrepreneurs. Another direction of studying the development of digital entrepreneurship is its influence on climate processes and the achievement of the Sustainable Development Goals. Thus, dynamic processes and changes in the environment of the digital economy are important factors of institutional transformations, which change the environment and are influenced by its changes at the same time. This leads to the necessity for researching the conditions and specifics of the above processes from the position of the management of competitiveness.

2 Materials and Methods

The methodological framework of this paper is based on the systems approach. According to this, the peculiarities of the management of entrepreneurship's competitiveness in the digital economy markets were considered from the position of identification of the main interactions, which ensure the development of the digital economy and facilitate the growth of the competitiveness of entrepreneurial structures in digital markets. The foundation of systemic interactions was considered from the position of the economic theory and macro-economics, which study in detail the parameters and conditions of competition in markets, allow for the characterisation of the level of intensity or concentration of the market share of economic subjects and describe consequences for enterprises that emerge because of the violation of market proportions. This methodology, set on the theory of management, allowed identifying the directions and opportunities for managerial influence of digital transformation on the processes of competitiveness management and characterising sectors and directions in which digital innovations have the largest potential from the position of the competitive advantages formation.

This research is interdisciplinary since it combines—within one scientific problem—several visions and approaches. According to this, from the position of elaboration on the set problem, attention should be paid to the works of well-known scholars in the sphere of the theory of market economics and competition [6, 11], digital transformation and development of the digital economy [2, 3, 9, 15], development of competition in the digital economy [8] and management of competitiveness of economic subjects in the digital economy [1, 7, 10, 12, 13]. A separate direction in the context of the research is the works connected with climate-responsible activities of entrepreneurial structures under the conditions of the digital economy [4, 5, 14].

As for the level of disclosure of the problem, it should be noted that the dynamics of the development of digital markets and parameters of competitiveness predetermine the need for additional studies aimed at precision detailing of the specifics of competition in digital markets, the impact of digital transformations on the competitiveness of entrepreneurial structures and substantiation of the features of competitive struggle under the given conditions. According to this, the main purpose of this research is to substantiate the main features of managing the competitiveness of entrepreneurial structures that function in digital markets.

3 Results

Competition is the key condition of the existence of a market. It reflects the character of interrelations between sellers and buyers, defines the pricing and turnover terms and forms the proportions of the distribution of resources and benefits. Competition is often treated as an important factor of economic development, which is manifested in the competitive struggle between economic subjects and the constant search for ways to win in it. Thus, the notion of competitiveness is an important criterion that characterises the level of success of a product or economic entity in the competitive struggle.

Despite the significant technical and institutional features, the digital economy has the same key features as the traditional economy. It is characterised by the market type of pricing, buyer–seller interactions and price fluctuations, caused by a quick change in demand or innovative breakthrough. However, the digital economy also has features that differentiate it from traditional markets. According to this, the models of the management of competitiveness, which are used under the conditions of the analogue economy, must always be checked for suitability for usage in the digital economic environment.

The initial point for analysing the management of competitiveness of entrepreneurial structures in the digital economy is the determination of the specifics of digital markets that influence the parameters of competition and competitive struggle. Such specific features define the competitive dynamics, influence the character of pricing and the cost of entering or leaving the market, and define the level of the market's competitiveness. Depending on the designated conditions or the use of a method, the competitive struggle could be characterised by important features (Table 1).

In many digital markets, traditional tools of competitive struggle have a very limited effect. Thus, for example, under the conditions of null prices, the tools of pricing competition cannot work in the usual form and require modification or full rejection from their use. The possibility for quick and cheap scaling of business processes on digital markets decreases the role of organisational tools of management in the competitive struggle. On the other hand, the specifics of digital markets could distort the character of competitive struggle and, accordingly, reduce the capabilities of the tools of competitiveness management.

According to this, an important factor of the competitive struggle in digital markets is the so-called "bonus for innovations", available for innovative companies. Through the achievement of technological leadership in a certain sphere, they possess—for some time—substantial competitive advantages and find themselves in the vacuum of anti-monopoly regulation. This allows for a quick implementation of the policy of market monopolisation through acquisitions of rivals or potential rivals. In such conditions, an important tool of competitiveness management is the necessity for synchronisation of the speed of business processes with the speed of digital markets' development.

Even though most of the key factors of the competitive struggle of digital entrepreneurs are based on innovations and information improvements, traditional tools of competitiveness management are also widely used by companies in the digital sector. Attention should be paid to those connected with the organisation of business processes and management of intellectual capital and innovations. In such conditions, any innovations could be treated as the key factor of competitive struggle.

The management of entrepreneurship's competitiveness in the digital economy is, in its nature, rather complex, since it covers different directions and levels of competition. Competition in the digital economy is not considered separately from digital

Table 1 Specific features of digital markets

Specific feature	Characteristics
Multilateralism of markets	In certain markets, a digital product is a platform that unites different groups of market participants, e.g., creators of content, consumers and advertisers
Strong network effects	A large number of digital product users under certain conditions, due to the network effect, could lead to the transformation of competitive markets into monopolies
Economics on scale	Digital markets are peculiar for high constant expenditures with low variables. This allows for a quick increase in audience coverage and the use of current assets to enter other markets
Dependence on Big Data	The processes of management require the use of a large array of data for substantiation of decisions
Effect of attachment	Long use of one digital platform makes it difficult to switch to another platform because of developments in the form of reputation, rating, etc.
Role of intellectual property rights	The information character of digital products often requires special permits (patents, licences) for using it
Zero prices (or close to zero prices)	Due to advertising sales and the possibility to gather data on consumers, seven out of the ten largest digital companies offer products and services for free (zero price)
Usage of breakthrough innovations	Due to the high level of innovativeness, breakthrough innovations are very often not subject to any regulation but create significant advantages in the cost and speed of transactions between market subjects
Integration of business models	Vertical integration and digital conglomerations, extended to digital markets, predetermine the risks of anti-competitive behaviour

Source Formed by the authors based on [8]

ecosystems. It includes a more complex interaction than just a competition for better conditions of functioning. Eco-systems are a form of interaction of entrepreneurial subjects, as a result of which a certain value is created. The digital ecosystem is based on a certain technological platform, around which other participants of the ecosystem—"complementors"—interact. A specific feature of ecosystems in the digital economy is the presence of cross-competition [10].

Cross-competition between entrepreneurs, the market and the economic sector appears because market subjects belong to different sectors and spheres of activity. They manage their competitiveness at a wide spectre of markets at the same time. In the digital economy, such competition could be implemented between platforms (e.g., iOS from Apple versus Android from Google), between a platform and "complementors" (e.g., integration of other services into the Microsoft operating system) and between complementors (when they compete for the same segment of consumers). All the above levels of competitive struggle take place simultaneously with the competition between ecosystems. Such a type of competition was called "Moligopoly" competition [9].

The characterised conditions determine the dynamic development and adaptation of managerial tools with the processes of digital transformation. In such conditions, certain factors become of higher priority, while the role of other factors is reduced. It is believed that the focus only on the achievement of digital maturity in the context of competitiveness management is wrong. A much better result is achieved through the synchronisation of talent, culture and organisational structure with the digital environment. According to such a vision, digital transformation ensures the transition of entrepreneurial structures to new conditions of functioning, modifying the organisational culture and creating—on its basis—its digital eco-system. In the context of this issue, the European Union suggest going beyond the limits of a company and forming—as a factor of competitiveness management—the necessary digital competencies of society [15]. In these realities, competition management acquires a more creative character.

Another factor that differentiates the tools of competitiveness management in the digital economy is the life cycle of a company's development. The necessity to pass the given phases of development in the traditional economy rather complicates the potential for the competitive struggle for new entrepreneurial structures. Unlike this, in the digital economy, access to global digital platforms and successful implementation of the open science concept allows young entrepreneurs to obtain more opportunities for overtaking the leaders. The key factor of such success could be better adaptation, flexibility or sense of the market. A vivid example of implementing such opportunities is numerous start-ups.

Another aspect of competitiveness management is determining the influence of digital innovations on the companies' competitive position. According to this, the focus is made on the necessity to ensure complex innovations, the combination of different technologies, the creation of new markets and opportunities and the improvement of value creation chains. These factors allow forming new values, which, in their turn, are the key factors of competitiveness. Thus, the digital economy interacts with traditional economic spheres and ensures the formation of new competitive advantages in all markets.

The key parameters of competitiveness, which form due to the digital transformation of the economy are as follows:

- Improvement in effectiveness and efficiency of business processes;
- Better perception of the needs and experience of customers;
- Increase in the level of skill and competence, including entrepreneurial skills;
- Development of the values of the company's organisational culture;
- Creation of new opportunities due to access to new digital platforms and opportunities [12].

Thus, radical digital transformations constantly create new values and opportunities and propose new methods of relations with customers, ways to develop value changes and forms of communication that are the basis for the competitive advantages of entrepreneurial subjects [7]. A separate source of competitive advantages in these conditions is the capabilities of certain management tools, e.g., monitoring or analysis.

The digital economy defines only a certain component of managerial processes and competitiveness. It uses the basis of traditional spheres, which cover communications, analytical tools and systems and mechanisms of management of business processes, etc. That is why the achievement of a high level of competitiveness and support for its sustainability requires complex managerial solutions that use digital technologies not as a goal but only as a tool [1].

Competition is traditionally treated as a factor of struggle, within which any means are justified. However, under the conditions of an increase in environmental risks and aggravation of climate problems, climate-responsible entrepreneurship becomes an inseparable component of the economic system. Moreover, under the conditions of the digital economy, responsible entrepreneurship often accepts the role of the tool of solving a certain social and global problem. Such an approach to the implementation of digital entrepreneurship is characterised as "digital sustainability", which is considered together with such concepts as social, sustainable and institutional entrepreneurship [5]. Climate-responsible digital entrepreneurship uses innovative solutions to manage the risks in natural and industrial systems [14]. Their integration with digital platforms allows modelling real processes and assessing them from the position of possible damage, losses or other models of the interaction between humans, nature and business.

However, while positively assessing the possibilities of the influence of digital entrepreneurship on the resolution of current climate problems, scholars emphasise that such interaction is not always harmonious and is very often based on a compromise [4]. Acceptance of such compromises requires systemic solutions and gradual transformations of the behaviour of market subjects. Mutual understanding will allow improving the world's capabilities in the resolution of environmental and climate problems, due to the availability of technological and value-based tools. In such conditions, coordination of the values of the digital economy development and the values of world development forms a potential for the creation of additional competitive advantages, based on the information and institutional tools.

4 Conclusion

The development of the digital economy is closely connected with the competition. It is based on the interaction of demand and offer, pricing parameters and determination of the market power of a market participant. The digital economy is based on principles that are similar to the traditional economy. Achievement of competitiveness is one of the criteria of success for entrepreneurs in digital markets. However, its level is formed given the specifics of digital markets, which include network effects, multilateralism, zero prices, use of breakthrough technologies, etc. In the digital economy, competition is not treated separately from the eco-system, so the management of competitiveness in this sphere is formed given the cross-competition and moligopoly. According to this, the management of competitiveness is implemented within isolated digital platforms, between certain subjects or between entire ecosystems.

From the position of the management of competitiveness by digital entrepreneurs, it is important to use the capabilities of digital solutions rationally.

Digital transformations are considered very influential tendencies, which affect business. Given that climate problems are, perhaps, the largest threat to the existence of mankind, implementation of climate-responsible digital entrepreneurship is a very important and topical task, the resolution of which creates additional opportunities for the formation of stable competitive advantages of enterprises in the digital economy.

References

1. Alexandrova E (2020) Digital economy in competitiveness of modern companies. In Antipova T, Rocha Á (eds) Digital science 2019. DSIC 2019. Advances in Intelligent Systems and Computing, vol 1114. Springer, Cham. https://doi.org/10.1007/978-3-030-37737-3_11
2. Bajgar M, Calligaris S, Calvino F, Criscuolo C, Timmis J (2019) Bits and bolts: the digital transformation and manufacturing. OECD Science, Technology and Industry Working Papers, 2019/01, OECD Publishing, Paris. https://doi.org/10.1787/c917d518-en
3. Calvino F, Criscuolo C (2019) Business dynamics and digitalisation. OECD Science, Technology and Industry Policy Papers, 62, OECD Publishing. https://doi.org/10.1787/6e0b011a-en
4. Dwivedi YD, Hughes L, Kar AK, Baabdullah AM, Grover P, Abbas R, Andreini D, Abumoghli I, Barlette Y, Bunker D, Kruse LC, Constantiou I, Davison RM, Wade W (2022) Climate change and COP26: are digital technologies and information management part of the problem or the solution? An editorial reflection and call to action. Int J Inf Manag 63:102456. https://doi.org/10.1016/j.ijinfomgt.2021.102456
5. George G, Merrill RK, Schillebeeckx SJD (2021) Digital sustainability and entrepreneurship: how digital innovations are helping tackle climate change and sustainable development. Entrep Theory Pract 45(5):999–1027. https://doi.org/10.1177/1042258719899425
6. Kim YR, Liu A, Williams AM (2022) Competitiveness in the visitor economy: a systematic literature review. Tour Econ 28(3):817–842. https://doi.org/10.1177/13548166211034437
7. Nikulin RA (2018) Increase of the company competitiveness in the digital economy. J Soc Sci Res. Academic Research Publishing Group, vol 1, pp 367–372
8. OECD (2022) OECD handbook on competition policy in the digital age. https://www.oecd.org/daf/competition/oecd-handbook-on-competition-policy-in-the-digital-age.pdf. Accessed: 04 Dec 2022
9. Petit N (2020) Big tech and the digital economy: the moligopoly scenario. Oxford University Press, Oxford
10. Petit N, Teece DJ (2020) Taking ecosystems competition seriously in the digital economy: a (preliminary) dynamic competition/capabilities perspective (hearing on competition economics of digital ecosystems, 3 Dec 2020), 7. https://one.oecd.org/document/DAF/COMP/WD(2020)90/en/pdf. Accessed: 04 Dec 2022
11. Porter ME (1990) The competitive advantage of nations. Harvard business review. https://hbr.org/1990/03/the-competitive-advantage-of-nations. Accessed: 04 Dec 2022
12. Rossato C, Castellani P (2020) The contribution of digitalisation to business longevity from a competitiveness perspective. TQM J 32(4):617–645. https://doi.org/10.1108/TQM-02-2020-0032

13. Sizova IY, Semenova EM, Zakharov AV, Sotnikova EA, Zviagintceva YA (2020) Managing company competitiveness in the digital economy. In: Popkova E, Sergi B (eds) Scientific and technical revolution: yesterday, today and tomorrow. ISC 2019. Lecture Notes in Networks and Systems, vol 129. Springer, Cham. https://doi.org/10.1007/978-3-030-47945-9_41

14. Sokolov AG, Abramov VM, Istomin EP, Korinets EA, Bolshakov VA, Vekshina TV (2020) Digital transformation of risk management for natural-industrial systems while climate change. In: IOP conference series: materials science and engineering, vol 940. International Scientific Conference "Digital Transformation on Manufacturing, Infrastructure and Service" 21–22 Nov 2019, St. Petersburg, Russian Federation, vol 940, p 012003. https://doi.org/10.1088/1757-899X/940/1/012003

15. Vuorikari R, Punie Y, Carretero Gomez S, Van Den Brande G (2016) DigComp 2.0: the digital competence framework for citizens. Update Phase 1: the Conceptual Reference Model. EUR 27948 EN. Luxembourg (Luxembourg): Publications Office of the European Union; 2016. JRC101254

The Model of Organisation of Green Entrepreneurship and the Climate-Responsible Management of Production Factors in the Digital Economy Markets

Alexey V. Tolmachev⊙**, Platon A. Lifanov**⊙**, Nataliya V. Ketko**⊙**, and Anastasia I. Smetanina**⊙

Abstract This paper dwells on the characteristics of the models for organising green entrepreneurial digital business initiatives and the managing climate-responsible digitalisation of production factors in countries of Latin America. Parameters of the implementation of start-ups are shown; at the modern stage, they are the main form of digital green entrepreneurship in the studied region. It was discovered that the key advantages of achieving success in the implementation of start-ups are connected with the implementation of accelerators, which are created by the government, at the level of public–private partnership and the private level—co-working, which provides space for the business environment. The main research methods are the method of deconstruction, statistical method and method of systematisation. The scientific novelty of this paper consists in the elaboration on the aspects that characterise the capabilities of digitalisation in the sphere of main climate-responsible entrepreneurship and the sectors of industry.

Keywords Digitalisation · Organisation of green entrepreneurship · Production factors · Accelerators · Start-ups · Coworking · Climate-responsible management

JEL Classification D23 · D24 · D83 · O32 · O33 · O35

A. V. Tolmachev
Kuban State Agrarian University, Krasnodar, Russia

P. A. Lifanov · N. V. Ketko
Volgograd State Technical University, Volgograd, Russia
e-mail: gsa-buch@list.ru

A. I. Smetanina (✉)
Institute of Scientific Communications (ISC-Group LLC), Volgograd, Russia
e-mail: luxury_economy@mail.ru

1 Introduction

Before the COVID-19 pandemic, many countries in Latin America demonstrated significant progress in the economy, which positively influenced the social sector and their position in the international economic space. The problems caused by the pandemic (economic crisis, inflation, growth of credit interest rates and stoppage of production due to limitations imposed by national governments) have a negative effect on the positions of this region's countries in the world market. However, the existing course toward intense industrial development and entrepreneurial activity are the priority that determines further prospects in the sphere of development of sectors that manufacture competitive products, which provide export revenues and important climate-responsible production management trends are gradually forming in the region under study, contributing to balanced development [12]. Implementation of digital tools is an important component of the acceleration of entrepreneurial activities; it ensures the preconditions for managing the production factors in these countries. The main traditional production factors, which are considered by economic sciences, include natural resources, human resources and capital. The new factors are entrepreneurial potential and data [20], climate-responsible production. Entrepreneurial potential is the basis of the functioning of medium, small and micro-companies and large corporations [1, 3, 5, 13]. It is also one of the key production factors. The intensity and level of the implementation of ICT in the formation of these factors and the functioning of the entrepreneurial environment determine the creation of a stronger foundation for the economic growth, the greening of the industry of countries of Latin America.

This paper's goal consists in the description of specific features of the digital provision of the green entrepreneurial sector and the sphere of production factors management in countries of Latin America, associated with the climate-responsible behaviour of subjects. For this, the following tasks have been developed: to determine the features of existing models of organizing business activities using ICT in individual states of the region under study, including identifying aspects of green business initiatives; highlight the key indicators of climate-responsible management of production factors in the context of digitalization at the level of Latin American countries.

For this, the features of the existing models of the organisation of entrepreneurial activities with the use of ICT in the selected countries of the region are determined, and also the aspects of green business initiatives are identified; the key indicators of climate-responsible management of production factors in the context of digitalization at the level of Latin American countries are determine.

2   Materials and Method

The main characteristics of the digital economy's influence on the production environment of the studied countries are studied in this paper on the basis of analysis of the materials of theoretical and practical works [2, 4, 6, 10]. The comprehensive research of these materials and the systematisation of statistical and analytical data allowed identifying the key specific aspects of the implementation of digitalisation in these spheres.

The method of deconstruction was used to divide scientific and complex studies to distinguish the sought characteristics of the digital provision of entrepreneurship and the production sphere. The statistical method was used to collect statistical indicators of the implementation of ICT in production and entrepreneurship. The method of systematisation allowed identifying the comprehensive characteristics of digitalisation within the studied spheres, which ensures the functioning of Latin America's countries' economies.

The system of indicators, which were studied, includes the following:

- Global Startup Ecosystem Index (GSEI) [8], which allows identifying the positions of countries in the global ranking of start-ups, which are the initial forms of entrepreneurship that have various models depending on the specialisation of business. GSEI is evaluated with the use of ranks.
- Business Score Startup (BSS) [8], which identifies the effectiveness of business during the organisation of start-ups. This indicator is assessed in points.
- Quality Score Startup (QlSS) [8], which shows the level of the quality of start-ups that are used by entrepreneurship. This indicator is also calculated in points.
- Quantity Score Startup (QnSS) [8], which demonstrates quantitative dissemination of start-ups (also measured in points).
- IT integration industry Index (ITIII) [7], which determines the ranking positions of the national production sector in the sphere of digital technologies implementation, i.e., achievements of countries in the sphere of digitalisation of production factors (rank).

3   Results

The assessment of the parameters of the formation of start-ups, which are the key form of entrepreneurship under the conditions of the digital economy and the basis of the formation of various models of business management, is performed in Table 1.

According to the data from Table 1, the studied countries of Latin America have middle positions by the implementation of start-ups. The best results in all indicators are observed with Brazil (26th position).

The entrepreneurial sector of Chile was able to ensure a very high level of effectiveness of start-ups' functioning (2.56 points). The large financial success of Chilean

Table 1 Indicators of the implementation of start-ups in the selected countries of Latin America (as of October 2022)

Indicator	Value						
	Brazil	Chile	Mexico	Argentina	Colombia	World (max)	World (min)
GSEI, rank	26	34	35	37	44	1 (USA)	100 (Kyrgyzstan)
BSS, points	1.4	2.56	2.43	2.03	2.13	3.8 (Sweden)	0.02 (Somalia)
QlSS, points	5.91	2.64	2.8	2.4	0.91	164.15 (USA)	0.05 (Somalia)
QnSS, points	2.17	1.4	0.79	1.24	5.36	27.56 (USA)	0.05 (Senegal, Kosovo)

Source Compiled by the authors based on [8]

start-ups is due to the use of the business accelerator, financed by Chile's government since 2010 for the growth of the economy and increase in the effectiveness of the business environment. The accelerator of the eco-system of Chilean start-ups (Start-Up Chile) includes the following: providing the participants (entrepreneurs) with coworking space (on a free basis or with low rental rates), start-up capital for the implementation of the selected business models of start-ups; free training on the basics of entrepreneurship (including in the sphere of digitalisation), conducted by various participants of the educational and technological environment by the terms of competition. According to [6], the provision of the accelerator's services without the financing of starting efforts of entrepreneurs does not facilitate the achievement of effectiveness. It is possible to note the stable effectiveness of Chilean start-ups Chile, which is demonstrated by stable and high values over various periods of the use of the given government initiatives on the acceleration of their implementation [4].

With the high effectiveness of start-ups' activities, Chile is ranked third in Latin America by their number. The largest number of this form of business (144 start-ups) is concentrated in the sphere of Big Data analysis and software (development and services) [18].

A specific feature of the implementation and functioning of start-ups in Chile is the fact that their maximum concentration is found in the country's capital—Santiago. The level of the digital economy and living standards of Santiago is much higher compared to other Chilean cities [8]. This fact was influenced by the 2020 pandemic, as a result of which the capital remained the centre of innovations, and four other cities, which had had similar indicators of development, slowed down in their progress. This might be considered a threat to economic indicators and indicators of digitalisation in Chile.

The main model of entrepreneurship that uses digital technologies and is connected with start-ups in Chile is crowdfunding (voluntary cooperation of people for the financing of certain initiatives, which is performed primarily in the remote form). Based on this model, the most well-known start-ups in Chile function: NotCo (Santiago) and Migrante (Santiago).

The high level of environmental pollution (water, air, soil) in Santiago is associated with large-scale consumption due to the concentration of the population and business in the Chilean capital. There is also a high negative impact on the climate in industrial and agricultural regions (in the context of animal husbandry). In order to solve this problem, the government adopted legislative acts regarding: a ban on the production and distribution of plastic, single plastic bags; prohibition of the use of non-environmentally friendly production technologies; implementation of responsibility for the recycling of tires, non-working computer equipment, batteries by manufacturers and importers [19].

This business initiative, which operates within the framework of the NotCo startup (Santiago) [11] is focused on the use of the latest imported equipment, which: does not create high volumes of defects that require processing and disposal; can be connected to various types of energy sources; is energy-saving, which eliminates excessive fuel consumption and environmental pollution. Separately, it should be noted the company's contribution to reducing the negative climate impact through the production of meat substitutes, dairy products, plant-based meat products. This approach allows to reduce the cultivation of farm animals, which increases the negative impact on the environment (water, soil, air) NotCo's business initiative operates in Latin America, so today it implements green business solutions in large areas.

The main features of the models of organisation of the entrepreneurial sector in the sphere of digital start-ups in Brazil are as follows [8].

- Services based on the use of digital technologies. The well-known start-ups in this sphere include iFood (Sao Paolo), which deals with the delivery of food from restaurants, cafes, supermarkets, etc.; Nuvemshop (Sao Paolo), which provides services for the development and maintenance of digital tools of creation and doing business in electronic platforms.
- E-commerce with the use of digital platforms. The most effective start-ups in this sphere include MadeiraMadeira (Curitiba) (household goods). This digital platform implements several measures for climate-responsible management, such as focusing on sustainable packaging, saving resources in the field of process management, cooperation with manufacturers with certificates of origin of raw materials (mainly for furniture products) [9].

It should be noted that Brazilian start-ups are characterised by a high level of organisation quality (at the level of a group, management (founder) and external agents). Unlike Chile, in which the main concentration of start-ups is in the capital, Brazilian start-ups are found in different cities.

According to [2], the best results are demonstrated by Brazilian start-ups, which take into account the following parameters in their organisation: provision of necessary resources (investment support, grants for development, own financial assets); the focus on latest market tendencies and indicators of demand; high standards of coordination of interaction at the internal and external levels; motivation, training and professional development of the team participants.

As for the models of organisation of the entrepreneurial sector, which is oriented toward digitalisation, in Mexico, the key role in their dissemination, similarly to other countries of Latin America, belongs to start-ups, which further develop into separate companies or structural departments of the existing companies. The quality of organisation of interaction in Mexican start-ups is not sufficient, which is caused by the absence of the practice of using governmental initiatives on the acceleration of their implementation. Still, there are certain government programmes in the sphere of attraction of international investors, which finance the business initiatives of start-ups and rental services at low prices [8].

The models of organisation of entrepreneurship with the use of digital start-ups in Mexico include the following [8].

- E-commerce (digital trade platform). Here the best development was achieved by the start-up Kavak (buy and sell of used cars, Mexico City);
- Services in the financial sphere. The best-known Mexican start-ups in this sphere are Bitso—an electronic platform that provides services on buy and sell of cryptocurrencies (Puebla); Clip—a digital platform for business environment participants' accepting all forms of electronic payments (Mexico City).

In 2022, Argentina raised the level of its digital economy and the creation and functioning of business initiatives, which ensure its development (start-ups). Various models of entrepreneurship that use start-ups are found in four cities.

- Buenos Aires, which is ranked 56th in the world by the provision of an eco-system of creation and functioning of start-ups. The following models of start-ups are found in Buenos Aires: online educational services (74); digital marketing technologies and services in the sphere of sales and promotion (48); e-commerce (digital trade platforms) (38); financial services on electronic payments (including cryptocurrency) (10) (examples: start-ups Ualá and Defiant). Successfulness of these business initiatives of the digital economy in Buenos Aires is predetermined by the activities of five accelerators that are based on public–private partnerships; fourteen co-workings, which provide space for the implementation of the business initiatives (offices and infrastructure) [14];
- Cordoba, which digital business initiatives developed in the last three years due to the quick growth of external international investments. The models of digital start-ups in Cordoba include digital services on car rental (28); crowdfunding (18); educational services in the electronic form (14) [15];
- Mendoza, which has start-ups that use the models focused on financial services, entertainment and leisure in social networks;
- Rosario, in which start-ups develop within private business initiatives. The models of Rosario's start-ups include crowdfunding, healthcare services, e-commerce and financial digital banking (example: Let'sBit) [16].

Assessment of the functioning of entrepreneurship based on digital tools in Colombia showed that the main models of organisation in this sphere are booking flights [start-up LifeMiles (Bogota)]; sales and services for digital applications on financial accounting for micro business in countries of Latin America [start-up

Treinta (Bogota)]; sales and services of digital applications on the search and sales (rent) of real estate [start-up La Haus (Medellin)] [17].

The analysis showed that the business sector of Mexico, Argentina, and Colombia has a significant range of prospects in the field of climate-responsible process management.

4 Discussion

It is possible to see that digital support for entrepreneurship led to the emergence of such a business initiative as digital start-ups (including those related to climate-responsible management), which became popular in the selected countries of Latin America. The effectiveness and quality of the formation of start-ups are connected with the level of the eco-system, which is formed at the level of public–private partnership. The main models of digital start-ups, which became most popular in the region, include e-commerce (digital trade platforms), digital services in various spheres of the economy and crowdfunding. The level of the eco-system of business initiatives of entrepreneurship can form based on public–private partnerships and based on private financing only. The latter can be easily achieved under the conditions of the openness of the economy and the existence of a sufficiently high level of protection of external investments. The emergence of new digital tools is an important precondition for the emergence of new models of organisation of start-ups in various spheres. Examples of start-ups of countries of Latin America demonstrate sufficient sustainability of these business initiatives in the face of serious challenges and risks in the economy and social sphere. Thus, this form of entrepreneurship is flexible and adaptive in the context of the modification of new models of business.

As shown in Table 2, the largest progress in the digitalisation of production factors (processes, use of human capital, resources management and finance) was achieved by Chile—from the 34th position in 2020 to the 34th in the 1st half of 2022.

The key characteristics of the management of digitalisation of production factors in Chile are as follows [10]:

Table 2 Rating indicators of the selected countries of Latin America in the sphere of digital support for production factors

Country	ITIII, rank		
	2020	2021	1st half of 2022
Brazil	48	49	43
Chile	40	39	34
Mexico	53	52	47
Argentina	52	59	53
Colombia	49	46	58

Source Compiled based on [4]

- Fair social management, connected with the implementation of ICT in the sphere of the processing industry, with the country's focus on employment.
- Wide robotization in the processing industry, which allows for the reduction of costs and an increase in product quality, which facilitates the growth of export. This sector keeps stable sales revenues, which account for 11.5–12% of the national GDP.

Other countries of Latin America, which are considered in this paper, use robotization, software and AI tools in the processing and food industry; however, they are not characterised by the policy of social justice in the context of the provision of employment given the implementation of digital tools.

The assessment of the management of digitalisation of production factors in the selected countries of Latin America showed that only Chile uses the policy of industry digitalisation that implies maintaining the employment level in the country (responsible industrial corporate management). It is worth noting that such an approach stimulated the achievement of a high world rating in the sphere of digital provision of export-oriented spheres of industry.

5 Conclusion

In this paper, the features of the formation of the models of entrepreneurial initiatives' organisation at the level of start-ups in the most developed countries of Latin America were studied. The success of the implementation of the models used is based on the parameters of the eco-system of start-ups, created by the members of the business environment and by the government—within a public–private partnership. It was proved that efficiency of the functioning of the studied business initiatives could be achieved only if there are such components as organisational, resource and professional readiness of start-ups' creators'; investments in functioning; creation of favourable conditions for activities (co-working). The presence of just one condition significantly reduces the prospects for the achievement of the expected effects of business initiatives, especially in the case of the lack of the required financial support.

Responsible corporate and climate management of the implementation of ICT in the industry is an important precondition for achieving stable indicators of effectiveness in the use of digital technologies in the main spheres and of economic growth on the whole. This approach to management implies that dismissed personnel, who work in the same sphere, can raise their professional level with the help of digitalisation programmes and help deal with the shortage of staff in other sectors, the achievement of environmental indicators of the functioning of production, administrative activities affects the receipt of additional economic and energy effects.

References

1. Akimova O, Volkov S, Kabanov V, Ketko N, Kuzlaeva I (2020) Regional entrepreneurship support infrastructure: Volgograd region case study. WSEAS Trans Environ Dev 16:397–412. https://doi.org/10.37394/232015.2020.16.40
2. Coda R, de Moraes GHSM, de Castro Krakauer PV, Junior JMP (2022) Startup founders' entrepreneurial profile in the Brazilian context. RPCA 16(1):15–36
3. Fadeeva EA, Denisov IV, Lifanov PA, Bagdasarian IS (2020) Factors of creation and functioning of AI: anthropogenic vs. social. Int Econ Policy Emerg Econ 13(5):464–470. https://doi.org/10.1504/IJEPEE.2020.110436
4. Flechas XA, Kazunari Takahashi C, Bastos de Figueiredo JC (2022) The triple helix and the quality of the startup ecosystem: a global view. Revista de Gestão. https://www.emerald.com/insight/content/doi/10.1108/REGE-04-2021-0077/full/html. Accessed 19 Nov 2022
5. Glotko A, Sycheva I, Petrova L, Tolmachev A, Islamutdinova D (2019) Environmental problems of processing industry in the agro-industrial complex of the region. J Environ Manag Tour 10(5):974–983. https://doi.org/10.14505/jemt.v10.5(37).04
6. Gonzalez-Uribe J, Leatherbee M (2018) The effects of business accelerators on venture performance: evidence from start-up Chile. Rev Financ Stud 31(4):1566–1603
7. IMD (2022) World digital competitiveness ranking. https://www.imd.org/centers/world-competitiveness-center/rankings/world-digital-competitiveness/. Accessed 19 Nov 2022
8. Lp.startupblink (2022) Global startup ecosystem index 2022. https://lp.startupblink.com/report/. Accessed 19 Nov 2022
9. Madeiramadeira (2022) Curitiba—Centro. https://www.madeiramadeira.com.br/lojas/parana-curitiba-centro Accessed 15 Nov 2022
10. Neira FG, García GF, Seguel MC (2018) ICT in Chile at the beginning of the fourth industrial revolution. Chinese Bus Rev 17(6):263–278
11. Notco (2022) No estamos aquí para iniciar una Revolución, ESTA ES UNA Revelación. https://notco.com/cl/sobre/sobre-nosotros. Data accessed 15 Nov 2022
12. Popkova EG, Shi X (2022) Economics of climate change: global trends, country specifics and digital perspectives of climate action. Front Environ Econ 1:935368. https://doi.org/10.3389/frevc.2022.935368
13. Shabaltina LV, Shchukina NV, Surkova OA, Smetanina AI (2022) A framework for reconstructive digital farming for areas with unfavourable climatic conditions for agricultural entrepreneurship. In: Environmental footprints and eco-design of products and processes, pp 215–222. https://doi.org/10.1007/978-981-19-1125-5_25
14. Startupblink (2022) The ecosystem of Bogota start-ups. https://www.startupblink.com/startup-ecosystem/bogota-co. Accessed 19 Nov 2022
15. Startupblink (2022) The ecosystem of Buenos Aires start-ups. https://www.startupblink.com/startup-ecosystem/buenos-aires-ar. Accessed 19 Nov 2022
16. Startupblink (2022) The ecosystem of Cordoba start-ups. https://www.startupblink.com/startup-ecosystem/cordoba-ar. Accessed 19 Nov 2022
17. Startupblink (2022) The ecosystem of Rosario start-ups. https://www.startupblink.com/startup-ecosystem/rosario-ar. Accessed 19 Nov 2022
18. Statista (2022) Number of start-ups in Chile as of February 2022, by industry. https://www.statista.com/statistics/1197926/chile-start-ups-industry/. Accessed 19 Nov 2022
19. Trade.gov (2022) Environmental technologies. https://www.trade.gov/country-commercial-guides/chile-environmental-technologies. Accessed 15 Nov 2022
20. Xu X (2021) Research prospect: data factor of production. J Internet Digit Econ 1(1):64–71

Foundations for Ensuring the Effectiveness of Climate-Responsible Entrepreneurship in the Markets of the Digital Economy

Aziza B. Karbekova⬤, **Abdil Tashirov**⬤, **Rustam E. Asizbaev**⬤, and **Platon A. Lifanov**⬤

Abstract We elaborate on the key aspects of achieving the effects of the use of digital technologies in the ecologically focused business environment of Middle East countries. It is proved that the presented directions for achieving effectiveness could be adapted in other countries, including developing ones, when implementing a national policy on the focus on digitalization in the green economy (with a range of stimuli), with factual support of government initiatives by entrepreneurship and determination of possible prospects for functioning. The tasks of this research are achieved with the help of system analysis, complex analysis and comparative analysis. This paper's approach implies the formulation of a list of features of achieving various environmental and economic effects from the use of modern tools of digitalisation in the formation of climate-responsible entrepreneurship.

Keywords Climate-responsible entrepreneurship · Digitalization · Digital economy · Decarbonization · Green economy · Solar energy

JEL Classification H23 · H25 · L94 · O13 · O38

A. B. Karbekova (✉)
Jalal-Abad State University Named After B. Osmonov, Jalal-Abad, Kyrgyzstan
e-mail: aziza-karbekova@mail.ru

A. Tashirov
OJSC "Guarantee Fund", Bishkek, Kyrgyzstan

R. E. Asizbaev
The Academy of Public Administration Under The President of The Kyrgyz Republic Named After ZH. Abdrahmanov, Bishkek, Kyrgyzstan

P. A. Lifanov
Volgograd State Technical University, Volgograd, Russia

1 Introduction

Climate-responsible entrepreneurship is one of the elements of responsible consumption and production, the course for which has been declared in the UN Sustainable Development Goals. This type of entrepreneurial activity is the basis of the concept of green economy development, according to which economic development is a dependent component of the external natural environment. A circular economy, similar to the green economy, implies the orientation at the participation of climate-responsible subjects of business, which are leaders in the cyclical use of resources, including technogenic resources. The role of entrepreneurs and enterprises, which are focused on the improvement of climate, becomes particularly relevant during the emergence of environmental problems the government cannot solve independently with the help of regulations and prohibitions.

Modern entrepreneurship, which offers environmentally-oriented products (services) to consumers or functions based on the focus on the preservation and improvement of climate, uses the latest digital technologies. The current course on climate responsibility of business is largely connected with the use of such ICT that ensure climate and economic effect. Studying this connection is important for the identification of new directions for the development of the entrepreneurial sector, especially in developing countries.

The purpose of this research is the identification of the directions for ensuring the effectiveness of climate-responsible entrepreneurial activities, which is focused on the use of ICT. Further in this research, we dwell on the rating of countries by the formation of green entrepreneurship and characterise the directions for the effectiveness of climate-responsible business that uses innovative digital technologies.

2 Materials and Method

We determine the key directions for achieving the effectiveness of a climate-oriented business sector in the sphere of the digital economy based on the analysis of the theoretical and empirical materials on this issue at the general level and in the context of certain regions and countries [1, 4–6, 8, 10–12, 18]. To find the list of specific directions for the countries of the given region (the Middle East), it is necessary to systematically study this issue. This will allow discovering the modern state of the given problems. Countries of the Middle East were selected for this research due to their entrepreneurship's significant achievements in the studied sphere.

The methodological framework of this research includes systemic analysis, which allows distinguishing the directions for effective climate-responsible entrepreneurship that functions with the use of ICT; comparative analysis, which is used to compare the estimate indices that identify the state of the effectiveness indicators;

complex analysis, which allows determining the directions for the effectiveness of climate-oriented business under the conditions of the usage of ICT.

3 Results

Let us dwell on the analysis of the positions of the Middle East countries by the main parameters of implementing green entrepreneurship in 2022 (Table 1). Also, within the further description of the countries' achievements in this sphere, we shall present the directions for the effectiveness of climate-oriented business that uses ICT.

Let us elaborate on the directions for effectiveness in the sphere of climate-responsible business, which is focused on the use of ICT in countries of the Middle East that demonstrate the highest achievements: Israel, the UAE, Saudi Arabia and Kuwait (Table 1).

Israel is the leader among countries in the region by the level of the general use of green technologies in entrepreneurship. This is due to the achievement of high results in the transition to renewable energy at the level of business subjects in all sectors. The contribution of Israel's entrepreneurship to decarbonisation and the use of purification technologies was not sufficient, however.

Table 1 Ratings of the Middle East countries by the indicators of green business in 2022

	Indicator	Value of indicators, rank						
		Israel	UAE	Saudi Arabia	Kuwait	Egypt	Turkey	Qatar
1	The general level of implementation of green technologies in business	30	41	51	58	59	69	73
2	Level of purification technologies that are implemented in processes and products	65	49	69	60	29	20	57
3	Level of CO_2 emissions from the activities of the business (minimum—1, the growth means an increase in emissions)	46	3	19	23	68	66	29
4	Level of the energy transition to renewable energy at the level of entrepreneurship	9	10	12	11	74	19	76

Source Compiled by the authors based on [13–17]

According to [18], the transition to renewable energy in Israel has been conducted since 2008 based on the programme documents, which regulate the refusal of the utility sector and economy of fossil energy resources, which are required for energy production. The legislation and programme foundations on the level of renewable energy implementation were changed a lot in Israel. In 2022, it was announced that Israel aimed at the growth of the use of solar energy by more than 50% by 2025, and the achievement of the level of renewable energy by 30 and 70% of coverage of energy needs through natural gas by 2030 [2]. Israel uses tax stimulation for companies that are involved in R&D and implementation of innovative technologies in the sector of renewable energy [18]. There is a corporate tax rebate, for the last five years, which equals 6% (regular tax rate is 23%) [9].

Israel uses stimulation measures for developing green energy at the level of enterprises. Among operators of the markets of renewable energy production, we can mention private companies [7]: EDF Renewables (32 MW per year); Enlight Energy (30 MW per year); YVS Renewable (50 MW per year); Zabar Solar (65 MW per year); Doral Energy (100 MW per year); Prime Energy (475 MW per year). The analysis showed that these companies function stably and effectively in the market of national energy, and some of them provide renewable energy to other countries. As for implementing green energy projects in other countries, it is worth mentioning the participation of Enlight Energy and NewMed, which are leaders in natural gas production in Israel [17]. Thanks to investments and clever management of gas production, these companies were able to reach energy independence from external supplies after the discovery of gas deposits near the Mediterranean coast. They were also able to ensure the organisation of these types of activities using public–private partnerships, with large projects on the delivery of renewable energy to countries of Africa and the Middle East.

In 2022, the United Arabs Emirates was ranked 41st in the world by the general level of implementation of green innovations in entrepreneurship. Though this is a medium level of effectiveness, the UAE's entrepreneurial sector was able to achieve high results in this sector (Table 1): decarbonisation (3rd position); energy transition to renewable energy in the economy (10th position).

These two directions are connected since the reduction of CO_2 emissions from the burning of fossil fuels is predetermined by the transition to renewable energy in the country. The results were achieved due to clear climate goals of the country, which were accepted and supported by the business: reduction of CO_2 emissions by 70% by 2030; the complete absence of CO_2 emissions in 2050. Companies in most of the industrial sectors conduct a transition to the use of solar energy without significant problems, but some enterprises that are not able to use solar energy due to technical features had to use R&D to find opportunities for transition to low-carbon hydrogen. Such a solution on the use of low-carbon hydrogen facilitates the achievement of climate neutrality of the economy [4, 10]. The main segments of entrepreneurship that implement a transition to the use of low-carbon hydrogen in the UAE at this stage include railroad and car transport, with possible options in the sphere of aviation [5].

Though Saudi Arabia is behind the UAE by the general level of the use of green technologies in the development of climate-responsible business, it still has certain achievements in the sphere of decarbonisation of companies' activities and implementation of projects on energy transformation toward green energy (Table 1). These achievements are connected primarily with the close cooperation at the level of the public–private partnership between the UAE and Saudi Arabia. This is predetermined by the climate and resource specifics of the two countries and traditions of cooperation at the level of the entrepreneurial sector (desert climate, gas deposits, solar energy). Despite the fact that up until 2020 the main source of electric energy generation in the countries of the Middle East was natural gas (the UAE—97.3%, with 2.7% accounting for solar energy; Saudi Arabia—2.7%, with 0.1% accounting for solar energy), 2020 saw the start of large-scale development of projects on electricity generation from solar energy, also with wind energy and nuclear energy. The problems caused by the COVID-19 pandemic reduced the rates of these projects implementation. It is worth noting the main projects at the level of the entrepreneurial sector [3, 6].

(1) Saudi Arabia:

- Project Al Faisaliah (Shuaibah) (electricity generation from solar energy, 600 MW per year), participating companies: ACWA, Al Babtain Holding Investment (national), Gulf Investment Corporation (Kuwait). Design work was started in 2020.
- Project Sudair solar power (electricity generation from solar energy, 1,500 MW per year), participating companies: ACWA, Badeel, PIF (national). Design and construction work was performed in 2020–2021, production began in 2022.
- Project Dumat Al Janda (electricity generation from solar energy, 400 MW per year), participating companies: EDF Renewables (France), Masdar (UAE). Design and construction took place in 2020–2021, production began in 2022.
- Project Rabigh (electricity generation from solar energy, 300 MW per year), participating companies: Al Jomaih Energy and Water Company (national company), Marubeni (Japan). Design and construction began in 2020, production is to be started in 2023.
- Project South Jeddah Noor (electricity generation from solar energy, 300 MW per year), participating companies: EDF Renewables (France), Nesma (national company), Masdar (UAE). Design and construction work took place in 2020–2021, production began in 2020.
- Project Sakaka (electricity generation from solar energy, 300 MW per year), participating companies: Al Gihaz, ACWA (national). Design and construction work took place in 2020, production began in 2021.

(2) United Arab Emirates:

- Project Barakah (electricity generation with the help of atomic energy, 1400 MW, participating companies: KEPCO (South Korea), ENEC (national company). Design and construction took place in 2020, production began in

2021; (electricity generation from solar energy, 2650 MW per year), participating companies: two Spanish companies, one Italian company, Masdar
(national company), one French company, one Chinese company, one American company and one company from Saudi Arabia). Design and construction
work took place in 2020, production started in 2021.

- Project AL Dhafra (electricity generation from solar energy, 2000 MW
  per year), participating companies: Masdar, Taqa (national), two Chinese
  companies, and one French company. Design and construction took place in
  2020–2021, production began in 2022.

The government of Saudi Arabia provides tax and other subsidies for companies
that participate in national programmes on the use of green technologies for the
production of green energy [2]. That is, the stimulation of green energy with the
use of new digital technologies (digital elements of solar, wind and atomic power
stations) is one of the decisive factors in favour of the choice of projects in this sphere
from the position of economic effectiveness for companies.

In Kuwait, the general rating of implementing green technologies in the sphere of
entrepreneurship is behind Israel, the UAE and Saudi Arabia (Table 1). A sufficient
level was achieved in the sphere of decarbonisation (23rd position), and transition
to green energy in business (11th position). According to [8], a large share of CO_2
emissions is produced by large desalination plants and electric power stations. Due
to specific climate conditions, Kuwait needs a lot of fresh water. At present, companies in Kuwait work on projects on decarbonisation, which ensured a transition to
renewable solar energy by 0.1% in 2022 [6].

Kuwait's goal is to reach 15% of renewable energy in the total energy balance of
the country by 2030. Kuwait's goals differ from the more ambitious goals of Israel,
the UAE and Saudi Arabia. However, even so, they are not easy to achieve, given
the traditional focus on the entrepreneurial sector on energy generation from fossil
fuels [1]. To deal with this problem, the government introduced certain stimulation
measures in recent years.

It is worth mentioning the tendency of the use of such type of renewable energy
in Kuwait as green hydrogen, which is produced with the use of new technologies of
electrolysis. Mass production of electricity generation based on green hydrogen is
expected to start in 2030, after an innovative Shagaya Renewable Energy Park will
have been constructed [12]. The entrepreneurial sector of Kuwait will participate
in competitions for the purchase of energy, which are conducted by the government
similarly to competitions in the markets of electric energy that is produced from other
sources. The project of this innovative energy park is financed through public–private
partnership.

We can state that Kuwait actively participates in the development of business initiatives in the sphere of climate-responsible entrepreneurship that is focused on implementing innovative digital technologies. Similarly to most countries of the Middle
East, preferential terms for the stimulation of such initiatives have a significant
positive influence on their implementation and further production.

4 Discussion

Analysis of the characteristics of the functioning of climate-responsible entrepreneur-ship, which uses the leading digital technologies, demonstrated that the most effective direction is the government's stimulating the implementation of green technologies to ensure sustainability and balance between economic and environmental develop-ment. The focus on this direction is peculiar for all countries that were studied in this paper. The main measures of stimulation include tax subsidies for business and other preferences in the sphere of entrepreneurship (rental discounts, loan discounts, etc.). The effectiveness of these measures is achieved for business and for the state under the condition of factual orientation toward the realisation of climate-responsible projects with the application of new digital technologies.

Analysis of the directions of public–private partnership and the level of stimulation of digitalisation in the sphere of the management of climate-responsible entrepreneur-ship as well as analysis of the materials of [11] demonstrate that a current aspect of achieving effectiveness is refusal from capitalisation that is aimed at an increase in effectiveness and cost of tangible assets in favour of capitalization that implies a wise proportion between tangible and non-tangible assets and an increase in their effectiveness. The focus on the effectiveness of capital that is connected with tangible assets was typical for most countries (including countries of the Middle East) during the period of economic growth and formal reaction to the necessity to solve climate problems. Such an approach was demonstrated by the example of the effective activ-ities of gas companies in Israel, which previously operated in the sphere of natural gas production and focused on economic growth and then began an energy transi-tion to green energy and worked on the achievement of economic effectiveness due application of innovative digital technologies in the sphere of solar energy.

An important direction of ensuring the effectiveness of green entrepreneurship with the application of green technologies is the realisation of partner projects that ensure the attraction of international investments and the achievement of environmental and economic effects for the participants and for countries on the whole.

5 Conclusion

In this paper, we analysed the characteristics and directions for implementing green technologies in the sphere of climate-responsible business, by the example of coun-tries of the Middle East. We discovered that such countries as Israel, Kuwait, Saudi Arabia and the United Arab Emirates were able to reach significant results in this sector, which led to them holding important positions in the green economy ratings. This concerns the implementation of renewable energy resources (wind energy, solar energy and atomic energy) and the realisation of projects on economy decarbonisa-tion. Two types of high-priority development of the green economy of the Middle

East countries are interconnected: reduction of CO_2 emissions in this region is mainly due to the transition of the energy sector to renewable energy.

Thus, it is possible to state that the described aspects of effectiveness are successfully implemented thanks to the considered countries' focus on international integration and support for the promotion of green economy products (in this case, energy) in other countries, with the attraction of substantial investments.

References

1. Alsayegh OA (2021) Barriers facing the transition toward sustainable energy system in Kuwait. Energy Strategy Rev 38. https://www.sciencedirect.com/science/article/pii/S22114 67X21001620. Accessed 06 Dec 2022

2. Chandak P (2022) Israel to generate 30% its energy need from renewable sources by 2030. https://solarquarter.com/2022/06/01/israel-to-generate-30-its-energy-need-from-renewa ble-sources-by-2030/. Accessed 06 Dec 2022

3. Chandak P (2022) Saudi Arabia launches new initiative to boost RE projects. https://solarquar ter.com/2022/03/17/saudi-arabia-launches-new-initiative-to-boost-re-projects/. Accessed 06 Dec 2022

4. Hanley ES, Deane J, Gallachóir BÓ (2018) The role of hydrogen in low carbon energy futures— a review of existing perspectives. Renew Sustain Energy Rev 82:3027–3045

5. Ibrahim MD, Binofai FAS, Mohamad MOA (2022) Transition to low-carbon hydrogen energy system in the UAE: sector efficiency and hydrogen energy production efficiency analysis. Energies 15:6663. https://www.mdpi.com/1996-1073/15/18/6663/pdf. Accessed 06 Dec 2022

6. Mashino I (2022) UAE and Saudi Arabia lead the decarbonization of the Middle East—accelerating business development in the region. Mitsui & Co. Global Strategic Studies Institute Monthly Report. https://www.mitsui.com/mgssi/en/report/detail/__icsFiles/afieldfile/2022/01/21/2112e_mashino_e.pdf. Accessed 06 Dec 2022

7. Mordorintelligence (2022) Israel renewable energy market—growth, trends, Covid-19 impact, and forecasts (2022–2027). https://www.mordorintelligence.com/industry-reports/israel-ren ewable-energy-market. Accessed 06 Dec 2022

8. Naseeb A, Ramadan A, Al-Salem SM (2022) Economic feasibility study of a carbon capture and storage (CCS) integration project in an oil-driven economy: the case of the state of Kuwait. Int J Environ Res Public Health 19(11):6490. https://www.mdpi.com/1660-4601/19/11/6490/htm. Accessed 06 Dec 2022

9. Nbn (2022) Taxation in Israel—general information. https://solarquarter.com/2022/06/01/isr ael-to-generate-30-its-energy-need-from-renewable-sources-by-2030/. Accessed 06 Dec 2022

10. Rissman J, Bataille C, Masanet E, Aden N, Morrow WR, Zhou N, Elliott N, Dell R, Heeren N, Huckestein B et al (2020) Technologies and policies to decarbonize global industry: Review and assessment of mitigation drivers through 2070. Appl Energy 266:114848. https://ene rgyinnovation.org/wp-content/uploads/2020/04/Technologies-and-policies-to-decarbonize-global-industry-review-and-assessment-of-mitigation-drivers-through-2070.pdf. Accessed 06 Dec 2022

11. Sabden O, Turginbayeva A (2017) Transformation of national model of small innovation business development. Acad Entrepreneurship J 23(2):1–14. https://www.abacademies.org/art icles/Transformation-of-National-Model-of-Small-Innovation-Business-Development-1528-2686-23-2-109.pdf. Accessed 06 Dec 2022

12. Shehabi M, Bassam D (2021) Opportunity and cost of green hydrogen production in Kuwait: a preliminary assessment. In: USAEE working paper, pp 21–536. https://papers.ssrn.com/sol3/papers.cfm?abstract_id=3995610. Accessed 06 Dec 2022

13. Technologyreview (2022) Carbon emissions. https://www.technologyreview.com/2022/03/24/1048253/the-green-future-index-2022/. Accessed 06 Dec 2022
14. Technologyreview (2022) Clean innovation. https://www.technologyreview.com/2022/03/24/1048253/the-green-future-index-2022/. Accessed 06 Dec 2022
15. Technologyreview (2022) Energy transition. https://www.technologyreview.com/2022/03/24/1048253/the-green-future-index-2022/. Accessed 06 Dec 2022
16. Technologyreview (2022) The green future index 2022. https://www.technologyreview.com/2022/03/24/1048253/the-green-future-index-2022/. Accessed 06 Dec 2022
17. Timesofisrael (2022) Israeli companies tout big plans to develop renewable energy projects in MENA. https://www.timesofisrael.com/israeli-companies-tout-big-plans-to-develop-renewable-energy-projects-in-mena/. Accessed 06 Dec 2022
18. Wu T, Li YT (2019) A review of the national green innovation system in Israel. IOP Conf Series: Earth Environ Sci 252. https://iopscience.iop.org/article/10.1088/1755-1315/252/4/042120/pdf. Accessed 06 Dec 2022

Government Support of Green Innovations: International Experience and Policy Recommendations

Vasily N. Tkachev⬤, Konstantin E. Manuylov⬤, and Elizaveta V. Kiseleva⬤

Abstract The research reviews the empirical evidence on the effectiveness of green innovation public policies and develops a set of recommendations concerning the optimal design of government stimulus programs based on the existing body of empirical studies, academic literature, official reports, and statistical data. The authors conclude that the public good nature of green innovations, noticeable initial costs, high levels of uncertainty and risk, and long payback period lead to market failures and underinvestment in their development and adoption, which necessitates the introduction of stimulus measures by the state. The authors find robust empirical evidence of a positive effect of government support on green innovations and suggest the following recommendations with respect to green innovation policy design: (1) various policy tools should be applied simultaneously and in coordination to exploit their synergetic effect; (2) governments and private companies should act in tandem with the former creating the optimal environment for the latter to realize their technological, managerial, and financial potential; (3) emerging economies with limited opportunities in developing green technologies should focus on absorption and diffusion of innovations obtained from other countries.

Keywords Green innovation · Government policy · Innovation policy · Government support · Sustainable development

JEL Classification Q580 · Q550

1 Introduction

The need for government support of sustainable businesses and technologies is recognized globally. Since water, air, and biosphere are public goods, market mechanisms fail to price in the full cost of their usage to society. Therefore, individual producers

V. N. Tkachev (✉) · K. E. Manuylov · E. V. Kiseleva
MGIMO University, Moscow, Russia
e-mail: tkachev_mgimo@mail.ru

and consumers tend to disregard this cost in their production and consumption decisions. Eco-aware businesses usually lose to unsustainable enterprises in terms of competitiveness because they bear more expenses and have to price similar goods higher. "Dirty" fossil fuel-based technologies also have the advantage that they are already installed, which creates additional entry barriers for sustainable competitors in the form of high fixed costs of developing new infrastructures and business models. Another market failure specific to innovation is that companies usually fail to fully appropriate the returns from their investments, which leads to under-investment in innovation [10]. These facts underlie the need for government intervention to internalize the positive externalities attributed to green innovations and the negative externalities arising from unsustainable business practices. Besides, public support of green investment can help achieve faster economic growth and, for developing markets, narrow the technological gap with developed countries [20].

Recognizing the importance of government support of green innovations, several initiatives that coordinate the efforts of national governments have been launched on the international level, such as the UNECE technical cooperation project "Building the capacity of SPECA countries to adopt and apply innovative green technologies for climate change adaptation."

However, the following important questions arise.

- How effective have these efforts been in promoting green innovations to date?
- What can be done to improve their efficacy further?

This paper aims to answer these questions by reviewing the existing body of evidence on the effectiveness of government support of green innovations and developing recommendations concerning the optimal parameters of sustainable innovation policies.

2 Methodology

The research is based on reports and data from international organizations and analytical agencies involved in green initiatives, academic literature, and official statistical data.

During the research, we applied the approach and definitions by the United Nations Economic Commission for Europe (UNECE). According to it, green innovation (eco-innovation) is defined as an innovation whose use is less environmentally harmful than its alternatives. The term innovation refers to new or significantly improved products, processes, and marketing or organizational methods [14].

3 Results

The difficulty of estimating the effect of various public initiatives on green innovation activity lies in the choice of representative indicators and their limited availability. The standard information source for innovation studies, the EUROSTAT/OECD Community Innovation Survey (CIS), is of limited use when identifying eco-innovations [16]. Besides, a common proxy for innovation activity—the number of patents registered—can be used to quantify the development of new technologies. However, it is poorly fitted to characterize the diffusion of existing technologies, which is also an important part of the innovation process. Therefore, researchers usually apply an extended set of methods, including surveys of companies. Despite this difficulty, a growing number of studies find a robust and positive effect of government initiatives on green innovations.

According to the research by Yu et al., financial constraints impede the capability of companies to develop green innovations; green finance policies can effectively ease these constraints [19]. However, as state-owned enterprises are more likely to receive the stimulus, the authors call for more support to be directed toward privately owned enterprises.

Wang et al. find a positive effect of government subsidies on the Chinese new energy vehicle industry [17]. They conclude that the effect is stronger for private-owned firms and the central area of the country.

Xiang et al. also demonstrate a positive impact of government subsidies on green investment and argue that government subsidies can encourage publicly listed companies to enhance their level of green innovation through debt financing and equity financing [18].

Another study finds a positive effect of government support on green innovation adoption in SMEs in Pakistan [7].

As the energy sector accounts for about two-thirds of CO_2 and other greenhouse gas emissions [14], energy-intensive industries play a crucial role in achieving low-carbon economic growth. A study of the effect of subsidies provided to energy-intensive firms in China to support green innovation estimates that government R&D subsidies increase the green innovation of energy-intensive firms, specifically their tendency and performance, by 107.3% and 54.1%, respectively, the impact being stronger for state-owned enterprises and small and medium businesses [4].

3.1 The Choice of Policy Instruments

Green innovations have not yet induced the self-sustaining and cost-reducing technological cycle. In many cases, they survive in the market only due to the massive government stimulus [14]. As public budgets become increasingly constrained, it is important to use public funds sparingly. Moreover, fostering green innovation should not significantly hinder economic growth.

Governments can now choose from a wide range of policy tools. Apart from tightening regulations, implementing new environmental standards, and improving the environment for sustainable businesses, they include a variety of financial mechanisms: green taxes on harmful environmental activities, subsidies, loans and grants for green investments and research, and tax rebates for meeting environmental standards and increasing demand for green technologies. However, when introducing government stimulus, it is important to understand which policy tools will be most effective in promoting green innovation.

During the Flemish CIS eco-innovation survey, a third of the participating companies mentioned current or expected regulations as a motive for introducing clean innovations, while about 15% were motivated by grants and subsidies [15].

According to another study, carbon pricing, which can be realized through a carbon tax or a cap-and-trade system, reduces the use of dirty technologies and creates an incentive for developing new sustainable technologies [2]. Future expectations of carbon prices provide a strong stimulus for companies to invest in R&D and accelerate the adoption of green technologies.

To increase the efficiency of R&D support allocation, governments can harness market mechanisms to extract information contained in markets. For example, Murray [9] concludes that a recommended way to provide public support for innovation is public co-investment with private partners, such as early-stage venture capital funds. Such design of government support multiplies the financial benefits of success for investors while maintaining incentives to make good investment decisions. Giebe, Grebe, and Wolfstetter [5] find that some degree of unnecessary government funding can be avoided if competition among applicants for R&D grants is introduced (e.g., in the form of auction mechanisms).

The alternative to direct support for R&D or innovation is indirect support, for example, through tax incentives, which are often praised for their non-discriminatory nature. However, as they support any formal R&D, their effectiveness in promoting green innovation decreases [14].

A growing body of evidence suggests that governments should deploy policy instruments simultaneously and in coordination instead of choosing from them.

Acemoglu et al. find that using a carbon tax alone leads to excessive consumption reduction in the short run while introducing subsidies alone makes them less effective compared with the simultaneous introduction of these tools [1]. Aghion, Hemous, and Veugelers utilize economic models of directed endogenous technological change to quantify this effect. They estimate that, if introduced separately, the carbon price would have to be about 15 times higher during the first five years, while subsidies would have to be, on average, 115% higher in the first ten years to provide an effect equal to their simultaneous introduction [2].

Another study stresses that public R&D support should be provided simultaneously with carbon pricing because it addresses the knowledge externality associated with creating new sustainable innovations and offsets the comparative advantage of the older unsustainable but already installed practices [16].

IEA [6] outlines four stages of the innovation cycle for new energy technologies and suggests a specific type of government intervention at each stage.

- For promising but not yet mature technologies: support research, large-scale demonstration, and addressing infrastructure and regulatory needs.
- For technologies that are technically proven but require additional financial support: more technology-specific incentives (e.g., feed-in tariffs) combined with regulatory frameworks or standards to create a market.
- For technologies close to competitive: technology-neutral incentives that are removed when market competitiveness is achieved.
- For competitive technologies: building public acceptance or adoption by identifying and addressing market and informational barriers.

The directed endogenous growth model mentioned above provides another important insight. If developed countries embark on an active course towards clean technologies and subsequently diffuse them to developing countries, the latter will become more active in implementing clean technologies even in the absence of green policy measures on their side [2].

A study by OECD [11] suggests that softer public instruments (e.g., developing a supportive economic environment, enhancing the quality and reliability of the information, strengthening knowledge and awareness, and increasing the level of commitment and addressing behavioral biases to support greener consumer choices) can be highly effective in complementing economic measures. Therefore, they must be carefully considered when designing comprehensive green development strategies.

3.2 Fundamental Versus Incremental Innovations

Though the ability to produce radical innovations can only be viewed as beneficial for countries and companies, it does not guarantee success in the contemporary globalized and highly competitive environment. The capacity to absorb advanced technology and diffuse it throughout the economy is also crucial for success. History provides several examples of emerging economies successfully adopting the technologies of leading industrialized countries and catching up with their productivity and income levels.

In this regard, governments can work in two main directions: supporting the diffusion of existing technologies or developing new ones. According to the study by Shu et al., government support more strongly mediates the effect of green management on radical product innovation than its effect on incremental product innovation [12]. A study by OECD states that carbon pricing fosters primarily incremental innovation [14].

3.3 Policy Implications

As different countries face varying environmental concerns and are characterized by different levels of economic, social, and political development, an individual, "tailor-made" approach is preferential in every case. However, existing theoretical concepts and reviews of practical experience make it possible to suggest the following universal recommendations.

- Governments and private companies should act in tandem, the former creating a framework and business environment for the latter to deploy their technological, managerial, and financial potential to full capacity.
- Policymakers should aim for the optimal level of green innovation support. The excessive stimulus may lead to distortion of market signals and wasteful investments. If the incentives are too weak, underinvestment in sustainable practices may lead to steep damage from climate change.
- Since the amount of available public funding is limited, governments must encourage investment from multilateral organizations, financial institutions, and the corporate sector [3].
- One important direction of green policy is the supervision of financial markets, which play a crucial role in channeling private investment. Though investor demand for sustainable instruments and policy shifts towards sustainability in some countries have already led to the emergence and dynamic growth of the sustainability sector in the financial markets, active government action is required to ensure its efficient and orderly development. Such action can include developing legal frameworks and guidance for financial institutions, providing risk assessment, disseminating knowledge about sustainable finance solutions, and improving digital and financial infrastructure and other measures [13].
- The effectiveness of individual policy instruments increases if they are introduced simultaneously and complement each other.
- It is recommended that governments decide on the appropriate measures of green innovation stimulus in cooperation with key stakeholders.
- An important policy measure is a public investment in basic and long-term research. Since fundamental green technologies have a public good character while representing a risky, uncertain, and long-term investment, they are unlikely to be developed by the private sector. Governments can also aim to improve the process of translating research into innovation by strengthening the links between science and business.
- Since the provision of targeted support to certain innovations can be risky because of their uncertain and random nature [8] and the lack of information on their maturity and future commercial potential, it is recommended that government support is provided based on performance rather than specific technologies or channeled to general-purpose technologies needed for a wide range of innovations (such as ICT, biotechnology, and nanotechnology) [14]. Due to the growing multidisciplinarity of research, investing in R&D to provide an underpinning for green

innovation will require a broad portfolio of investments instead of focusing only on energy and environmental R&D.

- Countries with limited opportunities to generate green technologies should focus on removing barriers to the absorption and diffusion of innovations from abroad. However, countries with a strong capacity to produce green innovations will also benefit from international cooperation because it makes it possible to share the costs of public investment, improves access to knowledge, and fosters technology transfer across countries.
- While public policies act as a strong stimulus for promoting eco-innovations, they need to be seen as merely one part of a full set of motives for the private sector to introduce green innovations. They will effectively work only in tandem with the demand for sustainable products and voluntary codes of conduct or sector agreements.

4   Conclusion

As the issue of climate change is becoming more pressing, governments have started paying increased attention to sustainable business practices. Green innovations have significant positive externalities but imply noticeable initial costs, high levels of uncertainty and risk, and long payback periods, which justifies and necessitates active public support for their development and diffusion.

This research contributes to the growing body of theoretical and empirical works exploring green innovation public policies. The main findings of the research include providing evidence of a visible and robust positive effect of government support on green innovations, as well as a number of recommendations concerning the optimal design of green innovation government support programs. The research results suggest that various policy tools should be applied simultaneously and in coordination. Governments and private companies should act in tandem with the former creating the optimal environment for the latter to realize their technological, managerial, and financial potential. Governments should provide funding for fundamental research while stimulating incremental innovations by private businesses. Emerging economies with limited opportunities to develop green technologies should focus on the absorption and diffusion of innovations obtained from other countries.

References

1. Acemoglu D, Aghion P, Bursztyn L, Hemous D (2012) The environment and directed technical change. Am Econ Rev 102(1):131–166. https://doi.org/10.1257/aer.102.1.131
2. Aghion P, Hemous D, Veugelers R (2009) No green growth without innovation. Bruegel Policy Brief, 2009/07. Retrieved from https://www.bruegel.org/sites/default/files/wp_attachments/pb_climatervpa_231109_01.pdf. Accessed 5 Dec 2022

3. Atalla G, Mills M, McQueen J (2022) Six ways that governments can drive the green transition. EY. Retrieved from https://www.ey.com/en_gl/government-public-sector/six-ways-that-governments-can-drive-the-green-transition. Accessed 5 Dec 2022
4. Bai Y, Song S, Jiao J, Yang R (2019) The impacts of government R&D subsidies on green innovation: evidence from Chinese energy-intensive firms. J Clean Prod 233:819–829. https://doi.org/10.1016/j.jclepro.2019.06.107
5. Giebe T, Grebe T, Wolfstetter E (2005) How to allocate R&D (and other) subsidies: an experimentally tested policy recommendation. Discussion Paper No. 108. Retrieved from https://sfb tr15.de//uploads/media/108.pdf. Accessed 5 Dec 2022
6. International Energy Agency (2010) Energy technology perspectives 2010. IEA, Paris, France. Retrieved from https://iea.blob.core.windows.net/assets/04776631-ea93-4fea-b56d-2db821bdad10/etp2010.pdf. Accessed 5 Dec 2022
7. Jun W, Ali W, Bhutto MY, Hussain H, Khan NA (2021) Examining the determinants of green innovation adoption in SMEs: a PLS-SEM approach. Eur J Innov Manag 24(1):67–87. https://doi.org/10.1108/EJIM-05-2019-0113
8. Milovidov VD (2018) Hearing the sound of the wave: what impedes one's ability to foresee innovations? Foresight STI Governance 12(1):76–85. https://doi.org/10.17323/2500-2597.2018.1.88.97
9. Murray G (1999) Early-stage venture capital funds, scale economies and public support. Venture Capital: Int J Entrepreneurial Finance 1(4):351–384. https://doi.org/10.1080/136910699295857
10. OECD (2011) Demand-side innovation policies. OECD Publishing, Paris, France. https://doi.org/10.1787/9789264098886-en
11. OECD (2011) Fostering innovation for green growth. OECD Publishing, Paris, France. https://doi.org/10.1787/9789264119925-en
12. Shu C, Zhou KZ, Xiao Y, Gao S (2016) How green management influences product innovation in China: the role of institutional benefits. J Bus Ethics 133(3):471–485. https://doi.org/10.1007/s10551-014-2401-7
13. Tkachev VN, Kiseleva EV, Fedyanina OV (2021) Financial sector growth, consolidation, and new technologies make it a powerful actor in tackling global environmental challenges. In: Zavyalova EB, Popkova EG (eds) Industry 4.0: exploring the consequences of climate change. Palgrave Macmillan, Cham, Switzerland, pp 375–388. https://doi.org/10.1007/978-3-030-75405-1_33
14. UNECE (2013) Innovation policy for green technologies: guide for policymakers in the transition economies of Europe and Central Asia. UN, New York, NY; Geneva, Switzerland. https://doi.org/10.18356/f1571ab8-en
15. Veugelers R (2012) Which policy instruments to induce clean innovating? Res Policy 41(10):1770–1778. https://doi.org/10.1016/j.respol.2012.06.012
16. Veugelers R (2016) Empowering the green innovation machine. Intereconomics 51(4):205–208. https://doi.org/10.1007/s10272-016-0603-1
17. Wang Z, Li X, Xue X, Liu Y (2022) More government subsidies, more green innovation? The evidence from Chinese new energy vehicle enterprises. Renew Energy 197:11–21. https://doi.org/10.1016/j.renene.2022.07.086
18. Xiang X, Liu C, Yang M (2022) Who is financing corporate green innovation? Int Rev Econ Financ 78:321–337. https://doi.org/10.1016/j.iref.2021.12.011
19. Yu C-H, Wu X, Zhang D, Chen S, Zhao J (2021) Demand for green finance: resolving financing constraints on green innovation in China. Energy Policy 153:112255. https://doi.org/10.1016/j.enpol.2021.112255
20. Zavyalova EB, Studenikin NV (2019) Green investment in Russia as a new economic stimulus. In: Sergi BS (ed) Modeling economic growth in contemporary Russia. Emerald Publishing Limited, Bingley, UK, pp 273–296. https://doi.org/10.1108/978-1-78973-265-820191011

Use of Climate-Smart Green Innovations by Sector of the Digital Economy

Ecological and Resource Potential of Interregional Commodity Exchange Logistics

Tamila S. Tasueva● and **Vera V. Borisova**●

Abstract The basis of stable socio-economic development of the region's economy is a balance of production and consumption of commodity and material values. Under the sanction pressure of the collective West on the Russian economy, the resource-saving and environmental orientation of logistics supply systems largely determine national security. One of the basic components of the new approach to supporting the balance of production and consumption of material resources should be anti-crisis management of resource conservation in the inter-regional exchange of goods. The research investigates the possibilities of crisis management of logistics of interregional commodity exchange, adequate to environmental and resource-saving requirements. The authors consider the processes and measures aimed at overcoming (limiting) the spread of the crisis situation in the inter-regional exchange of goods related to the use of material resources and green logistics. The authors substantiate the expediency of identifying the eco-resource potential of inter-regional commodity exchange logistics and using green logistics indicators and tools of standardization and certification of logistics processes in the supply chain for its assessment. It is shown that the fundamental shifts towards the ecological and resource orientation of logistics systems of inter-regional commodity exchange are associated with the depletion of all types of raw materials and driven by the need to find a possible alternative way of forming the ecological and resource potential. The question of the formation and development of the environmental and resource potential of the logistics of inter-regional commodity exchange in a situation of scarcity of raw materials is relevant due to the role of the environmental and resource component in the logistics of inter-regional commodity exchange. In conceptual terms, the problem associated with the identification of features of the environmental and resource potential of the logistics of inter-regional commodity exchange and the search for reserves of application of the toolkit of green logistics in supply chain management was identified.

T. S. Tasueva (✉)
Grozny State Oil Technical University, Grozny, Russia
e-mail: tamila7575@mail.ru

V. V. Borisova
Saint Petersburg State University of Economics, St. Petersburg, Russia

© The Author(s), under exclusive license to Springer Nature Switzerland AG 2023
E. G. Popkova (ed.), *Smart Green Innovations in Industry 4.0 for Climate Change Risk Management*, Environmental Footprints and Eco-design of Products and Processes, https://doi.org/10.1007/978-3-031-28457-1_18

175

This determines the relevance of this research. The authors conclude that the implementation of the reserves of environmental and resource potential of the logistics of inter-regional commodity exchange contributes to the organization of commodity exchange interaction between regions based on the introduction of the standards of green logistics and the use of resource-saving technical means and digital technologies. The authors confirm the hypothesis that the construction of logistic systems of interregional commodity exchange based on environmental and resource indicators will make it possible to increase their qualitative sustainability, balance the production and consumption of raw materials, and counteract the challenges of sanctions.

Keywords Green logistics · Logistic system · Interregional commodity exchange · Potential · Reserves · Ecological and resource indicators

JEL Classification M110

1   Introduction

The mechanism of logistics of interregional commodity exchange is implemented through the use of tools covering all functional areas of logistics and its elements, links, and subsystems. Logistics of interregional commodity exchange includes organizational and economic measures related to managing economic flows (primarily material flows) aimed at optimizing the total costs of commodity movement and supporting the balance of production and consumption of raw materials [1]. A number of scientists link the efficiency of regional logistics systems with the coordination of all participants in the supply chain to achieve a system-wide objective [2, 3], the transformation of financial [4] and information [5] flows, and the development of standards for logistics processes [6]. The generalization of different points of view showed that the content of the logistics system of inter-regional commodity exchange should be supplemented by the environmental and resource component [7].

The Chinese experience in supporting environmental protection measures and reducing the ecological footprint of logistics processes is of particular interest.

In February 2021, the State Council of China issued a "Guidance on the green and low-carbon cycle development economic system" [8].

In China, green logistics aims to reduce the environmental impact through the full utilization of logistics resources, the introduction of advanced logistics technologies, and the rational planning and implementation of logistics operations, such as transportation, warehousing, packaging, loading and unloading, handling, and distribution. The ultimate goal is sustainable development, combining economic, social, and environmental benefits [8].

The China Federation of Logistics and Purchasing has established a green logistics division to develop standards that promote energy conservation, reduce the carbon cycle, and comprehensively create a green supply chain ecosystem. It is necessary to highlight the following standards.

- "Composition and method of green logistics index accounting";
- "Green logistics enterprise assessment index";
- "Logistics";
- "Method of greenhouse gas emissions accounting at enterprises".

It is also necessary to note the standard that defines procurement controls for green manufacturing: GB/T39258-2020 National Standard "Green Supply Chain Management, Control of Procurement in Green Manufacturing Enterprises" [9].

China has developed 11 national standards (by type of raw material) to reduce emissions of harmful substances and greenhouse gases into the atmosphere during the movement of raw materials and supplies from the procurement market to the warehouses of the enterprise.

It is useful for Russia to use China's experience in introducing green standards into logistics processes. Russia has already implemented international environmental standards and standards for the emission of toxic gases from engines (vehicles are equipped with additional filters to reduce air pollution). Compliance with these standards has made it possible to reduce harmful emissions almost three times by using diesel fuel alone and ten times by using neutralizers.

Analysis of the supplier market and identification of all possible candidates for the supply of goods is based on the impact of the supplier company on the environment, its actions in terms of solving social problems, and effective management within its enterprise [10].

Environmental requirements for procurement are also fixed in Russian legislation [11, 12]. Many enterprises form their own questionnaires for ranking suppliers because there are no uniform requirements and state standards in Russia.

In Russia, attempts are being made to introduce digital platforms for remote assessment of the environmental performance of suppliers. Thus, Bidzaar ESG, a digital service of mass express evaluation aimed at ranking suppliers in terms of ESG criteria, has been developed. This service was created based on international standards for disclosure of non-financial reporting (Global Reporting Initiative (GRI), Sustainability Accounting Standards Board (SASB), etc.) in accordance with the draft of the open national standard ESG, prepared by Delovaya Rossiya with the recommendations of the Central Bank and the Russian green taxonomy. The developers of this service focus their attention on connecting small and medium businesses in Russia to the green agenda [13].

Green standards in logistics processes and resource-saving initiatives are voluntary in nature. They act in line with legal requirements and show how logistics activities affect the environment and how they fit into legal requirements in terms of environmental safety and resource and energy conservation.

It is assumed that the environmental standards of logistics processes and resource-saving initiatives of participants of inter-regional commodity exchange will reduce the adverse environmental impact at the following levels.

- Enterprise (intra-corporate logistics);
- Region (extended logistics supply chain);
- Global (international) supply chain.

The success in implementing the reserves of ecological and resource potential of the logistics of interregional commodity exchange largely depends on the coordinated work of all participants in the supply system.

2 Methodology

The methodological basis of this research is based on the concept of ESG logistics (ESG—Environmental, Social, and Corporate Governance), which implements the principles of environmental, social, and governance responsibility and combines multiple ideas of green, ethical, responsible, sustainable, and transformational development of logistics, which complement each other and were developed in the link of functional areas of supply, production, and sales logistics in the supply chain.

Principles and methods of building environmentally-oriented and resource-saving logistics systems receive new scientific interpretations in the context of interdisciplinarity and synergetic paradigm. Synergetics allows us to consider logistics supply chains from the perspective of the coexistence of economic entities with the natural world. Synergetic principles are complemented by a systemic approach that explains the interconnection and interdependence of participants in the logistics chain. The ecological and resource approach used by business entities must comply with the following principles.

- The timeliness of the information on the environmental assessment of processes in supply chains;
- The integrity and interdependence of the system elements;
- The combination of statics and dynamics for long-term forecasting of the development of a green supply chain;
- The consideration of industry specifics and the state of environmental policy in the region and the country.

3 Results

In the third decade of the twenty-first century, the speed of transformation of logistics processes is changing dramatically. The transformation of logistics activities has led to the widespread use of artificial material means and resource-intensive production and the depletion of non-renewable natural resources, leading to the deterioration of the environment. Such development of systems is characterized by specialists as a technogenic type of environmental and economic management [14].

Human activity in the context of developing an economic system of the technogenic type leads to the destruction of the biosphere, pollution, and the destruction of the ecology of territories.

A possible solution to the problem of environmental pollution in the development of logistics infrastructure is to turn to global practices, namely the concept of sustainable development.

The basis of this concept is the sustainable development of all areas of human activity by maintaining a balance between their environmental, economic, social, and cultural development, ensuring the efficient use of natural resources, energy-saving, and environmental technologies, supporting society, developing human resources, and preserving the specifics of cultural potential and the integrity of the environment and the biosphere for future generations [14].

The Russian economy and its logistics sector are only at the beginning of the way to apply ESG principles in their activities. Sustainable environmentally-oriented infrastructure is expensive and low-profit. Simultaneously, experts estimate the environmental potential of Russian technologies at more than three trillion rubles.

Analyzing the possibility of implementing the reserves of environmental and resource potential of the logistics of the inter-regional exchange of goods, we investigated the logistics processes of its components. Particular attention in the research focuses on the following logistics functions-activities: procurement, transportation, warehousing, inventory management, and the involvement of secondary material resources in economic turnover [15]. On the one hand, these functions-activities (logistical processes) are a driving force and a tool for solving environmental, economic, social, and cultural problems. On the other hand, these processes have a negative impact on the environment and leave a significant negative ecological footprint. Therefore, the search for ways to reduce the ecological footprint of logistics processes is particularly important in the organization of inter-regional supply systems.

ESG logistics is becoming increasingly popular in various industries and activities. The assets of ESG-based companies exceed $80 trillion, and the assets of investment funds exceed $30 trillion. The assets under the management of ESG-based investment funds total more than $30 trillion. The assets of ESG-administered funds exceed $30 trillion. The number of financial institutions working based on ESG principles is growing, infrastructure is developing (special sections at exchanges, rating agencies, and consultants), and regulation and supervision are changing.

Business entities prefer to work on the principles of ESG logistics with environmental commodities, transparent sources of origin of funds, and the implementation of socially significant goals.

The environmental component is one of the three elements of the ESG concept.

According to expert estimates [16], "Russian companies are good at formulating ESG principles, including environmental risk management principles, and developing appropriate regulations and strategies. However, only a small number of business entities implement and monitor ESG principles.

The development and implementation of green standards for logistics processes aim to ensure the reproducibility of the current best practices (benchmark) of logistics functions and operations by formalizing them to reduce the ecological footprint. The development of green standards for logistics processes in supply chains is preceded by an analysis of the best domestic and foreign practices and situational analysis of the current situation in the inter-regional exchange of goods in terms of the impact of logistics on the environment.

Recommendations on the implementation of green standards for logistics processes in the interregional exchange of goods are based on the interaction of participants in the supply chain (SC), which involves the formation of an end-to-end environmentally-oriented logistics process aimed at the joint creation of a reference project (model) of a complete environmentally-oriented logistics process that integrates the interests of all participants of SC, makes it possible to coordinate their environmentally-oriented and resource-saving actions, identifies opportunities and resource reserves of resource-saving and improvement of the ecological orientation of processes, and excludes logistic functions and operations that are detrimental to the environmental parameters of SC.

The business community, government agencies, public organizations, associations, and consumers are interested in realizing the environmental and resource potential of logistics. The same structures can initiate the development and implementation of green standards in logistics, which helps reduce the ecological footprint in the performance of logistics processes, increases their efficiency, and emphasizes the binding requirements for environmental protection.

4   Conclusion

The current logistics management system of interregional exchange of goods is integrated with the environmental and resource-saving management of enterprises involved in the supply chain. This system is aimed at developing and implementing science-based norms and standards of resource consumption and green standards of logistics processes.

The performance of logistics processes in the supply chain is associated with an active technogenic impact on all objects of the environment—the atmosphere, hydrosphere (surface and underground), subsoil, flora, and fauna. This leads to environmental degradation. Due to the sanction war and geopolitical events, the following changes in supply chains are observed.

- The performance of logistics processes occurs in the context of the introduction of new additional logistics facilities;
- Supply contracts are being renegotiated;
- Energy and freshwater needs are increasing;
- The use of natural resources, etc., is becoming increasingly complex.

Under such conditions, the realization of the reserves of environmental and resource potential of logistics is of particular importance. The introduction of green standards for performing logistics processes can reduce the environmental load.

Currently, some Russian companies (e.g., Severstal, Evraz, Sibur, and Rosatom) use environmental performance indicators to assess management performance.

The analysis showed that the requirements of foreign and Russian environmental standards differ in some cases. In particular, this is related to the level of availability of specific technologies for manufacturers and consumers. For example, it is difficult for Russian businesses to implement requirements related to waste disposal because of the low level of development of the system of collection of waste products. The approach to forming requirements for compliance with environmental legislation is markedly different.

Assessment and improvement of the environmental characteristics of logistics functions and operations in supply chains create conditions for information openness in providing information about the ecological footprint.

In developing green standards for logistics processes, it is advisable to use the international standards of the ISO series (International Standard Organization) and the international standards ISO 14000. These standards formulate useful provisions for logistics activities related to the organization of environmental management and environmental auditing. They contain principles, recommendations, and tools for organizing environmental activities at the enterprise and are recommendatory in nature. A separate document contains the ISO 14001 standard "Environmental Management System. Requirements and Guidelines for Use." Based on the provisions of this document, we have recorded the following basic requirements for an environmentally oriented supply chain.

- The goal of developing a green supply chain guides the implementation of green standards by all supply chain participants in all functional areas of logistics (supply, production, and sales) and for all key logistics activities;
- In the functional areas of logistics activities, companies must implement and comply with green standards for logistics processes, including procedures for determining significant environmental impacts;
- Local environmental goals, objectives, and performance indicators (KPIs) must be developed in supply logistics, production logistics, and sales logistics, comparable with the general environmental goal of supply chain development, considering significant environmental impacts and legislative and other requirements;
- The participants of the supply chain develop an action plan for the implementation of green standards in logistics processes, identifying those responsible for its implementation and determining the budget for this purpose;
- In accordance with the requirements of the introduction of green standards in logistics processes, it is advisable to carry out staff training aimed at resource conservation and the formation of environmental awareness among employees;
- Green standards for supply chain logistics processes are subject to monitoring and periodic audits to determine if they meet the stated criteria;
- Periodical reports on the fulfillment of the criteria stated in the green standards.

The basic principles of environmental logistics are aimed at the following.

- Rational use of natural resources;
- Minimal use of raw materials and packaging that cannot be recycled or disposed of safely;

- Application of economically justified and environmentally safe transportation and storage of material resources;
- Increasing environmental awareness and responsibility of personnel;
- Introduction of innovative technologies to reduce the environmental load;
- Maximum use of production waste, containers, and packaging as recyclable materials or their environmentally safe disposal;
- Full and rational use of the company's resources [17].

To summarize, the main objectives of the implementation of the environmental and resource potential of the logistics of the inter-regional exchange of goods are associated with the following.

- Use of the system of separate collection of industrial waste, waste containers, and packaging with their further targeted use;
- Introduction of new technologies for the use of secondary raw materials to produce material resources suitable for production activities;
- Use of natural energy in production to minimize environmental pollution, including the competent use of climatic features of the region, etc.;
- Improvement of existing environmental legislation, which provides not so much punishment for violations of the rules for separate waste collection and uneconomical use of natural resources, as tax or subsidy incentives for industries and enterprises that use environmentally safe and innovative technologies;
- Use of environmentally friendly packaging materials, especially in cases where recycling is impossible or difficult.

Acknowledgements The reported study was funded by RFBR, project number 20-010-00141\22 "Formation of the Institutional Framework of the Region Infrastructure in the Digital Economy."

References

1. Afanasenko ID, Borisova VV (2018) Digital logistics: textbook. Peter, St. Petersburg, Russia
2. Borisova V, Tasueva T, Budyakov A (2020) Configuration of logistic supply systems in the oil and gas industry of Russia. In: Proceedings of SCTMG 2020: international scientific conference "social and cultural transformations in the context of modern globalism". European Publisher Ltd., Grozny, Russia. Retrieved from https://www.europeanproceedings.com/files/data/article/10040/12025/article_10040_12025_pdf_100.pdf. Accessed 12 Feb 22
3. Shulzhenko TG (2020) Methodological approach to the reengineering of logistics business processes in the transport chains with the implementation of smart contracts. Manag Sci 10(2):53–73. https://doi.org/10.26794/2404-022X-2020-10-2-53-73
4. Parfenov AV, Naimin C (2021) Logistics and supply chain management in cross-border e-commerce. In: Shcherbakov VV, Smirnova EA (eds) Logistics and supply chain management. Saint Petersburg State University of Economics, St. Petersburg, Russia, pp 109–114
5. Rodkina T (2016) RFID technology in Russian logistics: real achievements and challenges. Russian Logistics J 1(2):48–53
6. Pechenko NS (2019) Customs regulation of logistics foreign trade processes. Proc St Petersburg State Univer Econ 6(120):184–189

7. Borisova VV (2022) ESG criteria for the development of logistics infrastructure in the region. Logistics Supply Chain Manag 6(19):26–30
8. State Council the People's Republic of China (2021) The state council on accelerating the establishment and improvement: guidance on the green and low-carbon cycle development economic system. Retrieved from http://www.gov.cn/zhengce/content/2021-02/22/content_5 588274.htm. Accessed 10 Feb 2022
9. State Administration for Market Regulation (2021) Interpretation of the national standards series of green/environmental supply chain management. Retrieved from https://www.samr. gov.cn/bzjss/bzjd/202104/t20210406_327602.html. Accessed 19 Feb 2022
10. Tasueva TS, Elibaeva PT (2020) Application of resource-saving capacity of logistics at oil and gas companies. Herald of GSOTU. Humanitarian, Soc Econ Sci 16(2), 30–34. https://doi.org/ 10.34708/GSTOU.2020.75.52.005
11. Russian Federation (2011) Federal law "On the procurement of goods, works and services by certain types of legal entities" (18 July 2011 No. 223-FZ, as amended 14 July 2022). Legal Reference System "ConsultantPlus", Moscow, Russia. Retrieved from https://www.consultant. ru/document/cons_doc_LAW_116964/. Accessed 19 Feb 2022
12. Russian Federation (2013) Federal law "On the contract system for the procurement of goods, works and services for state and municipal needs" (5 Apr 2013 No. 44-FZ, as amended 15 Nov 2022). Legal Reference System "ConsultantPlus", Moscow, Russia. Retrieved from http://www.consultant.ru/document/cons_doc_LAW_144624/df3ace 0ea577a92ea8b71c0d4363fbbe79da7160/. Accessed 19 Feb 2022
13. HSE University (2021) Mass remote ESG certification methodology for companies: developed by National Research University Higher School of Economics specially for Bidzaar company. HSE University, Moscow, Russia. Retrieved from https://bidzaar.com/start/wp-content/upl oads/2021/11/Методология_ESG_оценки_поставщиков_Bidzaar.pdf. Accessed 19 Feb 2022
14. Borisova V, Pechenko N (2021) Sustainable development of logistic infrastructure of the region. E3S Web of Conf 295:01042. https://doi.org/10.1051/e3sconf/202129501042
15. Tasueva TS, Budyakov AN (2021) Digital technologies in the purchasing practice of an oil and gas corporation. In: Dyatlov SA, Miropolsky DY, Selischeva TA (eds) State and market: mechanisms and institutions of Eurasian integration amidst increasing global instability. Saint Petersburg State University of Economics, St. Petersburg, Russia, pp 480–486
16. Russian Institute of Directors (2021) ESG issues in the practice of Russian public companies. In: The study was conducted jointly with Sber. Retrieved from https://nokc.org.ru/wp-content/upl oads/2021/05/rgur-27-05-2021-v1-prezentacziya-issledovanie-esg_rid_noks-1.pdf. Accessed 10 Feb 2022
17. Borisova VV, Lysochenko AA, Vorobyev GA, Pavlenko II, Avsharov AG (2021) Digital technologies in public procurement logistics. In: Popkova EG, Sergi BS (eds) Modern global economic system: evolutional development vs. revolutionary leap. Springer, Cham, Switzerland, pp 1394–1402. https://doi.org/10.1007/978-3-030-69415-9_154

Specific Features of State Regulation and Telecommunication Infrastructure of Climate-Responsible Entrepreneurship in the Digital Economy Markets

Tatiana N. Litvinova📵

Abstract In this paper, the authors formulate the peculiarities of the formation of such elements of the digital economy as state regulation and telecommunication infrastructure of the climate-responsible business sector of selected Asian countries. The statistical data on the formation are compared and the provisions of the analytical materials on this issue are studied. The scientific novelty of the considered study is due to the improvement of the provisions regarding modern differences and similar parameters for achieving results in the field of activities of climate-responsible business entities of the digital economy of Asian states. The analysis performed showed the inclination of the state regulation of the considered countries toward the creation of complex and bureaucracy-oriented legislative approaches and the absence of flexibility, which is required for the achievement of effect from the activities of the participants of Asian countries' digital economy markets. Certain tendencies of internal market protection and the insufficient consideration of opportunities for international integration in the sphere of ICT are peculiar to the three studied countries; they appear under the influence of the government's participation in the regulation of this sphere.

Keywords State regulation · Telecommunications · Investments · Digital economy · E-commerce · Big data · Wireless broadband · Climate-responsible business sector

JEL Classification F12 · G31 · K21 · K24 · L11 · O32

1 Introduction

Improvement in the technological characteristics of products (services) and expansion of the assortment are connected with the quick implementation of digital tools, which are drivers of economic growth and competitiveness in the national and international markets. Enterprises that have financial capacities developed the digital

T. N. Litvinova (✉)
Volgograd State Agricultural University, Volgograd, Russia
e-mail: litvinova1358@yandex.ru

economy of their countries and became the leading participants of the sectoral markets, ensuring the identification of key areas of climate-responsible behaviour for the market. Such enterprises function mainly in developed countries and emerging economies that are characterised by quick economic growth, which is based on the effectiveness of managing investments in digital innovations. The activities of enterprises (corporations), which are leaders in certain spheres depend on a range of conditions that affect the mechanism of digital tools implementation. Among such factors, an important place belongs to state regulation of the digital economy, including the order of operations of buying and selling of products (services). International economic development depends on the government's support for e-commerce, especially with large-scale character, aimed at promoting environmentally friendly products, and involves maintaining the foundations of responsible process management. The formation of telecommunication infrastructure, which is an element of the digital economy, forms preconditions for investments in certain territories and countries. They are performed based on the level of the business environment and the government's readiness to ensure its continuous functioning and to ensure the system of investment protection. Accordingly, the government's participation is important in the legislative and organisational spheres of the management of digitalisation's isolated sphere. In some countries, the role of the government is minimal in this direction; there are also opposite examples of direct regulation (support or restraint) of the activities of digital economy market participants. Specifics of the state regulation of the ICT sphere and formation of the telecommunication infrastructure of Asian countries pose a large interest from the position of determination of features of their functioning including in the context of a focus on climate-responsible behaviour.

In this paper, the authors seek the goals of determining the level of state regulation of the sphere of digitalisation and investing in telecommunication infrastructure, as well as presenting the characteristics of two elements (factors) of the formation of the studied countries' digital economies.

The purpose of the study is to determine the specific characteristics of the formation of the climatic, technological, and economic foundations of the telecommunications environment and state regulation of the digital economy in the studied countries of Asia. In order to achieve the target, we have identified the following tasks, among which are: to identify the level of state regulation of the digitalization sphere, investment in telecommunications infrastructure based on the activities of a climate-responsible business sector; to present the characteristics of the specifics of two elements (factors) of the formation of the digital economy of the countries under study.

2 Materials and Methods

A significant scientific and practical contribution to the study of the implementation of the considered tools of the Asian countries' digital economies was made by the authors of [4, 8–10, 13–17]. The above scholars consider the indicators and specific aspects of the regulatory policy and the level of telecommunications of the studied countries and identify the problems and advantages that are peculiar to the given territories. Taking into account the provisions of their works, it is necessary to emphasize the importance of further research of these problems in the context of the focus on modern specific features of the elements of the digital economy.

A range of methods used in this paper includes the following. The statistical method is utilised to select the data; the method of comparison is utilised to compare the indicators; the index method allows assessing the level of formation of indicators. Within the assessment of the character of state regulation of ICT, its influence on the main tools of the digital economy is studied. Also, analysis of telecommunication infrastructure implies the evaluation of the dynamics of investments in this sphere and the dynamics of formation of the main types of the countries' telecommunications.

3 Results

The analysis of state regulation of the ICT sphere and investments in the telecommunication sphere of the selected countries of Asia was performed. State regulation was assessed with the help of the indicator of the formation of the regulatory framework.

The indicators of the digital economies of China, Indonesia and Japan are presented in Table 1.

Analysing the data from Table 1, it is possible to see that the simplest system of legal regulation of the digital sphere (the system that is most attractive for enterprises) has been formed in China. The year 2019 saw a range of changes, which came into effect in 2021, that complicated doing business in the national digital sphere and influenced the main indicators in 2022. This is particularly true for e-commerce at the level of "seller-consumer". The choice of e-commerce as a sphere that is most dependent on legal regulation is due to the fact that the three selected countries were among the top 10 countries by online sales in the world in 2021 and the first half of 2022 (Table 1). According to Ref. [4], China is the world leader in the volume of sales in e-commerce, which became the basis of the country's economic growth in 2013–2020. The growth of the volumes of e-commerce in 2021—the first half of 2022 also led to an increase in the national GDP. According to Ref. [4], e-commerce facilitates the increase in the socio-economic level of regions and country on the whole and becomes the foundation for developing and implementing new digital technologies, which have larger value added and can raise the competitive positions of China. This is true for industrial robots and AI, which are actively implemented in certain high-precision sectors and are an important part of the export.

The relatively high position of China in terms of the organization of e-commerce (one of the top thirty countries) also implies efforts in the field of climate-responsible behaviour on the part of enterprises associated with its development. The analysis showed that the constant growth in the production of goods, and their implementation on electronic trading platforms led to the fact that by the end of 2018, online sales accounted for 18% of total retail sales [2]. At the same time, by the end of 2022,

Table 1 Indices of the formation of the digital economies of the selected Asian countries

Indicator	Value of indicators, rank								
	China			Indonesia			Japan		
	2020	2021	2022 (January – June)	2020	2021	2022 (January–June)	2020	2021	2022 (January–June)
1. Regulatory framework index, including the state of the dependent tool of the digital economy	18	15	16	51	50	49	44	48	47
1.1. Implementation of e-commerce, Index, including	19	22	25	50	48	51	16	15	16
1.1.1. Sales revenue, index	1	1	1	11	9	9	4	4	3
2. Investment in telecommunications, index, including the state of the dependant tool of the digital economy	36	37	34	61		8	52	53	32
2.1. Wireless broadband	24	23	20	42	42	47	2	2	2

Source Constructed based on [1, 7, 12]

this indicator is expected to grow by 100% and amount to about $14.5 billion per year [6]. This increase in the production of consumer goods and the corresponding development of e-commerce affect the rapid growth of solid waste. Their increase is typical for regions with difficult weather conditions, requiring special attention to the protection of goods with the help of reliable packaging. This fact affects the use of bulk packaging. Also, an additional fact is the packaging to give the product consumer appeal. The materials used for packaging generate a high level of solid waste.

In view of the need to reduce solid waste, technologies and new materials have begun to be introduced at the state level and large e-commerce entities (electronic trading platforms) that would protect goods from mechanical damage. The intro-duction of such innovations implies a reduction in the mass of packaging, while at the present time the consumer attractiveness (design) of packaging elements has begun to fade into the background [2]. Climate-responsible logistics in the field of e-commerce has been a priority in recent years. The introduction of solutions to reduce packaging has had a positive effect in terms of reducing the economic burden on the logistics sector. Namely, due to the reduction in the mass of packages of goods, the territory of warehouses and transport was increased, which made it possible to increase turnover with the existing logistics potential. Accordingly, such behaviour of climate-responsible entrepreneurship, dictated, among other things, by the government, had an impact on obtaining an additional synergistic effect.

Despite China being the leader in the volumes of e-commerce, the level of its implementation is behind other countries in terms of simplicity and quality of doing. As shown in Table 1, there was a decrease in this indicator in China. The changes were caused by the change in state regulation in this sphere. The year 2019 saw the adoption of anti-monopoly law, which regulated the quick growth of the market leaders, such as JD, Alibaba and PDD [13]. The measures on the reduction of subsi-dies for raw materials were adopted, which led to a decrease in the enterprises' sales revenue. This indicator for JD reduced by 34%, Alibaba—by 65% and PDD—by 74% [13]. The optimisation measures that were ensured at the level of legislation include the growth of minimum wages of delivery personnel; support for micro, small and medium companies in the sphere of e-commerce, connected with the focus on the reduction of regional imbalances in the sphere of population's incomes and employ-ment. We can state that support for the most vulnerable layers of workers is carried out by states using administrative methods as part of the need to counter the negative socioeconomic factors that arise in rapidly developing economies [10].

As for the problems of the legislative framework, it is necessary to note the complex customs procedures of product exports, which take two–three weeks. These processes have not been put in digital form to the level that would allow an increase in the speed of operations processing.

State regulation of the digital economy is peculiar for a range of problems, which include the digitalisation of processes and low level of automatization of customs registration operations. These problems influence the level of organisation and quality of China's e-commerce.

As for the level of investments in telecommunications in China, it is possible to note a gradual improvement of this factor of the digital economy, which led to the improvement of the state of the main type of the country's telecommunications— Wireless broadband (24th position in 2020, 23rd—in 2021, 20th—in 2022). The growth of the level of Internet quality, speed and coverage is proof of the effectiveness of investments in its development in the past three years.

According to empirical studies [16], improvement in wireless broadband led to an improvement in the level of economic digitalisation through the implementation of cloud computing and AI in large enterprises of the key export-oriented spheres. The analysis showed the best results in the activities of enterprises and high-tech companies in Eastern China, which is traditionally focused on leadership in digital-isation and innovations. Investments in wireless broadband also led to the growth of Chinese corporations' profits. An effect from the territorial implementation of wireless broadband was observed in the reduction of CO_2 emissions, achieved due to the optimisation of resources management (reduction by 4.3–4.4% in the first half of 2022 compared to the first half of 2021) [3]. Such an indicator demonstrates that further digital support for the economy positively influences the reduction of the negative climate impact of production through the implementation of energy-efficient digital technologies.

Unlike China and Japan, Indonesia demonstrates a low level of organisation of e-commerce (speed of services, payments, reaction to complaints, return of payments, etc.). The country's positions in the world ranking by the level of organisation and quality of e-commerce were different over 2020–2022. The reasons for this include the following.

- Problems of low legislative support for cybersecurity in the system of transactions for commodities. A high level of fraud in this sphere is caused by the absence of a clear legislative mechanism for protecting the cyberspace of the main trade platforms and payment systems. Such problems affect the trust of international customers [14];
- Problems of insufficiently effective regulation of unfair price competition in the sphere of e-commerce, which influences the participants' access to the market [17]. According to Ref. [8], national legislation treats such violations of the competitive environment as "burning money", connected with selling commodi-ties for a price that is lower than the market price or even the cost, to attract customers, compete with rivals and then raise prices when there are no rivals left. Legislative norms on the punishment of unfair sellers for such actions focus mainly on the sphere of foreign companies' activities; there is no regulation regarding national companies. This negatively influences the interaction between fair and reliable sellers and national consumers and reduces their capabilities.

A high level of formation of e-commerce is accompanied by a substantial level of sales, which are focused on the international market with the observation of competi-tive requirements and national market with certain violations of competitive terms. It should be noted that there are prospects for shaping approaches to climate-responsible business activities in the context of the rapid growth of the state's e-commerce.

Investments in the telecommunication sphere in Indonesia demonstrated an increase, but this did not positively affect the level of wireless broadband, which improved only at the level of certain regions with high digitalization [9]. Accordingly, the telecommunication infrastructure of Indonesia is peculiar to regional imbalance, which negatively influences the level of employment of territories that do not have effective wireless broadband coverage.

Similarly to China, Japan has rather strict legislative requirements for the functioning of large digital platforms (e.g., Rakuten) which tend to monopolise the national e-commerce market. Compared to China, these requirements are more complex; the legislation has many norms that allow for control of prevention of monopolisation or so-called price collusion on main products (services) which are sold in electronic form. However, Japan is not characterized by authoritarianism, for preventing measures are applied only to the conclusion of agreements and non-legal deals that pose a threat to the national e-commerce market. There is an example of the successful development of Rakuten, which created and manages the second-largest digital platform in the country. The company integrated into the financial services market, the smartphones market and the telecommunication technologies sector [15]. It is worth noting the successful climate-responsible initiative of Rakuten associated with the transition to energy management associated with the use of renewable energy sources (94.77%), which reduces the negative impact on the environment. As a result of this measure, pollution of the environment is ensured, since fossil fuels are not used [5].

An increase in the level of investments in telecommunications in Japan did not lead to an increase in the ranking of wireless broadband but supported its leading positions (2nd place in 2020–2022). The quality, speed and regional balance of wireless broadband coverage allowed for ensuring access to the services of exchange and transfer of information but did not affect the implementation of the digital economy opportunities that appear due to an improvement in this type of telecommunications. Japan's digital economy is not peculiar for a high level of large-scale start-ups (with investments of more than $1 billion) and does not have significant results in the sphere of cybersecurity (45th position in the world ranking in 2021–2022) and public–private partnership (41st position). These facts of insufficient manifestation of the Japanese economy's potential in the sphere of digitalisation are connected with the focus on traditionally low risks on projects, which formed at the level of various generations of the business environment [11]. The Japanese digital economy is also peculiar in expectations from the government regarding market self-regulation in the sphere of implementation of new ICT in sectors and enterprises that determine the growth of the economy.

4 Discussion

The study of transformations of state regulation and formation of the telecommunication infrastructure of the selected Asian countries allowed identifying the specific aspects of their formation, the main of which are as follows.

China

1. Large profits from e-commerce did not allow ensuring a high level of e-commerce in the context of the provision of quick service of delivery and return of payments for non-delivery of goods to foreign customers. This fact was predetermined by a range of legislative complexities of customs and payment state regulation of e-commerce in China. Thus, the quality of e-commerce organisation is at the medium level, given the problems of the regulatory and legislative character. State regulation of the digital economy focuses mainly on the resolution of internal problems and does not envisage the stimulation for the international integration of national products.
2. Investments in the telecommunications sphere led to the growth of the digital economy and the provision of a climate-responsible course for the development of industry, entrepreneurship in the field of logistics. This direction of digital support is an effective foundation for further innovations in the sphere of digitalisation of main export-oriented sectors.

The common specific characteristics of the two considered elements of the digital economy (state regulation and investments in telecommunications) include the growth of support for SMEs; reduction of the attractiveness of entrepreneurial climate for corporations; absence of legislative barriers for the development of telecommunication technologies.

Indonesia

1. The insufficiently effective legislative framework in the sphere of e-commerce is a barrier to increasing the quality of this tool; it reduces the opportunities of a competitive business climate in the country. The government's ignoring the problems of national participants of the e-commerce market may lead to the reduction of revenue in this sphere due to non-competitive actions of unfair players.
2. Focus on the regional imbalance in the provision of telecommunication infrastructure is connected with the inertia of local businesses and authorities, which affects the socio-economic sphere of the territories.

Japan has its national specifics of the formation of the digital economy and its elements. Despite the complexity and a large number of bureaucratic conditions, the system of state regulation of e-commerce does not seek the goals of authoritarian influence on the activities of corporations that account for a large share of the market but ensures a fair—from the position of Japanese economic traditions—protection of market players and consumers of products (services). The large size of the e-commerce market conforms to the significant level of its organisation (quality, service and speed of services), which is a sign of the effectiveness of the state regulation of this sphere. Given the influence of the global economic crisis, it might be necessary to adopt certain legislative changes, which would allow accelerating the activities of all market participants. This will enable to raise the level of GDP and the level of implementing digitalisation.

The formation of Japan's telecommunication infrastructure is peculiar to the traditionalist approach to the use of its tools for implementing new digital technologies at the level of the main sectors and spheres of life activities. Such an approach has remained unchanged in the course of the implementation of various ICT, which is connected with the protection from new risks and the absence of experience in dealing with new economic and technological risks.

5 Conclusion

Analysis of the two components of the selected Asian countries' digital economy allowed identifying the specific features and similarities in this sphere. The characteristic similarities include the complexity and bureaucratic load regarding the legal regulation of the e-commerce functioning (element that depends on the national legislator's approach); slowness of integration of ICT at the level of market participants. The differences in this category include the complex structure of the state regulation norms that was formed in the course of historical experience; the necessity to implement innovations in the management of competitive conditions, which are focused on the attraction of consumers, growth of market presence and competitiveness. Regarding the digital economy markets of China and Japan, we can note the introduction of a focus on climate-responsible behavior of business entities, which provides both environmental and economic benefits.

ICT markets of Asian countries have a high technological and economic potential, so there may appear new approaches to regulation, and there is a probability of acceleration of the entrepreneurial sector. This would be possible under the conditions of active interaction at the level of business and public authorities of all levels.

References

1. Austin R (2022) The top 10 e-commerce markets you should be targeting. https://localizejs.com/articles/the-top-10-e-commerce-markets-you-should-be-targeting/. Accessed 15 Nov 2022
2. Caiyi L, Xiaoyong L, Zhenyu L (2022) The nexus between e-commerce growth and solid-waste emissions in china: open the pathway of green development of e-commerce. Front Environ Sci 10. https://www.frontiersin.org/articles/10.3389/fenvs.2022.963264/full. Accessed 15 Nov 2022
3. Cambridge University Press (2022) Sustainable development report. https://dashboards.sdgindex.org/rankings. Accessed 15 Nov 2022

4. Feng Z (2022) Research on the contribution of electronic commerce to economic development: an empirical analysis based on artificial neural network. Hindawi Wireless Commun Mobile Comput 1. https://www.hindawi.com/journals/wcmc/2022/6271717/. Accessed 15 Nov 2022

5. Global.rakuten (2022) Environment. https://global.rakuten.com/corp/sustainability/enviro nment/. Accessed 15 Nov 2022

6. Globaldata (2022) China continues to lead global e-commerce market with over $2 trillion sales in 2022. https://www.globaldata.com/media/banking/china-continues-to-lead-global-e-commerce-market-with-over-2-trillion-sales-in-2022-says-globaldata/. Accessed 15 Nov 2022

7. IMD (2022) World digital competitiveness ranking. https://www.imd.org/centers/world-com petitiveness-center/rankings/world-digital-competitiveness/. Accessed 15 Nov 2022

8. Kenjiro J, Sudaryat S, Nova H (2022) Burning money" by e-commerce platform businesses and the relationship with selling loss based on business competition law In Indonesia. ULREV 6(1):46–56

9. Kharisma B (2022) Surfing alone? The Internet and social capital: evidence from Indonesia. Econ Struct 11(8). https://journalofeconomicstructures.springeropen.com/articles/10.1186/s40 008-022-00267-7#citeas. Accessed 15 Nov 2022

10. Maxyutova A, Kaliyeva S, Rakhmetova R, Meldakhanova M, Khajiyeva G (2022) Socio-economic factors in labour market regulation. Migr Lett 9(5):561–570. https://doi.org/10. 33182/ml.v19i6.2375

11. Mckinsey (2021) How Japan can make digital 'big moves' to drive growth and produc-tivity. https://www.mckinsey.com/capabilities/mckinsey-digital/our-insights/how-japan-can-make-digital-big-moves-to-drive-growth-and-productivity. Accessed 15 Nov 2022

12. Oberlo (2022) Ecommerce sales by country in 2022. https://www.oberlo.com/statistics/eco mmerce-sales-by-country. Accessed 15 Nov 2022

13. Schroders (2022) China e-commerce: is it time to look beyond regulatory pressures? https:// www.schroders.com/en/ch/asset-management/insights/markets/china-e-commerce-is-it-time-to-look-beyond-regulatory-pressures/#. Accessed 15 Nov 2022

14. Sufriadi Y (2021) Prevention efforts against e-commerce fraud based on Indonesian cyber law. In: 2021 9th international conference on cyber and IT service management (CITSM), pp 1–6. https://ieeexplore.ieee.org/document/9588900. Accessed 15 Nov 2022

15. Takigawa T (2022) What should we do about e-commerce platform giants? The Antitrust and regulatory approachesin the US, EU, China, and Japan. https://papers.ssrn.com/sol3/papers. cfm?abstract_id=4048459. Accessed 15 Nov 2022

16. Wang C, Zhang M (2022) The road to change: broadband China strategy and enterprise digi-tization. PLoS ONE 17(5). https://journals.plos.org/plosone/article?id=10.1371/journal.pone. 0269133. Accessed 15 Nov 2022

17. Wibowo AT, Prihartinah TL, Suherman AM, Sulistyandari (2022) The future of legal regulation related to predatory pricing practice in e-commerce Implementation in Indonesia. Webology 19(1):2605–2620

Features of the Development of Climate-Responsible Business in the Markets of the Digital Economy in the High-Tech Industry

Elena G. Popkova⬤, Victoria N. Ostrovskaya⬤, Elena N. Makarenko⬤, and Rustam E. Asizbaev⬤

Abstract The purpose of this research is the identification of key features of green digital business that is focused on functioning in the sphere of the high-tech industry. We analyse the main directions for the formation of the environmentally-oriented business sector in the sphere of high-tech productions, which are connected with digitalization. We consider and assess the advantages and shortcomings of these directions at the level of government and corresponding infrastructure. We also elaborate on the conditions that facilitate the effective implementation of the analysed directions for the ecologization of high-tech production. Research methods that were applied in this research include the method of sampling, comparative method and compositional method. The novel aspect of this research consists in the determination of the characteristics of the environmental formation of high-tech production at the level of modern business in the sphere of the digital economy.

Keywords Climate-responsible business · High-tech industry · Digital economy · CO_2 emissions · Electric energy

JEL Classification O32 · Q52 · Q53 · Q53 · Q55

E. G. Popkova (✉)
RUDN University, Moscow, Russia
e-mail: elenapopkova@yahoo.com

V. N. Ostrovskaya
Center of Marketing Initiatives, Stavropol, Russia

E. N. Makarenko
Rostov State Economic University, Rostov-On-Don, Russia

R. E. Asizbaev
The Academy of Public Administration Under The President of The Kyrgyz Republic Named After ZH. Abdrahmanov, Bishkek, Kyrgyzstan

1   Introduction

The business environment develops according to its strategies, which are connected with the satisfaction of consumer demand, and according to the conceptual programme parameters of national policy in a certain sphere. At present, modern consumers mostly support government programmes aimed at innovative environmental solutions in various spheres, including industry. This allows citizens of a given country receive access to the most competitive products that have climate-oriented characteristics. The export of such products stimulates international economic and technological integration. All high-tech productions use certain tools of digital support (either materials that are manufactured with the use of ICT and have environmental characteristics or products with digital elements, including high-tech products created with the use of the two above components).

The fact of growth of climate-responsible high-tech digital products in some countries and their low level in other countries is predetermined by specific features. They include the conditions of functioning of the given business (tax rates, rental discounts, business consultation, training of skilled personnel, immigration policy for labour resources, etc.), traditions in industry, society's support for ecologization at the national level, protection of investments (including international investments) and the existence of public–private partnerships in the sector of scientific and technological centres that ensure the support for national high-tech productions. Since the growth of high-tech productions that have climate-related advantages is beneficial for the government, business and infrastructure involved in R&D, it is possible to state that these parties should be interested in their support and development. However, not all countries work on the conditions for the creation of high-tech digital products with climate-responsible characteristics and high competitive advantages.

The purpose of this research is the identification of key features of green digital business that is focused on functioning in the sphere of the high-tech industry. For this, we identify and study countries that have advantages in the international market of climate-responsible business that works in the sector of the high-tech digital industry and analyse the features of the activities of the selected countries' business sector.

2   Materials and Method

The works that elaborated on the issue of formation of the conditions of green business in the high-tech industrial sector include [1, 3–7]. Since in this paper we deal with wide regional coverage, which includes achievements of the business environment of countries, which ensures the key world positions, we deem it necessary to perform a comprehensive assessment of this issue.

The tasks that were set for this paper implied the use of the following methods: the statistical method of sampling—to find the values of indicators that characterise the level of high-tech production and implementation of green technologies in the entrepreneurial sphere of the leading countries; the compositional method—to present the totality of features that were distinguished based on the study of scientific and practical data and materials of the conducted analysis. For this research, we selected countries in which innovative green business was able to ensure high technological indicators in the industrial sphere.

We analyse statistical materials for January–October 2022, and we study analytical characteristics of the business environment's functioning for the recent 3–4 years.

Further in this paper, we shall consider the ranking of digital business that works in the sector of high-tech production. To determine the leading countries, we compared the level of formation of the high-tech industry, which is focused on innovative information and communication technologies, and the level of the business environment's ecologization. Countries that achieved the highest values on both indicators shall be considered leaders in this area.

3 Results

Table 1 contains the rankings of the selected countries by the studied indicators. We selected twenty countries that are leaders in the sphere of high-tech industry and leaders in the export of high-tech products.

As is shown in Table 1, not all leaders in the sphere of digital high-tech production demonstrated achievements in the ecologization of entrepreneurial activities. Slovakia, Mexico, North Macedonia and Malaysia do not have positive results in this area. These countries are characterised by a large ranking gap between the indices that describe the formation of high-tech production with the use of ICT and the index of environmentally-oriented business.

Further in this paper, we shall dwell on the features of functioning of digital climate-responsible business in the markets of the digital economy in high-tech industry by the example of countries that are leaders in this area.

Let us consider Switzerland first. In 2022, the level of functioning of the Swiss high-tech sector is very high (2nd position in the ranking). The country is ranked 2nd by high-tech manufacturing. The level of export in this sector is medium (6.7% of total industrial exports, 29th position). Swiss companies in this sphere demonstrated a rather high position on climate-responsible management (14th position).

The main volume of high-tech production, which is oriented at climate-responsible management, accounts for the pharmaceutical sector, which manufactures products with the help of modern biotechnologies [9]. The participants of this sector include the following [9].

- Numab Therapeutics (investments equalled CHF 100 million in 2021);
- Anaveon (investments equalled CHF 110 million in 2021);
- Anjarium Biosciences (investments in equipment upgrade and innovative solutions equalled CHF 57 million in 2021);
- Alentis Therapeutics (investments equalled CHF 60 million in 2021);

Table 1 Rankings of countries in the sphere of high-tech production and green digital economy in 2022

Indicator/country	High-tech manufacturing		High-tech exports		High-tech imports		The green future index
	Value (%)	Rank	Value (%)	Rank	Value (%)	Rank	Rank
Singapore	74.7	1	29.4	5	24.9	6	29
Switzerland	67.3	2	6.7	29	5.9	109	14
Slovakia	61.5	3	8.8	21	12.3	21	46
Czech Republic	60.1	4	23.8	7	23.7	7	27
Hungary	59.8	5	14.9	10	15.8	15	24
Ireland	58.5	6	9	20	6.3	101	12
Germany	56.8	7	11.7	14	10.4	35	8
Republic of Korea	56.3	8	28.8	6	18.4	12	10
Japan	55.2	9	15	16	13.1	11	19
France	52.1	10	11.2	15	10	42	7
Mexico	50.3	11	16.9	9	19.5	9	54
Netherlands	49.8	12	13	12	13.6	18	3
Sweden	48.8	13	7.8	24	8.5	62	9
China	48.5	14	32.4	4	26.9	5	26
Denmark	48.3	15	5.8	34	6.6	99	2
North Macedonia	47	16	2.9	49	6.6	98	No data
Austria	45.8	17	7.3	25	7.9	76	23
Malaysia	45.5	18	46.9	1	29.4	4	65
Finland	44.6	19	4.6	38	7.5	80	6
USA	44.3	20	9.4	18	19.2	11	21

Source Constructed by authors with the use of [11, 12]

- Oculis (investments equalled CHF 52 million in 2021).

Due to the successful management, the above companies ensured the growth of overall liquidity (current assets/current liabilities ratio): 4.033 in 2019, 5.249 in 2020 and 7.046 in 2021. These companies have been increasing their profit and purchasing assets. Incomes from sales of products of the high-tech sector of Switzerland have been growing: CHF 4786 million in 2019, CHF 4883 million in 2020 and CHF 6658 million in 2021. These positive results were achieved thanks to clever R&D. The above companies' products are based on the use of innovative technologies that allow for the following [1, 9].

- Control over the technological process to prevent defects (anticipatory control system). Due to the elimination of production defects with the help of robotised controllers, the companies achieve the reduction of the negative burden on ecology that would have been substantial because of additional waste.

- Production of eco-friendly packaging for pharmaceutical products and equipment that does not create threats to the environment (natural materials that dissolve in water could be used as fuel during recycling).

Therefore, the key feature of ecologization of high-tech production in Switzerland is the use of technologies that prevent the creation of waste and pollution from waste processing (CO_2 emissions).

Like the pharmaceutical industry, the electronic industry is focused on modern digital technologies that are connected with the support for climate features of products. During the manufacture of electronic products, eco-friendly materials are used (they do not require special recycling). Productions in this sphere get subsidised insurance if a company installs equipment for the capture and removal of CO_2 [10]. Such an approach of insurance companies is based on public–private partnership, aimed at ecologization and growth of GDP from the activity of high-tech production.

It should be also noted that the Swiss authorities created a system of support for organisational and technical conditions of the functioning of climate-responsible businesses in the high-tech production sector [10, 13].

As for Ireland's experience in the functioning of climate-responsible business that deals with high-tech production, the following should be said.

- The agrarian sector, connected with fish production and vertical plantations (using wet peatlands, which account for 16.2% of the country's territory) accounts for 0.99% of the national GDP [7, 8]. This is not a large per cent of GDP, but companies in this sector are profitable and export-oriented. An important aspect in the development of this agrarian sphere is that the transfer of peatlands into lease allows saving their resource. The reduction of peatlands in Ireland started in the 1960s, when, together with their use for heating households, they were used for commercial purposes at mini power plants. Peatlands in Ireland are natural tools for the capture and removal of CO_2. The reduction of their areas became a threat to the environment. Also, the burning of peat deals 50% more damage to the environment compared to the burning of coal. Therefore, the government's transferring peats into the lease to agrarian companies is a measure to protect the climate from greenhouse gases. An example of digitalization in this sector is digital innovative sensors of climate regimes for growing and caring for plants and fish.
- The sector of medical equipment is represented by 450 companies of the national cluster [5]. Almost all companies of this cluster export their products, which ensures the international competitiveness of Ireland in this sphere. The companies were created in the late 1990s, with 40% of their capital provided by international owners (transnational corporations from the USA). The focus on the support for green initiatives in these Irish companies was influenced by US owners, who seek the goal of their products' conforming to the world standards in production, technological support and materials structure. Accordingly, technological lines were equipped with installations allowing controlling CO_2 emissions, which facilitated environmental protection.

Like the two above countries, Germany demonstrates positive results in the functioning of the high-tech industry with the participation of business that uses climate-oriented digital technologies.

The key sector here is the production of semiconductors, which is ranked 1st in Europe and is among the leaders in the world (digital chips). The following companies are the leaders in this sector [2].

- Intel factory in Magdeburg (American company, with investments of EUR 17 billion in 2020–2021, with expected investments of EUR 80 billion for the construction of new production facilities).
- Bosch (EUR 1 billion of investments in 2021–2022 for the construction of a new factory and the creation of more than 700 jobs.
- Globalfoundries (EUR 1.1–2.4 billion of investments in new production facilities in Dresden in 2022 and next years). Production of chips for 5G technologies is planned.
- Infineon (investments in renovation of the existing factory—EUR 1 billion).

The semiconductor industry creates a large climate footprint without the provision of optimisation measures. This applies to CO_2 emissions during production and the negative influence of used products waste from all modern digital devices (smartphones, tablets, etc.). To reduce the negative effect on climate, companies of this sector accepted responsibility in the sphere of ecologization, performing a transfer to the production of digital chips with the use of renewable energy sources. In this case, the focus on export is decisive in making decisions on the replacement of traditional fuel that is used during production by environmentally friendly fuel.

In France, the important sectors in the sphere of functioning of climate-responsible business are as follows [4].

- Aerospace production (drones, products for business aviation, small aviation and passenger planes). There are three aerospace clusters, in which activities are decentralised at the level of participating companies. The level of CO_2 emissions at the production state is high [6]. Effectiveness in this sphere is connected with investments in treatment facilities. Investing and management of decarbonisation are conducted at the level of cluster participants, which facilitates the joint resolution of environmental problems.
- Production of electric vehicles, which helps reach zero CO_2 emissions. In France and Norway, electric charging stations supply electric energy generated from renewable energy sources (nuclear power plants, hydroelectric power stations) [3].

4   Discussion

Based on the results of this research, we can distinguish the key features of the functioning of green business, which is involved in the high-tech digital sector. These features might be implemented in countries that face problems with ecologization in this sphere.

Attention should be paid to such direction as the implementation of anticipatory technologies in the sphere of control and purification processes, which ensure protection from waste and CO_2 emissions. This is stimulated by the insurance companies, government and foreign owners. This aspect should be implemented in countries that aim at the growth of high-tech production and have the potential and conditions for attracting investments in their ecologization. This requires a high level of investment protection (including foreign investments), the absence of corruption and the existence of attractive organisational and technical conditions for the administrative and production spheres. The realisation of this direction for high-tech production, which is connected with decarbonisation, implies the use of modern digital tools: robotization and artificial intelligence for control and detection of CO_2 emissions, digital sensors for mechanical removal of CO_2 emissions.

Another important aspect is the creation of organisational and technical conditions that would be attractive for the business environment and would be connected with the support for the initiatives of companies and climate-responsible high-tech clusters in the context of the provision of premises for office and production activities.

Another direction is also important: the entrepreneurial sector's focus on ecologization (including the use of renewable energy) in the context of opportunities for international integration, support for the export of products in markets that are interested in green products and climate-responsible processes of production (experience of Germany).

Countries that have legislation that supports cluster entrepreneurial activities should also focus on partnership cooperation in investing and organisational or ecological measures that raise the competitive advantages of high-tech products. Companies' participation in cluster structures allows them to achieve high results in the context of effectiveness in implementing climate-oriented measures given the acceptance of each concept of ecologization and active work according to the agreed partnership decision.

5 Conclusion

The performed research showed that the participation of the business environment in the resolution of climate problems becomes increasingly relevant. This applies also to the sectors of industry that ensure a high level of GDP and national export, namely high-tech industry. Many leading countries in the sphere of high-tech production do not achieve significant results in the context of the creation of additional environmental value (eco-friendly materials, resources, and energy resources that are used in production processes). Such results (positive or negative) largely depend on the specifics of doing business that were formed at the level of a country and on the influence of investors, owners and participants of the corresponding infrastructures (banking and insurance sectors).

The considered aspects of effectiveness that reflect on the specific features of ecologization of high-tech production can be implemented in countries that have attractive conditions for the organisation of activities (production, transport, office, financial and insurance). Further tendencies of ecologization and digitalization can lead to the emergence of new directions of effectiveness. This also applies to the introduction of new digital systems of monitoring emissions.

References

1. Gehring M, Meyer H-P (2022) Sustainability across the pharmaceutical value chain how Switzerland could take a leading role in promoting a greener approach. Conf Rep Chimia 76:800–804
2. Gtai (2022) Semiconductor industry in Germany. https://www.gtai.de/en/invest/industries/ind ustrial-production/semiconductors. Accessed 8 Dec 2022
3. Hausfather Z (2019) Factcheck: how electric vehicles help to tackle climate change. https://www.carbonbrief.org/factcheck-how-electric-vehicles-help-to-tackle-climate-change/. Accessed 8 Dec 2022
4. Jose LA Jr, Brintrup A, Salonitis K (2020) Analysing the evolution of aerospace ecosystem development. PLoS ONE 15(4). https://journals.plos.org/plosone/article?id=10.1371/journal. pone.0231985. Accessed 8 Dec 2022
5. McKernan D, McDermott O (2022) The evolution of Ireland's medical device cluster and its future direction. Sustainability 14:10166. https://www.mdpi.com/2071-1050/14/16/10166. Accessed 8 Dec 2022
6. Pierrat E, Rupcic L, Hauschild MZ, Lurent A (2021) Global environmental mapping of the aeronautics manufacturing sector. J Clean Prod 297. https://www.researchgate.net/pub lication/349746966_Global_environmental_mapping_of_the_aeronautics_manufacturing_s ector. Accessed 8 Dec 2022
7. Rowan NJ, Murray N, Qiao Y, O'Neill E, Clifford E, Barceló D, Power DM (2022) Digital transformation of peatland eco-innovations ('Paludiculture'): enabling a paradigm shift towards the real-time sustainable production of 'green-friendly' products and services. Sci Total Environ 838:3. https://www.sciencedirect.com/science/article/pii/S0048969722034258. Accessed 8 Dec 2022
8. Statista (2022) Ireland: distribution of gross domestic product (GDP) across economic sectors from 2011 to 2021. https://www.statista.com/statistics/375575/ireland-gdp-distribution-acr oss-economic-sectors/. Accessed 8 Dec 2022
9. Swissbiotech (2022) Swiss biotech industry highlights. https://www.swissbiotech.org/ind ustry/. Accessed 8 Dec 2022
10. Swissre (2022) Over USD 270 trillion in climate investments needed to meet 2050 net-zero targets. https://www.swissre.com/media/press-release/pr-20221007-USD-270-trillion-in-climate-investment-needed.html. Accessed 8 Dec 2022
11. Technologyreview (2022) The green future index 2022. https://www.technologyreview.com/2022/03/24/1048253/the-green-future-index-2022/. Accessed 8 Dec 2022
12. WIPO (2022) GII 2022 economy profiles. The following tables provide detailed profiles for 132 economies. https://www.wipo.int/global_innovation_index/en/2022/?utm_source= google&utm_medium=cpc&utm_campaign=Global+Innovation+Index+2022+%28EN% 29%3A+Search+Campaign&utm_content=search+ads&gclid=CjwKCAiAs8acBhA1EiwAg RFdw_3F9Z-tCM7aaTkY_dr5tA7XpcQIBTHDwivaiQEY4fvpH4cmePRTmRoCHRQQAv D_BwE. Accessed 8 Dec 2022
13. ZG.CH (2022) Zug: high-tech cluster. https://www.zg.ch/behoerden/volkswirtschaftsdirektion/ kontaktstelle-wirtschaft/clusters/high-tech-cluster. Accessed 8 Dec 2022

Climate-Resilient Smart Technologies—The Experience of the Largest Chemical TNCs

Natalia Yu. Konina⬤ and Elena V. Sapir⬤

Abstract The relevance of the research is determined by the growing importance of global warming and changes in the global economy on the transition to sustainable development. The chemical industry is one of the vital sectors of the economy. It is one of the three largest industrial emitters of greenhouse gases. The decarbonization of global chemical companies is important due to their key role in providing technologies, products, and services for related industries and contributing to energy security. The research analyzes the experience and prospects for applying climate-resilient innovations and technologies by the world's leading chemical companies. Chemical companies are faced with the urgent need to optimize production and improve safety, environmental friendliness, and strategic flexibility. Leading chemical multinationals change their strategies to catch up with green investments and gradually shift towards sustainable development. The transition to carbon–neutral chemical production is challenging for chemical transnational corporations (TNCs) due to the industry's dependence on petrochemical fossil raw materials. Of considerable interest is the analysis of the possibility of using smart innovations to decarbonize the production of chemical TNCs through the use of electrochemical technologies based on the low-emission electricity sector, as well as other innovative technologies. Chemical TNCs increasingly use AI in chemical research to develop new materials or chemical structures and develop new synthesis processes that can significantly reduce CO_2 emissions from the chemical sector.

Keywords Global warming · Climate change · Multinational companies · Transnational corporations · TNCs · Chemical industry · Chemical firms · Decarbonization · Smart technologies · Innovation

JEL Classifications F52 · F54 · Q2 · Q3 · Q4 · Q5

N. Yu. Konina (✉)
MGIMO University, Moscow, Russia
e-mail: nkonina777@gmail.com

E. V. Sapir
P. G. Demidov Yaroslavl State University, Yaroslavl, Russia

1 Introduction

The chemical industry is an important part of the global economy. At the present stage, the world economy is undergoing a significant transformation under geo-economic processes, digitalization of the economy, and the transition to sustainable development.

In 2021, the total global income of the chemical industry was about $4.73 trillion, the highest in the last 15 years. It is estimated that the industry directly employs about 15 million people, and another 105 million jobs indirectly depend on the chemical industry: for every person employed in chemical companies, there are another 7 jobs in other sectors of the global economy. More than 96% of all manufactured goods are made using chemicals that cannot be replaced by alternative materials. More than 80% of chemicals are sold to other companies and undergo numerous transformations in various value chains before reaching the final consumer [4].

New environmental conditions associated with technological, socio-economic, and climatic changes have led to significant changes in the operations of transnational corporations (TNCs) [10]. Demand for end products from chemical firms has been on the rise since the pandemic, but the cost of raw materials has become increasingly volatile, increasing pricing pressure. The chemical industry is characterized by accelerating consolidation, as evidenced by the mergers between major chemical transnational TNCs Bayer and Monsanto, Dow and DuPont, and Clariant and Huntsman.

Among chemical companies in developed countries, a number of major international companies and TNCs play a special role [8]. In the process of globalization, a small group of the largest chemical TNCs has been formed, which include such firms as Dow, Shell, Sabic and BASF, Sinopec, INEOS, Formoza Plastics, LG Chem, Linde, Mitsubishi Chemicals, LyondellBasell, Airliquide, Petrochina, etc. The leading chemical TNCs remain firms from developed countries, primarily from the USA and Germany [7], which are experiencing growing competitive pressure from firms in Asia and the Middle East. Chemical TNCs can be conditionally divided into three groups based on their position in the market and the strategies being implemented. The first group includes TNCs, specializing predominantly in basic chemicals and plastics, such as Exxon Mobil (USA), Shell Chemicals (UK), and Helm AG (Germany). The second category of chemical TNCs includes firms, for example, the German Evonik and the European AkzoNobel, DSM, and Solvay, specializing in special types of chemicals for certain consumers. The third group of TNCs is diversified companies, such as DowDuPont, BASF, Bayer, LyondellBasell, and Mitsubishi Chemical, producing a wide range of chemical products along the entire value chain. Companies in the chemical industry are actively engaged in innovation and digital transformation; digitalization is currently accelerating. The pandemic has shaken up customer demand, supply chain operations, employee engagement, and maintenance procedures, while demands for sustainability, personalization, and greater efficiency have risen substantially. Chemical companies are moving towards greater segmentation in how they manage their products, whether they are commodity products or innovative specialty products.

Being one of the world's most important industries, the chemical industry is the biggest industrial user of energy and one of the top three industrial contributors to greenhouse gas (GHG) emissions, along with steel and cement. For example, Germany's largest chemical multinational, BASF, is expected to have CO_2 emissions of between 19.6 million metric tons and 20.6 million metric tons in 2022 (20.2 million metric tons in 2021). The chemical industry can help decarbonize other industries; it supplies the materials used to make green hydrogen, solar panels, wind turbines, and many of the products on which the post-industrial society depends [3]. Thus, chemicals are a key factor in finding sustainable solutions that benefit society. As the global demand for chemicals continues to grow, the adoption of climate-resilient technologies and changes in emission-intensive chemical production technologies is of particular importance CO_2 [5].

2 Methodology

Raw material stewardship, climate change strategies, and the transition from conventional to renewable energy are the top three challenges of our time, all of which are being addressed by current efforts to achieve the sustainable use of carbon dioxide (CO_2) as a chemical feedstock. However, the ways in which CO_2 is used and utilized in industrial value chains are complex and very diverse. CO_2 can be used as a raw material for a wide range of products, such as plastics, transport fuels, and building materials. Identifying climate-resilient technologies is no easy task. Part of the challenge is identifying the appropriate CO_2 source for a particular application, as well as identifying appropriate carbon capture and pre-treatment technologies.

Effective adaptation of chemical firms to the challenges of climate change requires appropriate technological and institutional innovations. An effective decarbonization strategy for firms must adequately consider the economic impacts of climate change, as well as existing regulatory frameworks and consumer and investor behavior. The development and application of climate-resistant technologies by chemical firms is based on information about the formed public demand for the transition to a sustainable development trajectory and the necessary R&D based on large specifics of chemical technological processes. An important aspect is the influence of the external environment and the experience of other competing firms in creating climate-sustainable technologies that make it possible to deal with an increased carbon footprint.

The research is based on system analysis and in-depth research of activities of the biggest chemical TNCs connected with decarbonization. The case method is a convincing scientific methodology to check and confirm the validity of the scientific hypothesis.

The research addresses three specific research questions:

1. What circumstances influence the decarbonization of the largest chemical TNCs?
2. What factors influence the development of climate-resilient technologies by chemical TNCs?
3. Which climate-resilient technologies are most successfully applied by chemical TNCs?

3   Research Results

Sustainability and ESG issues are becoming major development areas for leading chemical companies, which are considered in the development goals and strategies being developed [19].

Climate-resilient smart technologies are primarily green innovation and the Internet of Things (IoT) technologies to develop green, durable, biodegradable, and green products for a sustainable future [1]. Climate-resilient smart technology encompasses all innovations that contribute to developing important products, services, or processes that reduce harm, impact, and environmental degradation while increasing the use of natural resources. Climate-resilient smart innovations are designed to combat climate change by minimizing carbon dioxide (CO_2) and other greenhouse gas emissions [20].

Several factors influence the largest chemical TNCs to make their operations and products more sustainable and achieve a low carbon footprint [13]. The EU's European Green Deal was unveiled in December 2019 with a set of proposals to reduce net greenhouse gas emissions by at least 55% by 2030, compared to 1990 levels, on the way to achieving climate neutrality by 2050. Decarbonization of chemical firms requires two key conditions for success: customer commitment, as goods will be more expensive than they are today, and government policy. In May 2022, Germany announced targets to cut CO_2 emissions by 65% by 2030, by 88% by 2040, and to near-zero emissions by 2045—a much more ambitious timeline than the EU's. The European chemical sector should cut greenhouse gas (GHG) emissions by 186 million tons by 2050. To achieve the 2050 Green Deal goals, chemical companies need to invest around one trillion euros from 2021 to 2050. The road of chemical firms to net zero is driven not only by stricter regulations, especially in Europe and Asia, and by investors increasingly looking at companies from an environmental, social, and governance perspective but also by anticipation of better business opportunities as customers in different markets increasingly demand products with a lower carbon footprint. Chemical TNCs with the most advanced ESG projects are looking to strengthen their competitive edge in a changing regulatory and investment environment that is becoming increasingly stringent for the most polluting companies. Sustainability issues are increasingly integrated into strategic business and financial planning, creating new challenges for investment policy [14]. For example, full decarbonization of chemical MNCs requires much more electricity from renewable sources at competitive prices than total wind energy currently generated in Germany.

Annual presentations to investors by major chemical firms such as BASF, Dow, and Air Liquide already show big demand for green investments, especially for renewable energy projects.

The peculiarity of the operations of chemical firms is that chemical products are a means of decarbonizing for a number of downstream processes that are difficult to reduce, and that the chemical industry itself is a sector that is difficult to reduce [12]. Chemicals play such a key role in society and industrial infrastructure today and in the future that companies need to transform.

In general, all chemical TNCs have more in common than differences in the approach of these companies to ESG. Many are focusing on climate targets because high energy consumption and carbon emissions are issues across the industry, especially for chemical products. Recurring pathways to achieve carbon reduction targets include introducing new low-carbon products and processes, increased use of renewable energy, and selective growth projects to renew product lines for promising applications (such as electric vehicle battery chemicals and recycled plastic for round packaging) [11]. Chemical companies are also increasingly promoting alliances with customers and suppliers to lower R&D and capital expenditures while better meeting customer requirements [12].

There are four climate-resilient technological paths for decarbonizing chemical TNCs, ranging from the more mature direct heat replacement of electricity and the use of hydrogen to technologically less mature but potentially more selective electrochemistry and plasma-based approaches [5]. There is significant technological complexity in reducing emissions associated with fuel combustion (i.e., consumed natural gas). Electrification is not always the solution for chemical companies that need heat above 1000 °C in their processes.

The most promising technologies for decarbonizing the activities of leading chemical firms are associated with the transition to electrochemical processes based on the low-emission electricity sector in the short and long term. The integration of electrochemical processes with the energy sector is of great importance to use the flexibility of the process to reduce energy costs in chemical production and provide valuable services to support the energy system.

When powered by renewable electricity, electrochemical processes have a smaller carbon footprint compared to conventional thermochemical methods. A promising direction for chemical TNCs to minimize their carbon footprint is electrolysis production using renewable energy sources.

Chemical TNCs can make a significant difference along the entire energy value chain by offering the technologies needed to convert renewable electricity, biomass, and waste into green hydrogen, green ammonia, e-methanol, e-fuels, and biofuels that will enable a sustainable future. In particular, BASF, Dow, Air Liquide, and other chemical and engineering companies help customers worldwide upgrade their refineries to produce bio-based diesel and SAF, and build low-carbon hydrogen plants.

Increasing use of the most advanced digital tools and knowledge management are important factors of the competitive advantage for chemical companies [16]. For example, automation speeds up R&D for new products from about two to three years to about four to six months so that factories can meet new needs much faster [9]. AI is applied in chemical research to develop new materials or chemical structures and develop new synthesis processes that increase sustainability [6]. The multivariate analysis allows scientists to more accurately determine the effect of individual ingredients in a mixture, improving the product's quality. Chemical firms implement digital twins to access a single view of the entire supply chain, from raw materials to production and market forecasts. Chemical companies also reinvent their business models and reshape supply chains, integrating their new suppliers into a digitally integrated ecosystem [17].

German chemical multinational BASF is a global leader in climate-resilient technology, with plans to scale up new decarbonization technologies after technology matures in the second half of the 2020s. BASF is taking steps to reduce the emissions associated with the production of its products. Despite current European energy problems, BASF confirms that it aims to cut greenhouse gas emissions by 25% by 2030 compared to 2018 to 16.4 million metric tons by 2030 and maintains its goal of achieving zero emissions worldwide by 2050 [2]. BASF releases an annual CO_2 emissions forecast for the BASF Group as part of its forecast with a range of plus or minus 0.5 million metric tons. In 2021, BASF reduced its CO_2 emissions by around 3% compared to 2020 despite significantly higher production volumes. To a large extent, this was due to the increase in the use of renewable energy sources. BASF is ready to offer customers the first products with net zero and lower carbon footprint [2].

There are several compelling examples of how chemical TNCs use climate-resilient smart technologies to optimize their operations. Carbon transparency is the basis for achieving zero greenhouse gas emissions. Thus, the largest digital European company Atos, in partnership with the German chemical TNC BASF, launch a digital solution for companies in the chemical and process industries to identify, monitor, and ultimately reduce the carbon footprint of chemical products. This recently launched partnership ecosystem will help decision-making for sustainable supplier selection based on the chemical carbon footprint data across the value chain [18].

BASF has high hopes for two climate-resilient innovative technologies: the electric cracker and methane pyrolysis. In addition to electrifying crackers, BASF is working on CO_2-free hydrogen production by pyrolysis of methane using a new process technology at its methane pyrolysis test facility in Ludwigshafen, Germany. In this process, methane is heated in a reactor to produce hydrogen with solid carbon pellets as a by-product. If heating is done with renewable electricity, it will be a CO_2-free process. According to the company, hydrogen from methane pyrolysis could be a key building block for CO_2-reduced ammonia and methanol production.

The most successfully climate-resilient technologies applied by chemical TNCs are based on renewable energy capacity. For German TNC BASF, the transition to renewable energy sources will be the main factor in reducing emissions until 2025. In 2021, renewable energy accounted for 16% of BASF Group's global electricity demand. By 2030, BASF predicts that 100% of its global electricity demand of the level 2021 will come from renewable sources.

When it comes to the use of renewable energy sources, BASF follows a "make and buy" strategy. This strategy includes investing in own renewable energy assets and purchasing green energy from third parties. To increase the share of renewables in energy supplies, BASF concluded several strategic partnerships. In 2021, BASF acquired a stake in the Vattenfall Hollandse Kust Zuid (HKZ) wind farm. Once fully operational, it will be the world's largest offshore wind farm, with a total installed capacity of 1.5 gigawatts. The project is expected to be fully operational in 2023. In May 2022, BASF and utility company RWE unveiled a plan to build a 2 gigawatt (GW) offshore wind farm off the northwest coast of Germany to provide BASF's Ludwigshafen plant with green power for CO_2-free manufacturing processes such as electric cracking, starting from 2030. According to BASF, this will reduce CO_2 emissions in Ludwigshafen by 2.8 million tons per year [2]. Additionally, BASF signed long-term purchase agreements for renewable energy with suppliers Ørsted and Engie.

About 50% of the steam demand at the Ludwigshafen site comes from steam generation processes that result in CO_2 emissions. The new approach is to produce steam using electricity. BASF works with Siemens Energy to improve energy efficiency by using electric heat pumps to produce CO_2-free steam from waste heat. This is the first BASF project for an acetylene plant that uses heat pumps and steam recompression to upgrade waste heat so it can be used as steam for the plant's steam network. The integration of this heat pump project will produce about 60 metric tons of steam per hour, also avoid CO_2 emissions by approximately 160,000 metric tons per year, and reduce annual cooling water consumption by more than 20 million cubic meters. An important climate-resilient project at BASF's main production site in Ludwigshafen is the development of an electrically heated steam cracker. Currently, cracker ovens are gas-fired and produce about 1 metric ton of CO_2 per metric ton of olefin. For CO_2-free hydrogen production, BASF is developing new processes, such as methane pyrolysis. In March 2022, BASF signed an agreement with SABIC and Linde to develop and test the world's first electrically heated steam cracker that will use electricity instead of natural gas to heat the ovens. If electricity comes from renewable sources such as wind and solar power, the new technology could cut carbon dioxide (CO_2) emissions by as much as 90%.

BASF plant Verbund in Antwerp is the largest chemical plant in Belgium and the second largest plant of BASF after Ludwigshafen. BASF aims to reduce site emissions from 3.8 million metric tons in 2021 to near zero by 2030. This could be made possible by importing green energy from offshore wind parks, coupled with the introduction of new low-emission technologies and the planned large-scale CCS project in the port of Antwerp. Zhanjiang in China is set to become the third largest factory of BASF with CO_2-neutral production. The advanced Verbund concept and the use of renewable energy sources will play a key role in significantly reducing CO_2 emissions compared to a gas-fired petrochemical plant.

In August 2022, BASF Schwarzheide GmbH and envia Mitteldeutsche Energie AG commissioned the first joint venture solar farm with an expected electricity generation of 25 gigawatt-hours per year, about 10% of the BASF production site in Lusatia's current annual electricity demand. In a very short time of six months, a huge solar farm with around 52,000 photovoltaic modules and a transformer station was built to supply green electricity to BASF's one of the first CO_2-neutral production sites.

By using green energy, low carbon steam, bio-based raw materials, and highly efficient processes, BASF has developed its own digital zero-emission and low-carbon products solution for approximately 45,000 products sold. The company expects demand for these products to outstrip supply over the medium term, with their market value more than offsetting higher production costs.

Between 2021 and 2025, BASF plans to invest about 1 billion euros in developing climate-resilient, low-emission smart technologies and scaling them up in pilot plants. This amount is included in BASF's current investment budget, with capital expenditure rising to around 2–3 billion euros over five years from 2026 to 2030.

Following the example of German TNC BASF, a number of multinational chemical companies have launched transformational sustainability initiatives. For example, the US TNC Dow has pledged to implement the principles of a circular economy in its business model, design 100% of its products and processes using sustainability criteria, including the principles of green chemistry, and reduce greenhouse gas emissions by 30% by 2030, including getting 60% of electricity from renewable energy sources.

Another German chemical TNC Bayer AG announced a strategic partnership with Microsoft to create a new cloud-based set of digital tools and data solutions for use in agriculture and related industries, offering new infrastructure and core capabilities to accelerate innovation, increase efficiency, and support sustainability across all value chains. Another climate-resilient technology is the Climate platform Bayer's Field View, which is now used on more than 180 million acres of farmland in more than 20 countries. This is in line with Bayer's ambitious goal of 100% digital sales support across Crop. Science by 2030 and accelerating its ability to deliver results-driven digital solutions to customers. Bayer is committed to setting a new industry standard for data-driven digital innovation.

4 Conclusions

To decarbonize, chemical TNCs have moved to implement climate-resilient innovations in all segments of the value chain. Chemical companies are taking significant steps to reduce their own greenhouse gas emissions and transitioning to a sustainable circular enterprise [15]. Another important area of climate-smart innovation is the reduction of Scope 2 emissions through the use of more sustainable energy sources. Of paramount importance are the efforts of chemical firms to reduce Scope 3 emissions by using more sustainable raw materials from the value chain.

This requires an innovative ESG-based approach that aligns with the company's true identity. This approach implies dedicated funding for disruptive technologies and wisely uses public funding to reduce R&D costs.

Climate-resilient innovation is essential to achieving chemical companies' plans to be more resilient and competitive and maintain sound balance sheets to guard against cyclical slowdowns. Considering a lot in common in goals and strategies, the extent to which the planned strategies will be implemented is determined by the level of leadership and operational efficiency of company management. Therefore, tying the remuneration of top managers to the success of climate-resilient innovations related to ESG performance is of great importance.

References

1. AL-Khatib AW, Shuhaiber A (2022) Green intellectual capital and green supply chain performance: does big data analytics capabilities matter? Sustainability 14(16):10054. https://doi.org/10.3390/su141610054
2. BASF (2021) BASF presents roadmap to climate neutrality. Retrieved from https://www.basf.com/global/en/media/news-releases/2021/03/p-21-166.html. Accessed 4 Dec 2022
3. Boulamanti A, Moya JA (2017) Energy efficiency and GHG emissions: prospective scenarios for the chemical and petrochemical industry. In: JRC science for policy report (JRC105767/EUR 28471 EN). Publications Office of the European Union, Luxembourg. https://doi.org/10.2760/630308
4. Geels FW (2022) Conflicts between economic and low-carbon reorientation processes: insights from a contextual analysis of evolving company strategies in the United Kingdom petrochemical industry (1970–2021). Energy Res Soc Sci 91:102729. https://doi.org/10.1016/j.erss.2022.102729
5. Griffin PW, Hammond GP, Norman JB (2018) Industrial energy use and carbon emissions reduction in the chemicals sector: a UK perspective. Appl Energy 227:587–602. https://doi.org/10.1016/j.apenergy.2017.08.010
6. Hermann E, Hermann G, Tremblay J-C (2021) Ethical artificial intelligence in chemical research and development: a dual advantage for sustainability. Sci Eng Ethics 27(4):45. https://doi.org/10.1007/s11948-021-00325-6
7. Konina N (2018) Evolution of the largest multinational German companies. Sovremennaya Evropa 2(81):49–59. https://doi.org/10.15211/soveurope220184959
8. Konina NY (2022) Contemporary European multinationals: changes in the balance of market power. Sovremennaya Evropa 2022(5):78–91
9. López FJD, Montalvo C (2015) A comprehensive review of the evolving and cumulative nature of eco-innovation in the chemical industry. J Clean Prod 102:30–43. https://doi.org/10.1016/j.jclepro.2015.04.007
10. Popkova EG, Zavyalova E (2021) Conclusion: future challenges related to socio-economic development and the institutional response to them. In: Popkova EG, Zavyalova E (eds) New institutions for socio-economic development: the change of paradigm from rationality and stability to responsibility and dynamism. De Gruyter, Berlin, Germany, pp 183–184. https://doi.org/10.1515/9783110699869-019
11. Prado V, Glaspie R, Waymire R, Laurin L (2020) Energy apportionment approach to incentivize environmental improvement investments in the chemical industry. J Clean Prod 257:120550. https://doi.org/10.1016/j.jclepro.2020.120550
12. Riedel NH, Špaček M (2022) Challenges of renewable energy sourcing in the process industries: the example of the German chemical industry. Sustainability 14(20):13520. https://doi.org/10.3390/su142013520
13. Rome A (2019) DuPont and the limits of corporate environmentalism. Bus History Rev 93(1):75–99. https://doi.org/10.1017/S0007680519000345

14. Sapir EV, Karachev IA (2020) Challenges of a new investment policy: investment promotion and protection. Financ: Theor Prac 24(3), 118–131. https://doi.org/10.26794/2587-5671-2020-24-3-118-131
15. Stephanopoulos G, Bakshi B, Basile G (2022) Reinventing the chemicals/materials company: transitioning to a sustainable circular enterprise. Comput Aided Chem Eng 49:67–72
16. Syahchari DH, Saroso H, Sudrajat DL, Herlina MG (2020) The effect of information technology, strategic leadership and knowledge management on the competitive advantage in the chemical industry. In: Proceedings of the ICIMTech 2020: international conference on information management and technology. IEEE, Bandung, Indonesia, pp 120–125. https://doi.org/10.1109/ICIMTech50083.2020.9211198
17. Winterhalter S, Weiblen T, Wecht CH, Gassmann O (2017) Business model innovation processes in large corporations: insights from BASF. J Bus Strateg 38(2):62–75. https://doi.org/10.1108/JBS-10-2016-0116
18. Wu C, Lin Y, Barnes D (2021) An integrated decision-making approach for sustainable supplier selection in the chemical industry. Expert Syst Appl 184:115553. https://doi.org/10.1016/j.eswa.2021.115553
19. Zavyalova EB, Popkova EG (eds) (2021) Industry 4.0: exploring the consequences of climate change. Palgrave Macmillan, Cham, Switzerland. https://doi.org/10.1007/978-3-030-75405-1
20. Zavyalova EB, Shumskaia EI, Kuzmin MD (2020) Smart contracts in the Russian transaction regulation. In: Popkova E, Sergi B (eds) Scientific and technical revolution: yesterday, today and tomorrow. Springer, Cham, Switzerland, pp 205–212. https://doi.org/10.1007/978-3-030-47945-9_22

The Role of Academic Leadership in Developing Green Innovation in Industry 4.0

Evgeniy S. Popov ⓘ

Abstract The paper aims to disclose the role of academic leadership in the context of the development of green innovation during the formation of Industry 4.0 based on the consumer-value approach, which is based on the principle of saving human labor and natural resources. The methodological basis of the research is represented by the universal consumer-somatic approach, which covers a range of social, educational, academic, and environmental fields. The author also uses the ecological-crisis approach, which makes it possible to reveal the source of contemporary problems in the field of the natural environment. The author reveals that academic leadership is an important point in the development and use of green innovation of Industry 4.0 through the development of the subjective factor of material production. The author examines the important points of higher education academia. The emphasis is placed on research and development, digitalization, and trans-professional identity. The role of objective and subjective factors in overcoming the ecological crisis was revealed from the position of use value. The author shows the role of academic leadership, which in the consumer-value approach has an impact on improving the efficiency of the subjective factor in the resource-using activities of a person by saving living labor and natural resources.

Keywords Law of use value · Academic leadership · Environmental crisis · Industry 4.0 · Digitalization · Trans-professional identity

JEL Classification A19 · J01 · Q50

E. S. Popov (✉)
Platov South Russian State Polytechnic University (NPI), Novocherkassk, Russia
e-mail: povove@mail.ru

1 Introduction

Under the current conditions of technological transition from Industry 3.0 to Industry 4.0, which, in fact, will end with the establishment of a quadruple industrial revolution, i.e., the replacement of the domination of automated production by the domination of cyber-physical systems; the current global problems (e.g., the further vectors of the higher education system, environmental problems, and the place of people in the areas of productive and non-productive labor) will ultimately solve themselves [1]. Despite the positive statements of the main apologist of Industry 4.0, economist K. Schwab, the scientific literature of the last decade focuses on conceptual search and evaluation of trends in the development of higher education, directions of design and development in solving the environmental crisis, and their overall interaction with the material production.

The point is that science and education are now universal social productive forces that do not directly provide human beings with material goods. However, being in the field of social consciousness, science and education have established themselves as factors of innovative development of production, but only through the embodiment in students, workers, and employees and in the objectification of the means of production, plants, and factories. In this regard, it is necessary to comprehend the new educational and scientific foundations in the conditions of the formation of Industry 4.0, which, in today's world, has acquired a focus on achieving academic leadership as a qualitatively different state of higher education at the regional and state levels.

We should also remember that Industry 4.0 is currently the main trend in strengthening the trend of turning material production into the technological application of science and education. Thus, this process affects the organization and effectiveness of the design and development of university teams, which raises an urgent issue in the development of green innovation in the context of the environmental crisis. Consequently, the rationale for academic leadership in the development of green innovation is important because environmental issues raise increasingly more questions for academia.

2 Materials and Method

The study of contemporary environmental problems is the most important research work, which indirectly but deeply penetrates social beings. Nevertheless, the Russian scientific literature on this issue has developed a theory of ecological crisis [2]. The most important feature of the theory is to reveal the causes of negative changes in the composition of the atmosphere, hydrosphere, biosphere, and structural ecosystem transformations of flora and fauna depending on certain resource-use activities of people. According to the ecological-crisis approach, when producing real-life activities, people seek to increase the consumption of commodities by reducing labor costs on restoring the ecological-resource equilibrium. Thus, increased consumption is achieved through resource depletion and environmental degradation.

In this context, we consider the cost-effectiveness of resource use, where the increase in consumption of a certain part of society and labor costs associated with the restoration of resources and environmental conservation are related to profits. Therefore, the rational interaction of production and consumption with overcoming the regional ecological crisis is possible only on the opposite of the law of value to the economic basis on which other forms of social life are raised.

As a law of labor economy, the law of use value establishes a condition for the efficiency of resource use, the economy of human labor, and material objects of nature with the provision of full welfare with preservation of the environment [3]. Accordingly, on this plane, the achievements of scientific and technological green innovations in Industry 4.0 as progressive techniques and technologies that save working time, public labor, and the natural environment are decisive for nature and humankind [4, 5].

Considering the theoretical and methodological toolkit of academic leadership, the author notes that the emergence of academism of higher education on a global scale is a factor in the development of digitalization and trans-professional identity [6–8]. In this way, academic leadership participates in the development of green innovation by focusing on increasing the effectiveness of graduates of higher education institutions and research and innovation activities. Thus, the further goal of this research is to reveal the specific role of academic leadership in the context of the consumer-value approach in developing green innovation in Industry 4.0.

3 Results

According to the materialist understanding of history, the existence of society is determined by the production as a process of appropriation of natural objects by individuals within the system of social relations. This means that the development of a person and society occurs in the context of purposeful resource-use activities, in the course of which the natural world is transformed into things useful for consumption. However, with further social development and a successive change in production modes, the forms of exploitation of natural resources have also changed, which has now led to threats of a global ecological crisis.

According to the UN Conference on Environment and Development in Rio de Janeiro (1992), the essence of the discussed threat is a mode of socio-economic development based on a system of profit production [9]. It is important to note that the agenda of the Earth Summit conference subsequently became the basis for the sustainable development concept. The sustainable development concept is a set of international rules and regulations aimed at minimizing a human impact on the environment and achieving sustainability (the balance between nature and people) by reducing pressure on the planet's ecosystem. Most countries have ratified this concept.

However, the concept that constitutes one of the contemporary theoretical directions of analysis and overcoming the environmental crisis, in fact, has not emerged from the value paradigm. In fact, in achieving Sustainable Development Goals, it proceeds from the value equivalence (equilibrium), from which in subsequent economic forms unfolds profit [3]. Therefore, in the scientific discussion, it is necessary to recognize the incompatibility of the value paradigm in overcoming the ecological crisis and the need to replace it with another paradigm based on the labor theory of consumer value.

In contrast to the value plane, the main source of society's innovative development in the consumer-value approach is the excess of labor productivity over labor costs. Thus, implementing this approach in resource management makes it possible to progressively overcome the ecological crisis through labor activity, according to which labor costs under production conditions should be less than the saved labor under consumption conditions. In other words, in practice, human resources should be used to outstrip the growth of production results in relation to the labor costs of resource extraction and the restoration of the ecological-resource equilibrium. Simultaneously, the increase in the contribution to the satisfaction of needs must be at the lowest cost of all kinds of the planet's resources.

Focusing on the consumer-value approach in overcoming the negative trend of the ecological crisis, it is necessary to emphasize that the achievement of the goal is possible with the use of advanced technology and equipment. Thus, when using the resources, two necessary moments are overcoming the ecological crisis and the development of green innovation in Industry 4.0. These principles must be defined in a unified approach. For this reason, it must be clearly understood that although there are approaches to a socially beneficial exit from the ecological crisis associated with an increase in labor costs to restore the ecological-resource equilibrium and reduce the mass of use values, the most effective way is the transition to green innovation in the field of resource extraction. The consumer-value character of green innovation makes it possible to achieve the socio-economic and environmental goal through the economy of social labor—an increase in substitutable living labor and a decrease in spent past labor, which includes the consideration of the costs of manufacturing, maintenance, and disposal (transformation resource-waste-resource) of green means of production, as well as the restoration of ecological-resource equilibrium [4].

Simultaneously, the measurement of green innovation in Industry 4.0 in the plane of use value as the proportion of released and expended labor determines the area of effective resource extraction through productivity growth and directs the design and development work and the educational sphere of society. Thus, green innovations as material factors of production are the result of the technological development of the design team. Design and research work should be focused on the effect of future savings of labor and resources, which makes it possible to achieve the desired result in the conservation of the natural environment through their implementation in material production.

In turn, the growth of the effectiveness of green innovation in Industry 4.0 as a major factor in overcoming the ecological crisis is possible primarily through comprehensive human development as an increase in the use value of its labor force, realizing the growth of resource efficiency by saving working time and resources, as well as a qualitative leap in its subjective aspect of material production (increasing the level of education, retraining, and development of spiritual, creative, physical, and psychological traits) [10].

In connection with the latter, a strategy of academic leadership as a special state of the educational sphere can be identified as a basis for comprehensive human development, which includes digitalization and trans-professional identity. Thus, if we consider the complex problem of the impact of academic leadership on the comprehensive development of the individual, we can conclude that the tendency of trans-professional identity in higher education academia becomes a condition for increasing the use value of labor.

Consequently, if the trans-professional identity of graduates of higher educational institutions is the formed social and professional preferences at the individual and collective levels, including lifelong education, and professionalism as a process of constant retraining in the context of the ability to integrate the acquired competencies in the working environment [11], then it that academic leadership becomes a prerequisite in the positive assertion of academicism of higher education actors as holders of the necessary cumulative knowledge, abilities, and skills that contribute to the achievement of the goal in the environmental and socio-economic areas of society.

In this context, the use value of the labor force of actors whose future labor replaces more labor of simple qualifications with the labor of higher qualifications in the workplace is increased. In this way, the usefulness of the objects of consumption created by graduates trained within the framework of academicism of higher education, acting in the future as workers involved in productive labor in economic essence, increases even while saving time and natural resources.

We should not forget that the formation of Industry 4.0 further increases the complexity of manufacturing techniques and technologies, increasing the circulation of information in society. In turn, this determines the need for constant renewal and acquisition of new knowledge by actors at the academic level. In this regard, within the framework of the formation of academic leadership, there is an ongoing process of digitalization of academia as the formation of a mixed model of education, making it more accessible in the sense that almost all educational programs will be available in video format, online format, and distance education. The latter forms an extensive practice of the transfer of academic knowledge. That is, on the one hand, actors acquire specialized fundamental knowledge about the laws of nature and society. On the other hand, actors get interdisciplinary knowledge of various fields of scientific knowledge, contributing to the expansion of labor practices. Ultimately, actors save time in updating and gaining additional knowledge about the world by instantly finding scientific literature and video and audio lectures, as well as by interacting with various experts in various fields of scientific knowledge.

We can conclude that moments of digitalization and trans-professional identity within academic leadership have a qualitative impact on education. These moments are primarily reflected in the academic actors of higher education that will act as specialists in certain labor activities who can increase the consumer-value efficiency of material production.

4 Conclusion

This research revealed the specific role of academic leadership in developing green innovation in Industry 4.0. The author showed that in the context of the consumer-value approach, the objective factors of production of green innovations should lead to savings of natural resources and human labor. Through the inclusion in academism of higher education, the subjective factor increases the use value of labor, which leads to an economy of public labor in general. It is also shown that the consumer-value approach is a fundamental paradigm in overcoming the ecological crisis.

References

1. Schwab K (2016) The fourth industrial revolution (trans from English). Eksmo, Moscow, Russia (original work published 2016)
2. Friedman VS (2017) The global ecological crisis: from the lecture course "The nature conservancy: biological foundations, simulation models, and social applications." URSS, Moscow, Russia
3. Elmeev VY (2007) Social economy of labor: general foundations of political economy. Publishing house of St. Petersburg University, St. Petersburg, Russia
4. Dolgov VG (1998) Management of scientific and technological progress: Consumer and value-added basis. Publishing house of Leningrad State University, St. Petersburg, Russia
5. Dolgov VG, Elmeev VYa, Popov MV (1991) Choosing a new course. Mysl, Moscow, Russia
6. Iordache-Platis M (2021) The academic leadership approaches of the professional and personal student development. In: Proceedings of the ECMLG 2021: 17th European conference on management, leadership and governance, Valletta, Malta, pp 202–211
7. Siddiqui AK (2016) Academic leadership through e-education in India. Int J Sci Res 5(5):801–804. Retrieved from https://www.ijsr.net/archive/v5i5/NOV163487.pdf. Accessed 12 Nov 2022
8. Vodenko KV (2021) Regionalization of state policy in higher education sphere in the context of academic leadership formation. Manag Issues 1:156–168. https://doi.org/10.22394/2304-3369-2021-1-156-168
9. UN General Assembly (1992) Rio declaration on environment and development. In: The United Nations conference on environment and development, 3–14 June 1992, Rio de Janeiro, Brazil. Retrieved from https://www.un.org/en/development/desa/population/migration/generalassem bly/docs/globalcompact/A_CONF.151_26_Vol.I_Declaration.pdf. Accessed 12 Nov 2022
10. Udovichenko AS (2022) Time for development. Scientific monograph. Individual entrepreneur Anglinova L. N., St. Petersburg, Russia
11. Vodenko KV (2022) Academic leadership and higher education in the context of the development of trans-professional identity in the conditions of digitalization and regionalization. Social'nye i Gumanitarnye Znania [Soc Hum Sci] 8(3):300–309

Managing the Development of Climate-Responsible Entrepreneurship in the Digital Economy Markets in Fintech

Natalia G. Vovchenko⬭, Konstantin A. Zenin⬭, Sergey P. Spiridonov⬭, and Victor P. Kuznetsov⬭

Abstract *Purpose* Finding the conceptual framework of climate-responsible entrepreneurship in the fintech market, describing the approaches and directions of its implementation, characterising these approaches, determining the real state of their realisation in the activity of fintech companies and elaborating on obstacles that hinder the potential of their use. *Design/Methodology/Approach* The methodological framework of this research is interdisciplinary, which is implemented through the interaction of the IT sphere, the concept of sustainable development and specifics of the activity in the financial sphere. Combining all components, we emphasize the possibility of the formation of a synergetic effect, which allows solving the current climate problems more rationally. To better describe the potential of such interaction, we use the system approach. It allows determining the directions for the fintech companies' influence on the Sustainable Development Goals and the fight against climate change. We also consider the development of general concepts of social responsibility and ESG investing and comprehensive specialised concepts of Green FinTech and Climate FinTech, which offer effective tools to achieve the goals of climate sustainability within the studied segment of the digital economy. *Findings* We consider and generalise approaches to the complex implementation of climate-responsible activity in the sphere of digital entrepreneurship in the sector of fintech solutions. We also identify the key concepts that reflect on the current processes most completely and have the largest potential to oppose climate change. We dwell on the list of key tools of the digital economy that ensure the fintech technologies' effect on

N. G. Vovchenko (✉)
Rostov State Economic University, Rostov-on-Don, Russia
e-mail: nat.vovchenko@gmail.com

K. A. Zenin
Essentuki Institute of Management, Business and Law, Essentuki, Russia
e-mail: eiubp@eiubp.ru

S. P. Spiridonov
Tambov State Technical University, Tambov, Russia
e-mail: spiridonov_sp@bk.ru

V. P. Kuznetsov
Minin Nizhny Novgorod State Pedagogical University, Nizhny Novgorod, Russia

© The Author(s), under exclusive license to Springer Nature Switzerland AG 2023
E. G. Popkova (ed.), *Smart Green Innovations in Industry 4.0 for Climate Change Risk Management*, Environmental Footprints and Eco-design of Products and Processes, https://doi.org/10.1007/978-3-031-28457-1_23

climate processes and characterise the main directions of fintech solutions' influence on climate change. We describe the directions for involving the largest Green FinTech and Climate FinTech companies with the resolution of the problem of climate change and characterise the problems of comprehensive use of the fintech potential in the achievement of sustainable climate goals. *Originality/Value* This research allows for the systemic consideration of the change of conceptual approaches to the support for climate responsibility of digital entrepreneurship in the fintech market, substantiating the directions and tools that facilitate the increase in fintech companies' impact on the state of climate change and generalising the characteristics of fintech companies' impact on the provision of climate sustainability through the formation and use of diversified solutions and approaches.

Keywords Sustainable development goals · Social responsibility · Climate change · Fintech · Digital economy · Climate responsibility · Digital entrepreneurship

JEL Classification O13 · O33 · O38 · Q01 · Q54 · Q55 · Q56 · Q58

1 Introduction

Climate-responsible entrepreneurship becomes more popular in the sphere of development of start-ups and investing in the digital economy. The growth of the interest of IT specialists in the problems of climate change and the search for possibilities to solve or assuage them with the help of innovative digital products forms a new direction of economic digitalisation. This powerful tendency influences different spheres of social and economic life, forming a synergetic effect and stable eco-systems, which can solve complex tasks in the sphere of sustainable development by joining the efforts of IT, financial institutes and climate initiatives.

Abstracting from the interests of the IT sphere, let us note that the goals of financial and climate organisations are traditionally considered poorly compatible. According to this, concepts that imply the financing of climate programmes are mostly based on institutional sources (national or international). Until recently, the interests of private financial establishments in the sphere of investing in climate projects were considered from the position of residual approach or declaration of values. However, at present, financial products or companies integrate climate goals in the system of financial indicators.

An important aspect that defines the development of climate-responsible entrepreneurship in fintech is the shift of values and principles in financial markets. According to modern approaches, the interests of financial establishments, together with profit-making, can include also other socially important aspects. As a consequence, concern regarding climate change is replaced with real interests and actions, especially in the spheres that have a long-term horizon of planning, in particular risk management, pension insurance, trading etc.

2 Materials and Methods

The methodology of this paper is interdisciplinary. It is realised at the intersection of several scientific and practical spheres that cover the concept of sustainable development and the conditions of functioning of financial markets and the digital economy, as well as innovative solutions. The system approach in this case is implemented through the study of mutual links and interaction between the elements of the studied spheres and the search for opportunities for their integration and directions for increasing the mutual positive influence. To substantiate the given aspects, we use secondary methods, which include an overview of scientific works on the given topic and elaboration on real examples of climate-responsible activity of fintech companies, as well as a description of different approaches to the development of sustainable climate-responsible entrepreneurship under the conditions of the digital economy in the market of fintech solutions. It allows determining the opportunities for achieving the results due to the rational interaction between the elements and discovering the directions of the influence on the state of the problem.

Given the complexity of the considered processes and their interdisciplinary character, we used the works of scholars and practitioners that studied the issue of sustainable development and digital technologies, the formation of the fintech market, etc. The main works used in this work contain detailed studies on the directions of transformation of the fintech sector in the context of the growth of the impact of climate change [5]; the influence of the financial processes and financial technologies on climate change [8, 10]; motives for the growth of interest in the Sustainable Development Goals among long-term investors [4]; formation of the concept of Climate FinTech and options of its implementation in the form of real projects and companies [1, 3, 7]; processes of fintech companies' acquiring the status of companies that are resilient to climate change [2]; directions and level of influence of fintech companies on climate change [11].

Given the differences in approaches to the problem of managing the development of climate-responsible entrepreneurship under the conditions of the digital economy in fintech markets, the main goal of this research is the generalisation and comparison of different views in the sphere of financial technologies of the climate character, as well as substantiation of universal directions for increasing the favourable influence of fintech solutions on the resolution of climate problems.

3 Results

Climate responsibility of entrepreneurship in the digital market and the traditional market is based on the desire for compliance with the following SDGs: SDG 7 "Affordable and clean energy", SDG 11 "Sustainable cities and communities", SDG 12 "Responsible consumption and production", SDG 13 "Climate action", SDG 14 "Life below water", SDG 15 "Life on land" and SDG 17 "Partnership for the goals" [13]. Despite the social significance of the declared goals and their rather wide implementation in the system of state regulation, the achievement of the SDGs has a recommendatory character and their stimulation has a limited effect.

In view of the main purpose of entrepreneurial activity, which implies the maximisation of profit, climate responsibility is considered in the context of voluntary reaction, image approach or systemic interest in compliance with the above goals. The main comprehensive approaches, which are intended to direct the activity of entrepreneurial structures to the achievement of the above goals, are corporate social responsibility and the concept of financing of environmental and social issues and the issues of corporate governance (ESG). These concepts offer the algorithms to integrate the SDGs into the system of management and envisage the possibility of monitoring and reporting on their achievement.

The use of these approaches in the digital markets of fintech forms a new construct, which has a larger potential for the influence on climate processes, due to the integration of the interests of business, sustainable development, IT and financial institutions. In this construct, entrepreneurship is the basic element that defines the character of relationships and key interests; the SDGs point at values and priorities; IT proposes technical solutions; financial interests connect them, providing the key resource for development. Thus, under the conditions of the digital economy and an increase in the impact of financial technologies on business and social processes, the new concepts that are intended at the study of the effect of financial processes on climate change through the active use of financial technologies are Green FinTech [11] and Climate FinTech [1]—particular sub-domains of the FinTech market.

The role of such sub-domains grows. According to the forecast of the audit company KPMG, among the key trends of fintech in 2022, one of the main ones is the acceleration of ESG investing among companies focused on climate change, decarbonisation and circular economy [6]. Thus, it is possible to state and climate-responsible activity of digital entrepreneurship in the fintech markets forms within a sustainable ecosystem, which unifies the SDGs, interests of financial markets and capabilities of innovative technologies in the digital economy.

Within the Green FinTech and Climate FinTech concepts, climate-responsible companies obtain technological solutions to implement financial services which allow for better research of the interaction between the business and environment and climate change, optimisation of financial flows for ensuring sustainable development and creation of new opportunities for direct investment in climate projects.

As of now, fintech is treated as a totality of complex ecosystems, most of which achievements are connected with the following tools:

- Financial inclusiveness;
- Blockchain, cryptocurrency and bitcoins;
- Financial services;
- Financing of entrepreneurship;
- P2P crediting;
- Distributed ledger technology and the Internet of trust [9].

An important role in the achievement of the SDGs belongs to blockchain technology, which is used in special apps, the organisation of decentralised markets of renewable energy, direct financing of climate and carbon loans, innovations in the sphere of financial tools, including green bonds, etc.

The potential of FinTech allows companies to effectively influence the climate and to achieve key business goals. The main directions of such influences are shown in Fig. 1.

As we see, the use of fintech influence to overcome the consequences of climate change is rather diversified. Within the proposed concept, financial functions which facilitate the improvement of the capabilities for the financing of climate projects or support for financial inclusiveness and justice are optimised; innovative solutions for the planning of long-term financial processes are developed; access to knowledge on the fight against climate change is improved; the database of more environmental solutions for the financial sector is formed. Due to this diversity of directions and opportunities, companies have different approaches to the identification of their role in sustainable processes and implement it according to their creative ways.

Diversification of the influence of fintech companies on climate change is presented in Table 1. It presents information on the largest specialised Green FinTech and Climate FinTech companies that are active players and motivate and offer solutions for fighting climate change.

As can be seen, fintech companies are active in the climate-responsible context. At the same time, their initiatives differ, covering the following directions:

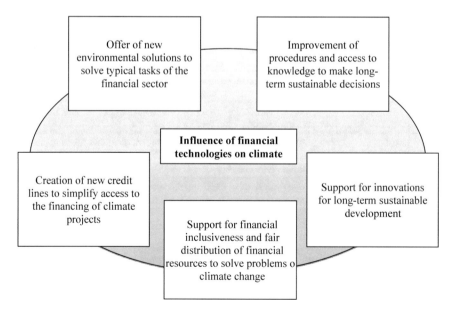

Fig. 1 Directions for the influence of financial technologies to overcome the consequences of climate change. *Source* Made by the authors based on [8]

Table 1 Characteristics of the influence of the most popular fintech companies on climate change

No.	Fintech company	Influence on climate
1.	Aspiration	Financing of environmental projects and refusal to provide services to oil corporations; transfer of a part of each transaction for planting trees
2.	TreeCard	Replacement of plastic or its reuse for the needs of the financial establishment, direction of 80% of profit for recovery of forests
3.	Starling Bank	Rejection of bank branches and paper; use of renewable energy sources, implementation of projects on planting trees
4.	Stripe	Offer of the possibility for quick transfer of money for the needs of climate sustainability, encouraging customers to perform such transfers
5.	Triodos Bank	Financing of only those companies that comply with the sustainable development principles
6.	Almond	Creation of a service that offers more environmental alternatives for the business in the sphere of supply, production and promotion of products
7.	Tred	Possibility to track carbon footprint based on payments by bank card
8.	Clim8	Services on saving money and recommendations on the financing of environmentally sustainable projects
9.	Trine	Business loans, including investing in the production of energy from alternative sources
10.	PensionBee	Refusal from the financing of oil, natural gas and tobacco business and arms trade
11.	Raise Green	Search for environmentally sustainable companies at the local level, allowing participation in investing in their activity and monitoring the influence of your investments on climate change
12.	Joro	Measuring the individual carbon footprint of each user and encouraging users to make more conscious decisions
13.	Treelion	Uses a blockchain-based platform to accumulate the potential of environmental digital products
14.	Miris	Offers a structure of green financing, which allows tracking and reporting the financial flows from the position of their impact on climate sustainability
15.	CO2X	Offers solutions to SMEs, which allows them to track their carbon footprint and receive green financial services

Source Made by the authors based on [3, 12]

- Refusal of financing of projects and companies with large climate footprint;
- Refusal of plastic in the operational activity;
- Accumulation and redirection of financial resources for the recovery of forests or development of alternative energy sources;
- Encouraging customers to finance climate projects;

- Creation of opportunities to choose more environmental solutions in the value chain;
- Creation of opportunities for tracking carbon footprint;
- Top-priority financing of environmentally sustainable projects;
- Stimulation for the financing of sustainable projects with the possibility of personalised determination of the influence of each investment on the general state of the environment;
- Possibility to track environmental and climate effectiveness of financial flows.

4 Discussion

Summing up the information on the direction of implementing the Green and Climate FinTech concepts, it should be noted that despite the growth of activity and development of this direction, there are still substantial risks within this sphere. These risks limit its potential. The main challenges are the problems with financing of climate projects through international funds and government institutions, limited financing of start-up projects of the corresponding direction through strict selection at early stages and the phenomenon of greenwashing, which is a type of climate PR of companies that imitate environmental activity. In this case, greenwashing is a factor of repulsion and a decrease in interest in climate projects due to the low level of real influence on environmental projects.

Besides, a separate direction of climate-responsible entrepreneurship in the fintech sphere is the development of its own operational processes according to the SDGs. Although the products of the digital economy, including the fintech sector, exists in the digital dimension, the manufacture of the products leaves a real footprint in the environment. Here we speak not only of emissions of fintech companies, which are gradually replaced with alternative energy sources but also of their suppliers' emissions.

In this context, the focus should be made on the following: despite the lower level of the use of physical infrastructure and the domination of the remote form of employment and application of cloud technologies, fintech companies also leave a visible carbon footprint. The main sources of climate pressure of fintech companies are the consumption of energy for services, centres of data processing, crypto mining, etc. This issue is especially acute in the case of fintech companies' using fossil fuels during this.

5 Conclusion

To sum up, we should note that climate change is getting more attention from financial institutions, especially in the spheres that use a long-term horizon of planning. This increases the necessity for the management of climate-responsible digital entrepreneurship in the fintech market. For the comprehensive use of the capabilities of this sphere in the reduction of the impact of climate change, the specialised concepts of Green and Climate FinTech were formed. They allow combining the capabilities of IT and the financial sphere to ensure the achievement of the SDGs. The main digital tools that are used for this are financial inclusiveness, blockchain, P2P crediting, DLT, etc. In this context, the directions within which fintech companies implement their climate-responsible activity are the facilitation of access to the financing of climate projects, development of innovative solutions that facilitate long-term financial planning given the impact on climate change, support for access to knowledge in the sphere of financial and production processes' influence on climate, and proposition of ready environmental solutions for the financial sector.

Based on this, fintech companies have the potential and implement projects in the sphere of climate responsibility, which ensures the readjustment of their operational processes, stimulates change in the priorities of the activity of partners and contracts, creates opportunities for the purposeful improvement in the state of climate sustainability and form new opportunities for the financing of climate projects. The current problems, which restrain the potential of fintech on the opposition to climate change, are based on the problems of limited financing, greenwashing and hindrance of the tendency for the reduction of fintech companies' pressure on climate through gradual replacement of energy sources.

References

1. Barclays (2021) Climate FinTech. Rise Insights report. Retrieved from https://rise.barclays/con tent/dam/thinkrise-com/documents/Rise-FinTech-Insights-climate-fintech-2021-DIGITAL. pdf. Accessed 12 Dec 2022
2. Blackshaw P (2021) Is fintech turning green? Value Walk, 1 Oct. Retrieved from https://www. valuewalk.com/is-fintech-turning-green/. Accessed 12 Dec 2022
3. Devs D (2022) 10 best green fintech companies and their sustainable financial solutions, 31 May. Retrieved from https://dashdevs.com/blog/top-10-fintech-companies-that-worship-sustai nability-social-change/. Accessed 12 Dec 2022
4. Eccles RG, Klimenko S (2019) The investor revolution. Shareholders are getting serious about sustainability. Harv Bus Rev (May–June 2019):106–116
5. Gomber P, Kauffman RJ, Parker C and Weber BW (2018) On the fintech revolution: interpreting the forces of innovation, disruption, and transformation in financial services. J Manag Inf Syst 35:220–265
6. KPMG (2021) Top fintech trends for 2022. KPMG Insights. Retrieved from https://home.kpmg/ xx/en/home/insights/2022/01/top-fintech-trends-in-h2-2021.html. Accessed 12 Dec 2022
7. McCreary A, Zhao Y, Chang A (2022) Climate FinTech. Mapping an emerging ecosystem of climate capital catalysts. Report. New Energy Nexus. Retrieved from https://www.newene rgynexus.com/wp-content/uploads/2021/01/New-Energy-Nexus-Climate-Fintech-Report.pdf. Accessed 12 Dec 2022
8. Mhlanga D (2022) The role of financial inclusion and FinTech in addressing climate-related challenges in the industry 4.0: lessons for sustainable development goals. Front Clim 4. https:// doi.org/10.3389/fclim.2022.949178

9. Mokhtar BSK (2022) A bibliometric analysis of fintech trends and digital finance. Front Environ Sci 9. Retrieved from https://www.frontiersin.org/articles/10.3389/fenvs.2021.796495; https://doi.org/10.3389/fenvs.2021.796495
10. Nassiry D (2018) The role of fintech in unlocking green finance: policy insights for developing countries. ADBI working paper, no. 883. Retrieved from https://www.econstor.eu/handle/10419/190304. Accessed 12 Dec 2022
11. Puschmann T, Hoffmann CH, Khmarskyi V (2020) How green FinTech can alleviate the impact of climate change—the case of Switzerland. Sustainability 12:10691. https://doi.org/10.3390/su122410691
12. Ściślak J (2022) Top 13 green FinTech companies—sustainable FinTech firms. Fintech business. Code & Pepper. https://codeandpepper.com/13-top-green-fintech-companies/. Accessed 12 Dec 2022
13. UN (2022) The sustainable development goals. United Nations. Retrieved from https://www.un.org/sustainabledevelopment/sustainable-development-goals/. Accessed 12 Dec 2022

Development of a Strategic Management System for the Development of the Oil and Gas Complex

Sergei B. Zainullin⬤, Penesiane E. O. G. Seri⬤,
Rosemary C. Nwachukwu⬤, and Oumar M. Ouattara⬤

Abstract The company's access to international markets in the fiercely competitive oil and gas sector is only possible if advanced management methods are used. The international market is extremely spacious, which creates significant opportunities for companies in the oil and gas sector. However, it also puts forward additional requirements for management. The export–import policy of the company in the oil and gas sector requires a specific approach to maintain the competitiveness of oil and gas products and companies directly from quantitative and qualitative parameters and more diligent adherence to marketing principles and methods. In the globalization context, foreign economic activity is the main component of the company's evolutionary development in the gas and oil sector. For a long time, the economic literature has been spreading the opinion that globalization is one of the strategies of oil and gas companies to conquer foreign markets.

Keywords Strategic planning · Strategic management · Oil and gas complex development · Ivory Coast oil and gas complex · Hydrocarbon industry

JEL Classification L22 · M21 · P28

S. B. Zainullin
Synergy University, Moscow, Russia

S. B. Zainullin · P. E. O. G. Seri (✉) · R. C. Nwachukwu
RUDN University, Moscow, Russia
e-mail: emmanuelonesine@yahoo.fr

O. M. Ouattara
Ufa State Petroleum Technological University, Ufa, Russia

1 Introduction

International marketing strategies are characterized by complex situations in which future conditions cannot be considered by methods based on extrapolation. Therefore, it is extremely important to analyze external conditions of activity and development prospects and the position of an oil and gas company in the competitive struggle to identify trends, threats, barriers, and opportunities that can significantly affect the functioning of an oil and gas company in the foreign market.

Arutyunova and Martynenko [1] suggest that strategies can be developed for oil and gas companies over some time to develop production and business activities for specific markets or market segments based on market conditions and the capabilities of the oil and gas company. This requires the formation of a comprehensive international plan that will cover all markets and products of an oil and gas company, which is especially important in terms of entering foreign markets. Thus, the strategy provides for the study of all alternatives and options in the field of foreign economic activity that relate to the long-term goals and their justification for making informed management decisions. According to Ishtuganov [3], in the future, the effectiveness of the development of such a strategy can be assessed depending on the degree of achievement of the goals of the gas and oil company.

2 Methodology

The methodology of this research is based on scientific works of national and foreign authors on the issues of the evolution of a management system for efficient development of the hydrocarbon industry in general and, in particular, that of the Ivory Coast. The different methods used to realize this research are cognition, comparison, formal logic and dialectic methods, analysis, and synthesis.

3 Results

International strategies are a comprehensive basis for achieving the fundamental objectives of an oil and gas company in the foreign market. The oil and gas company perceives each country's market as having excellent integrity with different needs and buyers, a different mentality, and different ways of decision-making. During international marketing, an oil and gas company should prepare a different entry and business strategy for each foreign market, which will contain an appropriate structure, as indicated by Ivlev and Savina [4]. The strategy of the oil and gas company is mainly aimed at preventing the negative effects of external factors.

According to Pupentsova et al. [7], it is particularly important to develop a foreign market entry strategy when expanding the number of markets where an oil and gas company conducts its business activities. At each stage of internationalization of an oil and gas company, various combinations of actions are used, each of which is associated with certain costs, risks, and effects.

The foreign market entry strategy of an oil and gas company refers to the level of development of the company's strategy because it determines the ways to use the resources of an oil and gas company more effectively, as well as a way to harmonize the potential of an oil and gas company with its objectives [9].

There are several main strategies for a company's access to international markets: direct and indirect exports, joint business activities, and direct investments.

According to this analysis, strategies differ depending on whether the firm exports domestic products or establishes production facilities in foreign markets.

Thus, in the foreign market, the company usually begins its activities with indirect exports, then develops towards direct exports and, finally, towards production abroad.

Kondratyev [5] suggests that one of the drawbacks of systematizing such strategies is the conditional ranking of the level of attractiveness and value without any criteria. The choice of the most responsive strategy requires some justification. Kondratyev proposes to concretize more basic strategies for entering the international market using the matrix "product price – incentive cost." This approach involves two profit strategies and two market entry strategies, depending on the promotion cost and the commodity prices. However, the authors do not provide for using quantitative criteria, offering only a list of conditions for applying a particular strategy.

Ivory Coast is known as the world's largest cocoa supplier country, which accounts for 40% of the global market [2]. Few people know that it is also a country producing 36,000 barrels/day of oil from fields on the Jacqueville shelf, which is operated by several world oil majors.

In 2018, approximately $890 million worth of oil was produced; about 23 million people live in the country; GDP per capita is $2278; for comparison, in Russia—$26,456 [10].

In September 2021, the Minister of Mining and Oil of Ivory Coast, T. Kamara, announced the discovery of an oil and gas field: on the deep-water shelf of block CI-101, 60 km from the coast during exploratory drilling conducted by the Italian company ENI; the depth of the well is 3445 m; drilling time—30 days; the volumes proven reserve include: up to 2 billion barrels of crude oil and up to 70 million m^3 of associated petroleum gas (APG) [6].

In 2019, Ivory Coast signed contracts with Eni and Total for the exploration of four oil blocks with investments of $185 million.

Oil production in the country is provided mainly on the shelf, near the border with Ghana. Ivory Coast revised its oil code in 2015 to attract new investors through production-sharing agreements (PSAs). The country has 51 allocated subsurface areas; four areas are under development, 26 are at the exploration stage, and 21 are free or under negotiation.

In 2014, Total mentioned the very promising results of its research on the Ivory Coast shelf. In recent years, the British company Tullow Oil has also announced significant discoveries.

Entering the Latin American market can be difficult due to political differences. Evidence of this can be seen in war-torn countries such as Afghanistan or Rwanda, where government turmoil makes it almost impossible for Exxon Mobil, its subsidiaries, and other similar corporations to gain any positions. If this were achievable, government corruption would be almost impossible. It is impossible to make a profit in any form. Another type of barrier that can hinder a company's entry into a market during its economic conquest is the different laws of other countries. Russian companies (e.g., Rostec and Rosneft) are the most active in the Latin American market.

The large-scale and sustainable cooperation between Russia and Venezuela was based on similar political and economic features of the two countries. In political terms, these are two countries with super-presidential power and similar political systems. The second feature that brings both countries closer is the raw materials economy and the specialization in gas and oil extraction. The USA has argued that Russian energy assistance was an important factor in Venezuela's political stability during the crisis. Thus, in January 2020, North American diplomacy, through the mouth of the special representative for Venezuela, Elliott Abrams, said that Russia acquires and resells 70% of Venezuelan oil, becoming, in effect, a financial "lifeline" for Caracas. This fact demonstrates the importance and value for Venezuela of cooperation with Russian oil corporations, primarily Rosneft.

Let us conduct a comparative analysis of the strategies of Russian oil and gas companies in Latin America:

Rosneft

Types of strategies and their goals:

- *Commodity market*: geographical diversification of sales channels and changes in the product basket in accordance with market trends.
- *Resource market*: the growth of exploration and production of hydrocarbon in Russia, Brazil, and Venezuela.
- *Technological*: improving the efficiency of management based on the database and digital counterparts of the company's property, developing technologies in the field of artificial intelligence, robotics, industrial internet, and uncrewed vehicles.
- *Integration*: creating own gas stations and refineries in Russia and Latin America; creating own thermal and electric stations.
- *Finance and investment*: steady accumulation of positive cash flow, logical reduction of financial liabilities, and payment of dividends in accordance with the best practices in the oil sector.

Gazprom

Types of strategies and their objectives:

- *Commodity market*: increase of new markets for its products and diversification of activities.
- *Resources market*: discovery of new oil and gas fields in Russia and abroad.

- *Technological*: development of nature-saving technologies and digital projects of oil and gas exploration and production.
- *Integration*: development of sales business and production of refined petroleum products; creation of own thermal and electric stations.
- *Finance and investment*: growth of free cash flow, growth of the company's market value.

Lukoil

Types of strategies and their goals:

- *Commodity market*: increase of new sales markets and diversification of refined products.
- *Resource market*: development of new deposits.
- *Technological*: increasing the oil recovery coefficient, the development of digital technologies in such areas as a digital twin, digital personnel, robotization of routine processes, and digital ecosystem.
- *Integration*: development of sales business, construction of new oil refineries, and creation of own thermal and electric stations.
- *Finance and investment*: growth of free cash flow and growth of stock returns.

Surgutneft

Types of strategies and their goals:

- *Commodity market*: an increase in types of refined petroleum products with high environmental and operational properties, diversification of product supply to new markets.
- *Resource market*: development of new deposits.
- *Technological*: improving the efficiency of exploration and development of oil and gas fields, digitization of fields, wells, and other production processes.
- *Integration*: creating own gas stations, extending related services, and creating own thermal and electric stations.
- *Finance and investment*: Growth of free cash flow, growth of stock return.

The long-term foreign economic strategies of Russian oil and gas companies have a pre-programmed aspect. In the short-term aspects, the strategies can be intuitive. The foreign economic aspect of the commodity market strategy of oil and gas companies is to increase new sales markets (countries of the world economy) and change the product basket in accordance with market trends, i.e., with the needs of foreign buyers.

Kondratyev [5] suggests that the product mix includes crude oil, natural gas, natural gas liquids, and refined oil and gas products. The external economics of the resource financing strategy also includes investments in the exploration and development of new oil and gas fields in Latin America.

The financial and investment strategy of the external integration economic sectors, as well as oil and gas companies, is "upstream" integration, which is manifested by the establishment of their own gas stations in the countries of the world economy and the construction and purchase of oil and gas processing plants and thermal and power stations. In terms of technology strategy, the foreign economic activity aims to import the latest digital technologies, robots, and drones to improve the efficiency of many business processes in the oil and gas sector.

The increase in the number of partner countries of Rosneft, Lukoil, and Gazprom and the maintenance of the number of partner countries of Lukoil at the same level confirm the first hypothesis regarding the strategy of foreign economic activity. An expansion of the number of nomenclature groups (commodity items) occurred for Rosneft and Surgutneftegaz corporations. Gazprom and Lukoil corporations experienced a drop in the number of exported commodity items, which suggests that the second hypothesis does not work for all companies.

Any models, concepts, and strategic goals are considerably adjusted by practical actions to implement them in order for Russian companies to enter the oil and gas market of Latin America. When implementing strategic tasks, the tools traditional for the professional activities of the Russian leadership could not be used. Various tools are used to achieve the goals, which are applied in accordance with Russia's political and diplomatic actions, particularly in the international arena, including the following:

- Centralized support for the activities of Russian companies;
- Use of "active events";
- Corruption of corporate and government managers in other countries;
- The use of informal (unofficial) structures and connections to promote the state strategy of Russia [8].

Efforts were directed to the formation of impact formation in two main directions:

- Manipulation through the main sectors of the economy or important national companies and "economic development of Russia's interests" (economic influence);
- Maintaining political division, autocrats, nationalists, and populists and bringing pro-Russian leaders to power who will consider Russia's interests in their policies (political influence).

Both sets of tools are adaptively applied and, as history shows, are quite effective.

The strategic goals of Russian oil and gas companies were achieved through controlled energy companies that had the task of subordinating the market of a particular country. A pro-Russian lobby was formed from local partners involved in energy trade, ready to promote Russia's interests. Russian-backed (often created) economic and political actors have also become the tools used by the Russian ruling elite to advance its interests in key geopolitical decisions on the international stage.

As part of a strategy to develop economic cooperation with Latin American countries, Russia is ready to offer lucrative contracts to individual representatives of the business and political communities in these countries, involving them in transparent agreements on energy trade.

Simultaneously, Russia has always adhered to the principle of "non-deterioration" of Gazprom's positions. Russia signed several intergovernmental documents that provided for the development of energy infrastructure "with the involvement of European and Russian investors, as well as loans and capital from third countries." It even proposed the creation of a "transnational consortium" in the gas trade in Europe. These international agreements legalized the consolidation of Russian capital through the mechanism of bringing Russian-controlled intermediaries in trade to its market.

Despite the alternation of controlled intermediaries, the strategic goal of Russian oil and gas companies is to continue to "strengthen Russia's role as a driving force in the creation of a new system of political and economic relations between the countries of Latin America". In the strategy of oil and gas companies to enter the Latin American market, the development of various forms of joint economic activities has become an instrument of economic development, the use of which is synchronized with other means of influence to achieve a multiplier effect.

The analysis also identifies the following challenges to achieving the strategic objectives of the Russian oil and gas industry in the Latin American market [11]:

- Monopolization of the market and development of management in consumer countries (conclusion of contracts on favorable terms for Russia, which are also profitable for political and business leaders of individual countries, formation of a system of privileged partners in trade with the Russian Federation, providing advantages to individual countries (companies) in access to the Russian market);
- Reforming the internal energy markets of consumer countries (adoption of legislation or its implementation aimed at liberalizing energy markets and limiting the abuse of monopoly position and monopolism in energy markets).

In relation to Latin America, the direct practical actions of oil and gas companies on the Latin American market were implemented in the following areas:

- Preservation and strengthening of technological dependence:

  (1) Maintaining dependence on Russian energy technologies, the supply of Russian components, and the use of Russian control systems;
  (2) Breaking the production chains that connected Russia and Latin America, implementing projects to "replace" components, and implementing new energy projects (gas pipelines, oil pipelines, and power supply networks).

- Obtaining operational control:

  (1) Control over the activities of companies through the buyout of energy companies, hobbies of the management of state-owned companies, the creation of debt obligations to the Russian capital, and the formation of joint companies to manage energy monopolies.
  (2) Control of the energy market for the preservation and increase of supplies of Russian (from Russia) companies to the Latin American market, blocking the diversification of supplies of resources or equipment.
  (3) Control over the organizational, technical, and software functioning of the market (control over the flow of information) [12].

The result of implementing such a strategy will be the formation of a system of dependencies that makes it possible to control energy policy.

Thus, since the deficit in the global energy markets will continue in the medium term, despite short periods of saturation of demand, the fundamental indicators of the market in 2022 look optimistic, which is positive for Russian oil and gas players who can also significantly succeed due to the normalization of the geopolitical situation that has emerged in the market.

Gazprom PJSC has historically been significantly undervalued due to geopolitical and sanctions risks, being the "hand of the Kremlin" in the eyes of international investors. Coupled with strong fundamentals of the gas market, which have already been reflected in the company's record revenue and profit, this could be a turning point for Gazprom shares, stimulating their recovery despite the uncertainty caused by the spread of the Omicron strain in the world. Caught in the "Mexican impasse" between the EC's internal consultations on energy and the Kremlin's verbal interventions in the export market, Gazprom was deprived of additional export volumes and a rally of spot gas prices in Europe. Nevertheless, the fundamental indicators of the gas market retain growth potential, and prices for financial transaction tax (FTT) are steadily rising against the background of an equally obvious gas shortage in Europe.

Although both of these scenarios involve too many "ifs…," which significantly increases the uncertainty in the spot gas market, Gazprom may turn out to be a key beneficiary of the conflict resolution. Thus, the gas monopoly benefits from the growth of long-term supply volumes and from the potential growth of exchange prices for gas in the European market. Since a possible compromise between the Russian and European leadership indirectly involves resolving the situation around Nord Stream 2 and the issue of further gas transit through Ukraine, the year 2022 may turn out to be a turning point for Gazprom, removing several political and operational risks. If we ignore the political background, we highly appreciate the potential for further growth of Gazprom's capitalization in the medium term.

Thus, the company's record revenue and profit for Q3 2021 and favorable price conditions in the gas market suggest further growth of financial indicators in the current and the next year. Although the further expansion of Gazprom's ambitious capital investment program still raises concerns about the generation of single-dealer platform (SDP), we note that Gazprom's efforts to reduce the debt burden coupled with a generous liquidity cushion provide a confident financial position for the company, forming a reserve for several years ahead. We note that throughout the COVID-19 pandemic, Lukoil's management has paid primary attention to maintaining high standards of corporate governance despite unfavorable market conditions and maintaining stable cash generation, which was ultimately reflected in the formation of dividend payments for the current year. Coupled with management's conservative approach to cost control and capital investment, an investor can obtain a balanced proxy for oil underpinned by significant resistance to commodity market fluctuations.

4 Conclusion

The outbreak of COVID-19, caused by a new strain, caused severe turbulence in the global oil market, and the price of Brent crude oil fell to its lowest level since the beginning of 2022. The recovery of the oil rebound remains uncertain due to possible work stoppages and other traffic restrictions. We believe that Russian companies can become a "safe haven" for the hydrocarbon sector during the turbulent period in the world markets in Latin America thanks to the development of its strategic management system. As an example of a good management system, we note that even though Lukoil has limited potential to increase production, which could become a deterrent in a free market, the company has made significant progress in the area of efficient capital management and cost optimization. Thus, the company looks particularly attractive under conditions of possible market restrictions, being a key beneficiary of the recovery of refining margins, which, in turn, allows for some financial stability in the face of lower oil prices. Thus, the company can achieve its objectives if the oil market correction leads to a further decline in oil prices. Russia has become one of the largest oil and gas companies in the world thanks to its strategic management system. Indeed, if Ivory Coast adopts these different development strategies and puts into practice the management methods of the Russian industries, the Ivorian hydrocarbon industries will make a big place in the international oil and gas market.

Acknowledgements This paper has been supported by Peoples' Friendship University of Russia Strategic Academic Leadership Program.

References

1. Arutyunova DV, Martynenko MA (2021) Exit strategy formation mechanism of the company to the global market. Manag Econ Soc Syst 2(8):5–14
2. Gnigu M (2019) Industrial risk management the case of cumulative damage of the GESTOCI contract. Unpublished master thesis, INP-HB University of Yamoussoukro, pp 80–84
3. Ishtuganov MB (2020) Mutual exemption from liability as a type of indemnity in the international oil and gas industry. Theoretical and practical foundations of scientific progress in modern society. In: Sukiasyan AA (ed) Proceedings of the international scientific and practical conference "Theoretical and practical foundations of scientific progress in modern society". Aeterna, Ufa, Russia, pp 57–60
4. Ivlev YuN, Savina VS (2020) About some ways of implementing the company's strategy of entering international markets. In: Velikorossov VV (ed) Innovations in the management of socio-economic systems (RCIMSS-2020). Ruscience, Moscow, Russia, pp 231–242
5. Kondratyev VB (2019) Global oil and gas market: basic trends. Russ Min Ind 6(148):24–29. https://doi.org/10.30686/1609-9192-2019-6-148-24-29
6. Ministry of Petroleum, Energy and Renewable Energies (2020) CIV 2020 energy report: final audit report. Abidjan, Cote d Ivoire. Retrieved from https://www.dgenergie.ci/fichiers_uploades/files/rapport/Livret%20du%20Bilan%20%C3%A9nerg%C3%A9tique%20CIV2020_Final.pdf. Accessed 10 Sept 2022

7. Pupentsova SV, Lysenko AN, Zhavoronkov DK (2020) Risk assessment of the company's strategy for entering the international market. Postgraduate student. Appendix J Bull Transbaikal State Univ 14(2):82–90
8. Pushkareva PP, Zakharov GV, Klimenko AO (2020) Review and classification of methods and strategies for industrial companies entering foreign markets. Humanitarian Sci Bull 7:122–130. https://doi.org/10.5281/zenodo.3959173
9. Rahimov IR, Omarov MM (2020) Key marketing strategies when companies enter international markets. In: Kuznetsov YuV, Malenkov YuA, Sokolova SV, Anokhina EM, Zhigalov VM, Kaisarova VP et al (eds) Current management issues: new management methods and technologies in the regions. Scythia-print, St. Petersburg, Russia, pp 342–344
10. Seka KO (2019) Optimization of trading activities at Abidjan-Vridi oil terminal. Unpublished master thesis, INP-HB University of Yamoussoukro, pp 30–32; 34–37
11. Sidorchak DS, Dyakonova MA (2021) The essence of the exit strategy of the enterprise to the international market. In: Proceedings of the 11th international scientific-practical conference: management of socio-economic development of regions: problems and ways of their solution. Kursk Branch of Financial University under the Government of the Russian Federation, Kursk, Russia, pp 146–148
12. Ukolova EV (2020) Specificities of the strategy and tactics of entering the international markets of American companies. Eurasian Union Sci 5–2(74):8–12

Risks and Financial Barriers to the Implementation of Sustainable Infrastructure Investment Projects in the Housing Sector

Svetlana B. Globa [ORCID]**, Evgeny P. Vasiljev** [ORCID]**, Viktoria V. Berezovaya** [ORCID]**, and Dmitry V. Zyablikov** [ORCID]

Abstract The paper aims to identify and study aspects that make it possible to improve the quality of functioning of utility infrastructure. The utility complex is one of the most complex elements of the functioning of the territory. To ensure and maintain a decent level of quality-of-life indicators, it is necessary to ensure an uninterrupted supply of heat, water, electricity, and gas to all types of buildings, guarantee their proper quality and safety, and minimize the negative impact on the environment. The relevance of this research is determined by the variety of problems in the functioning of utility infrastructure accumulated over the years of network use and those that have arisen during new construction. The authors use the methods of abstraction, analysis, synthesis, induction, deduction, and other methods to identify risks and financial barriers to introducing advanced environmentally friendly technologies at residential infrastructure facilities for the sustainable development of territories. The research object is the projects for the modernization of utility infrastructure considering the directions of sustainable development of the territory. The presented research classifies the problems of the development of utility infrastructure that affect the provision of conditions for the quality of life of the population, including the state of the used equipment and the compliance of technology with recent requirements in the field of digitalization, ecology, and safety; the state and quality of structures; the workload of fixed assets of enterprises and their moral and physical deterioration; the state of networks and the availability of reserves and other factors that cause an increase in the likelihood of human-made accidents. The risks of reducing the quality of utilities have been identified, among which operational, technical, economic, social, and legal risks are highlighted. It is noted that the state of utility engineering infrastructure largely determines the current characteristics and

S. B. Globa (✉) · E. P. Vasiljev · V. V. Berezovaya · D. V. Zyablikov
Siberian Federal University, Krasnoyarsk, Russia
e-mail: sgloba@sfu-kras.ru

V. V. Berezovaya
e-mail: VVBerezovaya@sfu-kras.ru

D. V. Zyablikov
e-mail: dzyablikov@sfu-kras.ru

© The Author(s), under exclusive license to Springer Nature Switzerland AG 2023
E. G. Popkova (ed.), *Smart Green Innovations in Industry 4.0 for Climate Change Risk Management*, Environmental Footprints and Eco-design of Products and Processes, https://doi.org/10.1007/978-3-031-28457-1_25

potential of the residential sector and utilities. The scientific novelty of this research lies in the actualization of the system of factors affecting the stable functioning of utility infrastructure to ensure a high quality of life for the population.

Keywords Risks · Public–private partnership · Project management · Concession · Investments · Financing

JEL Classification Q48 · Q56 · Q57 · Q53 · O33 · L97

1 Introduction

The activities of utilities affect the environment, which is also an important characteristic of the quality of life. Water from natural sources is used for domestic and industrial water supply purposes, and then, after treatment, is returned to the natural environment. However, wastewater that enters infrastructure facilities after use is often insufficiently treated due to the deterioration of laid communications and the use of outdated treatment technologies or equipment. Simultaneously, about half of the volume of discharge of polluted wastewater into the natural water bodies of Russia is carried out by housing and utility services enterprises. Thus, according to experts, the deterioration of utility networks in some regions reaches 60–70%; in remote areas and small settlements, it reaches up to 85%. Unfortunately, we constantly see reports in the news feed about communal accidents and their consequences.

Operating boiler houses of centralized heat supply systems emit many pollutants and harmful substances into the atmosphere. Household and industrial waste placed in organized landfills and illegal landfills pollute the soil due to the lack of high-quality organization of sorting, recycling, and safe disposal of waste.

All over the world, public–private partnership (PPP) shows effectiveness of use in the fields of transport, education, healthcare, industry, and many other sectors. In terms of the number of such projects in Russia, the leading sectors are utilities and energy (90%), the social sector (7%), transport (3%), and other sectors. In recent years, the need for new infrastructure facilities has increased significantly.

Despite the widespread practice of using concession and PPP mechanisms and models, their study shows the limitations of the approach to managing such projects in Russia: unclear distribution of roles, lack of guarantees for all parties to the project, issues of risk distribution, and difficulties with financing for various reasons. These disadvantages significantly reduce the possibility of the most effective use of the advantages of these projects.

Therefore, risk management as an element of the financial management of the project makes it possible to identify the main project risks (including potential ones) and reduce their impact on the financial result.

Due to better management, PPP involves the transfer of risk to a private partner. However, the public sector takes some responsibility for possible losses and minimizes the negative consequences for the private investor.

Decision-making by participants in residential infrastructure modernization projects takes place in conditions of partial uncertainty, which is characterized by a deviation from the planned result. The need to assess the possible consequences leads to the concept of risk. As an element of financial management in the project, the risk management system should have certain goals and actions and exclude uncertainty.

It is also important to consider this issue in the methodological context of the assessment of different types of risks. This assessment makes it possible to identify the threat at the stage of its inception and, accordingly, to prevent negative consequences. In other words, the availability of more flexible methods will make it possible to assess financial risks with different risk profiles and degrees of risk.

2 Methodology

The study of the problem is determined by the works of scientists on the functioning of housing and utility services. The literature analysis has shown the lack of a common understanding of the complex of problems and solutions to this urgent problem. Thus, it is constantly in the focus of attention of regional and municipal authorities and economists.

The issues of risk management and financing in PPP projects are presented in the studies of Adamiya and Tretyakova [1], Galazova [9], Kovaleva et al. [14], Petrova [22], Shuliuk [27], and others.

The problems and advantages of using concession and PPP methods in the implementation of infrastructure projects are studied by Ahmed et al. [2], Dzgoeva and Savelchev [8], Marcelo et al. [18], and others.

Features and prospects for the use of public–private partnership schemes in Russia and other countries are studied by Akpoghome and Nwano [3], Atadjanova [4], Chernov [6], Chumichkin [7], Azimul Haque [5], Khatskevich and Tatarinova [10], Khmurova et al. [11], Khusnutdinova [12], Korechkov [13], Kreukels and Spit [15], Kuzmin et al. [16], Lukyanenko [17], Morozova et al. [19], Omelyanenko and Kovtun [20], Pernsteiner et al. [21], Prieto et al. [23], Prizhennikova [24], Sabetska et al. [25], Shaimukhametov [26], Stepanov and Legostaeva [28], Wang et al. [29].

Recently, large cities saw a tendency of mass housing construction without assessing the compliance of the current and prospective availability of utility infrastructure. This led to an increase in disproportions and an increase in the deterioration of existing networks and accidents.

Thus, the following types of territories can be distinguished from the point of view of the need and scope of modernization of utility infrastructure:

1. Existing territories of residential development. These territories often have a high degree of physical and moral deterioration of the networks. They are characterized by high density and non-consideration of engineering communications, as well as the insufficient capacity of municipal infrastructure.

2. New micro districts and territories of complex development. These territories are often characterized by insufficient free power and the high cost of connecting to utilities.
3. Remote areas and territories that are not provided with utility infrastructure. These territories are characterized by aggressive use and consumption of natural resources (e.g., heating with wood, coal, or burning of petroleum products), lack of technologies for cleaning, sorting, and disposal of waste, and environmental pollution.

Utilities include cold and hot water supply, sewerage, heat, gas, and electricity. Thus, it can be stated that the utility complex is one of the most difficult elements of the territory's functioning. It is necessary to ensure the uninterrupted supply of heat, water, electricity, and gas to all types of buildings, ensure their proper quality and safety, and minimize the negative impact on the environment. This requires a large amount of repair, cleaning, and preventive work on utility networks and facilities, uninterrupted functioning of emergency services, and a high management level of utility facilities.

The main constituent parts of the utility infrastructure are the networks of water management and energy complexes, including water supply and sanitation systems, heat supply, electricity, and gas supply.

Let us highlight the most pressing problems of the development of utility infrastructure that affect the quality of life of the population:

- Morally and technically obsolete equipment and technologies that do not meet today's requirements in the field of environmental protection and environmental safety;
- Obsolete structures built 40–50 years ago, requiring repair, reconstruction, modernization, and introduction of digital technologies and automation;
- Overloading of existing pipelines, sewage treatment plants, and energy sources due to deterioration of the main and auxiliary equipment, limiting their potential capacity and throughput;
- A high percentage of internal corrosion of pipes, causing contamination and deterioration of water quality after its purification;
- The lack of capacity reserve of networks and their high wear and tear, leading to a high level of losses and unaccounted water consumption;
- Low performance and imperfection of technologies for cleaning, processing, and utilization of used water resources and the resulting sediment;
- Constantly arising unauthorized dumps of household and industrial waste, which must be eliminated and placed in landfills, practically not amenable to sorting or recycling;
- High volumes of accumulated garbage on the streets due to the melting of the accumulated snow cover during the winter period, as well as high dust in the summer due to untimely and incomplete cleaning of the streets and improper arrangement of lawns and coatings;
- The lack of a high-quality storm sewerage system, leading to non-normative ingress of the storm and meltwater into the household sewerage system;

- The lack of a unified approach to the separate collection of waste, their sorting and processing;
- The increasing complexity of repairs of public infrastructure networks due to the high density of buildings, the impossibility of long-term restriction of traffic on the sections of pipe passage;
- The increasing likelihood of human-made accidents due to hazardous utility infrastructure facilities (gas pipelines, etc.) in protected areas.

3   Results

Failure to comply with the high level of public services can lead to negative social consequences. Therefore, it is important to manage the risks of reducing the quality of utilities, among which we highlight:

- Operational risks expressed in possible interruptions in the provision of utilities due to poor quality and incorrect performance of work;
- Technical risks arising from the malfunction or use of obsolete and worn-out equipment;
- Economic risks manifested in the incurrence of losses in the implementation of activities and affecting the financial stability of enterprises;
- Social risks arising from the poor-quality performance of work or prolonged interruptions in the supply of utility resources;
- Legal risks arising from the imperfection of the legislative framework.

Thus, the main principles of the formation of a program for the development of engineering utility infrastructure for the sustainable development of the region include the following:

- The use of technologies that ensure the normative quality of resources entering the buildings (water, gas, heat, and energy) and reducing the accident rate of networks and environmental and sanitary-epidemiological safety of the territory;
- Application of energy-efficient and energy-saving materials and technologies;
- Development of alternative energy sources that minimize environmental pollution;
- Accounting for emerging agglomerations and the development of satellite cities.

Investments are needed to ensure equal access to quality utilities for all residents and create new resource supply systems.

The development of utility infrastructure helps overcome territorial disproportions and ensures equal conditions for the comfortable living of residents of various territories of the region.

The level of utility infrastructure development largely determines the activity of production and the social sphere. It also acts as a factor in the formation of the territorial structure of the economic complex of the region.

4   Conclusion

The research showed that most of the problems of the current state and prospective development of utility infrastructure systems are interrelated and affect the ecological state of the environment and the level of quality of life of the population. The degree of development of utility infrastructure is one of the dominant factors in improving the quality of life of the population. Thus, it is important to ensure the uninterrupted and trouble-free functioning of utility infrastructure systems, ensuring the preservation and safety of the environment through the use of environmentally friendly technologies at all stages of the life cycle.

Acknowledgements The study was carried out within the framework of the research grant of the Krasnoyarsk regional fund to support scientific and technological activities on the topic "Development of models of financial support for investments in the utility infrastructure of the region, taking into account the best Russian and world practices and features of the spatial and territorial development of the Krasnoyarsk Territory," No. KF-835, agreement on the procedure for targeted financing No. 226 of April 20, 2021.

References

1. Adamiya TT, Tretyakova GV (2020) Risk management in public-private partnership projects: application of current approach and its improvement. Innov Investment 1:35–38
2. Ahmed Y, Atan I, Sipan B, Binti H, Hashim A (2020) Public-private partnership strategy for housing provision in Abuja, Nigeria. Int J Sci Technol Res 9(1):1958–1965. Retrieved from https://www.ijstr.org/final-print/jan2020/Public-Private-Partnership-Strategy-For-Housing-Provision-In-Abuja-Nigeria.pdf. Accessed 27 Sept 2022
3. Akpoghome TU, Nwano TC (2019) Public-private-partnership (PPP) in Nigeria. KAS Afr Law Study Libr 6:482–501. https://doi.org/10.5771/2363-6262-2019-4-482
4. Atadjanova D (2021) Theoretical and methodological fundamentals of public-private partnership. Innov Econ 4(8):11–18. https://doi.org/10.26739/2181-9491-2021-8-2
5. Azimul Haque AHM (2020) Public-private partnership: problems and prospects. https://doi.org/10.6084/m9.figshare.13150694.v4
6. Chernov AV (2020) Project finance in Russia: the status, international practices, and public-private partnership. Finan Credit 26(3):630–643. https://doi.org/10.24891/fc.26.3.630
7. Chumichkin DA (2020) Results of the use of PPP mechanisms in modern Russia. Econ Soc 7(74):623–626
8. Dzgoeva DT, Savelchev LA (2021) On the development of public-private partnership in the field of housing and communal services. Adm Consult 11:154–162. https://doi.org/10.22394/1726-1139-2020-11-154-162
9. Galazova S (2021) Financing of public-private partnership projects based on "Smart technologies." In: Popkova EG, Sergi BS (eds) "Smart technologies" for society, state and economy. Springer, Cham, Switzerland, pp 1696–1703. https://doi.org/10.1007/978-3-030-59126-7_185
10. Khatskevich EM, Tatarinova LY (2020) Public-private partnerships in Siberia: implementation results and barriers to development. ECO J 50(1):29–44. https://doi.org/10.30680/ECO0131-7652-2020-1-29-44
11. Khmurova V, Mykolaichuk I, Kandahura K, Sylkina Y, Sychova N (2021) Strategy for the development of public-private partnership in the context of global changes. Sci Horiz 24(8):108–116. https://doi.org/10.48077/scihor.24(8).2021.108-116

12. Khusnutdinova ER (2019) Perspective directions of the development of the public-private partnership in Russia. Colloquium-J 13–11(37):156–157
13. Korechkov YuV (2020) Entrepreneurial activity in the regional economy: based on public-private partnership. Soc Polit Res 4(9):95–105. https://doi.org/10.20323/2658-428X-2020-4-9-95-105
14. Kovaleva TM, Milova LN, Khvostenko OA, Popova EV (2021) Finance of public-private partnership in the Russian Federation. In: Ashmarina S, Mantulenko V, Vochozka M (eds) Engineering economics: decisions and solutions from Eurasian perspective. Springer, Cham, Switzerland, pp 240–246. https://doi.org/10.1007/978-3-030-53277-2_27
15. Kreukels AMJ, Spit TJM (2008) Public-private partnership in the Netherlands. J Econ Soc Geogr 81(5):388–392. Retrieved from https://www.researchgate.net/publication/229584413_Public-private_partnership_in_the_Netherlands. Accessed 27 Sept 2022
16. Kuzmin E, Zinatullina E, Mezentseva E (2021) Regional distribution of public-private partnership projects: case of Russia. In: E3S Web of conferences, vol 301, p 01005. https://doi.org/10.1051/e3sconf/202130101005
17. Lukyanenko OYu (2018) The development methods of national public administration in public-private partnership. Bull Kemerovo State Univ Ser Polit Sociol Econ Sci 1:14–20. https://doi.org/10.21603/2500-3372-2018-1-14-20
18. Marcelo D, House RS, Mandri-Perrott C, Schwartz JZ (2019) Do countries learn from experience in infrastructure public-private partnerships? Public-private partnerships practice and contract cancellation. J Infrastruct Policy Dev 3(1):56–75. https://doi.org/10.24294/jipd.v3i1.1084
19. Morozova LS, Khavanova NV, Vetrova EA, Sulyagina JO (2020) Scientific basis of the organisation of public-private partnership. Rev Inclusiones 7(S2–1):535–545
20. Omelyanenko V, Kovtun G (2022) Revitalization of old industrial territories on the basis of public-private partnership. East Eur Econ Bus Manag 2(35):141–147. https://doi.org/10.32782/easterneurope.35-20
21. Pernsteiner S, Schaffhauser-Linzatti M, Karl R (2008) Reporting on public-private partnership projects. In: Proceedings of the 65th international Atlantic economic conference, 9–13 Apr 2008. Warsaw, Poland
22. Petrova I (2021) Innovative landscape as a form of public-private partnership. Intellect XXI 5:63–67. https://doi.org/10.32782/2415-8801/2021-5.13
23. Prieto R, Duvall T, Swain K, Placilla M, Diwik J (2009) Perspectives on public private partnerships
24. Prizhennikova AN (2019) Public-private partnership in Russia: features of the implementation forms. Colloquium-J 5–6(29):40–46
25. Sabetska T, Stefanyshyn L, Hryhoriv S (2021) Management of public-private partnership projects in the field of capital construction. Black Sea Econ Stud 63:52–57. https://doi.org/10.32843/bses.63-8
26. Shaimukhametov EI (2021) Specificity and features of PPP in the fuel and energy sector. Student Herald 4–5(149):61–63
27. Shuliuk B (2021) Corporate financing of public-private partnership projects: assessment of financial opportunities and risks. Finan Credit Act Probl Theory Pract 3(38):78–85. https://doi.org/10.18371/fcaptp.v3i38.237422
28. Stepanov MS, Legostaeva AA (2022) Public-private partnership project management system in the Republic of Kazakhstan: analysis of problems and perspectives. Central Asian Econ Rev 1:86–100. https://doi.org/10.52821/2789-4401-2022-1-86-100
29. Wang X, Yin Y, Xu Zh (2022) The influence of trust networks on public-private partnership project performance. In: Proceedings of the institution of civil engineers—management, procurement and law. https://doi.org/10.1680/jmapl.21.00025

Alternative Energy Investments as a Part of Global Smart Green Innovations

Olga V. Khmyz⑩ and Margarita I. Solomenkova

Abstract Concerns about the global environmental situation are leading to a search for new climate solutions, one of which is to stimulate an energy transition at the global level by expanding the use of alternative energy sources and introducing smart energy-saving technologies. Significant financial resources are needed for this purpose. Therefore, it is vital to assess the prospects for alternative energy in the post-COVID economy by analyzing the current state of this segment of the global energy market and the impact of the COVID-19 pandemic on it. The research uses statistical data from the World Bank, the International Energy Agency, and Bloomberg Information and Analysis Agency. The authors apply the methods of retrospective and comparative analysis and the least squares method. The analysis showed the growing importance of alternative energy, especially in developed countries, the serious impact of global energy inflation, and the favorable prospects for renewable energy sources in general. The study has drawn conclusions about the possibility of increasing investment in alternative energy.

Keywords Alternative energy · Renewable energy sources · Global energy transition · Investments in alternative energy · Climate agenda · Smart green innovations

JEL Classification G12 · C12

O. V. Khmyz (✉) · M. I. Solomenkova
MGIMO University, Moscow, Russia
e-mail: khmyz@mail.ru

M. I. Solomenkova
e-mail: solomenkova@icloud.com

© The Author(s), under exclusive license to Springer Nature Switzerland AG 2023   247
E. G. Popkova (ed.), *Smart Green Innovations in Industry 4.0 for Climate Change Risk Management*, Environmental Footprints and Eco-design of Products and Processes, https://doi.org/10.1007/978-3-031-28457-1_26

1 Introduction

The global green agenda and continuation of the climate improvement work, begun by developed countries in the last century and stimulated by the adoption of the Paris Agreement [13] and its developing regulations, has not stopped at the beginning of the second decade of the new century, and in the post-pandemic period as well. One of the most promising areas of world energy development for improving the global environmental situation seems to be stimulating the global energy transition to alternative energy [15].

The accelerated transformation of the global energy system from fossil to alternative energy sources covers renewable energy sources, the integration of renewable energy systems and technologies, and measures to increase energy efficiency and strengthen the electrification of end users (primarily energy-efficient smart homes and transportation [1]) using renewable energy sources and reducing the use and production of fossil energy sources [6]. In today's global financial economy, the global energy transition requires a gradual reduction of investment in fossil fuels (in 2020, global investment in the oil and gas sector has shrunk by almost 30% [9]) and an increase in the amount of financial investment in alternative energy. The funds required are estimated [2] to be over $100 trillion. Such significant financial resources are needed to implement innovative energy technologies and abandon traditional (fossil) energy sources.

Therefore, the paper aims to analyze the current state of investment in alternative energy and determine the near future of such investments.

2 Methodology

Understanding alternative energy as an energy sector with resources or processes that are replenished (restored) at regular intervals, we can distinguish five types of energy: solar, wind, geothermal, bio, and hydropower [11]. Additionally, there are certain enabling technologies that play an important role in the transition to green energy. The scaling up of renewable energy and the integration of solutions for electrifying energy services, such as transportation and clean cooking solutions, provide enormous social and environmental benefits and climate change mitigation and make it necessary to channel financial flows into building infrastructure for alternative energy production. For example, smart green innovations can improve the quality of life of the nearly three billion people who lack clean cooking methods and the 759 million people who lack access to electricity [14]. In doing so, offering advanced cooking solutions to billions of people would reduce global soot emissions by 50–90%, greatly limiting global warming.

The main ways to invest in alternative energy are as follows:

- Shares of companies whose business is based on the production of alternative energy technologies;

- Shares in new alternative energy projects and ethical financial firms;
- Units of a mutual fund or index fund with a large basket of securities of companies generating alternative energy;
- General energy or infrastructure funds, private investment, venture capital, etc.

Notably, shares of alternative energy companies have posted significantly higher total returns over the past decade: 422.7%, or nearly seven times higher than nearly 60% of fossil fuel companies. Over five years, the rate of return is lower but three times higher than that for fossil fuel companies [8]. The lower value of the standard deviation of the portfolio return also suggests an advantage of investing in clean energy companies despite the impact of COVID-19.

Over the past twelve years, the volume of global investment in alternative energy has generally shown an upward trend, but it has varied from year to year. Thus, there was an 11% drop in 2018, with a recovery in 2019 [5]. In part, this is due to lower capital costs because more renewable energy capacity could be purchased for the same amount of investment. Another factor was a change in policy in major markets, such as China, where a government restriction on the number of solar projects eligible for a feed-in tariff in mid-2018 led to a sharp drop in solar investment. Nevertheless, globally, the decline in Asia was partially offset by an increase in Europe.

Despite the generally negative effects of the COVID-19 pandemic on the global economy, average annual investment in renewable energy remained stable in 2019–2020 compared to 2017–2018. Notably, due to the dramatic cost reductions in solar and wind technologies associated with increasing competitive innovation and production expansion, the same level of investment in 2019–2020 has resulted in more capacity expansion than in the previous period [4].

The year 2021 saw another record increase in renewable energy capacity (in annual terms) of nearly 6%, although that year still saw supply chains disrupted in the COVID-19 pandemic, construction delays still occurring, and prices for raw materials and metal rising rapidly, and reaching highs. This was especially noticeable in the world market of traditional energy carriers. The increase in oil and gas prices led to the global energy crisis of early 2022. The current difficult situation (especially in developed countries) has once again demonstrated the need to accelerate the transition to clean energy sources and the importance of alternative energy in general. New power generation capacity from solar, wind, and other renewable energy sources increased to record levels worldwide in 2021 and will grow even more [10]. By 2030, the global renewable energy market will reach $177.6 billion, more than doubling compared to 2020. Nevertheless, much of this will depend on the financial resources expended, especially by developed (European) countries.

3 Results

To confirm the obtained estimate of the high degree of investment attractiveness of the global alternative energy sector and related companies, based on statistical data

Table 1 Intermediate indicators LSM (T = 12)

Coef.	Estim.	Std. error	t-stat.	P-mean
Const.	− 26,867.8	6342.6	− 4.2361	0.0039***
l_CO$_2$	46.1693	469.426	0.0984	0.9244
l_GDP	− 104.261	271.506	− 0.3840	0.7124
l_Popul	1321.95	518.196	2.5511	0.0380**
Elecpw	− 0.112513	0.216527	− 0.5196	0.6193
Dep. var. mean	286.0000	Std. dev. dep. var		50.20503
Sum res. squares	4486.771	Std. model error		25.31733
R-squared	0.838175	Adj. R-squared		0.745703
F (4,7)	9.064125	P-mean (F)		0.006706

Source Calculated by the authors
Asterisks indicate that explanatory variable is statistically significant

from public sources (World Bank website) for the period from 2010 to 2021, we build econometric models and forecast the trend of global investment in alternative energy.

For the calculations in the GRETL program, we select the following factors:

- Invest (Global investment in renewable energy, B\$);
- CO_2 (CO_2 emissions, kt);
- GDP (current US\$);
- Elecpw (Electric power consumption, kWh per capita);
- Popul (Total population).

The importance of the factors was proven [3]. We obtain the following values for the dependent variable Invest (Table 1).

At the 5% confidence interval, only the logarithm factor from Population is significant. The adjusted coefficient of determination was as high as 0.75. To get away from large values of the factors CO_2, GDP, and Popul without prejudice to the simulation, we will reduce their digit capacity through the introduction of multipliers: 10^{-6} for CO_2, 10^{-12} for GDP, and 10^{-6} for Popul. Let us build a model of the same kind to make sure that the transition to data with other factor digits is correct. Since the qualitative indicators remain the same, the transition is correct. Thus, we continue modeling.

Based on the analysis of the time dynamics of the factors, we determine the need for introducing non-linear factors. The 2016–2021 dynamics for the CO_2 and Elecpw factors change significantly from the 2010–2015 dynamics. Logarithms can be used to compensate for large values. However, in this case, these actions are not justified because the use of the logarithm is dictated not by the magnitude of the factor but by the nature of the dynamics, which is close to a simple linear relationship for the factors CO_2, GDP, and Popul and does not require the use of logarithms, unlike the

Table 2 Final LSM indicators

Coef.	Estim.	Std. error	t-stat	P-mean
Const.	− 761.362	151.226	− 5.0346	0.0005***
Popul.	0.141585	0.0204253	6.9318	< 0.0001***
Dep. var. mean	286.0000	Std. dev. dep. var.	50.20503	
Sum res. squares	4776.204	Std. model error	21.85453	
R-squared	0.827736	Adj. R-squared	0.810509	
F (1,10)	48.05029	P-mean (F)	0.000040	

Source Calculated by the authors
Asterisks indicate that explanatory variable is statistically significant

dynamics of the Elecpw factor. The rest of the factors can be used without using any functions. The results confirm the changes made.

Let us check the model for multicollinearity of factors, which can worsen qualitative indicators. For the inflation factor method (VIF), the minimum possible value equals 1.0; values higher than 10.0 may indicate the presence of multicollinearity: CO_2—8.603, GDP—11.205, Popul—9.033, and l_Elecpw—2.366. From the above data, we can see that although the VIF values for a number of factors are at the limit of 10, the conventionality of this limit does not provide a basis for stating the fact of multicollinearity.

Unfortunately, the small sample size does not allow us to use asymptotic tests for heteroscedasticity that are valid for large samples. Thus, we use a step-by-step removal of the factors with the worst parameters. After removing the factors CO_2, l_Elecpw, and GDP, the adjusted coefficient of determination increased (Table 2).

The result is a one-factor model (pairwise regression) with excellent performance. The free term (constant) and the coefficient before the Popul factor are significant even at the 0.01% significance level. The coefficient of determination is 0.81.

Consequently, simplifications to the pairwise regression gave the best option. The most important factor for predicting investment is to predict the size of the population. That is, if we predict the size of the population, we can calculate the corresponding value of the investment by the equation Investment = − 761.362 + 0.141585 * Population, where the size of the population should be in millions of people.

Since the paired model showed a better result, we can build a trend, indicating its equation and the values of its coefficient of determination.

Statistical data on the amount of global investment in alternative energy is obtained for the last (pre-pandemic) decade. Thus, the forecast for 2022–2026 will be optimal and accurate (Fig. 1).

The refined trend shows that investment flows into alternative energy have an upward trend, which will continue for at least the next three years. This once again confirms the feasibility and reliability of investment in this sector.

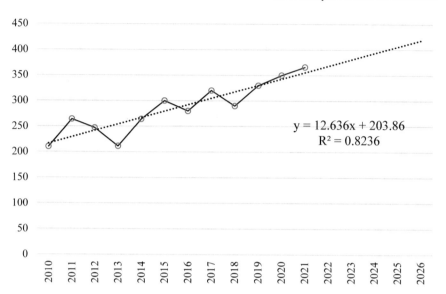

$$y = 12.636x + 203.86$$
$$R^2 = 0.8236$$

Fig. 1 Investments in renewable energy sources, 2010–2026, $ billion. *Source* Compiled by the authors

4 Conclusion

The analysis of the situation and trends in the contemporary world energy market has shown that despite the COVID-19 pandemic and geopolitical factors, the importance of alternative energy for the global economy continues to grow.

Based on the conducted research and its goals and objectives, we can make the following conclusions.

Alternative energy is a promising and even necessary direction for reducing the effects of climate change, recommended to investors for directing their financial investments. There are several ways to do this:

- To buy shares in renewable energy generating companies located around the world;
- To directly finance alternative energy projects or developments;
- To invest in alternative energy through funds, IPOs, or buy green bonds [12].

The global market for renewable energy sources is expanding; the increase in their capacity is renewing records every year. The increasing annual growth rate of alternative energy capacity reflects the fact that it has become an attractive investment proposition. Investment in renewable energy has grown from less than $50 billion annually in 2004 to about $300 billion in the pre-pandemic years and is expected to reach $1.4 trillion in 2022, with accumulated investment in renewable energy exceeding $2.4 trillion in 2022 [7]. That said, the current political uncertainty makes any prediction difficult. Nowadays, Western countries aim to move away from fossil fuels as quickly as possible, which will likely be the trend over the next few years. In the Russian Federation, the government also assists development under the Support Program, with substantial investments being made in solar, wind, and hydropower.

The leaders in financing alternative energy are China and developed countries (the USA, Germany, and the UK). The position of developing countries in this indicator should strengthen by 2025 due to the reduction of the cost of solar and wind energy to a competitive level. The accelerated pace of the energy transition and the increased competitiveness of alternative energy is evidenced by the significant reduction in the cost of energy over the past decade. Overall, renewable energy accounts for the majority of investments in transitional technologies and smart green innovations, with solar and wind being the most significant investment streams. Nevertheless, the financial flows into alternative energy production technology must be increased even further to achieve the goals of the Paris Agreement.

However, there is a risk of not meeting projections due to rising raw material and logistics prices in 2021–2022, which could increase costs over time. Chances for improvement exist with the timely application of additional hedging mechanisms, cost sharing, or strategic allocation of equipment.

In addition to achieving carbon neutrality and improving global environmental and climatic conditions, alternative energy contributes to reducing unemployment, improving global health, and increasing the GDP of countries that develop the corresponding energy capacities. Additionally, shares of alternative energy companies are characterized by higher returns and minimal volatility compared to fossil fuel companies, which makes them profitable for investors. Simultaneously, despite the promising prospects, investors face several risks and barriers to making investment decisions. Among these risks, the most relevant in the current environment are varieties of political and regulatory risks. However, with competent and timely actions, investment risks can be leveled because the built models and the calculated trend of global investment in alternative energy has an upward direction at least until 2026. This indicates the stability and predictability of the alternative energy market and, accordingly, is an argument in its favor when making investment decisions.

In general, the trend of investment in alternative energy has been prevalent for at least the last 12 years, despite the emergence of new options to support the transition to zero-carbon emissions, such as investments in energy storage projects, electric vehicle production, nuclear power, hydrogen projects, etc.

References

1. Antunes ME, Chamberlain L (2022, May 12) 3 urban energy innovations with global implications. World Economic Forum. Retrieved from https://www.weforum.org/agenda/2022/05/3-urban-energy-innovations-with-global-implications/. Accessed 21 Sept 2022
2. Bennett S, Le Marois J-B, Orgland N (2021, July 23) Ten years of clean energy start-ups. IEA, Paris, France. Retrieved from https://www.iea.org/articles/ten-years-of-clean-energy-start-ups. Accessed 23 Sept 2022

3. Chernyak O, Chernyak Y, Fareniuk Y (2017) Forecasting of global new investment in renewable energy. In: Proceedings of HAICTA 2017: 8th international conference on information and communication technologies in agriculture, food and environment, Chania, Greece, pp 272–278. Retrieved from https://ceur-ws.org/Vol-2030/HAICTA_2017_paper33.pdf. Accessed 25 Sept 2022

4. Climate Policy Initiative (CPI) (2021) Global landscape of climate finance 2021. Retrieved from https://www.climatepolicyinitiative.org/wp-content/uploads/2021/10/Full-rep ort-Global-Landscape-of-Climate-Finance-2021.pdf. Accessed 25 Sept 2022

5. Frankfurt School-UNEP Centre and Bloomberg NEF (2020) Global trends in renewable energy investment, 2020. Frankfurt School—UNEP Collaborating Centre, Frankfurt am Main, Germany. Retrieved from https://www.fs-unep-centre.org/wp-content/uploads/2020/06/GTR_ 2020.pdf. Accessed 25 Sept 2022

6. Gaulin N, Le Billon P (2020) Climate change and fossil fuel production cuts: assessing global supply-side constraints and policy implications. Climate Policy 20(8):888–901. https://doi.org/ 10.1080/14693062.2020.1725409

7. Hall S (2022, July 7) These charts show record renewable energy investment in 2022. World Economic Forum. Retrieved from https://www.weforum.org/agenda/2022/07/global-renewa ble-energy-investment-iea/. Accessed 23 Sept 2022

8. IEA and Centre for Climate Finance & Investment (2021) Clean energy investing: global comparison of investment returns. Centre for Climate Finance & Investment of the Imperial College Business School, London, UK. Retrieved from https://iea.blob.core.windows.net/ assets/ef1d6b50-66a6-478c-990e-ee227e2dd89b/Clean_Energy_Investing_-_Global_Compar ison_of_Investment_Returns.pdf. Accessed 23 Sept 2022

9. International Energy Agency (IEA) (2020) Database: fuel supply. Retrieved from https://www. iea.org/reports/world-energy-investment-2020/fuel-supply. Accessed 21 Sept 2022

10. International Energy Agency (IEA) (2022, May 11) Renewable power is set to break another global record in 2022 despite headwinds from higher costs and supply chain bottlenecks. Retrieved from https://www.iea.org/news/renewable-power-is-set-to-break-another-global-rec ord-in-2022-despite-headwinds-from-higher-costs-and-supply-chain-bottlenecks. Accessed 25 Sept 2022

11. International Renewable Energy Agency (IRENA) (2022) World energy transitions outlook 2022: 1.5 °C pathway. IRENA, Abu Dhabi, UAE. Retrieved from https://irena.org/publicati ons/2022/mar/world-energy-transitions-outlook-2022. Accessed 25 Sept 2022

12. Khmyz OV, Oross TG, Prudnikova AA (2023) Factors of attractiveness of green bonds as a financing tool for countering adverse climate change. In: Popkova E, Sergi B (eds) Current problems of the global environmental economy under the conditions of climate change and the perspectives of sustainable development. Advances in Global Change Research, vol 73. Springer, Cham. https://doi.org/10.1007/978-3-031-19979-0_4

13. United Nations (2015) Paris agreement. UN, New York, NY. Retrieved from https://unfccc.int/ sites/default/files/english_paris_agreement.pdf. Accessed 12 Sept 2022

14. World Bank (2021, June 7) Report: universal access to sustainable energy will remain elusive without addressing inequalities. Retrieved from https://www.worldbank.org/en/news/press-rel ease/2021/06/07/report-universal-access-to-sustainable-energy-will-remain-elusive-without-addressing-inequalities. Accessed 25 Sept 2022

15. World Economic Forum (WEF) (2021) Fostering effective energy transition, 2021: insight report, April. Retrieved from https://www3.weforum.org/docs/WEF_Fostering_Effective_Ene rgy_Transition_2021.pdf. Accessed 19 Sept 2022

Smart Digital Innovations in the Global Fashion Industry and a Climate Change Action Plan

Natalia Yu. Konina🄳

Abstract The increasing importance of global warming and the demand to minimize carbon footprint determines the relevance of this research. The global fashion sector comprising textile, garments, and offline and e-commerce retailers, significantly impacts climate issues: 10% of global greenhouse gas emissions come from the global fashion sector, with the potential for further negative impacts. Moreover, the global fashion sector generates 2.1 billion metric tons of greenhouse gases annually. This paper is devoted to minimizing the negative climate impact of the global fashion industry by adopting solutions based on Industry 4.0. The research focuses on a combination of technologies and organizational efforts required to improve the environmental impact of the fashion sector. The research aims to determine the existing and applied technologies connected with Industry 4.0 that can significantly improve the climate impact of the global fashion industry. The research tasks suggest a detailed comparative analysis of the existing and forthcoming new technologies based on digital advances and their large-scale implementation in the fashion sector. The possibilities for decarbonization of the fashion industry start with fiber production. The existing technologies, if implemented correctly, can significantly diminish greenhouse emissions while creating or growing fibers. The author investigates the conditions necessary for the implementation of these technologies. The research confirms that the existing technologies (including AI, RFID tags, big data analytics, and forecasting) can minimize carbon dioxide emissions. Textile and clothing factories need to use green energy. Big data can radically improve behavior forecasting and reduce extra carbon dioxide emissions. The leading research method is systemic and comparative analysis based on statistical data. The analysis indicates that significant changes in consumer behavior and approaches are required to implement Industry 4.0 to minimize the negative fashion impact on the climate. The practical significance of the research is determined by the extensive and growing impact of the global fashion industry on climate warming and the potential to improve the situation.

N. Yu. Konina (✉)
MGIMO University, Moscow, Russia
e-mail: nkonina777@gmail.com

Keywords Industry 4.0 · Digital innovations · Fashion industry · Sustainability · Decarbonization · Global warming · Fashion · Textiles · Fibers

JEL Classification L67 · Q53 · Q55

1 Introduction

The fashion industry is one of the most important industries with a significant economic impact worldwide. Recently, the fashion industry's significant contribution to global warming and environmental issues has attracted the attention of buyers. Significant innovations in the fashion industry have focused on green products. Smart innovations, such as the Internet of Things (IoT), virtual reality (VR), augmented reality (AR), artificial intelligence (AI), and 3D printing, offer new opportunities by transforming the traditional methods of product design, manufacturing, marketing, and sales [7].

The increasing importance of global warming and active attention to these issues from key economic actors determines the relevance of this research. The global fashion sector, including textiles, apparel, offline retail, and e-commerce, significantly affects environmental and climate issues. By some estimates, the fashion industry accounts for up to 10% of the world's CO_2 emissions, 20% of the world's industrial wastewater, 24% of insecticides, and 11% of pesticides used [14].

Climate stabilization requires reliable, fast, and sustainable global action to reduce greenhouse gas emissions and achieve zero net CO_2 emissions. The IPCC says the world needs to cut carbon emissions by up to 58% by 2030 to stay within 1.5 °C of the global temperature rise.

According to McKinsey, the global fashion industry emits 2.1 billion metric tons of greenhouse gases annually, which is 10% of the world's total annual greenhouse gas emissions, more than the combined emissions of Germany, France, and the UK [1]. The textile industry is the second most polluting industry globally after the oil industry, accounting for about 2.1 billion tons of greenhouse gas emissions (more than international flights and sea transport combined) [13]. By 2050, the fashion industry will use up to 25% of the global carbon budget [14]. There is an urgent need to tackle the adverse impact of the fashion industry on sustainability through approaches to a circular economy, which is defined by three main areas of activity, namely: reuse, recycling, and the reduction in use [4]. Consumers want profound changes in the approach to minimize the fashion industry's negative impact on the environment in general and on global warming in particular [12]. Achieving this goal requires coordinated efforts by firms and society, as well as changes in consumer behavior.

2 Methodology

The research is devoted to the problems of practical solutions based on Industry 4.0 to reduce the impact of the global fashion industry on global warming.

The research aims to determine whether existing and applied technologies associated with Industry 4.0 can radically improve the impact of the global fashion industry on the climate and the environment.

The research objectives involve a comparative analysis of the climate impact of existing and future digital technologies and their practical implementation in fashion. This research aims to analytically test the following hypotheses:

1. The growing influence of Industry 4.0 technologies on the current state of the fashion industry;
2. The existing prospects for improving the fashion industry in terms of climate policy;
3. The need for a set of measures and broad international cooperation to reduce the negative impact of the global fashion industry on the climate.

An integrated approach is applied to analyzing aspects of global warming and changes in the global fashion industry. The leading research method is systemic and comparative analysis. The authors applied a systemic approach to consider several diverse elements previously studied separately. The author used a targeted sample approach when selecting company cases because it offers the most representative cases. The sampling criterion was whether the company had implemented at least one of the 4IR technologies, which impacted its carbon footprint. To identify common cases, the author used a broad web search.

The practical significance of this research is determined by the significant and growing influence of the global fashion industry on climate warming and the potential for improving the situation.

The most important theoretical and practical issue of the fashion industry's influence on global climate change is to determine whether implementing smart innovations can diminish the fashion industry's negative impact on the climate. In conclusion, based on the conducted analysis, the author highlighted the main outcomes.

3 Results

The influence of the fashion industry on climate begins with the production of raw materials, particularly textile fibers and yarns. The existing technologies, if properly applied, can significantly reduce greenhouse gas emissions from the creation or growing of fibers.

The $3 trillion global fashion industry uses significant amounts of raw materials (particularly textile fibers and yarns, primarily polyester and cotton), water, and electricity to produce 100 billion garments and accessories annually. About 60% of produced garments are thrown away within a year. McKinsey conducted a study showing that today's fashion industry produces 10% of greenhouse gas emissions, putting the industry in second place on the impact of global warming after energy [1]. The environment suffers significantly due to ill-conceived overproduction, suboptimal logistics, and fast fashion trends. The acceleration of fashion leads to an even more significant surplus in apparel consumption, which means a considerable increase in production volumes and related pollution [5]. It is estimated that the global middle class could be 5.4 billion by 2030, up from 3 billion in 2015. Each North American consumer buys 37 kg of textiles annually. In Western Europe, this figure is estimated at 22 kg. The consumption of textiles and clothing in Africa, the Middle East, and India is around 5 kg per capita.

Demand for clothing and other lifestyle-defining commodities can be expected to increase for middle-income people. If the consumption of fashionable clothes continues at the current level, three times, more natural resources will be required by 2050 compared to what we used in 2000. Current "fast fashion" means that the garment is produced at a rate of 50 cycles per year, compared to two cycles of traditional fashion [10].

Fast fashion, popular since the early 2000s, has pushed overconsumption and a concise life cycle of purchased garments into the mainstream through its ability to offer a wide variety of fashion items at low prices [9].

In the highly competitive apparel, fashion, and footwear world, integrated manufacturing technologies are rapidly evolving in line with Industry 4.0 standards [2]. The rapid development of new technologies makes possible innovations in manufacturing that optimize existing systems and lead to new processes. Key digital technologies (e.g., new mobility, AI, AR, VR, 3D, cloud computing, business intelligence, and social media) transform the global marketplace creating unexpected market risks and opportunities. Many companies from different sectors of the global economy have already undergone a digital transformation, with their business models influenced by the twin forces of digital disruption and digital globalization [6].

The IoT, VR, AR, AI, and blockchain implemented as standalone and combined solutions are the cornerstone of Industry 4.0, with IoT leading the fashion industry 4.0, providing various economic benefits for companies [8].

Market forces dominate the global textile, apparel, and footwear industries; a small number of large companies dominate, while the transnational companies leading in fashion are adept at leveraging established global value chains. Global competition between brands and retailers amid the active use of digital technologies puts downward pressure on companies' prices and profits.

Chemical fibers include synthetic and artificial fibers. However, synthetic fibers include fibers such as polyester, nylon, acrylic, and others. Currently, synthetic fibers dominate in fiber production (~ 69%). The rest is made up of natural fibers and blends of natural and synthetic fibers. The primary textile fiber is polyester, growing from 3.37 million tonnes in 1975 to 57.1 million tonnes in 2020 [16]. Simultaneously, the production of polyester for textiles in 2015 alone resulted in 706 billion kg of greenhouse gas emissions, equivalent to the annual emissions of approximately 185 coal-fired power plants. A polyester shirt has a higher carbon footprint than a cotton shirt (5.5 kg vs. 4.3 kg). The carbon footprint of a polyester t-shirt is 20.56 kg CO_2 throughout its entire life cycle. The consumer use phase yields the highest carbon footprint at 30.35%. Polyester fiber manufacturing is the second-largest carbon footprint source (28.94%). Spinning is an essential part of the carbon footprint in the t-shirt production phase. The dyeing, finishing, and pretreatment stages depend on different chemicals and energy sources. The carbon footprint of the transportation stage depends on the transport mode and the trip's distance [11].

The most significant decarbonization technologies relate to the materials used. The ability of startups to innovate and attract the attention of large apparel brands is fundamental. London-based Post Carbon Lab wants to further reduce emissions by converting fabrics into carbon-absorbing surfaces. Designers and researchers Dian-Jen Lin and Hannes Hulstaert use photosynthetic coating technology, which involves introducing microorganisms such as seaweed. The resulting textiles with living organisms can absorb climate-changing carbon dioxide [17].

Among the achievements of Industry 4.0 that can affect the carbon footprint of the global fashion industry, the following ones have the most significant potential in terms of raw materials:

- Alternative natural fibers;
- Genetically modified fiber crops;
- Bio-leather.

One of the promising technologies of the fourth industrial revolution is the development of alternative natural fibers and precision farming for fiber crops. These technologies can produce textiles with superior properties from renewable and biodegradable raw materials, including textile fibers from forest plants. Through green chemistry and enzyme science, innovative companies have solved the problem of the stiffness of natural fibers such as flax, hemp, or jute. Hemp requires much less water than cotton; it also proliferates, and its roots aerate the soil, making it fertile for future crops. Moreover, hemp provides about three times more fiber per acre than cotton.

Lyocell fibers made from cellulosic pulp from wood or bamboo are one of the available solutions for fiber availability. Lyocell fibers are five to six times stronger than cotton and garments made from cotton. To promote further wood-based specialty fibers, the biggest global producer Lenzing (Austria) released new features on its E-Branding Service platform. While the sustainability of the production of wood fiber is highly dependent on wood sources and chemical treatments used, these materials can provide high environmental performance. The benefits of wood-based fibers come with small risks associated with land use and deforestation.

Another technology of the Fourth Industrial Revolution essential for decarbonization is genetically edited fiber crops. Genome editing technologies increase the yield of fiber crops. It can manipulate the physical properties of yarns and fabrics. The complex genome characterization of allotetraploid cotton is challenging. However, recent studies have shown that CRISPR/Cas9-mediated mutation of cotton genes is possible, although the heritability of these gene modifications requires further research. With cotton accounting for 2.4% of the world's cropped area and consuming 24% and 11% of global insecticide and pesticide sales, respectively, further improvements in cotton grown could significantly impact sustainability [15].

The development of cotton genome editing opens the possibility of applying genome editing to other natural fiber plants. Genetic modifications in cotton have reduced insecticide use and increased yields. Nevertheless, it has unintended consequences such as loss of biodiversity and monopoly on seed supplies. This technology can solve the problems of declining yields due to soil erosion, water storage, and overuse of agrochemicals while providing value-creation opportunities for industry leaders and significant cotton-exporting countries such as China, India, and the USA. Fiber plant varieties can be adapted on a large scale to increase yields, reduce maintenance costs, and increase farmer income, positively impacting economic growth and GDP in cotton-producing countries while reducing supply-side risks [3].

Open-source platforms with commercial plant genome data and editing technologies are vital to smart innovations for improved, more resource-efficient textile fibers.

An interesting technology of the Fourth Industrial Revolution is biofabricated leather, i.e., the production of skin without using animal skins based on the use of biological tissue grown in the laboratory from self-created collagen cells. Collagen is refined and processed through a simplified tanning process that uses fewer chemicals. There is no waste because the size and shape are determined by design; physical properties such as the variable topography of the sheet are customizable. This process is faster and cleaner, resulting in an ethical product with less environmental impact. Biofabricated leather can compensate for the reduced supply of traditional leather. The production of biofabricated leather is an emerging development with a long-term commercial perspective. Biofabricated leather, the latest option available in a long list of skin substitutes and alternatives, is being developed by Modern Meadow, a US biotech company. The company has received over $180 million in investment. It ferments specific types of yeast to grow collagen—one of the main components of traditional leather—and then processes this protein to create a leather material that does not involve livestock in its supply chain [3].

Leather is the most economically important byproduct of the meat industry. The environmental impact of the skin is related to the effects of industrial agriculture and leather processing, the latter being highly toxic. For example, levels of toxins associated with tanning in the Gange River near Kapoor, India, are around 6.2 mg per liter, up from the government-set limit of 0.05 mg per liter. It is estimated that 430 million cows will need to be killed annually by 2025 to meet the world's fashion needs, with potentially significant benefits from lab-grown materials. Laboratory-grown or biotech leather can be lighter, thinner, and more substantial. Current technology makes it possible to develop new products in which laboratory-grown leather can mimic the properties of even rare or extinct animals [3].

A significant part of textiles' carbon emissions is connected with fabric finishing, dyeing, and printing with chemical dyestuffs. Moonlight Technologies (USA) tries to change the way goods are produced by smart carbon-negative, plant-based technologies for fabrics, working with plant-based chemistry and sustainably derived, non-toxic, biodegradable colorants for hard goods and textiles.

Digitalization allows designers and brands to find almost any textile on the net, helping to build a circular textile economy. There are many new material and textile digital libraries that help market players to refine their search by demonstrating how the fabrics look and feel in reality. Using metaverse and digital technologies, brands have started to embrace virtual solutions that allow users to try on clothes virtually. By feeding basic information into the app, such as gender, height, and weight, consumers can digitally visualize how the clothing will look and get a feel of the textile the cloth is made of. This can reduce the necessity to have sample pieces resulting in less deadstock.

The embodiment of AR makes it possible to try on clothes, shoes, and accessories using the camera and evaluate how well a particular cloth will look on a person. Large brands and stores have increasingly resorted to this technology in recent years. US retailer Walmart recently bought Zeekit, which develops virtual fitting rooms, and such technologies are starting to appear from Farfetch, IWC, Massimo Dutti, Allbirds, and others. Gucci has made it possible to try on shoes right in the application. Behind these developments is the Belarusian company Wanna, which took on AR projects in 2017. Cooperation with Gucci in 2019 was their first, yet very successful, experience working with virtual fashion.

The carbon footprint of any textile product depends on the longevity of the consumption and its reuse. The fibrous product can be reused, recycled, incinerated, or disposed of in landfills. Smart innovations are used to identify and reuse fibers. TrinamiX, a Germany-based provider of mobile spectroscopy solutions, developed a technology that makes it possible to identify and better sort more than 15 textile types and compositions. A good solution to minimize consumption and connected CO_2 footprint is the recent launch of online renting platforms. The famous British fashion chain Marks & Spencer has launched its versatile and largest 78-piece A/W'22 collection on the biggest UK rental platform Hirestreet.

Depending on the geography, there is a demand for entirely different types of clothing and sizes. Marketplaces are good solutions. One of the new sustainable marketplaces for carbon-neutral fashion items is launched by Verte Mode (USA) to change the way goods are produced and enable consumers to transform their shopping behaviors.

Carbon-neutral fashion means less waste and less transportation. It makes it possible to automatically choose the suitable types and sizes of clothes, depending on the demand, and deliver them to specific stores. For example, in the Netherlands, they buy more clothes with long sleeves, and in China with short ones. The next step is to use localized models for different countries, considering the peculiarities of the figure of the local population. AI solutions for optimal logistics and distribution utilized by famous fast fashion brands are a good example. Less transportation brings less carbon footprint.

4 Conclusion

In the new emerging architecture of the world economy, global warming and sustainable development are playing an increasingly important role. Prospects for existing technologies ranging from AI smart factories, RFID tags, AI analytics, and forecasting can minimize waste and improve the efficiency of fashion companies and their carbon footprint. Textile and garment factories can and should switch to clean energy.

While no alternative to natural fibers can replace cotton or take over a significant market share, combining these materials can significantly reduce the industry's environmental footprint. The technology is already improving primary synthetic fiber polyester production and reducing the carbon footprint. Nevertheless, further research and investment are needed. Significant opportunities to accelerate sustainable production are associated with artificial cellulose fibers such as Tencel-Lyocell. Big data can dramatically improve consumer response and behavior prediction. The next frontier has to do with AR, VR, and virtual clothing.

References

1. Berg A, Magnus K-H, Granskog A, Lee L (2020, Aug 26) Fashion on climate: report. McKinsey & Company. Retrieved from https://www.mckinsey.com/industries/retail/our-ins ights/fashion-on-climate. Accessed 28 Sept 2022
2. Bertola P, Teunissen J (2018) Fashion 4.0. innovating fashion industry through digital transformation. Res J Text Apparel 22(4):352–369. https://doi.org/10.1108/RJTA-03-2018-0023
3. Esposito M (2018) White paper: driving the sustainability of production systems with fourth industrial revolution innovation. World Economic Forum, Geneva, Switzerland. Retrieved from https://www3.weforum.org/docs/WEF_39558_White_Paper_Driving_the_ Sustainability_of_Production_Systems_4IR.pdf (Accessed 28 September 2022)
4. Jia F, Yin S, Chen L, Chen X (2020) The circular economy in the textile and apparel industry: a systematic literature review. J Clean Prod 259:120728. https://doi.org/10.1016/j.jclepro.2020. 120728
5. Kong H-M, Ko E, Chae H, Mattila P (2016) Understanding fashion consumers' attitude and behavioral intention toward sustainable fashion products: focus on sustainable knowledge sources and knowledge types. J Glob Fash Market 7(2):103–119. https://doi.org/10.1080/209 32685.2015.1131435
6. Konina N (2021) Introduction: at the dawn of the fourth industrial revolution-problems and prospects. In: Konina N (ed) Digital strategies in a global market. Palgrave Macmillan, Cham, Switzerland, pp 1–12. https://doi.org/10.1007/978-3-030-58267-8_1
7. Konina NY, Sapir EV (2021) Geo-economic aspects of the "green economy" in industry 4.0. In: Zavyalova EB, Popkova EG (eds) Industry 4.0. exploring the consequences of climate change. Palgrave Macmillan, Cham, Switzerland, pp 337–352. https://doi.org/10.1007/978-3-030-75405-1_30
8. Konina N, Dolzhenko I, Siennicka M (2020) The evolution of fashion consumer perception in post-industrial Era. In: Kovalchuk J (ed) Post-industrial society. Palgrave Macmillan, Cham, Switzerland, pp 223–233. https://doi.org/10.1007/978-3-030-59739-9_18

9. Lu S (2018, Sept 4) Changing trends in world textile and apparel trade. Retrieved from https://www.just-style.com/analysis/changing-trends-in-world-textile-and-apparel-trade_id134353.aspx. Accessed 28 Sept 2022
10. Marques A, Moreira B, Cunha J, Moreira S (2019) From waste to fashion—a fashion upcycling contest. Procedia CIRP 84:1063–1068. https://doi.org/10.1016/j.procir.2019.04.217
11. Moazzem S, Crossin E, Daver F, Wang L (2018) Baseline scenario of carbon footprint of polyester t-shirt. J Fiber Bioeng Inf 11(1):1–14. https://doi.org/10.3993/jfbim00262
12. Moran CA, Eichelmann E, Buggy CJ (2021) The challenge of "Depeche Mode" in the fashion industry—does the industry have the capacity to become sustainable through circular economic principles, a scoping review. Sustain Environ 7(1):1975916. https://doi.org/10.1080/27658511.2021.1975916
13. Nature Climate Change (2018) The Price of fast fashion. Nat Clim Chang 8(1):1. https://doi.org/10.1038/s41558-017-0058-9
14. Rana S, Pichandi S, Karunamoorthy S, Bhattacharyya A, Parveen S, Fangueiro R (2015) Carbon footprint of textile and clothing products. In: Muthu SS (ed) Handbook of sustainable apparel production. CRC Press, Boca Raton, FL. Retrieved from https://www.routledgehandbooks.com/; https://doi.org/10.1201/b18428-10. Accessed 1 Nov 2022
15. Shirvanimoghaddam K, Motamed B, Ramakrishna S, Naebe M (2020) Death by waste: fashion and textile circular economy case. Sci Total Environ 718:137317. https://doi.org/10.1016/j.scitotenv.2020.137317
16. Statista (2021) Production of polyester fibers worldwide from 1975 to 2021. Retrieved from https://www.statista.com/statistics/912301/polyester-fiber-production-worldwide/. Accessed 29 Sept 2022
17. Win TL (2020, Oct 14) Polyester from pollution? Fashion's next generation goes green. Thomson Reuters Foundation. Retrieved from https://news.trust.org/item/20201014125237-kvnf3/. Accessed 28 Sept 2022

Adopting Ecological Mindset in the Process of Teaching English for Professional Purposes: Project-Based Approach

Lyudmila S. Chikileva and Elena A. Starodubtseva

Abstract The research aims to demonstrate the importance of students' ecological awareness and the ways to develop it in the process of educating professionals for the green economy. The research is based on a project-based approach. Bloom's Taxonomy was the methodological basis of the research in the context of ecological awareness. The authors used the survey to obtain data and create learners' profiles. The method of semantic interpretation was applied to analyze the data. Participating in the project based on an interdisciplinary approach makes it possible to form various types of competencies, develop critical thinking, and pay special attention to metacognition and different types of intelligence. It becomes possible due to project-based teaching, the purpose of which can vary depending on the teacher's and learners' objectives. The main result is that students become self-aware of their role in the country's future economy as ecologically minded professionals who can contribute to the green economy and the preservation of nature. The research has proved that project-based teaching is applicable to building students' ecological awareness, developing critical thinking, understanding their future role, and contributing to the green economy.

Keywords Green economy · Framework program · Higher-order thinking skills · Project-based teaching · Metacognition

JEL Classification A22 · D83 · I2G0 · J24 · M14 · O10

L. S. Chikileva (✉) · E. A. Starodubtseva
Financial University under the Government of the Russian Federation, Moscow, Russia
e-mail: lchikileva@fa.ru

E. A. Starodubtseva
e-mail: eastarodubtseva@fa.ru

© The Author(s), under exclusive license to Springer Nature Switzerland AG 2023
E. G. Popkova (ed.), *Smart Green Innovations in Industry 4.0 for Climate Change Risk Management*, Environmental Footprints and Eco-design of Products and Processes, https://doi.org/10.1007/978-3-031-28457-1_28

1 Introduction

Nowadays, Russia is facing not only such traditional challenges as the need to diversify the economy and reduce dependence on the raw materials sector but also the decline in global business activity due to the COVID-19 pandemic, geopolitical factors, the accompanying drop of prices and decreased demand for fossil fuels, which can become long-term. As some countries have begun a large-scale transformation of their economic systems in response to changes in the global climate and environmental crisis [1, 5–9], Russia risks lagging behind the leading economies of the world in its economic and technological development if it does not take measures to respond timely and properly to these events. The way of getting out of the crisis is to determine Russia's economic development in the coming years or even decades.

It is important to consider that the ideas of the Green Course are to stimulate the economy through investments in infrastructure and create new jobs with an emphasis on green technologies. The Green Course of Russia is a framework program for the long-term development of Russia up to 2050. University students are exactly those people who are bound to realize this program. Therefore, it seems to be of great importance to develop their skills to make them capable of implementing smart green innovations [14]. It will be possible due to developing their higher-order thinking skills (HOTS), interdisciplinary education, and metacognition. These skills can be acquired in the lessons of English for Professional Purposes. If students are inspired to adopt a sustainability mindset, they will be able to achieve sustainability goals after graduation when they start their professional careers. Students can be given a green quiz to get an idea of their own footprint. After answering the questions, they will find out how sustainable their lifestyle is and what improvements may be possible. Project-based teaching can be used to prepare students to become ecologically minded professionals.

2 Materials and Methods

To realize the research goal, we used the following strategies, approaches, and methods: Bloom's Taxonomy, creating learners' profiles, questionnaires, the Likert scale, and a project-based approach. We asked students to fill in questionnaires, which helped carry out self-assessment and multiple intelligences self-assessment. Their answers were measured according to the Likert scale. A total of 145 students participated in the survey. They were university students who studied English as a foreign language. To define their naturalistic intelligence, we asked them to agree or disagree (using answers from strongly agree to agree, neutral, disagree, and strongly disagree) with several statements. Then the students were to record their scores in the table and answer the question "What are the three categories in which you scored the highest?" There were two stages in the survey: at the beginning of the term and its end, after participating in the project "My Ecological Impact," aimed at developing linguistic, economic, ecological, and metacognitive competencies. The aim of the project "My Ecological Impact" was to investigate environmental footprints and find out further possibilities for taking part in eco-industrial projects.

The students were given the project "My Ecological Impact," aimed at developing linguistic, economic, ecological, and metacognitive competencies. First, they were to answer the questions, record their scores in the table, and determine the category in which they scored the highest. This procedure was repeated after the experiment and showed the development of all types of intelligence, competencies, and expertise.

The project's main aim was to investigate environmental footprints and possibilities of participating in future eco-industrial projects.

3   Results

For many years, experts in the field of education have sought to develop a clear and accessible theory that would help teachers to develop students' thinking skills in a more effective and systemic way, as well as to assess these skills. The most famous model describing the process of thinking is Bloom's Taxonomy. This model consists of thinking skills structured from the most basic to the most advanced level. Bloom's Taxonomy is a tool that helps track the performance of all types of student intelligence, its development, and assessment.

The taxonomy includes remembering, understanding, application, analysis, synthesis, and evaluation. Remembering and understanding are most widely represented in the learning process: students learn some material and reproduce it.

The weak points of the original classification of Bloom's pedagogical goals are considered to be the absence of a level responsible for productive and creative activity and the lack of actions involving problem-solving or project activities.

In the late 1990s, a group of scholars, including Lorin Anderson, Bloom's former student, compiled an updated version of Bloom's taxonomy, which is more consistent with the educational practice of the twenty-first century. The new version includes the highest component—creativity. This skill combines what is already known to create something new, for which students generate, plan, and produce. The new taxonomy distinguishes between the knowledge of "what" (i.e., the content of thinking) and the knowledge of "how" (i.e., the procedure used for solving problems).

According to Anderson and his colleagues, meaningful learning provides students with the knowledge and access to the cognitive processes they will need to solve problems successfully [2]. We can illustrate it with the following example: different questions can be aimed at developing and assessing different levels of thinking. Simple questions involve knowing certain facts and operating on a specific topic (information). This type of questions is often present in traditional forms of control: tests, terminological dictations, etc. The purpose of clarifying questions is to provide feedback on what has been said. Clarifying questions are also asked to obtain information that is not in the message but implied. Such questions usually begin with the following wording:

- "That is, you say that ...?";
- "If I understand correctly, then ...?";
- "Do you really think that ...?".

Interpretive or explanatory questions usually begin with the word "why." They can be aimed at establishing cause-and-effect relationships (e.g., "Why do you think so?"). If the student knows the answer to this question, then it turns from an interpretative one into a simple one. Therefore, this type of question works when there is an element of independence in the answer to it. Creative questions can be formulated in the conditional mood, which implies an element of assumption and an impulse to fantasy. Evaluative questions aim to clarify the criteria for evaluating certain events, phenomena, and facts (e.g., "How does franchising differ from a startup?"). Practical questions aim to establish the relationship between theory and practice (e.g., "Where can you use statistics or accounting in everyday life?").

Analyzing numerous classroom scenarios, we can see that many teachers do not consider promoting equity. As a rule, only about 25% of students are involved. The situation can be changed if we use project-based tasks, presentation of teamwork, cooperative learning, etc. A teacher is supposed to ensure a learner-centered approach that encourages active learning, critical thinking, and problem-solving.

We should consider all language learners whom we teach. Knowing them, we can change teaching and assessment methods to create an educational environment where everyone is encouraged to study despite their gender, level of English, and individual traits. When we start teaching a new group of students, we can find out their goals and what motivates them to achieve them. Starting teaching, it is necessary to remember that students' types of intelligence should be considered.

Psychologist Howard Gardner introduced his theory about several types of intelligence in 1994. He also introduced the concept of multiple intelligence, which includes all its types. According to G. Gardner, school and university systems deprive students of whole-range opportunities to learn something that is not part of the curriculum, which leads to a lack of success in the future and the inability to realize their potential [11].

We should definitely mention that the term "intelligence" is used differently compared with the classical definition. According to the Cambridge dictionary, intelligence is the ability to understand and learn well and to form judgments and opinions based on reason. Not all of Gardner's types of intelligence are based on reason, though his desire to distinguish different students' inclinations are worth considering. Let us look at them briefly.

A fairly common type of intelligence can be called logical-mathematical. Students possessing this type of intelligence do not experience any difficulties with mathematics and everything related to it. They easily remember phone numbers, dates, or other numerical values; they also quickly solve examples and problems. The main characteristic of such people is the ability to structured analysis, simple installation of cause-and-effect relationships, and systematization of all components of life. We can develop and maintain logical and mathematical intelligence with the help of regular training. Solutions to various tasks, puzzles, or mathematical examples are perfect. It would not be superfluous to allow the student to apply the skills in everyday life.

The next type of intelligence distinguishes its owners with a special love for linguistics and everything that concerns it. It helps kids learn to speak, listen, write, and read quickly. These students adore literature; they like to study foreign languages, create something, and participate in interviews. As a rule, they become sociable people. Education for such students is best carried out in a group form. It is important that the student can discuss anything with others. Simultaneously, communication becomes the main activity for development; reading can be no less useful.

According to Gardner, one of the most useful of the eight main types of intelligence is interpersonal communication [11]. Students can easily interact with each other, make a good impression, and become good leaders. The main way to develop this intelligence is communication. Students should be provided with the opportunity to participate in various discussions and performances on stage, as well as participate in the organization of events.

Students with visual-spatial intelligence have no problems with geographical orientation or remembering new places. They are good at choosing colors and matching clothes. They can easily solve problems related to geometry, drawing, and painting. Their great advantage is the skill of clear visualization. Like any other type of intelligence, according to Gardner, visual-spatial intelligence can be developed [11]. It may be useful for students to draw, visualizing real or fictional objects. We can ask them to find their own solutions for various problems.

For students with the body-kinetic type of intelligence, their own body becomes the main way to express their feelings. They like to dance, play sports, craft various objects, and perform in front of other people. These activities help them to be healthy and achieve success in many areas. They may also have a character trait expressed in a vivid desire to compete.

The existential or intrapersonal type of intelligence is one of the most special types that help get a high level of awareness, easily understand feelings, talk about difficult issues, and draw the right conclusions. It is very important to allow these students to express their opinions, listen to them, and support some ideas. Only such an attitude will help them develop their intelligence to the desired level.

Those who have musical intelligence are very actively drawn to music. They quickly learn to play musical instruments, are fond of famous musicians' works, tend to attend concerts, and turn on loud music. These people easily remember melodies and sing songs without paying much attention to what others will say about their singing.

Naturalistic is another type of intelligence. It makes a person capable of certain types of activities [11]. The main interest of such people is the surrounding world. They enjoy interacting with nature. Experiments with natural phenomena often become one of the main tasks for students with naturalistic intelligence, for example, the germination of plants. They like being alone with nature. Students with such intelligence often take measures to prevent pollution and the planet's destruction.

As all types of intelligence can be developed. With the help of our project, we aimed to develop the following skills:

- Visual-spatial—students will go outside, watch, and create videos;
- Interpersonal—students will learn cooperatively, working in small groups;
- Intrapersonal—students will think and raise their awareness individually by finding and selecting new material;
- Musical—students can perform songs connected with flora;
- Linguistic—students will have a debate, make a presentation, and improve their English reading, writing, listening, and speaking skills;
- Logical-mathematical—students will collect and analyze statistics and present charts;
- Naturalistic—students will work out different measures to preserve nature and become aware that everyone can and must start preserving the environment not only as a citizen but as professional as well.

The desired outcomes would be for the teacher to act as a facilitator and observer, to support but not be overprotective, and to encourage students to work independently. For students, the outcome is to be creative and self-reflective about their learning styles and strategies, as well as to develop metacognition.

Metacognition is defined as thinking about thinking. According to Anderson, metacognitive learners are aware of their own thinking process and know which learning strategies to use at different stages of their learning process [3]. Metacognition follows the sequence of thinking and reflective processes. They are as follows:

- Preparing and planning for learning. It is important for students to think ahead about their learning process with certain learning goals in mind. Teachers can facilitate this process by providing students with specific and achievable learning goals. Successful students can select learning strategies and use metacognitive skills based on their learning situations.
- Selecting and using learning strategies. The ability of learners to identify and use learning strategies to fit a particular learning goal indicates that this learner is metacognitively aware. Teachers should ensure that learners are introduced to various strategies and techniques. Students need to be trained on how to best use these strategies. Moreover, students should be aware that no strategy will fit all their learning needs.
- Monitoring strategy. It is necessary to be able to monitor which learning strategy is used. Students should do periodic self-checks to see whether the strategy they selected is still the best one for the learning task. Teachers can help students learn to control the use of their strategies by periodically pausing and reflecting on the learning process.
- Combining various strategies. The ability to use metacognitive skills in combination is another characteristic of metacognitive learners. Such learners can connect, sequence, and coordinate multiple strategies to achieve the desired learning outcome. Teacher-facilitators can make sure that students are aware of the multiple strategies that are available to them for specific activities.
- Evaluating strategy use and learning. Part of the metacognitive learning process involves being able to assess whether the learning is happening effectively.

Teachers can facilitate such self-evaluation by asking prompting questions that will help learners connect the dots between their intended learning goals and learning outcomes. The examples of such questions are as follows:

- "What am I trying to accomplish?";
- "What strategies am I using?";
- "How well am I using them?".

Lessons of English are important for teaching metacognitive skills. English teachers can help students learn and practice metacognitive strategies, conduct self-reflection, and self-assessment, and become better learners [11].

As for project-based teaching, it is necessary to point out that it must be well-planned. The students were given the project "My Ecological Impact," aimed at developing different competencies—linguistic, economic, ecological, and metacognitive. First, the students were to answer the questions, record their scores in the table, and determine the category in which they scored the highest. This procedure was repeated after the experiment and showed the development of all types of intelligence, competencies, and expertise.

The main goal of the project "My Ecological Impact" was to investigate environmental footprints and possibilities to take part in future eco-industrial projects. It started with investigating the latest ecological projects in Russia [4, 12]. It is reported that the government will finance the construction of eco-industrial parks in 2022–2024. As a result, new production facilities will be able to process more than 850 thousand tons of waste annually [12]. Most students were unaware of the project "Ecology" aimed at environmental education, environmental protection, and preservation of biological diversity. They were also unaware that the project is carried out in the following areas: waste disposal and recycling, landfill elimination, conservation of forests and reservoirs, reduction of emissions into the atmosphere, and development of ecological tourism [4]. There is no doubt that anyone can plant a tree, clean up rubbish at the seaside, on the shore, and in the forest, or sort household waste. However, some aims can be specified for students majoring in Economics and Management. Students can find their own ways, make their own contributions to realizing ecological goals, make economic calculations, and suggest taking managerial measures. One of the main advantages of the project is that it is based on an interdisciplinary approach.

The students learned about new possibilities for realizing their professional knowledge—"there are to appear eco-industrial parks in several regions. As a result, new production facilities will appear on the territories of such parks, including for the processing of tires, plastic, chemical waste, and many others, and new jobs will be created at the enterprises, and the production of products in demand on the market will be mastered" [10].

The main question the students had to answer was "What can be done to improve the ecology, and what am I able to do?" This question is very important as it develops HOTS—analyzing, evaluating, and creating. Not only did the students recognize the existing practice, but they also developed their critical thinking and creative abilities.

The answers were different and promising. For example, answering the question of how to save electricity, the students said that they could do the following:

- Turn off phones, computers, and TV sets from time to time and spend week-ends without these devices;
- Arrange a candlelit dinner;
- Avoid charging gadgets overnight;
- Maintain a good condition of electrical appliances: clean the kettles from scale and the dust collector in the vacuum cleaner from garbage;
- Replace regular light bulbs with fluorescent ones.

The students were mostly interested in recycling. They visited the museum in the ecological center "Sborka," where they saw more than 800 items made from waste. The students also decided to join "Paperbattle," aiming to make gathers of wastepaper. Demonstrating their professional knowledge, the students calculated the ecological and economic results of these measures. They proposed the following:

- If possible, give up disposable paper and plastic dishes;
- Go shopping with a cotton shopper;
- Give boxes a second life: store seasonal clothes and shoes in them or bring boxes to the recycling points;
- Hand over unwanted books to the library or bookcrossing points;
- Giving preference to durable things, which will allow one to be surrounded by high-quality items and produce less garbage.

It is necessary to note that some students who study Economics and Business have become the winners of startup projects' contests and received a grant to implement their projects. The students set a goal to find new high-tech solutions for environmental management, reducing the loss risks of the country's natural assets and combating the effects of global warming. The students who participated in project-based activities have great potential. They have their own vision of the country's environmental problems and opportunities for their solution.

The results of the experiment are presented in Table 1 and in Fig. 1.

The table contains the percentage of answers to the questionnaire based on the Likert scale with the range of answers from "strongly agree" to "strongly disagree." The questionnaire was presented to the students before the project and after it to get evidence of whether their attitude towards ecological problems has changed or not. Their answers show a positive tendency towards ecological awareness; more students appreciate the beauty of nature and are bothered by pollution. Of particular interest is the change in the students' willingness to minimize pollution in any way they can (cumulative changes 36.0%) and sort out waste before throwing it away (cumulative changes 20.8%).

Considering the responses of students before their participation in the project, it should be noted that the majority of their answers to the first statement reflect a neutral attitude to nature (65.6%). Few students strongly agree with this statement (3.4 and 20.7%, accordingly). As for the second statement, the results are different. The attitude towards pollution is mostly negative: 26.2 and 31.9% of the respondents

Table 1 Results of the questionnaire survey before and after the experiment

No.	Questions	Strongly agree, %		Agree, %		Neutral, %		Disagree, %		Strongly disagree, %	
		Before	After	Before	After	Before	After	Before	After	Before	After
1	I appreciate the beauty of nature	3.4	10.1	20.7	36.4	65.6	48.3	5.5	3.3	4.8	1.9
2	I am bothered by the pollution	26.2	31.9	55.9	57.4	13.8	9.3	3.4	1.1	0.7	0.3
3	I try to minimize pollution in any way I can	8.9	11.5	12.4	45.8	61.4	35.8	7.6	5.5	9.7	1.4
4	I like to learn more about nature while reading books and watching films	14.5	25.9	9.6	12.0	52.4	41.7	21.3	18.5	2.2	1.9
5	I sort out waste before throwing it away	6.2	12.1	31.0	45.9	55.2	39.4	6.2	1.8	1.4	0.8

Source Compiled by the authors

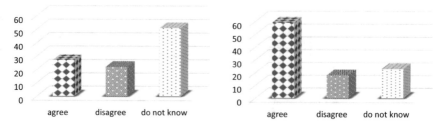

Fig. 1 Percentage of students' answers to the question "Will you do your best to find solutions to ecological challenges in your future professional activity?" before and after the experiment. *Source* Compiled by the authors

strongly agree and agree with the second statement. The answers to the third statement revealing students' efforts to minimize pollution show that most students are neutral (61.4%); some students strongly agree or agree with this statement (12.4 and 8.9%). As for statements 4 and 5, it is apparent that the attitude of respondents is mostly neutral (52.4 and 55.2%).

According to Fig. 1, students' willingness to actively solve ecological issues at their future working place has increased dramatically (from 27 to 59%). They have become aware of their responsibility for the environment not only as citizens but also as future professionals who can make their impact at their working places. They have recognized the opportunities; the majority are determined to develop and realize their ecological mindset in their professional activity.

4 Conclusion

The results of the chosen project showed the development of naturalist intelligence, as well as other types of intelligence, including logical-mathematical, intrapersonal, interpersonal, visual-spatial, and linguistic.

The students focused on what they learned and on how they learned it. As they were given an opportunity to choose how to learn, students had to think about their ways of learning; they could use various learning strategies. According to the Transformative Learning theory [13], the students were able to better understand and become conscious about their way of thinking and viewing the surrounding world. The students were exposed to HOTS:

- Examining, categorizing the species (analyzing);
- Deciding what items should be eliminated and left for comparison, debating while preparing the project (evaluating);
- Planning and formulating the results, presenting them in different ways: pictures, photos, PowerPoint presentations, and video (creating).

Language learners have adopted an ecological mindset. They have become personalities caring about the environment and evaluating their own impact to preserve it as ecologically minded university graduates who will be able to contribute to the green economy.

Acknowledgements The authors are grateful to the reviewers for their evaluating of the article.

References

1. Alsayegh MF, Rahman RA, Homayoun S (2020) Corporate economic, environmental, and social sustainability performance transformation through ESG disclosure. Sustainability 12(9):3910. https://doi.org/10.3390/su12093910
2. Anderson LW, Krathwohl DR (eds) (2001). Longman, New York, NY
3. Anderson N (2002) The role of metacognition in second language teaching and learning. ERIC digest. ERIC Clearinghouse on Languages and Linguistics, Washington, D.C.
4. Autonomous Non-Profit Organization "National projects" (n.d.) National project "Ecology." Retrieved from https://национальныепроекты.рф/projects/ekologiya. Accessed 12 Nov 2022
5. Bhattacharyya A (2019) Corporate environmental performance evaluation: a cross-country appraisal. J Clean Prod 237:117607. https://doi.org/10.1016/j.jclepro.2019.117607
6. Bitencourt CC, de Oliveira Santini F, Zanandrea G, Froehlich C, Ladeira WJ (2020) Empirical generalizations in eco-innovation: a meta-analytic approach. J Clean Prod 245:118721. https://doi.org/10.1016/j.jclepro.2019.118721
7. Boiral O, Heras-Saizarbitoria I, Brotherton M-C (2019) Improving corporate biodiversity management through employee involvement. Bus Strateg Environ 28(5):688–698. https://doi.org/10.1002/bse.2273
8. Chang T-W (2020) Corporate sustainable development strategy: effect of green shared vision on organization members' behavior. Int J Environ Res Public Health 17(7):2446. https://doi.org/10.3390/ijerph17072446
9. Chen F, Ngniatedema T, Li S (2018) A cross-country comparison of green initiatives, green performance, and financial performance. Manag Decis 56(5):1008–1032. https://doi.org/10.1108/MD-08-2017-0761
10. Financial University under the Government of the Russian Federation (2022, Oct 19) Finalists of the "Create Your Own Startup" contest presented their startup project at Russian Energy Week. Retrieved from http://www.fa.ru/News/2022-10-19-omela.aspx. Accessed 11 Nov 2022
11. Gardner HE (2006) Multiple intelligences: new horizons in theory and practice. Basic Books, New York, NY
12. Government of the Russian Federation (2022, Nov 7) The government will finance the construction of pilot eco-industrial parks in the regions. Retrieved from http://government.ru/docs/46984/. Accessed 12 Nov 2022
13. McGonial K (2005) Teaching for transformation: from learning theory to teaching strategies. In: Speaking Teach Center Teach Learn 14(2). Retrieved from http://arrs.org/uploadedFiles/ARRS/Life_Long_Learning_Center/Educators_ToolKit/STN_transformation.pdf. Accessed 11 Nov 2022
14. Melnichuk MV (2020) On the issue of assessing the human capital competencies. Mod Sci Actual Probl Theory Pract. Series: Economics and law 7:56–60. https://doi.org/10.37882/2223-2974.2020.07.25

The Impact of ESG Strategy on Brand Perception of Fuel and Energy Companies

Alina G. Mysakova ⓘ and Kristina S. Zakharcheva ⓘ

Abstract This article regards the impact of Environmental, Social and Governance factors (ESG) on the brand image of companies in the fuel and energy sector. It is highlighted that due to the lack of standard definitions or evaluation models, empirical studies use a wide variety of data and approaches in assessing the dependence on ESG factors of a company's brand value. The article reveals the most substantial factor for the fuel and energy companies brand value - the environmental factor as well as the authors' idea of the importance of each ESG factor in the perception of the company's brand by key stakeholder groups.

Keywords ESG investments · Sustainability · Brand value · Energy company · Brand perception · Sustainable development

JEL Classification M31 · Q29

1 Introduction

The prolonged COVID-19 pandemic has increased people's concern about the quality of their health and well-being, which has raised issues related to the environment. The public demand for environmental friendliness is increasing. Brands actively implement ESG in business practices: launch eco-campaigns and socially responsible initiatives, develop appropriate reporting ESG standards, and protect the environment. Among researchers and practitioners, there is a growing consensus that ESG factors significantly affect the company's market value. Initial research suggests that financial markets reward companies with a good ESG score, while a lower ESG score

A. G. Mysakova (✉) · K. S. Zakharcheva
MGIMO University, Moscow, Russia
e-mail: a.g.mysakova@gmail.com

K. S. Zakharcheva
e-mail: kriszakharcheva@gmail.com

© The Author(s), under exclusive license to Springer Nature Switzerland AG 2023   277
E. G. Popkova (ed.), *Smart Green Innovations in Industry 4.0 for Climate Change Risk Management*, Environmental Footprints and Eco-design of Products and Processes, https://doi.org/10.1007/978-3-031-28457-1_29

may be a sign of higher risk, given that the company is managed less efficiently than other companies in the same industry.

This research examines the influence of environmental, social, and governance (ESG) factors on the brand image of companies from the fuel and energy industry. However, it should be emphasized that due to the lack of standard definitions or evaluation models, empirical studies use a variety of data in assessing the impact of ESG factors on the brand value of companies and different approaches.

This research focuses on fuel and energy companies to address a brand review from the point of view of different stakeholders for three main reasons. First, the energy industry is crucial due to its large size and a high number of stakeholders. According to the EU Science Hub, at the global level, the direct contribution of the energy industry to GDP in the big countries was 15–20% in recent years; it supported more than 60 million jobs worldwide, half of which were in the fossil fuel industries.

The second reason is that the energy industry has a high value of multiplier effect on the economy based on its direct economic impact and its significant indirect and induced impacts. The direct impact is generated by revenues from selling different types of energy resources, employees' salaries, and private investments. The indirect contribution includes government spending for supporting the activity of energy companies, capital investments by all industries directly involved in upstream and downstream industries, and, ultimately, the supply-chain effect generated by direct purchases of fuel and energy products. The induced impact is based on the spending by those who are directly or indirectly employed by this industry.

The third reason is that the companies from the energy industry have one of the most profound impacts on the environment through air pollution, climate change, water pollution, thermal pollution, and solid waste disposal. The emission of air pollutants from fossil fuel combustion is the major cause of urban air pollution. Burning fossil fuels is also the main contributor to the emission of greenhouse gases, making up to 68% of global greenhouse emissions.

The awareness of the importance of ESG strategy in creating value for companies and society has led companies to strengthen their efforts in integrating ESG principles into their activities and provide the necessary data (non-financial disclosures) required by socially responsible investors. The public report on ESG issues and participation in ESG ratings is a method by which companies demonstrate a commitment to transparency and proactive management approaches to address externalities.

The importance of ESG factors in investment decisions has increased in recent years due to the growing interest of investors and regulators in ESG or socially responsible investments (SRI). As for the purpose of this research, three main strands of literature were examined:

- The impact of ESG factors on portfolio performance;
- The relation between ESG factors and stock price;
- The role of ESG ratings in incorporating firms' ESG information.

According to the first strand of literature, considerable work has been performed in measuring the impact of sustainable investments on portfolio performance [5, 8, 18].

Indeed, ignoring ESG factors may have negative effects on portfolio performance [17].

Research papers on the influence of ESG factors on stock prices provide the following results: several authors [3, 4, 10, 12, 14–16] studied how investors perceive firms' attitudes toward ESG and how they incorporate this factor in their investment decision that in its turn affected stock prices. There is evidence that companies that meet sustainability requirements have better market performance [6, 9], even though ESG factors may impact differently according to specific businesses and sectors. Friede et al. [13] analyzed more than 2000 studies, beginning from 1970, exploring the effects of ESG on financial results. Many of these studies (62%) highlight evidence of the positive effects of ESG factors. Giese and Nagy [16] built a financial model according to which the value of a company is a function that includes ESG scores and shows a lag between changes in ESG scores and their impact on the financial results. Their work shows that stock markets react more sensitively to ESG information for companies that do not have extreme ESG scores, i.e., neither very low nor very high, and that stock markets show a stronger reaction to improvements in ESG ratings rather than to drops in ESG performance.

The third strand of literature investigates the role of the ESG index in incorporating ESG information. ESG indexes are necessary to carry out research exploring the contribution of ESG factors on investments or portfolio performance and stock prices. Most studies [2, 19, 22] have relied in their analysis on ESG ratings. Nevertheless, scholars are aware that the ESG indexes market is still at an early stage, resulting in possible misleading research results.

Dorfleitner et al. [7] provided evidence of a lack of convergence of ESG measurement conception. In turn, Escrig-Olmedo et al. [11] state that ESG indexes do not fully incorporate sustainability principles in their methodology. Krüger [20] points out a "measurement error" due to the non-quantitative nature of ESG standards: each ESG rating has its own method to evaluate ESG performance, which leads to a considerable divergence among different ESG ratings related to the same company. Berg et al. [1] point out that the discrepancy in ESG ratings depends on using different introductory and methods while evaluating ESG performance: agencies refer to different ESG categories, use different methods to assess these categories, and give different importance to each of them.

The authors consider the following hypotheses:

- Hypothesis 1: There is a positive correlation between the score in ESG rating and investors' interest in investing in the company;
- Hypothesis 2: Young investors pay more attention to the way a company implements its ESG strategy;
- Hypothesis 3: The E (environment) factor is the major ESG factor affecting the brand value of the fuel and energy companies.

2 Methodology

This research analyzes how ESG investments affect the brand perception of fuel and energy companies to test the hypotheses that provide companies with competitive advantages and sustainable development in the conditions of decarbonization.

The research methods are based on the fundamental principles of branding and ESG components. The research used the scientific works of scientists in the field of management, economic analysis, ESG investments, and sustainable development strategies. The empirical base includes officially published reports and statistics on the websites of analytical and rating agencies.

The study on the influence of each ESG factor on the brand value of oil and gas companies used methods of logical and statistical analysis, as well as the synergetic principle.

3 Results

The contemporary ESG agenda is based on the Sustainable Development Goals (SDGs). Unfortunately, this fact does not mean that every company is actively working on each SDG. ESG is based on 17 SDGs, which the UN General Assembly formulated in 2015. SDGs include hunger and poverty eradication, gender equality, quality education, responsible consumption and production, peace, justice, and others. SDGs are formed at the intersection of three sectors: environmental, social, and economic.

The strategy of sustainable development of an oil and gas company is an understanding of how to best use assets, processes, products, services, and innovations to achieve a sustainable future.

Nowadays, ESG is an element of corporate strategy that allows oil and gas companies to achieve their business goals through the creation of new products, the transformation of business models, increasing the value chain efficiency and reducing risks. This is the standard model for providing financing: when making decisions about financing and subsidizing, investors, banks, and governments look at the criterion of ESG maturity. It is also a form of non-financial reporting that helps businesses create a positive image among customers and partners.

ESG is a key value for the investment attractiveness of an oil and gas company. Investors take ESG factors into account in their decision-making to manage risks and ensure a sustainable and long-term return on investment. They want to know what actions the company is taking in this direction before investing their funds in it: does its management consider its social obligations in its development strategy and prioritize them at the expense of short-term profits—all of this constitutes a responsible investment.

The information field that the company creates is monitored by investors, financial analysts, and rating agencies. The way the company declares its ESG activities should

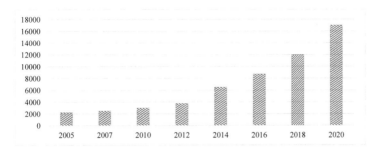

Fig. 1 US sustainable investments have surged 456% since 2010 ($b). *Source* Compiled by the authors based on the materials of the US SIF Foundation [23]

be consistent with its overall business strategy and as close as possible to how it implements its goals in this direction.

Investors' interest in ESG is growing every year. For example, in 2020, American investors invested 456% more in ESG companies than in 2010 (Fig. 1).

ESG investments in 2018 amounted to $30.7 trillion, which is 34% more than in 2016. Responsible investment is growing rapidly in all regions worldwide, especially in the USA, Europe, and Japan. SRI investments are projected to reach $50 trillion by 2040. It is also worth noting that despite the serious crisis in the financial market in 2020, caused by the contraction of business activity as a result of the spread of COVID-19, ESG investments have shown a record increase. It is also predicted that economic recovery in most countries will follow ways to increase investment in projects that comply with ESG principles (e.g., the EU Green Deal package of measures).

However, despite the rapid growth in socially responsible investment, this volume is insufficient to meet the goals of the Paris Agreement. According to research by the Organization for Economic Cooperation and Development (OECD), this requires annual investments (SRI investments) in the amount of $6.9 trillion by 2030, which is seven times higher than the stated above growth projections. This is supported by the fact that paying attention to ESG concerns does not compromise returns but rather increases them.

To stimulate the growth of ESG investment, the UN created the Principles for Responsible Investment (PRI) organization, which aims to promote and simplify the integration of ESG principles into the investment decision-making process. Currently, one of the key tasks of PRI is the unification of approaches to assessing ESG factors and their introduction into the process of forming credit ratings, which investors also use when analyzing investment opportunities. This is an important step since now there are many ESG ratings (e.g., MSCI ESG Rating, Sustainalytics, FTSE4Good Index, and the Dow Jones Sustainability Index) with different methodology, which sometimes makes it impossible to get an accurate and real assessment of the investment attractiveness of a company.

Therefore, on the one hand, ESG rating agencies contribute to reducing information asymmetry. On the other hand, ESG rating valuation might be misled by different kinds of assessment standards. However, investors have been proven to manage to profit from the information coming from these ratings in evaluating stock prices, thanks to the increasing volume of information on the company's ESG profile. Thus, there is a positive correlation between the effectiveness of the company implementing the ESG strategy and investors' interest in investing in the company. However, it is not so evident if the scores in concrete ESG ratings are regarded.

According to a recent study, 55% of the surveyed companies primarily work to reduce greenhouse gas emissions during production. Another 37% are worried about reducing electricity consumption.

ESG investments directly affect the company's image. The brand is inseparably linked with the issue of sustainable development; its brand mission coincides with the mission of activities in this area.

The most striking example is BP, which integrated the sustainable development agenda into the brand's DNA in the transformation process: "Our goal is to rethink energy for people and our planet. We want to help the world reach net zero and improve people's lives."

"Brand-led" sustainability is an approach in which the issue of sustainable development is a part of the brand's DNA. Moreover, it is embedded directly in the brand's mission. In this case, the brand directs and defines all activities of the company and additionally helps differentiate in the market.

In this case, the portfolio of relevant initiatives and brand communications, in general, work to maintain the brand's mission and focus on sustainable development [21]. That is, the deeper the topic is integrated into the brand, the less investment and effort will be required for additional communication of the agenda.

Over the past year, users have become more attentive to brands' position on important social issues. This is especially true for zoomers—Generation Z representatives. According to the surveys of Clear Channel UK and JCDecaux UK, on average, 58% of customers say they pay attention to the brand's position on social issues, and 56% of respondents consider social responsibility and environmental friendliness to be the key characteristics of a trustworthy brand. Another 55% of respondents pay attention to ethical production.

Nevertheless, insincerity is the main mistake that oil and gas brands make when planning environmental campaigns. In an attempt to earn "social approval points," brands can launch initiatives they do not believe in and campaigns whose values they do not share.

Potential customers feel such insincerity well. Greenwashing is a marketing technique in which non-ecological and even harmful goods and services are consciously positioned as environmentally friendly. Creating the image of eco-activists, brands using greenwashing do not benefit the environment.

Small initiatives can become the first steps toward fuel and energy companies' ESG investments. It is necessary to be sincere and avoid greenwashing. Customers will not take positively a campaign that openly manipulates facts. Experts and social media users will inevitably find out the truth, and the brand's reputation will suffer.

The third hypothesis of this research concerns the influence of ESG factors on the oil and gas companies' market value. An oil and gas company is a sensitive industry in matters of environmental protection. Shareholders are concerned about environmental practices, which will directly affect stock prices. Therefore, it seems necessary to carefully assess environmental indicators. As for the impact of the social factor, all investments in social initiatives are considered as costs for the company. Therefore, the impact of the social factor on the market value of companies is negative. It can be assumed that the management factor has a positive effect on the market value of oil and gas companies. Moreover, it is necessary to improve the quality of the company's management, including accounting and transparency measures, as well as relations within the company. However, the synergistic effect of ESG factors has the most important impact on the market value of oil and gas brands.

4   Conclusions

This research contributes to a better understanding of the ESG factors' influence on the oil and gas companies' market value in the conditions of decarbonization. It also answers three hypotheses.

The issue of sustainable development is global changes in what society and consumers expect from the business and not just another trend. Therefore, for branding, this topic is more than an answer to the actual needs of the target audience. It is a starting point for brand transformation and an opportunity to differentiate not only with communications but with real actions that embody the brand's promise.

Oil and gas companies often refuse to invest in social facilities in favor of profit. Nevertheless, it has been shown that projects in the field of environmental protection, social sector, and management contribute to sustainable profit and contribute to long-term success.

In general, the obtained data confirm the relationship between the effectiveness of ESG and the market value of companies. According to the first research hypothesis, to establish the dependence of the brand's position in the ESG rating, it is necessary to conduct further research on the correlation of each rating and the value of shares and create a single tool that considers financial and non-financial indicators.

According to the second research hypothesis, ESG factors represent value for the investment attractiveness of an oil and gas company. Investors consider all ESG initiatives when making decisions. Over the past year, consumers have become more attentive to the activities of brands on important social and environmental issues, which is especially reflected in Generation Z.

Nevertheless, there are differences between environmental, social, and state factors. Given that the chosen research area is the oil and gas industry, the hypothesis was put that the impact of the environmental factor on the value of companies would be more significant. Moreover, investors will be receptive to the initiatives of companies related specifically to environmental issues. However, it turned out that financial implications also play an important role as investors consider the costs associated with environmental initiatives of oil and gas companies.

The overall results of this research are consistent with the goal of the influence of ESG factors on the increase in the value of the oil and gas brand. The synergistic effect of ESG factors has the most important impact on the market value of oil and gas brands.

Thus, the obtained results provide new insight into the impact of each ESG factor on the market value of oil and gas companies, providing shareholders with a useful tool for long-term success and sustainable profit generation.

References

1. Berg F, Koelbel JF, Rigobon R (2019) Aggregate confusion: the divergence of ESG rating. Forthcoming Rev Finance. https://doi.org/10.2139/ssrn.3438533
2. Boze B, Krivitski M, Larcker DF, Tayan B, Zlotnicka E (2019) The business case for ESG. Rock Center for Corporate Governance at Stanford University, Stanford, CA. https://ssrn.com/abstract=3393082. Accessed 31 Oct 2022
3. Cao J, Titman S, Zhan X, Zhang WE (2019) ESG preference, institutional trading, and stock return patterns. https://doi.org/10.2139/ssrn.3353623
4. Chan Y, Hogan K, Schwaiger K, Ang A (2020) ESG in factors. J Impact ESG Investing 1(1):26–45. https://doi.org/10.3905/jesg.2020.1.1.026
5. Ciciretti R, Dalò A, Dam L (2019) The contributions of betas versus characteristics to the ESG premium. CEIS Working Paper No. 413. https://doi.org/10.2139/ssrn.3010234
6. Dimson E, Karakas O, Li X (2015) Active ownership. Rev Financial Stud 28(12):3225–3268. https://doi.org/10.2139/ssrn.2154724
7. Dorfleitner G, Halbritter G, Nguyen G (2015) Measuring the level and risk of corporate responsibility. An empirical comparison of different ESG rating approaches. J Asset Manag 16:450–466. https://doi.org/10.1057/jam.2015.31
8. Dorfleitner G, Utz S, Wimmer M (2013) Where and when does it pay to be good? A global long-term analysis of ESG investing. In: Proceedings of the AFBC 2013: 26th Australasian finance and banking conference, Sydney, Australasian, pp 1–29. https://doi.org/10.2139/ssrn.2311281
9. Eccles RG, Ioannou I, Serafeim G (2014) The impact of corporate sustainability on organizational processes and performance. Manage Sci 60(11):2835–2857. https://doi.org/10.2139/ssrn.1964011
10. Erhardt J (2020) The search for ESG alpha by means of machine learning—a methodological approach. https://doi.org/10.2139/ssrn.3514573
11. Escrig-Olmedo E, Fernández-Izquierdo MÁ, Ferrero-Ferrero I, Rivera-Lirio JM, Muñoz-Torres MJ (2019) Rating the raters: evaluating how ESG rating agencies integrate sustainability principles. Sustainability 11(3):915. https://doi.org/10.3390/SU11030915
12. Evans JR, Peiris D (2010) The relationship between environmental social governance factors and stock returns. https://doi.org/10.2139/ssrn.1725077
13. Friede G, Busch T, Bassen A (2015) ESG and financial performance: Aggregated evidence from more than 2000 empirical studies. J Sustain Finance Investment 5(4):210–233. https://doi.org/10.1080/20430795.2015.1118917
14. Gary SN (2016) Values and value: University endowments, fiduciary duties, and ESG investing. J College Univ Law 42(1):247–309. Retrieved from https://www.nacua.org/docs/default-source/jcul-articles/volume-42/42_jcul_247.pdf?sfvrsn=250b66be_10. Accessed 31 Oct 2022
15. Gary SN (2019) Best interests in the long term: fiduciary duties and ESG integration. Univ Colorado Law Rev 90:731–801. https://doi.org/10.2139/ssrn.3149856
16. Giese G, Lee L-E, Melas D, Nagy Z, Nishikawa L (2019) Foundations of ESG investing: how ESG affects equity valuation, risk, and performance. J Portfolio Manage 45(5):69–83. https://doi.org/10.3905/jpm.2019.45.5.069

17. Glossner S (2021). Repeat offenders: ESG incident recidivism and investor underreaction. https://doi.org/10.2139/ssrn.3004689
18. Hübel B, Scholz H, Webersinke N (2020). Permutation tests for stock index performance: evidence from ESG indices. https://doi.org/10.2139/ssrn.3528309
19. Khan M, Serafeim G, Yoon A (2015) Corporate sustainability: first evidence on materiality. Account Rev 91(6):1697–1724. https://doi.org/10.2139/ssrn.2575912
20. Krüger P (2014) Corporate goodness and shareholder wealth. J Financ Econ. https://doi.org/10.2139/ssrn.2287089
21. Ponomareva EA, Nozdrenko EA, Mysakova AG (2022) Sustainable development and industry 4.0 determinants in communication strategies in fuel and energy complex. In: Zavyalova EB, Popkova EG (eds) Industry 4.0. Palgrave Macmillan, Cham, pp 395–407. https://doi.org/10.1007/978-3-030-79496-5_36
22. Schoenmaker D, Schramade W (2018) Principles of sustainable finance. Oxford University Press, Oxford, UK. Retrieved from https://ssrn.com/abstract=3282699. Accessed 31 Oct 2022
23. White L, Whieldon E (2021) The ESG trends that will drive 2021s—podcast. S&P Global. Retrieved from https://www.spglobal.com/marketintelligence/en/news-insights/latest-news-headlines/the-esg-trends-that-will-drive-2021-8211-podcast-61980796. Accessed 31 Oct 2022

Foreign and Russian Experience in the Application of Climate Resilient Digital Technologies in the Oil and Gas Industry

Anastasia V. Sheveleva⬤ and Elana A. Avdeeva⬤

Abstract At the current stage of the development of world trends and the oil and gas industry, digital technologies play an important role by contributing to significant improvement of processes in the organizational and production areas of companies. Digital technologies have been actively used in the oil and gas industry in recent years, as they contribute to reduced production costs, increased productivity, and energy efficiency. Simultaneously, digitalization reduces CO_2 emissions and the negative impact on the environment. Analysis of the experience of leading foreign oil and gas companies (BP, Chevron, ConocoPhillips, Eni, ExxonMobil, Shell, Equinor, and Total) and Russian oil and gas companies (PJSC NK Rosneft, PJSC Gazprom, and PJSC Lukoil) shows that the most climate resilient digital technologies in the oil and gas industry include digital fields, digital twins, artificial intelligence, robots and drones, and Internet of Things. The research aims to analyze the experience of using digital technologies of Industry 4.0 by international oil and gas companies to combat climate change and find possible ways of regulation and management based on the environmental factor of ESG regulation. To achieve this purpose, the authors applied methods of analysis and synthesis, which made it possible to identify a list of climate-resilient digital technologies and formulate basic recommendations that contribute to the creation of a competitive advantage for oil and gas companies achieved through the integrated application of organizational, corporate, managerial, and technical components. Based on the research results, the authors conclude that the approaches to the application of these climate-resilient digital technologies in foreign and Russian companies in the oil and gas industry have similarities and differences. Digital and ESG transformations build synergy only when unified approaches to control and recording are in place. An updated configuration of such a system within the environmental cluster is proposed. In this regard, the analysis of foreign and Russian experience in the application of climate-resilient digital technologies based on the environmental factor of ESG regulation in the oil and gas industry seems to

A. V. Sheveleva (✉) · E. A. Avdeeva
MGIMO University, Moscow, Russia
e-mail: a_sheveleva@rambler.ru

E. A. Avdeeva
e-mail: elavtraum@gmail.com

© The Author(s), under exclusive license to Springer Nature Switzerland AG 2023
E. G. Popkova (ed.), *Smart Green Innovations in Industry 4.0 for Climate Change Risk Management*, Environmental Footprints and Eco-design of Products and Processes, https://doi.org/10.1007/978-3-031-28457-1_30

be of proper relevance. The results can be used in the operations of international oil and gas companies that have introduced or are introducing digital technologies to protect the environment, tackle climate change, and reduce environmental damage.

Keywords Digital technologies · Climate resilient digital technologies · Industry 4.0 · Oil and gas companies · Climate agenda · ESG transformation · Digital transformation

JEL Classification O130 · O440 · Q540

1 Introduction

Global challenges occurring at the current stage of the world's development are introducing significant changes in the regulation of activities at the government level and directly at the industry level.

The UN Framework Convention on Climate Change and the Paris Agreement on Climate Change has imposed additional environmental requirements on international companies and, therefore, contributed to the creation of new regulatory mechanisms. In 2021, the IEA published its Net Zero by 2050: A Roadmap for the Global Energy Sector.

In the face of global international challenges, such as coronavirus infection (COVID-19) and geopolitical aspects and transformations of 2022, the transition of the oil and gas industry to the daily implementation and application of contemporary approaches based on digital technologies has accelerated.

Effective and rational implementation of Industry 4.0 technologies in foreign oil and gas companies have created certain practices that can also be used by Russian oil and gas companies. "The international market is closed for Russia, so we will focus on new partners and their requirements, for example, China. Though there was a time when the Russian advantage was cheapness, now it is quality. If it is not possible to participate in the international competitions, then it is necessary to create our own rating-economic and climatic," said the Special Representative of the President of the Russian Federation on Climate Issues Ruslan Edelgeriev at the St. Petersburg International Economic Forum of 2022 dedicated to the study of the Russian potential in carbon units trade [1]. Therefore, ESG transformation in Russia, where the indicator is digitalized by 95%, can be implemented with reference to foreign practices but with the use of domestic approaches and specific features.

In this regard, the analysis of the use of climate-resilient digital technologies by foreign and Russian oil and gas companies is quite relevant.

2 Methodology

The authors have studied various sources on ESG and the digital transformation of the oil and gas industry, which made it possible to identify existing points of view and build the authors' vision that digital technologies are used by foreign and Russian oil and gas companies as a way to achieve climate-resilient goals.

The creation of the digital economy, the transition to Industry 4.0, and the integration of Industry 4.0 technologies into the processes ensuring the sustainable development of industrial companies, including in the oil and gas sector, are considered in the works of foreign and Russian authors: Abd-Rabo and Hashaikeh [2], Avdeeva [3, 4], Bobylev et al. [5], Daneeva [6], Dementieva and Zavyalova [7], Lomachenko [8], Popkova and Zavyalova [9], Sheveleva and Zagrebelnaya [10], Tyaglov et al. [11], Zavyalova and Krotova [12], Zavyalova et al. [13].

Several papers highlight the use of digital technologies to combat climate change: Bettini et al. [14], Lapão [15], Sheveleva and Cherevik [16], Sheveleva et al. [17].

It should be noted that the above works do not comprehensively address ESG and the digital transformation of oil and gas companies to combat climate change. This issue remains to be substantially unexplored. In this regard, the authors suggest that the introduction of digital technologies in the fight against climate change based on the Environmental Cluster (ESG regulation), as well as the creation and approval of a unified methodology for assessing the results of this approach, will contribute to the effective improvement of business processes in oil and gas companies.

The application of analysis and synthesis methods has made it possible to identify the main climate-resilient digital technologies in the oil and gas industry, which have been used in foreign practice for a long time and are only starting to be used by Russian companies.

Based on the research results, the authors have drawn relevant conclusions and formulated their recommendations.

3 Results

One of the 17 UN Sustainable Development Goals for 2030 is Goal 13: Climate Action.

Since 1990, global carbon dioxide emissions have increased by almost 50%. Simultaneously, this rate grew faster from 2000 to 2010 compared to the three previous decades. Limiting global warming to an increase of 1.5 and 2 °C will require rapid, far-reaching, and unprecedented changes in all spheres of life.

However, digitalization is not a new phenomenon for the oil and gas industry. According to audit company Ernst and Young, the first computer technologies have been introduced by oil and gas companies for reservoir modeling since the 1960s, for processing field data since 1973, and the first three-dimensional seismic models were built in the 1990s, leading to a 40% reduction in the cost of finding new fields on average. As a result, the growth of proven oil and gas reserves increased by 2.5 times. In turn, this led to a production rate increase. At the present stage, oil and gas companies are already using digital technologies in a variety of their activities.

In this regard, it is important to note the widespread use of digital technologies that contribute to improving the indicators of climate-resilient background. Simultaneously, international organizations contribute to more stringent regulations in the oil and gas industry in terms of greenhouse gas emissions, primarily carbon dioxide (CO_2) and methane (CH_4).

The largest international oil and gas companies are aware of the urgency of the climate change issue and have already started introducing Industry 4.0 technologies into their operational activities to reduce greenhouse gas emissions. Only successful implementation practices define the climate-resilient digital technologies that can be implemented within a pilot project and be scaled up to fit other facilities and companies.

The key technological areas of Industry 4.0 in the oil and gas industry are big data, chatbots, robots (including drones), optical character recognition technology, artificial intelligence, industrial Internet of Things, virtual and augmented reality, and blockchain.

According to the international research company Start Us Insights, artificial intelligence (AI) and the Internet of Things (IoT) are often used in the oil and gas industry. Efficiency improvement in all activity areas is based on cloud technologies, big data analytics, preventive maintenance, and production management system. AI supports oil well imaging and robots on oil platforms. Blockchain solutions form a transparent pricing chain. Virtual and augmented reality contribute to ensuring work safety and implementation of remote interaction in work and training processes.

Given that fleet aging and equipment wear in the production cycle of vertically integrated oil and gas companies is estimated to exceed 65–70%, contemporary practices, trends, and forms of application of innovations and promising digital technologies in hydrocarbon production, transportation, and processing facilities can be divided into two main groups. Newly commissioned facilities have advanced equipment from the start and are designed with the use of optimal approaches focusing on obtaining the maximum economic effect, including based on energy efficiency. The second group includes the development and modernization of previously existing facilities that have been operating for a long time in various areas of the fuel and energy sector, primarily in the electricity and oil and gas sectors, based on the introduction of the best available technologies for their efficient use.

According to the research of consulting agency Wood Mackenzie, the most relevant technologies for the oil and gas sector are big data, cloud technologies, remote control, and IoT. In addition to exploration and mining, the most promising areas for introducing digital technologies are oil refining and petrochemicals, commerce, and logistics.

Contemporary methods of data analysis based on AI, machine learning, and neural networks allow for a fundamentally new approach to solving current problems. However, their implementation requires dealing with BigData.

BP, Shell, ExxonMobil, ConocoPhillips, Chevron, SaudiAramco, Sinopec, and other major foreign oil and gas companies actively use IoT in their cross-technology digital solutions for remote monitoring, predictive analytics, creating models of digital twins, and minimizing costs, including the energy costs. The industry members use the services of the largest technology providers in the information and telecommunications area: Amazon, Cisco, PTC, Microsoft, IBM, and Intel.

The companies often use the following cloud platforms: Amazon Web Services (BP and Shell), Microsoft Azure (Shell, Chevron, Equinor, and ExxonMobil), Google Cloud Platform (Total), and IBM Cloud. Chevron migrates its applications to Microsoft Azure, including Platform-as-a-Service (PaaS), Internet of Things (IoT), big data, and Infrastructure-as-a-Service (IaaS). The companies expect more CAPEX and OPEX efficiency. As part of its strategy, Chevron has created a cloud readiness accelerator program to speed up the transition to Microsoft Azure.

The largest oil and gas companies widely use sensors and invest in relevant technological companies. According to BP, more than 99% of the company's oil and gas wells have sensors that continuously generate data flows used to analyze conditions at each site, optimize equipment performance, and identify maintenance needs.

To successfully use 3D printing throughout the oil and gas sector, it is necessary to determine the effectiveness of its implementation, as well as the parts and components that are profitable to produce. While 3D printing improves the efficiency of supply chains, additive manufacturing poses a number of regulatory challenges; the certification of 3D printing materials is a possible barrier to the widespread use of this technology in the oil and gas sector.

Robotics and automation of offshore oil platforms during the exploration and mining stages can reduce carbon dioxide emissions, increase safety and revenues, and reduce capital costs by 30% and operating costs by 50%.

For the benefit of the refining and petrochemicals segment, oil and gas companies are implementing refinery plant digitalization projects. From the point of view of an oil and gas company, the systems of integrated management and smart operator are the most relevant for this segment.

Russian oil and gas companies are slower in terms of digitalization in general and the introduction of climate-resilient digital technologies in particular. Only when the Russian government adopted the digital economy program in 2017, the largest oil and gas companies PJSC NK Rosneft, PJSC Gazprom, and PJSC Lukoil started to incorporate digitalization issues into their strategies, plans, and programs, including those related to sustainable development and climate change.

The analysis of foreign and Russian experience of digital technologies application in the oil and gas industry allowed us to identify the most promising of them (Fig. 1).

In general, oil and gas companies use a set of digital technologies in their activities, and these technologies are often interconnected. The application of digital technologies by oil and gas companies is aimed at increasing efficiency and maximizing value, improving operations safety, decarburization, and achieving climate-resilient goals.

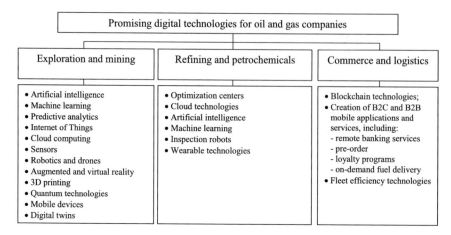

Fig. 1 Promising digital technologies for oil and gas companies. *Source* Compiled by the authors based on the research results

Climate-resilient digital technologies for ESG are hardly used by companies (35%) or are only used for some tasks (50%). The main obstacles to the use of digital solutions within the ESG activities of companies are the following:

- Lack of necessary competencies (45%);
- Insufficient development of ESG practice (44%);
- Lack of understanding of the need for ESG and digitalization synergy (40%);
- Poor cooperation within the company in terms of the application of digital ESG solutions (40%);
- Insufficient development of digital practices (11%) [18].

The results of the application of digital technologies by leading foreign and Russian oil and gas companies can be summarized as follows (Table 1).

According to consulting company Accenture, organizations integrating digital technologies into their operating models are growing twice as fast as those falling behind in the digital world.

Simultaneously, an ESG indicator is by 95% a digital indicator. It is calculated by various methods in different organizations, including MSCI ESG Rating; FTSE4Good Index Series, ISS ESG, World Wildlife Fund (WWF) Environmental Information Transparency Ratings, S&P Global Ratings ESG, sustainable development indices of the Russian Union of Industrialists and Entrepreneurs and Moscow Exchange, and the National Rating Agency.

The calculation methodologies are often not disclosed in full. Based on the ESG Rating Methodology (non-credit ratings to assess the company's exposure to environmental, social, and corporate governance risks) of the National Rating Agency, it was revealed [19] that there are discrepancies even in the same document while the specific features of the oil and gas industry operations in the field of exploration

Table 1 The results of the application of climate-resilient digital technologies by leading foreign and Russian oil and gas companies

Digital technologies/Result	Digital deposit	Digital twin	Artificial intelligence	Sensors and fiber optics	Cloud technologies	Robots and drones	Blockchain technologies	3D printing	Big data technologies	Internet of Things	AR/VR
Reduced energy consumption	✓		✓		✓			✓		✓	✓
Reduced harmful emissions	✓	✓	✓			✓				✓	
Increased power supply safety	✓	✓	✓			✓				✓	
Minimized costs	✓	✓	✓			✓	✓	✓	✓	✓	✓
Productivity and profit growth	✓	✓	✓	✓	✓	✓			✓	✓	✓
Improvement of business processes under ESG regulation, E Cluster (Environmental)	✓|	✓|	✓|	✓|	✓|	✓|	✓|	✓|	✓|	✓|	✓|

Source Prepared by the authors based on the results of the study and mining, oil refining and petrochemicals, and commerce and logistics are not considered.

To implement climate-resilient digital technologies under ESG transformation in the oil and gas industry, it is necessary to create a unified reporting system with a common methodology—by clusters (ESG rating), by industry, and throughout the country.

In addition to the introduction of a common tool and calculation methodology, the following promising areas of climate resilient and ESG regulation digital technologies of the oil and gas industry are recommended:

- Injection of CO_2 to maintain reservoir pressure at the later stages of development or at the exhausted gas fields for the extraction of residual gas reserves and CO_2 burial;
- Cycle process instead of simple gas injection into formations for "storage";
- Gasification of process vehicles (LPG or CNG/LNG);
- Study of the prospects of production of aviation condensed fuel from associated gas;
- CO_2 to polymers, in particular, polyol polymers production from CO_2 (e.g., Cardyon technology from Covestro);
- Methane from CO_2. Direct methane generation by CO_2 hydrogenation, including above-critical conditions.

In addition to the introduction of a common tool and calculation methodology, the following promising areas of climate-resilient digital technologies in exploration and mining activities of the oil and gas industry are recommended:

- Complex and green robotic works on exploration, design, and survey with UAVs and space sounding (field works in forest zones with minimized tree felling, wireless sensor probes and information systems for data collection, drones for a seismic survey);
- Early decision-making technologies in prospecting and exploration drilling (predictive assessment of the feasibility of continuing drilling according to the analysis of sludge by a gas analyzer, i.e., molecular geochemistry at the initial depths of drilling);
- 4D seismic monitoring at mainland fields through a system of stationary fiber-optic sensors in observation wells, especially horizontal ones (preserved and inoperable emergency wells);
- Area modeling of flow properties and formation piezoconductivity based on seismic data and subsequent use of results in designing a development system;
- Multi-well deconvolution technology of Polykod startup with PolyGon software complex for the analysis and interpretation of dynamic data of production rates and pressures in the development of oil and gas fields;
- Study and assessment of the practical application of electrokinetic and electroosmotic phenomena in low permeability porous media.

In addition to the introduction of a common tool and calculation methodology, the following promising areas of climate-resilient technologies in refining and petrochemicals areas of the oil and gas industry are recommended:

- Low-carbon fuels and GTL oils of the best quality made from CO_2;
- Application of supercritical water vapors (or other solvents) for highly efficient processing of heavy oil residues;
- Production and use of polycyclopentadiene in the Russian industry;
- Investigation of potential uses of graphene and its oxide in oils and lubricants, oil refining processes, bituminous materials, and drilling fluids.

In recent years, the largest international oil and gas companies have accumulated some experience in successfully implementing certain projects in this area. There have also been failed projects. It is advisable to adopt proven solutions.

A particular technology should be chosen according to the set of design conditions (climatic, infrastructure, and production conditions) and the selected goals.

4 Conclusions

The competitive advantage of international oil and gas companies is due to the combined application of organizational, corporate, and technical components based on Industry 4.0:

- Introduction of contemporary monitoring, regulation, and control systems with a clear recording of indicators and deviations;
- Creation of an environment to ensure unhindered interaction by creating working groups at the corporate and industry level;
- Formation of mutually beneficial partnerships within the framework of the topical research;
- Continuous improvement of management processes through the introduction of innovations, digital technologies, and software products;
- Application of other digital approaches to accelerate the processes aimed at identifying optimal solutions.

It is necessary to transform the ESG rating methodology (E, Environmental Cluster) to cover the full cycle of production activities (from the design and construction stage to the implementation of financial and logistics components), which will contribute to the creation of a common climate resilient digital pattern (like IFRS and RAS or Indicators and Criteria for G, Corporate Governance Cluster).

When forming a common climate-resilient digital pattern, it is necessary to create working groups in the production and auxiliary business areas to consolidate and record all parameters.

Adopting a common methodology is possible by introducing a legislative instrument at the state level.

The above measures will help international oil and gas companies to implement digital and ESG transformation more effectively to combat climate change and minimize its impact.

Acknowledgements Many thanks to the conference organizers for an excellent organization and an interesting program. Several most pressing issues were raised there. The conference made it possible to explore the views of Russian and foreign colleagues and discover their experience on the considered problem within a short time. Special thanks to the reviewers for their attention to this research and valuable recommendations.

References

1. Ecology of Russia (17 June 2022) The Russian environmental and climate efficiency rating may appear in 2023. Retrieved from https://ecologyofrussia.ru/rossiyskiy-reyting-ekologicheskoy-i-klimaticheskoy-effektivnosti-mozhet-poyavitsya-v-2023-godu/. Accessed 11 Nov 2022
2. Abd-Rabo AM, Hashaikeh SA (2021) The digital transformation revolution. Int J Humanities Educ Res 3(4):124–128. https://doi.org/10.47832/2757-5403.4-3.11
3. Avdeeva EA (2021) Energy management and certification systems for international oil and gas companies (on the example of the USA and Saudi Arabia). Financ Econ 5:223–228
4. Avdeeva EA (2021) China's competitive advantages in the field of energy efficiency: case of China National Petroleum Corp. Risk Manage 2(98):5–15
5. Bobylev SN, Chereshnya OY, Kulmala M, Lappalainen HK, Petäjä T, Soloveva SV, Tynkkynen VP et al (2018) Indicators for digitalization of sustainable development goals in PEER program. Geogr Environ Sustain 11(1):145–156. https://doi.org/10.24057/2071-9388-2018-11-1-145-156
6. Daneeva YO (2019) Digitalization in oil and gas sector: path to sustainability. Financ Econ 11:120–123
7. Dementieva A, Zavyalova E (eds) (2020) Corporate governance in Russia: Quo vadis? De Gruyter, Berlin. https://doi.org/10.1515/9783110695816
8. Lomachenko TI (2020) Creative strategy of digitalization as a factor of safety and sustainable development of the oil and gas complex. J Altai Acad Econ Law 10–1:64–68
9. Popkova EG, Zavyalova E (2021) New institutions for socio-economic development: The change of paradigm from rationality and stability to responsibility and dynamism. De Gruyter, Berlin. https://doi.org/10.1515/9783110699869
10. Sheveleva AV, Zagrebelnaya NS (2020) The practice of introducing digital technologies in the oil and gas industry. In: Isachenko TM (ed) Modern international economic relations in the post-bipolar era. MGIMO University, Moscow, pp 244–250
11. Tyaglov SG, Sheveleva AV, Rodionova ND, Guseva TB (2021) Contribution of Russian oil and gas companies to the implementation of the sustainable development goal of combating climate change. IOP Conf Ser: Earth Environ Sci 666:022007. https://doi.org/10.1088/1755-1315/666/2/022007
12. Zavyalova EB, Krotova TG (2021) Methods for achieving SDG 1, poverty eradication. In: Popkova EG, Sergi BS (eds) Modern global economic system: evolutional development vs. revolutionary leap. Springer, Cham, pp 1987–2001. https://doi.org/10.1007/978-3-030-69415-9_218
13. Zavyalova E, Lee JS, Ostrovskaya E (2020) Corporate social responsibility. In: Dementieva A, Zavyalova E (eds) Corporate governance in Russia: Quo vadis? De Gruyter, Berlin, pp 173–187
14. Bettini G, Gioli G, Felli R (2020) Clouded skies: how digital technologies could reshape "Loss and Damage" from climate change. WIREs Clim Change 11(4):e650. https://doi.org/10.1002/wcc.650

15. Lapão LV (2020) Climate change, public health impacts and the role of the new digital technologies. Eur J Public Health 30(Supplement_5). https://doi.org/10.1093/eurpub/ckaa16 5.652
16. Sheveleva AV, Cherevik MV (2022) Digital technologies in the oil and gas sector and their contribution to UN climate action goal. In: Zavyalova EB, Popkova EG (eds) Industry 4.0: fighting climate change in the economy of the future. Palgrave Macmillan, Cham, pp 307–315. https://doi.org/10.1007/978-3-030-79496-5_28
17. Sheveleva AV, Tyaglov SG, Khaiter P (2021) Digital transformation strategies of oil and gas companies: preparing for the fourth industrial revolution. In: Konina N (ed) Digital strategies in a global market: navigating the fourth industrial revolution. Palgrave Macmillan, Cham, pp. 157–171. https://doi.org/10.1007/978-3-030-58267-8_12
18. SKOLKOVO Moscow School of Management (24 Oct 2022) How digitalization helps ESG business transformation. Retrieved from https://www.skolkovo.ru/expert-opinions/kak-cifrov izaciya-pomogaet-esg-transformacii-biznesa/. Accessed 15 Nov 2022
19. National Rating Agency LLC (2022) ESG ratings methodology (non-credit ratings to assess the company's exposure to environmental, social and corporate governance risks) (Approved by Order No. PR/07-06/22-1 of the General Director of 7 June 2022). Moscow, Russia. Retrieved from https://www.ra-national.ru/sites/default/files/analitic_article/Methology_ESG ratings_corp.pdf. Accessed 17 Nov 2022

Achieving Energy Efficiency Goals in the Public Procurement

Julia A. Kovalchuk⊙ **and Igor M. Stepnov**⊙

Abstract The energy efficiency course is an important component of the sustainable development of countries, ensuring the solution of a set of tasks: saving energy resources; involving business in the environmental agenda; changing construction norms and rules; optimizing the characteristics of industrial, commercial, and non-commercial real estate objects, considering the requirements for saving resources (energy, heat, water, etc.) and the use of energy-efficient materials; increasing responsibility for the efficient use of resources; supporting energy saving initiatives in the society; forming new consumption patterns. The research aims to systematize and evaluate the directions of including the principles of energy conservation and energy efficiency in the elements of the public procurement mechanism. The research methods include the analysis of Russian and international practices (USA, China, EU, and Japan) of the inclusion of public procurement in the implementation of national development programs and strategies, including the solution of energy efficiency problems. As a result, the research identifies the main components in the subjects of contracts, the customers of which are state and municipal institutions of different sectors of the economy (educational, medical, cultural and leisure, sports, social service and protection institutions, etc.): replacement of non-energy efficient appliances and indoor lighting, modernization of heating, water supply, and sanitation, installation of automatic systems accounting and control over the consumption of various resources in buildings and infrastructure facilities, and modernization of street lighting in cities and other localities. The original conclusion is made that public expenditure savings and the implementation of the dominant energy saving in terms of solving technical, economic, and socially significant tasks allow us to testify to the effects of improving public procurement management and the development of energy, construction, and other industries.

J. A. Kovalchuk (✉) · I. M. Stepnov
MGIMO University, Moscow, Russia
e-mail: fm-science@inbox.ru

I. M. Stepnov
e-mail: stepnoff@inbox.ru

Keywords Energy saving · Public procurement · Business responsibility · Building's reconstruction · Energy-efficient devices · Green public procurement

JEL Classification H57 · Q47 · Q55

1   Introduction

Public procurement is an important element of the functioning of the economic system of society and an instrument of economic policy [13] that ensures the creation of public goods and promotes business development and fair competition in various industry markets [17]. Due to the implementation of competitive procedures and basic principles of public procurement (legal grounds for regulating the procurement market within the framework of the state contract system [2], openness and transparency [16], the development of a competitive environment [11], etc.), they achieve minimization (optimization) of public expenditures or savings of state, regional, and local budgets and form a useful effect [12] in the context of the implementation of planned government expenditures.

Public procurement is characterized by the fact that its modernization in almost all countries is based on the need to introduce their own solutions that help eliminate bottlenecks in the implementation of the above-mentioned principles of the functioning of the contract system (including promoting the participation of small businesses in procurement, reducing corruption, the spread of electronic procurement, etc.) and on the inclusion of the best foreign practices to increase transparency and openness of procurement.

This also applies to such an important task as energy conservation when the need for energy resources is growing, and almost all countries need a more active transition to clean and environmentally friendly resources. Public procurement can be an excellent driver for solving this problem.

2   Methodology

The research methodology is based on the application of methods of comparative analysis of international practice in the field of accounting for energy saving and energy efficiency parameters, which forms an assessment and a set of various tools for improving the mechanism of procurement activities of authorities [1, 4, 20] in recognition of uncertainty in energy markets and imperfect competition [9] in conjunction with the sustainable development goals [10].

To determine the potential of using the principles of energy conservation in public procurement, it is important to consider the experience of countries that have gone from adopting strategic initiatives by the state to understanding the public importance of solving energy conservation issues.

In the USA, since 1975, the Federal Energy Management Program has been functioning with the result of a 49% reduction in energy intensity. For public facilities, guidelines for construction in accordance with the best practices of sustainable development and energy efficiency have been established since 2018. These guidelines contribute to optimizing the use of materials, reducing energy consumption by buildings and reducing energy consumption costs, managing water consumption (drinking and industrial water, storm drains), and reducing environmental impacts during the construction and operation of buildings. An important component of public procurement is the conclusion of contracts for modernizing buildings and infrastructure facilities to achieve water and energy efficiency.

Japan established an Energy Saving Center in 1978 and adopted an energy conservation law in 1979. In addition to the requirements for commercial and residential facilities, public procurement also provides for the certification of buildings according to energy efficiency standards. The current goal is to reduce electricity consumption by 13% by 2030, which will require increasing the energy efficiency of buildings by 35%. Only for the infrastructure solutions of Tokyo city, the introduction of LED (light emitting diode) lighting up to 100% is provided.

From 2003 to 2013, the "Intelligent Energy of Europe" program was implemented in Europe, which contributed to the introduction of energy efficiency criteria into the mechanism of public procurement for the construction, repair, and maintenance of buildings, which could lead to energy savings of 50% or more [6], including the purchase of energy-saving materials, energy-efficient (by at least 20%) and more environmentally friendly public transport, and the development of generation and use of energy from renewable sources. Further, the "Horizon 2020" program from 2014 to 2020 and the current "Horizon Europe" program from 2020 to 2027 are integrative [15]. These programs, as part of the implementation of common approaches to financing research and innovation for a low-carbon economy, are aimed at supporting the development of energy-efficient technologies and their introduction to markets. Public procurement under the current program is carried out for the construction and reconstruction of energy-efficient buildings [8], including residential and industrial facilities, the construction of heating and cooling systems, and the involvement of small businesses in the field of energy generation and transmission, including alternative energy and co-financing of energy-efficient projects in the public–private partnership format.

In China, the implementation of the "Made in China 2025" strategy is aimed at modernizing production towards environmentally friendly projects that are inseparable from improving overall energy efficiency in the manufacturing industry [19]. Accordingly, public procurement is aimed at facilitating the solution to this task, including the work of a specially created Society for Resource Conservation and Environmental Protection to support the sustainable development of economic production activities while preserving natural ecosystems [18].

The reference to the experience of these countries is connected with the statement of the fact of concern about the energy resources spent in high-income countries. Thus, the highest level of primary energy consumption per person was observed in 2020 in Canada (361.1 GJ), Norway (356.0 GJ), and the USA (265.2 GJ) [5].

Simultaneously, there is an increase in energy consumption due to the development of industry in developing countries—for example, over the past ten years, the average annual growth in energy consumption per person was 3.3% in China and 3.5% in India. In developed countries, consumption decreased by an average of 0.5% per year.

3 Results

Within the framework of achieving the goals of reducing the energy intensity of the gross domestic product in Russia, an important component was the adoption of the Law on Energy Saving and on Improving Energy Efficiency in 2009. This law was implemented to determine the amount of state budget funds to finance energy-saving programs in the amount of at least 21% and in the introduction of energy surveys for federal and municipal institutions to determine the optimal resource (energy, water, thermal, etc.) characteristics of buildings and real estate objects. An essential and consistent step in this process of achieving energy-saving goals in the country should also be considered inclusion in the subject of public procurement.

The Russian Federation has implemented the practice of considering the priorities of the state in strategies and doctrines (e.g., the Environmental Safety Strategy of the Russian Federation until 2025), which:

- Does not mean their consolidation in the form of mandatory norms in procurement legislation [14];
- May have an impact on the adoption of decisions by the authorities in solving strategic tasks of territorial development;
- Only encourages the application of relevant principles (for example, the principle of environmental friendliness) in procurement.

The Russian Federation has been implementing the state program "Energy Conservation and Energy Efficiency Improvement for the Period up to 2020" since 2010. Currently, as part of the development of the public procurement mechanism, work continues on the dissemination of practices for achieving energy conservation and energy efficiency in public sector organizations (in educational institutions, medical organizations, cultural, leisure, and sports institutions, institutions of social service and protection, etc.). Most of these purchases are purchases of municipal customers, which reflects a purposeful solution to the problem of expanding the application of energy efficiency principles for municipal-level facilities. Only in 2021, according to experts of energy service companies, the total value of all concluded state contracts in the field of energy efficiency reached 57.5 billion rubles, which is almost three times higher than in 2020.

The main measures to be carried out within the framework of public procurement are the replacement of non-energy efficient appliances and indoor lighting, modernization of heating, water supply, and sanitation systems, installation of automatic accounting systems, and control of consumption of various resources in buildings and infrastructure facilities. About 75% of the contracts in the framework of public procurement were concluded by educational organizations—kindergartens, schools, colleges, and universities. Together with their main activities for teaching children and adults, these educational organizations, thanks to budgetary funds, can modernize the conditions of activity and their current expenses for electricity, heating, and water supply. These measures are aimed at solving the problem of energy saving and, in the future, will create the effect of saving current costs for the operation of buildings owned by the state and municipal and intended primarily for non-commercial activities.

Given the climate in Russia, an important aspect is the modernization of the heating system in buildings of various uses. Thus, almost 25% of all contracts within the framework of public procurement were concluded for the modernization of the heating system, which included the installation of thermostats on heating devices and water heating systems made of polymer pipes—this option is more energy efficient than metal pipes, as well as durable, practical, and economical.

Tenders for the modernization of street lighting in cities and other settlements have become another important direction for the formation of a public effect in the implementation of energy-saving principles in public procurement. As a rule, in practice, this means replacing lamps and lighting equipment to ensure a low level of electrical energy consumption while maintaining an intense luminous flux, durability in operation, and ease of maintenance and replacement of burned-out lamps.

Types of lamps that can be used for street lighting include gas-discharge, LED, and incandescent lamps, including halogen and fluorescent lamps. Since the main criterion for street lamps is their efficiency, incandescent lamps are excluded from the list due to low efficiency—their luminous efficiency is no more than 13 lumens per 1 W.

The solution was to use LED lamps when installing street lighting because they create a stable luminous flux, differ in brightness and color reproduction, and are durable and energy-intensive. Nowadays, up to 50% of lamps used in street lighting in Russia are LED lamps. Every year, it is planned to increase its share in street lighting through public procurement.

The important characteristics contributing to their use in social institutions also include the fact that LED lamps do not heat up and can be supplemented with light and motion sensors. Depending on their purpose, lamps of different power are used to light the following objects:

- Facade and architectural illumination—with a power of 8–60 W;
- Parks, parking lots, and railway platforms – with a capacity of 60–120 W;
- Stadiums and city squares—with a capacity of 120–240 W.

In general, the optimization of the selection and placement of lighting devices makes it possible to reduce the installed capacity by up to 20% on the streets and roads. The replacement of obsolete gas-discharge, sodium, and fluorescent lamps provides real energy savings, which makes it possible to reduce electricity consumption by at least 15% and increase the lighting level by 22%. Replacing outdated lamps provides noticeable energy savings. Due to a longer period of use (up to 5–8 years instead

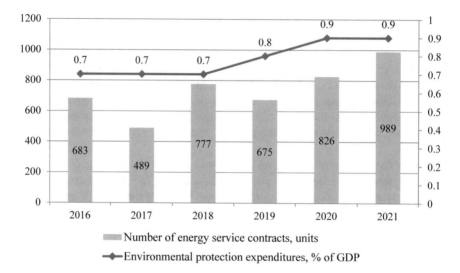

Fig. 1 Indicators of environmental friendliness and energy efficiency in Russia. *Source* Compiled by the authors based on data from the Association of Energy Service Companies [3] and the Federal State Statistics Service of the Russian Federation [7]

of 1–1.5 years), the reliability of lighting increases, and the costs of their operation decrease.

Given the prevalence of green public procurement in developed countries, unfortunately, such a trend has not yet been observed in Russia. Even though public spending on environmental protection has increased tenfold over the past decade and has been growing for the past three years (Fig. 1), and ecology has become one of the priorities of the country's development, there is no clear indication in the public procurement system of the allocation of criteria for environmental friendliness when choosing suppliers.

The explanation of this position may be as follows. First, this is the lack of a legal foundation based on which the public procurement system can orient itself in its work with green procurement, as well as the insufficiency of defining environmental rules and regulations. One of the first steps is introducing a list of goods for public procurement in the Russian Federation in 2023. For this list, the share of secondary raw materials used in production will be indicated. This should stimulate the expansion of the market for products involving the use of recycling and the development of ideas of the circular economy and increase the demand for such products from government customers. Environmental safety should become one of the public procurement principles along with their transparency or competitiveness because this will also contribute to the renewal of the country's economy.

An indirect confirmation of the attention of state customers is the use of the mechanism for concluding energy service contracts, the number of which has also been growing in recent years (Fig. 1). Energy service contracts are concluded with suppliers who have won a competition in public procurement for energy-saving measures (maintenance, design, acquisition, financing, installation, commissioning, operation, maintenance, and repair of energy-saving equipment at one or more customer facilities). Simultaneously, the contract price is determined based on the expected amount of energy savings, and the payment for the services of the energy service company corresponds to the percentage of savings achieved. This approach is important for promoting the principles of energy efficiency in public procurement and allows for significant government savings.

Undoubtedly, an accompanying and important tool for promoting the principles of energy conservation and energy efficiency in society are educational programs that are conducted to inform specialists of state and municipal institutions and citizens about the need to save energy, water, and heat resources. Within the framework of public procurement, this practice significantly contributes to achieving energy efficiency goals with the use of contemporary educational practices, changing human behavior itself in the context of the need to reduce environmental impact.

Therefore, based on the research results, it is necessary to highlight the consideration of energy conservation and energy efficiency as important components of the sustainable development of countries, ensuring the solution of a set of tasks and allowing them to be presented as criteria for concluding contracts in public procurement, namely:

- Saving energy resources;
- Development of alternative energy;
- Reduction of CO_2 emissions;
- Involvement of business in the environmental agenda;
- Optimization of energy characteristics of buildings, real estate, and infrastructure facilities;
- Changes in the rules and regulations for the construction of industrial, commercial, and non–commercial facilities, taking into account the requirements of saving resources (energy, heat, water, etc.);
- Increasing responsibility for the inefficient use of resources;
- Support the society's initiatives in the field of energy conservation and the formation of new consumption models.

4 Conclusions

Saving public finances should not remain the only indicator of the effectiveness of public procurement. The desire to improve the procurement mechanism in relation to the implementation of the dominant energy saving in terms of solving technical, economic, and socially significant tasks, as well as achieving a balance of interests of the state customer and potential suppliers in the framework of competitive procedures affects the development vectors of almost all sectors of the country's economy. The implementation of the trend towards environmental friendliness in public procurement becomes a driver for innovation to create environmentally friendly products and services and solve environmental problems, for example, reducing greenhouse gas emissions. The application of energy efficiency principles contributes to the development of the public procurement system and the most important industries that ensure economic growth: energy (based on the development of alternative energy), the construction industry (based on the use of new economical and energy-efficient structures), and various industries that produce energy-saving materials, equipment, and appliances and form progressive energy-saving standards for new and reconstructed buildings and structures.

References

1. Akram MW, Mohd Zublie MF, Hasanuzzaman M, Rahim NA (2022) Global prospects, advance technologies and policies of energy-saving and sustainable building systems: a review. Sustainability 14(3):1316. https://doi.org/10.3390/su14031316
2. Anchishkina O (2017) Contract institutions in the Russian economy: the sphere of state, municipal, and regulated procurement. Vopr Ekon 11:93–110. https://doi.org/10.32609/0042-8736-2017-11-93-110
3. Association of Energy Service Companies (2021) Russian energy service market, 2021. Retrieved from https://www.escorussia.ru/media/catalog/2022/08/Краткий_обзор_за_2021_г.pdf. Accessed 11 Dec 2022
4. Aydin E, Correa SB, Brounen D (2019) Energy performance certification and time on the market. J Environ Econ Manag 98:102270. https://doi.org/10.1016/j.jeem.2019.102270
5. BP (2021). BP Statistical review of world energy. Retrieved from https://www.bp.com/en/global/corporate/energy-economics/statistical-review-of-world-energy.html. Accessed 11 Dec 2022
6. D'Agostino D, Tzeiranaki ST, Zangheri P, Bertoldi P (2021) Data on nearly zero energy buildings (NZEBs) projects and best practices in Europe. Data Brief 39:107641. https://doi.org/10.1016/j.dib.2021.107641
7. Federal State Statistics Service of the Russian Federation (2021) Environment/Environmental protection costs in the Russian Federation. Retrieved from https://rosstat.gov.ru/folder/11194. Accessed 11 Dec 2022
8. Fregonara E, Ferrando DG, Tulliani J-M (2022) Sustainable public procurement in the building construction sector. Sustainability 14(18):11616. https://doi.org/10.3390/su141811616
9. Genc TS, Thille H, Elmawazini K (2020) Dynamic competition in electricity markets under uncertainty. Energy Econ 90:104837. https://doi.org/10.1016/j.eneco.2020.104837
10. Ghalambaz S, Hulme CN (2022) A scientometric analysis of energy management in the past five years (2018–2022). Sustainability 14(18):11358. https://doi.org/10.3390/su141811358
11. Gorokhova DV (2020) Public procurement in the Russian Federation: retrospective and development. Financial J 12(2):57–68
12. Khan N (2018) Prelims. In: Public procurement fundamentals. Emerald Publishing Limited, Bingley, pp i–xvi. https://doi.org/10.1108/978-1-78754-605-920181008
13. Melnikov VV (2022) Efficient public procurement and the role of competition. J Inst Stud 14(3):119–131. https://doi.org/10.17835/2076-6297.2022.14.3.119-131
14. Shadrina EV, Vinogradov DV, Kashin DV (2021) Do environmental priorities of the state affect the practice of public procurement? Public Adm Issues 2:34–60
15. Slepak VY, Pozhilova NA (2021) Legal foundations of funding fundamental science projects within the Horizon Europe programme. Kutafin Law Rev 8(3):423–442. https://doi.org/10.17803/2313-5395.2021.3.17.423-442
16. Smotritskaya II (2019) Contract procurement system in the context of Russian reforms. Bull Inst Econ Russian Acad Sci 6:9–25. https://doi.org/10.24411/2073-6487-2019-10067

17. Stepnov IM, Kovalchuk JA, Gorchakova EA (2019) On assessing the efficiency of intracluster interaction for industrial enterprises. Stud Russ Econ Dev 30:346–354. https://doi.org/10.1134/S107570071903016X

18. Sun Z, Zhang J (2022) Impact of resource-saving and environment-friendly society construction on sustainability. Sustainability 14(18):11139. https://doi.org/10.3390/su141811139

19. Wei Z, Han B, Pan X, Shahbaz M, Zafar MW (2020) Effects of diversified openness channels on the total-factor energy efficiency in China's manufacturing sub-sectors: evidence from trade and FDI spillovers. Energy Econ 90:104836. https://doi.org/10.1016/j.eneco.2020.104836

20. Yang C, Masron TA (2022) Impact of digital finance on energy efficiency in the context of green sustainable development. Sustainability 14(18):11250. https://doi.org/10.3390/su141811250

Practical Application of the Energy Justice System for the Assessment of the Compliance of Energy-Related Strategies and Projects in the Arctic Zone with the Principles of Sustainable Development

Anna B. Krasnoperova ⓘD

Abstract The research aims to analyze the integration of sustainable development and energy justice principles in the processes and initiatives related to the exploitation of energy resources in the Arctic region, comment on the issue of ensuring fairness and equity as far as the process of infrastructural expansion in the considered region is concerned, and outline further recommendations to establish energy justice framework. During the research, the author uses methods of scientific knowledge such as qualitative and quantitative analysis, synthesis, a comparative approach to the study of specific cases, and statistical data analysis. The research shows that the principles of sustainable development play an increasingly prominent role in the exploitation and development of energy resources. The considered concept is particularly important in the Arctic region for several reasons. First, harsh climatic conditions lead to the fact that local communities become very vulnerable to changes in the surrounding ecosystem caused by the implementation of energy projects and the development of energy infrastructure. Second, the energy sector is the most visible component of the Arctic economy, which largely determines the socio-economic well-being of local residents. Simultaneously, one can hardly deny that the implementation of oil- and gas-related initiatives often results in significant social and economic concerns and trade-offs. Therefore, it is essential to incorporate the principles of energy justice, sustainability guidelines, corporate social responsibility frameworks, and other similar tools in the business strategies and operational practices of energy companies operating in the Arctic zone.

Keywords Sustainable development · Energy justice framework · Fossil fuels · Renewable energy sources · Distributional justice · Procedural justice · Justice as recognition · Restorative justice · Arctic region · Offshore energy projects

JEL Classification Q01 · Q40

A. B. Krasnoperova (✉)
MGIMO University, Moscow, Russia
e-mail: krasn_anna97@mail.ru

1 Introduction

The transformation of energy systems aimed at ensuring universal access to affordable, efficient, and sustainable energy services primarily based on renewable energy sources has recently become the issue of utmost importance. Despite the fact that the share of renewables in the global energy mix has reached 24% in 2016 and is continuing to increase, the availability of environmentally clean energy services is still limited. Furthermore, approximately 840 million people worldwide do not have access to electricity at all, with the said figure projected to increase to 8% of the global population by 2030 since annual population growth outpaces the current rate of improving energy provision [7].

That being said, the necessity of expanding energy access while simultaneously promoting the use of renewable energy sources and increasing energy efficiency is explicitly expressed in the UN Sustainable Development Goals (SDGs). Apart from SDG 7, which is directly related to energy in its various forms and the provision of energy services, the need to accelerate energy transitions is addressed in SDG 13, which outlines the role of low-carbon energy system transformation in climate change mitigation. What is more, several international documents, including the Paris Agreement, emphasize the importance of fostering access provided through the development of renewable energy solutions [2].

To resolve energy-related issues, international organizations and national governments adopt measures to implement low-carbon energy transitions and promote wider access to energy services [4]. Nevertheless, the said measures may result in social injustices toward certain categories of potential stakeholders, in particular indigenous communities and local residents. Thus, the justice dimension of energy transition initiatives may not be taken into due consideration, leading to further exacerbation of social and economic problems.

2 Methodology

The theoretical and methodological basis of the given research is composed of scientific research conducted by members of the academic community and recognized experts in the field of sustainable development and justice-related issues in the energy sector, such as Dementieva [3], Delina [2], Dworkin [17], Harsem [6], Heffron [9, 10], Lee [21], Kadenic [8], Krotova [18, 20], McCauley [9], Nerini [4], Noble [12], Ostrovskaya [21], Parlee [13], Popkova [14], Pichkova [20], Sidortsov [15], Sovacool [16], Studenikin [19], and Zavyalova [3, 14, 18–21] as well as the relevant reports issued by IEA, IRENA, and UNSD.

Throughout the course of the research, the author applies such methods of scientific research as qualitative and quantitative analysis, synthesis, a comparative approach to the study of specific cases, and analysis of statistical data.

3 Results

The energy system in the Arctic is based on fossil fuels, most notably gas and oil, the resources of which are abundant in the said region. That being said, in recent years, projects based on renewable energy sources have become increasingly prominent due to the development of technologies, equipment, and infrastructure, as well as due to the mounting need to mitigate global climate change [6].

According to the 1982 United Nations Convention on the Law of the Sea (UNCLOS), five countries of the Arctic, namely, Canada, Denmark, Norway, Russia, and the USA, have the right to exploit the natural resources of the region. Thus, the said countries are directly engaged in implementing energy-related initiatives in the Arctic. Nevertheless, other Arctic and non-Arctic countries and large international companies also take a prominent part in scientific research programs and direct energy-related operations, such as natural resources excavation, designing renewable energy sites, and infrastructure development [10]. Consequently, these diverse economic and political actors undertake the responsibility for a complex and highly vulnerable environment, which has a population of more than four million people, including a substantial indigenous population, with local inhabitants being highly dependent on the region's unique ecosystem [11].

One can hardly deny that the increasingly active implementation of energy-related projects in the Arctic results in substantial social and environmental consequences. Energy development significantly affects the ecological situation, further triggering the negative changes caused by global warming, such as sea ice melting, which may lead to the displacement of local communities and disruption of the natural habitat of the species on which indigenous peoples rely for their livelihoods [15]. Moreover, apart from environmental concerns, energy-related initiatives have direct social implications. There is no avoiding the premise that the development of energy infrastructure generates certain social and economic benefits, such as expanding energy access, making energy services more efficient, and creating job opportunities. Nevertheless, the said benefits are accompanied by substantial costs, the allocation of which does not necessarily answer the criteria of fairness and equity. In the situation of rapid infrastructural change, the opinions voiced by local residents are not taken into due consideration; the ability of the affected communities to take an active part in the decision-making process remains highly questionable [16]. What is more, in case certain energy projects, especially the ones based on fossil fuel excavation, are closed due to a variety of reasons, the process of value creation rapidly becomes reversed, which results in the social and economic decline of local communities.

In this regard, the notion of establishing an energy justice framework, namely, of ensuring wider availability and affordability of energy services and observing the due process in the production, transportation, and use of energy, providing all potential stakeholders with the right to engage in policy-making, recognizing and respecting disparate forms of knowledge, cultural and traditional differences, and addressing the so-called "restorative justice," that is to say, the dimension of energy justice that focuses on the mitigation of negative economic (for instance, job loss) and environmental consequences [9], becomes the issue of utmost importance for the Arctic region. Even though the given aspect of Arctic development has been underestimated for a substantial period, with economic and technological issues predominating over social and ethical points, the concept of energy justice has become increasingly prominent in scientific research over the last few years [14]. The practical implementation of the framework described above poses significant questions.

3.1 Recent Improvements in the Distributional Dimension of Justice and Their Potential Drawbacks

The first dimension of the energy justice framework is the so-called distributional justice, which is a spatial concept focused on energy-related issues in connection to a particular locality. One can hardly deny that a certain place has a direct impact on the livelihoods and daily routines of local communities [5]. Thus, distributional justice explores how the development of energy infrastructure affects local residents, their economic well-being, and social relations. The said aspect of energy justice appears to be of particular significance in the Arctic because the indigenous population in the given region is particularly vulnerable as far as their dependence on the surrounding environment is concerned.

There is no avoiding the premise that the dimension of energy justice described above has recently seen certain improvements. Most of the recent projects devoted to fossil fuels excavation are implemented offshore (for instance, energy resources exploration in the Beaufort Sea of Canada, which is undertaken by such companies as Chevron, BP, and Exxon Mobil), therefore reducing potential threats for local communities [8, 12]. Furthermore, certain Arctic countries are currently engaged in activities aimed at the transition towards renewable energy sources, which are widely considered to pose less danger to the fragile ecosystem of the region than traditional gas and oil [19]. One of the latest examples of such projects is the proposed development of a wind farm in Finnmark county, predominantly populated by the Saami people [8].

Notwithstanding the fact that the aspects of distributional justice are increasingly featuring in the policies implemented by the governments of Arctic countries and in the business models of the companies engaged in energy-related activities, the considered notion is still far from being resolved. While the development of renewable energy sources may be beneficial from an environmental standpoint, the said process often causes other negative consequences, such as disruption of the reindeer population and natural habitat of other species, on which indigenous communities are largely dependent. What is more, offshore energy projects also present a potential threat of ecological catastrophe, such as an oil spill, which would have a crucial impact on local residents. Development of energy infrastructure in any form also results in increased industrialization, shipping, etc., affecting indigenous communities [17].

3.2 Addressing Procedural and Recognition Dimensions of Justice: The Lack of Due Consideration

The second component of the energy justice framework is the procedural justice concept, which emphasizes the necessity of engaging all potential stakeholders in the policy-making process and in other procedures related to the production and use of energy in a fair and equitable way. Closely linked to this concept is the third dimension of energy justice, that of justice as recognition, which calls for proper recognition and respect for indigenous populations and non-indigenous local communities, the various forms of knowledge these local residents possess, their livelihoods, and their vulnerability in terms of energy infrastructure development [18].

As it is clearly perceived from a number of case studies, the extent to which local communities are involved in large-scale development projects directly affects the social and economic consequences of the said projects overall and the affected communities in particular. That being said, local residents are often given the ability to participate in the decision-making process only after the given project has started rather than at an earlier phase of planning and designing infrastructure development. Therefore, local knowledge, which may often be highly valuable for the developers, and social and economic concerns of the local communities do not receive due consideration.

Simultaneously, in recent years, international companies engaged in energy-related activities in the Arctic, as well as the governments of the Arctic countries, have made certain progress toward wider recognition and inclusion of local communities [20]. For instance, as a number of energy projects implemented in Canada demonstrate, multiple indigenous communities are actively involved in the said developments. Furthermore, the social justice side of the projects is considered along with economic and environmental aspects [13]. Nevertheless, the ability of the local residents to influence the policy-making process remains somewhat questionable. While local knowledge may be consulted, it tends to have little impact on the keystone decisions in the area of energy development. Thus, the participation of the Arctic communities, especially of the indigenous population, in the development of energy-related projects appears to be substantially limited.

3.3 Restorative Aspect of Energy Justice: Employment Concerns in the Energy Sector

Another dimension of the energy justice framework, which may not have received extensive coverage in the scientific literature, but which nevertheless seems to be highly significant, is the so-called "restorative justice." This term refers to the mitigation of negative economic (e.g., loss of jobs) and environmental consequences of the transition towards a more sustainable energy system primarily based on renewable sources [9]. There is no avoiding the premise that the given issue, particularly its economic aspect, is of utmost importance in the Arctic region, where a substantial proportion of local residents is engaged in the energy sector. Local communities, particularly the indigenous population, often find themselves in a disadvantaged position because, due to the lack of specific education and sometimes due to certain cultural and social prejudices, these categories of people are mostly employed only at the low-level, underpaid positions in energy development projects. Furthermore, with the transition towards a low-carbon energy system and given the postponement of a number of fossil fuels excavation projects caused by the decline in oil prices, even these employment opportunities are likely to be lost, which will have a pronounced negative impact on the socio-economic well-being of the affected communities.

As one can clearly perceive based on the examples from the recent past, the governments of the Arctic countries currently do not necessarily step in to address the negative effects mentioned above. A common problem in the Arctic region is the "resource curse." The "resource curse" is the economic phenomenon characterized by the disproportionate allocation of capital, including financial, social, and human capital in a certain resource extraction sector, causing a lack of investment in other areas, in particular in innovative ones, and leading to substantial economic and social problems due to the high volatility of resource industries. Thus, should the oil and gas sectors come to decline, urgent state intervention should be needed to mitigate social and economic concerns faced by the local communities [1, 3].

4 Conclusion

As far as establishing an energy justice framework is considered, current policies implemented by the governments of the Arctic countries and the companies engaged in energy-related activities in the region are mostly aimed at promoting the distributional dimension of energy justice, that is, at mitigating the potential threats posed by the development of energy infrastructure for the particular localities crucial for the livelihoods of the indigenous peoples and non-indigenous Arctic communities [21]. Thus, environmental and social consequences of the energy projects based offshore or focused on renewable energy sources do not receive necessary consideration because these activities are assumed not to have a direct impact on local residents. One can hardly deny that the negative implications of these projects, albeit not so prominent, are still highly substantial for the inhabitants of the Arctic region. Therefore, should the currently applied approach be replaced with a thorough evaluation of the potential drawbacks of all the energy initiatives without any exceptions, that would address the social injustices and environmental concerns more fairly and equitably.

Another issue that should be considered when promoting the energy justice framework in the Arctic is ensuring the wider participation of the affected communities in the decision-making process. Mobilization of local knowledge and inclusion of the local residents from the earliest stages of planning and designing energy-related projects would be beneficial for the considered communities and the companies or governmental authorities implementing these projects. On the one hand, wider participation of the indigenous communities and non-indigenous Arctic populations in the policy-making process would result in better representation of the said categories of people and, consequently, in a more fair and equitable approach to the implementation of energy-related initiatives. On the other hand, local knowledge often proves to be invaluable for the companies and governmental authorities engaged in energy infrastructure development due to the highly specific climatic conditions in the region. Thus, local communities and indigenous populations should receive wider access to the decision-making process, particularly in the earlier phases of energy-related projects; the disparate forms of knowledge, including local knowledge and various viewpoints voiced by all the potential stakeholders, should receive due recognition and respect.

Finally, there is no denying that the transition towards low-carbon energy sources, which has become increasingly prominent in the Arctic region in recent years, is likely to cause such adverse economic effects as loss of jobs in the energy sector and overall economic decline. The population of the Arctic appears to be largely dependent on the fossil fuels industries, with the said notion being particularly substantial in regards to non-indigenous local residents, who constitute a substantial proportion of the Arctic communities and who are mostly employed on the gas and oil excavation projects and energy infrastructure development. The decline of the said industries would be a problem of utmost significance for the Arctic region. Therefore, the governments of the Arctic countries should consider the issue of designing and implementing such social and economic programs that would provide alternative employment opportunities, support the development of other economic sectors in the region, and ensure the social and economic well-being of the communities negatively affected by energy system transformations.

References

1. Andreeva E, Kryukov V (2008) The Russian model: merging profit and sustainability. In: Mikkelsen A, Langhelle O (eds) Arctic oil and gas: sustainability at risk? Routledge, Abingdon-on-Thames, pp 240–288
2. Delina LL, Sovacool BK (2018) Of temporality and plurality: an epistemic and governance agenda for accelerating just transitions for energy access and sustainable development. Curr Opinion Environ Sustain 34:1–6. https://doi.org/10.1016/j.cosust.2018.05.016
3. Dementieva A, Zavyalova E (eds) (2020) Corporate governance in Russia: Quo vadis? De Gruyter, Berlin. https://doi.org/10.1515/9783110695816
4. Fuso Nerini F, Tomei J, To LS, Bisaga I, Parikh P, Black M, Mulugetta Y et al (2018) Mapping synergies and trade-offs between energy and the sustainable development goals. Nat Energy 3:10–15. https://doi.org/10.1038/s41560-017-0036-5
5. Gieryn TF (2000) A space for place in sociology. Ann Rev Sociol 26(1):463–496. https://doi.org/10.1146/annurev.soc.26.1.463
6. Harsem Ø, Eide A, Heen K (2011) Factors influencing future oil and gas prospects in the Arctic. Energy Policy 39(12):8037–8045. https://doi.org/10.1016/j.enpol.2011.09.058

7. IEA, IRENA, UNSD, WB and WHO (2019) Tracking SDG7: the energy progress report, 2019. Washington, DC. Retrieved from https://iea.blob.core.windows.net/assets/74db56e5-92a9-448a-ae9e-7eda20df541b/2019_Tracking_SDG7_Report.pdf. Accessed 11 Dec 2022

8. Kadenic M (2015) Socioeconomic value creation and the role of local participation in large-scale mining projects in the Arctic. Extractive Ind Soc 2(3):562–571. https://doi.org/10.1016/j.exis.2015.04.010

9. McCauley D, Heffron R (2018) Just transition: integrating climate, energy and environmental justice. Energy Policy 119:1–7. https://doi.org/10.1016/j.enpol.2018.04.014

10. McCauley D, Heffron R, Pavlenko M, Rehner R, Holmes R (2016) Energy justice in the Arctic: implications for energy infrastructural development in the Arctic. Energy Res Soc Sci 16:141–146. https://doi.org/10.1016/j.erss.2016.03.019

11. NOAA (2019) Arctic report card. Retrieved from https://arctic.noaa.gov/Report-Card/Report-Card-2019. Accessed 12 Dec 2022

12. Noble B, Ketilson S, Aitken A, Poelzer G (2013) Strategic environmental assessment opportunities and risks for Arctic offshore energy planning and development. Mar Policy 39:296–302. https://doi.org/10.1016/j.marpol.2012.12.011

13. Parlee BL (2015) Avoiding the resource curse: Indigenous communities and Canada's oil sands. World Dev 74:425–436. https://doi.org/10.1016/j.worlddev.2015.03.004

14. Popkova EG, Zavyalova E (2021) New institutions for socio-economic development: the change of paradigm from rationality and stability to responsibility and dynamism. De Gruyter, Berlin. https://doi.org/10.1515/9783110699869

15. Sidortsov R, Sovacool B (2015) Left out in the cold: energy justice and Arctic energy research. J Environ Stud Sci 5:302–307. https://doi.org/10.1007/s13412-015-0241-0

16. Sovacool BK (2014) What are we doing here? Analyzing fifteen years of energy scholarship and proposing a social science research agenda. Energy Res Soc Sci 1:1–29. https://doi.org/10.1016/j.erss.2014.02.003

17. Sovacool BK, Dworkin MH (2015) Energy justice: conceptual insights and practical implications. Appl Energy 142:435–444. https://doi.org/10.1016/j.apenergy.2015.01.002

18. Zavyalova EB, Krotova TG (2021) Methods for achieving SDG 1, poverty eradication. In: Popkova EG, Sergi BS (eds) Modern global economic system: evolutional development vs. revolutionary leap. Springer, Cham, pp 1987–2001. https://doi.org/10.1007/978-3-030-69415-9_218

19. Zavyalova EB, Studenikin NV (2019) Green investment in Russia as a new economic stimulus. In: Sergi BS (ed) Modeling economic growth in contemporary Russia. Emerald Publishing Limited, Bingley, pp 273–296. https://doi.org/10.1108/978-1-78973-265-820191011

20. Zavyalova EB, Pichkova LS, Krotova TG (2021) The role of business and government cooperation in preventing and mitigating global biogenic challenges. In: Popkova EG, Sergi BS (eds) Modern global economic system: evolutional development vs. revolutionary leap. Springer, Cham, pp 2229–2244. https://doi.org/10.1007/978-3-030-69415-9_244

21. Zavyalova E, Lee JS, Ostrovskaya E (2020) Corporate social responsibility. In: Dementieva A, Zavyalova E (eds) Corporate governance in Russia: Quo vadis? De Gruyter, Berlin, pp 173–187

Impact Produced by Climate-Resistant Smart Technologies on Energy Saving and Energy Efficiency of Large Industrial Facilities

Yana A. Saltykova 📵

Abstract Each large industrial facility works toward saving energy and increasing its production energy efficiency. Notably, as a result of the world energy crisis, this trend has become highly relevant. The need to develop energy-efficient technologies permeates all business processes, affecting final product quality and cost. Climate technologies play a significant role. They determine air circulation efficiency and sustainability of air circulation at industrial sites, thus enhancing labor safety and reducing risks for occupational diseases. Additionally, new smart climate technologies that consider energy efficiency reduce the expenses to manage them and speed up the operation. In this regard, solutions for climate-resilient smart technologies used at large industrial facilities and their impact on energy saving and efficiency, which generally affects final product costs, are of particular interest. In the context of the industrial crisis, the use of smart climate technologies is of particular relevance because reducing product cost positively affects product competitiveness. The use of effective technologies reduces staff illness, thereby improving labor efficiency.

Keywords Industry · Smart technologies · Climate technology · Labor productivity · Investments · Efficiency

JEL codes O14 · O16 · O32 · O33 · L23

1 Introduction

In the context of the world economic crisis and energy shortage, the use of technologies to improve energy production efficiency is becoming more relevant.

Industrial facility and manufacturing plant operation is more strongly associated with energy consumption, which generates the need to find solutions for reducing

Y. A. Saltykova (✉)
MGIMO University, Moscow, Russia
e-mail: saltykova.ya@mail.ru

its consumption and has an immediate impact on the reduction of final product cost and the growth of the enterprise's competitiveness.

In addition to affecting the cost of final products produced by industrial enterprises, energy efficiency mitigates the negative impact produced by light and heat sources on employees working in workshops and industrial sites. Workshops and buildings of industrial facilities are designed to be obligatorily supplied with climatic systems [4].

To understand the climate technology concept and its impact on the energy efficiency of the enterprise, it is advisable to consider the concept in more detail. Thus, energy efficiency can be determined as an efficient energy distribution contributed by energy saving in workshops or management premises [7].

The given industry is at the interface of economics and engineering. It is associated with a significant reduction in final product cost and better microclimate in workshop premises [14].

The research aims to assess the impact of climate-resistant smart technologies on energy saving and energy efficiency of large industrial facilities.

2 Materials and Method

The author used the comparative historical method, regression and visual data analysis, comparison and analogies, statistical empirical data processing, expert-analytical method, and other methods.

3 Results

Energy efficiency is promoted by a set of measures, the structure of which is the following [11]:

- Choosing minimum and maximum values for optimal microclimate parameters;
- Infiltration reduction (infiltration air consumption);
- Increased air distribution efficiency on premises;
- Local air conditioning;
- HVAC system decentralization (heating, ventilation, air conditioning);
- Zoning of the system fundamental solution on the cardinal points;
- Use of preheating and cooling;
- Waste and natural heat and cold disposal;
- Combining microclimate support systems with other systems;
- Improvement of technical system automation tools.

It can be clearly seen that the most important role in the energy efficiency of construction is given to heating and cooling systems, local air conditioning, and indoor climate. Using energy-intensive climate technologies is a priority.

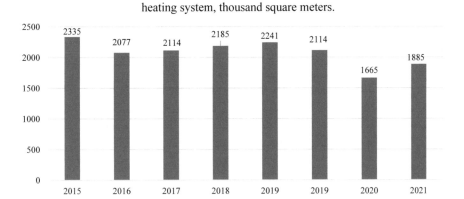

Fig. 1 Pattern of annual energy consumption to ensure climatic equipment operation at the industrial workshop. *Source* Compiled by the author based on the PJSC Avitek data [16]

With that said, climatic technologies comprise a set of air-conditioning, heating, sanitary, and utility equipment provided in all workshop buildings and management premises [13].

In fact, climate technologies include air conditioners, ventilation, air supply systems, and other engineering structures that contribute to a healthy environment in workshop buildings, support staff health accordingly, and improve operation convenience and quality in each room.

Specifically, climate technologies are one of the powerful sources of energy that industrial enterprises consume because spacious areas of workshops are equipped with these technologies, while most of them consume much energy. Thus, the pattern of annual energy consumption to ensure climatic equipment operation at the PJSC Avitek rolling plant is shown in Fig. 1.

It takes much energy to create a workshop climate. It accounts for 30–40 million rubles per year in value terms, which is quite a high cost even in the context of large enterprises. That is why many manufacturers tend to introduce energy-intensive technologies to equip their workshops, thus reducing costs. To achieve this goal, finding solutions to reduce energy consumption and increase the energy efficiency of climate technologies has recently become significantly more relevant. In this regard, induction or automated smart technologies are being developed, contributing to the uninterrupted operation and maintaining optimal conditions for workshop and industrial site production environments. To understand the significance of the impact of smart climate technologies on the energy efficiency of the workshop, it is necessary to consider their characteristics. The main difference between the smart and the traditional climatic system is the capability of the first one to remove air mass volumes so that the air would be normalized and got fresh [15]. More details on the advantages of smart climate technologies over traditional ones are shown in Fig. 2.

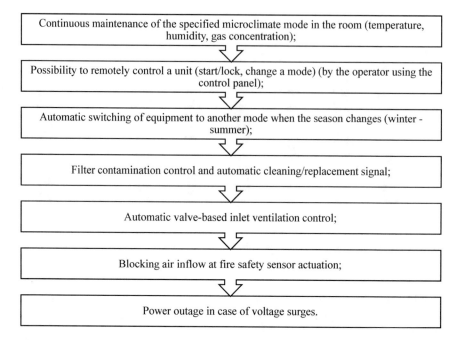

Fig. 2 Advantages of smart climate technologies over traditional ones. *Source* Compiled by the author based on [15]

Autonomous industrial heating systems show the greatest efficiency, which, in fact, ensures the complete independence of enterprises from city utilities and an independent choice of temperature mode depending on the industrial needs [6]. Smart industrial heating systems are divided into infrared (radiant), convective, and radiant-convective. With their ability to disperse air throughout the workshop, their energy efficiency is several times higher than that of the traditional climate systems. In traditional systems, the air is distributed in the upper part of industrial premises, which reduces heating and increases energy consumption for workshop air conditioning and heating [10].

In turn, radiant (infrared) heating systems heat objects (floor, equipment) isolated and heat people (heat is felt on skin and clothes) in areas where this is actually required, which significantly reduces energy consumption and increases the energy efficiency of production. The smart system is based on the sunlight warming effect; the difference is only that the radiation spectrum works in the infrared range. Simultaneously, it has no harmful ultraviolet light.

Having studied the latest technologies and their application in industrial enterprises, it was concluded that gas radiant heating systems are the most energy-efficient ones among smart climatic technologies [12]. It should be clarified that the systems are based on providing heating and air conditioning for workshop buildings that are stand-alone energy systems and run on natural gas, which significantly reduces electricity and heat energy costs for enterprises. This is due to the fact that the cost of natural gas required to generate a thermal energy unit (specifically, radiant) is significantly lower than the cost of electricity consumed by traditional systems. Compared to radiant water heating systems, the use of gas technologies eliminates the need to build a boiler plant for heating water, as well as pipelines and pump installation for its circulation from the boiler plant to the corresponding energy radiator and vice versa.

Let us compare the economic indicators of the use of a gas radiant heating system to traditional ones. The data were taken from the Sibshvank JSC website, which is a developer of these systems. An example provides the equipment of the plant workshop with an area of 700 m^2 with the gas radiant heating system. Economic indicators of gas radiant heating systems in comparison with traditional ones are as follows:

- Engineering expenses: gas radiant heating system—500 thousand rubles, traditional heating systems—300 thousand rubles.
- Installation expenses: gas radiant heating system—1000 thousand rubles, traditional heating systems—1700 thousand rubles.
- Heating expenses: gas radiant heating system—18,000 thousand rubles per year, traditional heating systems—28,000 thousand rubles per year.
- Energy power expenses associated with building air conditioning and ventilation: gas radiant heating system—1200 thousand rubles per year, traditional heating systems—3200 thousand rubles per year.
- Payback period: gas radiant heating system—0.5 years, traditional heating systems—2 years [2].

The analysis proves that using gas radiant heating systems significantly reduces energy costs. In the example given above, the savings can be easily calculated as early as at the system installation stage because, despite the higher design costs, the cost of an installation system is much lower than that of traditional systems. Savings for the plant amounted to 500 thousand rubles. Over the year, the company got an effect of 12 million rubles, which was associated with a significant decrease in energy costs when installing climatic systems in the workshop. Alongside that, the climate in the workshop improved, and the personnel sickness rate decreased. The latter is another resource to reduce enterprise costs because higher labor productivity and lower salary costs for employees replacing the diseased have a positive effect on the total plant costs.

The operation of the considered systems is improved by equipping them with innovative operating panels that enable remote adjustment of workshop heat exchange and air ventilation.

Smart-type climate systems also run on infrared scanning and automatically recognize the spots for temperature change [8]. In this regard, large developers recommend installing air handling units with a plate recuperator that provides air heating and air conditioning in ventilated rooms by drying and filtrating induced air. To automate the solution, a Freon thermal insulated air compressor condensing unit can be used.

Local sources of steam, moisture, and harmful substance emissions are removed by local exhaust hoods built into clean room facilities. Air is supplied to the rooms through filtered boxes.

Automatic regulators are installed on the inflow and exhaust system to maintain pressure drops between rooms during the operation of the ventilation system. Automation monitors and regulates the entire ventilation system in accordance with the specified parameters.

In general, this system provides the required temperature, humidity, air pressure, and sanitary and hygienic environmental standards in clean areas and adjacent plant premises.

Additionally, introducing a smart climate system significantly increases energy efficiency, reducing energy consumption and plant production costs.

However, the use of these systems is to be carefully analyzed for return on investment and the possibility of being used in each specific production, concerning its specifics and energy intensity. Thus, the implementation of a climate system at one of the production sites should be accompanied by revenue, cost, and energy consumption trend analysis.

An example of introducing the Tion climate system into the Aluminum Technologies LLC industrial site workshop will be considered. It should be clarified that fresh air flowing into the workshop building was controlled manually before the innovative project introduction. After the workshop climate control unit was improved by installing the automation system equipped in a small box and connected to the Internet; carbon dioxide concentration, temperature, and humidity are controlled automatically. As a result, energy is efficiently managed, preventing it from releasing outside and waste consumption in the workshop. Additionally, due to the replaced internal filters of the plant climate system, ozone is generated to destroy all microorganisms, disinfect the air, and decompose to oxygen at its output. Due to this, the filters are not clogged and need less frequent replacement. The company has significantly reduced energy costs by implementing these systems. The cost-effectiveness analysis is as follows:

- Elimination of heat and energy losses—7750 thousand rubles per year;
- Elimination of electricity losses—1350 thousand rubles per year;
- Labor cost saving for personnel responsible for climate system operation, including social contributions—624 thousand rubles per year;
- Energy-output ratio reduction—15%;
- Unit cost reduction—10%;
- Reduction of losses from staff illness—580 thousand rubles per year.

Considering that the total cost of the system upgrade amounted to 2 million 800 thousand rubles, the company recovered all costs as early as in the first year of its operation and earned additional income from energy savings in the amount of 7 million 504 thousand rubles.

Thus, as a result of the project implementation, the indicators of cost, production unit costs, energy capacity, and energy efficiency will change. However, it should be borne in mind that the trend towards green energy and clean technologies contributes to increased product competitiveness. That is, this is the case when the company increases its own environmental safety rating, strengthening its importance in the world market.

This is particularly important for industries that are large international market players. The focus here is on metallurgy, the oil and gas industry, and the defense industry. Importantly, representatives of these industries generally have a broad production infrastructure, each element requiring building an eco-friendly air atmosphere.

Therefore, enterprises will benefit from switching to the use of smart climate solutions, which will affect economic and social indicators. Simultaneously, when choosing smart climate technologies, companies must use advanced assessment systems based on expert opinion and device performance. Assessment parameters can be rated, while the importance of each indicator should be selected based on expert estimates [1].

Parameters for assessing smart climatic systems should include the following:

1. Energy saving;
2. Energy efficiency;
3. Payback period;
4. Autonomous climatic system parameter control;
5. Service life;
6. Maintainability;
7. Automated temperature, humidity, and pressure setting;
8. Provision of controllable digital information to the independent unit;
9. Noise level;
10. Coverage area.

These criteria must be given qualitative characteristics, after which each factor is rated on a ten-point scale. As a result, the choice falls on the device that will reach the maximum score.

For illustrative purposes, three devices will be considered to select a more suitable one for the Avitek JSC workshop located in Kirov [16]. Table 1 shows climate system characteristics with the parameters specified.

Based on the given equipment parameters, these parameters are further assessed by experts on a ten-point scale. Interestingly, the experts are Avitek labor protection specialists consisting of three people. Each of them rates on the following scale:

- 1–3—low rating;
- 4–7—medium;
- 8–10—high.

On this basis, the final assessment of the selected technologies for equipping the workshop with a climate system is further calculated. The mean assessment scores are given in Table 2.

According to the assessment results, the advantage is taken by the Chinese innovation because it gained a higher score. However, if the importance of each parameter in the overall assessment system is considered, energy saving and energy efficiency

Table 1 Climate system characteristics

Assessment parameter	Refengineering [5]	INTEKH-Climate [3]	Xian LIB (China) [9]
1. Energy saving (loss of energy in % to the overall supply) (%)	3–5	15–17	7–8
2. Energy efficiency, energy consumption, kWh of operation	50	77	45
3. Payback period (according to the developer) (years)	1.5	2	1
4. Autonomous climatic system parameter control	Full	Partial	Full
5. Service life (years)	5–7	7–10	10
6. Maintainability	High	Medium	Low
7. Automated temperature, humidity, and pressure setting	Available	Partial	Full
8. Provision of controllable digital information to the independent unit	Available	Partial	Available
9. Noise level (dB)	30	35	25
10 Coverage area (m^2)	2000	5000	5000

Source Compiled by the author based on the internal Avitek JSC reporting on new technology introduction

take leading positions. Simultaneously, Refengineering shows higher rates according to these criteria. Additionally, a Chinese company obtained a higher score due to its high coverage area rates, but the workshop where the installation will be placed is 1800 m^2, which is why this criterion is not so important for the system. Based on the aforementioned, it becomes more logical to choose the Refengineering production climate system. Additionally, when selecting a contractor, the system transportation, installation, and warranty maintenance costs are considered, making it possible to make a final decision.

However, the preference is mainly given to the most energy-efficient, silent, and autonomously controlled installations.

4　Conclusion

The analysis proves that the climate system impacts energy efficiency and energy saving at large industrial companies because they immediately affect electrical and

Table 2 Climate system parameter assessment for workshop operation

Assessment parameter	Refengineering	INTEKH-Climate	Xian LIB (China)
1. Energy saving (loss of energy in % to the overall supply)	8	4	6
2. Energy efficiency, energy consumption, kWh of operation	8	5	9
3. Payback period (according to the developer)	5	4	10
4. Autonomous climatic system parameter control	10	8	10
5. Service life	5	7	8
6. Maintainability	10	8	5
7. Automated temperature, humidity, and pressure setting	10	8	10
8. Provision of controllable digital information to the independent unit	10	8	10
9. Noise level	7	5	8
10. Coverage area	7	10	10
Total	80	67	86

Source Compiled by the author

thermal energy costs, minimizing them and providing good conditions for workshop operation. Additionally, sustainable smart climatic technologies reduce energy consumption by automatically distributing heat and air, thereby preventing energy loss, minimizing heat loss, and distributing air to only those areas where it is needed indeed. On this basis, studying new sustainable smart technologies in climate systems that can be adapted to industrial business operations is becoming more attractive and profitable. Their use will affect the enterprise's energy efficiency, its cost, and the environment in general, particularly staff health and production ecological cleanness, which is also a factor in the enterprise's world market competitiveness.

References

1. Aivazyan SA, Golovan SV, Karminsky AM, Peresetsky AA (2011) On approaches to rating scale comparison. Appl Econometrics 23(3):13–40. Retrieved from https://cyberleninka.ru/art icle/n/o-podhodah-k-sopostavleniyu-reytingovyh-shkal. Accessed 1 Nov 2022
2. Climate and Heating Systems "Schwank" (n.d.) Heating, cooling, and ventilation for your building. Retrieved from https://schwankstore.ru/. Accessed 1 Nov 2022
3. Climate Company "INTECH-Climate" (n.d.) A climatic equipment catalog. Retrieved from https://www.airventilation.ru/Katalog-kondicionerov.htm. Accessed 1 Nov 2022
4. Climate Company "Standard Climate" (n.d.) Industrial plant ventilation. Retrieved from https://www.airclimat.ru/Ventilyatsiya-tseha.htm. Accessed 1 Nov 2022

5. Company "Refengineering" (n.d.) Industrial refrigeration equipment production. Retrieved from https://refeng.ru/proizvodstvo-oborudovaniya. Accessed 1 Nov 2022
6. Dmitrienko NI, Stadnik MS, Kashnikova AA (2022) Features of using gas radiant heating. In: Proceedings of the XIV international student scientific conference. Sochi, Russia. Retrieved from https://scienceforum.ru/2022/article/2018028913/. Accessed 1 Nov 2022
7. Dyurmenova SS, Makhov AY (2020) Ways to improve building energy efficiency. Young Sci 31(321):18–21. Retrieved from https://moluch.ru/archive/321/72917/. Accessed 1 Nov 2022
8. Evstratov VV (2020) Developing of technical equipment for automated room temperature control system. Young Sci 52(342):17–20. Retrieved from https://moluch.ru/archive/342/pdf/1714/. Accessed 1 Nov 2022
9. Focus Technology Co Ltd (n.d.) Xi'an LIB environmental simulation industry. Retrieved from https://ru.made-in-china.com/co_libindustrym/. Accessed 1 Nov 2022
10. Gas radiant heating—a radical solution to improve energy efficiency and competitiveness at industrial enterprises. Retrieved from https://isup.ru/articles/6/9846/. Accessed 1 Nov 2022
11. Lysyov VI, Shilin AS (2017) The ways to increase energy efficiency of buildings and structures. J Refrigeration Air Conditioning 2:18–25. https://doi.org/10.17586/2310-1148-2017-10-2/3-18-25
12. Maintenance portal "Dom-i-remont" (2020) Radiant heating: disadvantages and advantages of the mini-sun use. Retrieved from https://dom-i-remont.info/posts/otoplenie/luchistoe-otoplenie-nedostatki-i-preimushhestva-pomoshhi-mini-solncza/. Accessed 1 Nov 2022
13. Ministry of Construction, Housing and Communal Services of the Russian Federation (2021) Rulebook SP-56.13330.2021 "Production Buildings Code" SNiP 31-03-2001 (Approved by Order on December 27, 2021 No. 1024/pr). Retrieved from https://mooml.com/d/normativno-pravovye-dokumenty/proektirovanie-inzhenernye-izyskaniya/52465/. Accessed 1 Nov 2022
14. Rubtsova MV, Semenova EE (2021) Accounting for the influence of the building shape on its energy efficiency. Eng Constr Bull Caspian Region 2(36):10–15. Retrieved from https://cyberleninka.ru/article/n/uchet-vliyaniya-formy-zdaniya-na-ego-energoeffektivnost. Accessed 1 Nov 2022
15. VentingInfo (2020) Production ventilation—types of systems, requirements. Retrieved from https://ventinginfo.ru/sistemyventilyacii/ventilyatsiya-proizvodstvennyh-pomeshhenij-vidy-sistem-trebovaniya. Accessed 1 Nov 2022
16. Zaitsev AI (2021) Avitek PJSC production capabilities. In: Report "On the functioning of the environmental management system of the enterprise for 2020." Kirov, Russia. Retrieved from https://www.vmpavitec.ru/upload/iblock/d9a/w1e84az6xqle9qzrey4q57s4kny228rf.pdf/. Accessed 1 Nov 2022

ESG Investments of Oil Companies: Foreign and Russian Experience

Ksenia A. Voronina [ID]

Abstract The research reveals the peculiarities of ESG investments of oil companies in Russia and abroad. The author presents the structure of ESG investments and describes the main factors restraining the growth of such investments in the oil industry. Additionally, the author provides a brief analysis of the development and implementation of ESG investments in the world practice as a measure to control climate change based on Industry 4.0. Moreover, the author underlines the necessity of liquidation of formal indicators and determination of unified ESG standards for oil companies. The reasons for the reduction of foreign oil companies' investments in projects related to fossil fuels are analyzed. The leading Russian and foreign oil companies and their compliance with ESG principles, as well as their activities in the field of ESG investments, are considered. The research outlines the reasons, directions, and incentives for oil companies to increase ESG investments. The author supposes that for Russia, the insufficient compliance of oil companies with ESG requirements may become a reason for limiting access to the development of Arctic resources. The research suggests that the ESG concept may become a new system of protectionism for EU countries. A comparative analysis between Russian company Rosneft and Norwegian company Equinor is presented, the correlation between the environmental and financial performance of oil companies is revealed, and some statistical data for full and detailed disclosure of the research topic is given. The research results are of practical importance and can be considered by oil companies as part of their activities in the field of ESG investments.

Keywords ESG investments · ESG standards · Responsible investing · Oil companies · Energy transition · Climate change

JEL codes E220 · Q540

K. A. Voronina (✉)
MGIMO University, Moscow, Russia
e-mail: k.voronina@inno.mgimo.ru

1 Introduction

Following the priorities and commitments in the area of sustainable development, promotion of ESG agenda, and doing business on the principles of environmental friendliness, social responsibility, and high quality of corporate management require oil companies to make appropriate investments.

ESG investments of oil companies are aimed at reducing their negative impact on the environment, ensuring effective interaction with society, increasing transparency of doing business, and improving certain operational and production performance indicators. Otherwise, the efficiency of oil companies decreases, and their reputation deteriorates.

Foreign oil companies consider ESG investments primarily as one of the measures of controlling climate change based on Industry 4.0, increasing the level of corporate and social responsibility, transparency of companies, and profitability of their investment portfolio.

The principles of ESG investments are still in their infancy in Russia. The main participants of this process are exporting companies, especially oil companies, which often simply have to meet Western standards and requirements to sell their products in foreign markets. In today's conditions, ESG investments of oil companies are strongly influenced by economic and political factors.

In this regard, a comparative analysis of ESG investments of Russian and foreign oil companies becomes even more relevant.

2 Methodology

To comprehensively study the problems, features, directions, and incentives of ESG investments of Russian and foreign oil companies, the author has studied several sources on the research topic, which made it possible to determine the existing points of view and give the author's proposals on this issue.

The essence and significance of ESG investments, the theoretical basis of responsible investing based on historical arguments and a detailed analysis of ESG factors, and the practical applicability of different approaches to maximize returns on ESG investments have been widely discussed in the studies of foreign authors, including Hill [5], Silvola and Landau [16], and Swedroe and Adams [17].

The measures of controlling climate change based on Industry 4.0, the methods of achieving sustainable development goals, the contribution of Russian oil and gas companies to the implementation of sustainable development principles, and the peculiarities of the energy transition are presented in the studies of some Russian authors, including Sheveleva et al. [15], Tyaglov et al. [18], Tyaglov et al. [19], Zavyalova and Krotova [20], Zavyalova and Popkova [21], Zavyalova and Studenikin [22].

Foreign and Russian experience of applying ESG principles in developing strategies of oil and gas companies and the prospects of ESG investments in Russia are studied in the works of Kurnosova [6] and Petrov et al. [8].

To understand states' interest in the global energy transition and identify the correlation between ESG investments and the achievement of sustainable development goals, the energy strategies of states, in particular, the energy strategy of the Russian Federation until 2035 [1], were considered.

So far, there has been no comparative analysis of ESG investments of Russian and foreign oil companies.

As part of the application of a systemic approach, the author conducted a comparative analysis of the Russian oil company Rosneft and the Norwegian energy company Equinor, which revealed the correlation between the economic and environmental performance of companies and made it possible to identify the incentives, directions, problems, features, and prospects of ESG investments of oil companies in Russia and abroad.

Using the methods of generalization and synthesis, the author made the appropriate conclusions.

3   Results

At present, the desire of oil companies to comply with certain environmental, social, and governance standards is explained by the increasing awareness of countries of the problems associated with climate change based on Industry 4.0. In the context of the global energy transition, the integration of ESG principles into the activities of oil companies is considered as one of the factors for their future success. The ESG agenda plays a significant role in the sales of oil companies' production in foreign markets.

The problem that affects Russian and foreign oil companies is the lack of unified ESG standards that oil companies should strive to meet. The existence of formal indicators of oil companies' compliance with ESG standards leads to the fact that companies set convenient standards for themselves, misleading investors and authorities that focus solely on various companies' reports and their strategies, which usually reflect only positive aspects and do not cover such accidents as oil spills. Therefore, it is necessary to clearly define the criteria (standards) for oil companies as part of their compliance with the ESG agenda and change, in the interests of the state and investors, the methods of control over oil companies so that all factors of the company's activities are considered. At the present stage, the compliance of companies with the principles of ESG is assessed by a rating among companies. However, such an assessment is currently not accurate enough due to the lack of unified indicators, while the indicators assessed by Russian and foreign companies are different.

There are lots of incentives for foreign oil companies to make ESG investments. Among them, it is necessary to highlight the increasing confidence on the part of state authorities, increasing investment attractiveness, gaining access to participation in state programs and oil sector tenders, the formation of an environmental image within the framework of access to external markets, strengthening the competitiveness of an oil company, obtaining tax benefits and preferences, a simpler procedure for obtaining or extending oil production licenses, reducing the number of state inspections, and reducing costs when selling products abroad.

For Russian oil companies, such incentives are much less, which leads to the fact that ESG investments for such companies are a forced measure that requires high financial costs but allows these companies to sell their products and implement their projects abroad.

Singling out the criterion E (Environment) among other ESG investment criteria, it is necessary to specify that in the short term (between 2021 and 2025), the Norwegian international energy company Equinor takes the lead, while the British-Dutch oil and gas company Shell becomes the leader from 2026. Equinor plans to make the most of its green investments in 2023, when the figure is to reach $4500 million [7].

According to Rating-Agentur Expert RA GmbH (RAEX-Europe), an independent credit rating agency that evaluated companies according to three ESG criteria, the Russian company Rosneft took the leading position among Russian oil and gas companies in December 2021. The second place was taken by Lukoil [13].

According to the ESG ranking of the world's largest oil and gas companies by capitalization as of September 2021 (broken down by ESG criteria), Total, Royal Dutch Shell, BP, and Rosneft have the highest rates. In 2021, Rosneft made the most significant part of its ESG investments in Component G (Governance) [2].

As of October 2022, Rosneft continues to lead the ESG rating among Russian oil and gas companies, focusing on developing social and environmental programs, as evidenced by RAEX-Europe data. The leader is followed by major Russian companies Novatek and Lukoil [14].

As for the directions of ESG investments for Russian oil companies, on average, the largest amount of investment falls on the development of corporate governance, followed by the social aspect and the environmental component.

To analyze ESG investments and determine the features and problems of such investments, as well as the incentives, directions, and prospects of ESG investments of oil companies, the author considered two companies: the Russian oil company Rosneft and the Norwegian oil company Equinor. Both companies operate with state participation. After analyzing the reports of Russia's leading ESG oil company, Rosneft, it is possible to compile a series of systematized data and describe in detail the company's activities in this area. Below, a comparative analysis of the Russian company Rosneft and the Norwegian company Equinor will be given on a number of grounds.

According to the sustainability reports of Rosneft, the level of capital investment in environmental protection for the period 2018–2021 is characterized by a mixed trend. This indicator amounted to approximately 47 million rubles in 2018 and 52 million rubles in 2019. The indicator decreased to 44 million rubles in 2020 and reached the highest indicator for the considered period (55 million rubles) in 2021. This type of investment corresponds to the investments in category E (Environment).

The lowest level of investment was registered in 2020, which can be explained by the severe consequences of the COVID-19 pandemic for the oil industry [9, 12].

Simultaneously, the reduction of the total greenhouse gas emissions indicators of Rosneft for the period 2018–2021 is not directly related to the amount of increased investment in Category E (Environment), except for 2021, when the company's greenhouse gas emissions significantly reduced to 72.9 million tons of CO_2-equivalent, and investments in Category E (Environment) increased to 55 million rubles [10, 12].

As for the S (Social) component, Rosneft's report presented indicators on expenditures on social programs, financing the social sector of the regions, and charity. In 2020, spending on social financing was minimal and amounted to 45 billion rubles due to the negative impact of the COVID-19 pandemic when the oil industry was shaken by lower oil prices and future uncertainty [11].

Although Rosneft makes relatively significant financial investments to improve corporate governance, the reports do not contain data on investments in component G (Governance) in monetary terms. The reports reflect data on the personnel structure by categories (workers, specialists and employees, and managers), the personnel structure by gender, and indicators regarding anti-corruption.

The greatest results in increasing ESG investments were achieved by those Russian oil companies that are actively cooperating with foreign oil companies within the frameworks of implementation of joint projects. For foreign partners, the ESG investment indicator is, in many respects, a conscious measure of environmental conservation.

The average low ESG ratings among Russian oil companies may be one of the reasons for the prohibition of their oil resources development projects in other countries, particularly in the Arctic, as well as the sale of production in foreign markets. Undoubtedly, there are other reasons for restraining the activity of Russian oil companies, including sanctions of Western countries and economic and political issues. Simultaneously, Russian oil companies wonder whether the ESG concept is a new system of protectionism for Western countries.

The Norwegian oil company Equinor should be highlighted among foreign oil companies with state participation. The company's data on compliance with ESG criteria are analyzed to conduct a comparative analysis between the Russian company Rosneft and the Norwegian Equinor. Equinor is selected for the study because it is the company that, among other foreign oil companies, discloses the largest amount of ESG and sustainability data.

According to Equinor's sustainability reports, the company invests the most in category G (Governance). Each year, the company increases its investments in improving corporate governance, data transparency, and corporate responsibility, as well as in the development of business integrity. In 2021, the highest level of investments in category G (Governance) for the period from 2018–2021 was recorded—311 billion rubles [4].

Data on the level of ESG investments in category E (Environment) are not disclosed in Equinor's annual reports. The company specifies the directions of eco-investments and forecasts the amount of investment in renewable energy and green

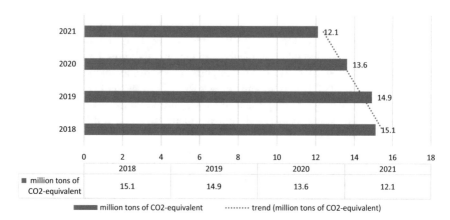

Fig. 1 Total greenhouse gas emissions of Equinor in 2018–2021 (million tons of CO_2-equivalent). *Source* Compiled by the author based on [3]

technologies. Thus, Equinor published data that investments in renewable energy and low-carbon technologies will amount to about $23 billion for the period from 2021 to 2026; the company will increase the share of capital expenditures in the same area to 50% by 2030 [4].

Considering Equinor's total greenhouse gas emissions statistics from 2018 to 2021, there is a clear downward trend in emissions, which suggests that Equinor's green investments have a positive trend. Equinor's total greenhouse gas emissions data for the period 2018–2021 are presented in Fig. 1.

In 2021, Equinor's investments in social programs increased by an average of 100 million rubles compared to 2020 and 2019 [4].

Thus, Equinor directs the largest part of its investments to category G (Governance), then investments are allocated to the development of renewable energy sources and green technologies, which corresponds to category E (Environment), and the smallest amount of investment is allocated to the social sphere—category S (Social).

Table 1 presents a comparative analysis between the Russian Oil company Rosneft and the Norwegian company Equinor according to five criteria:

1. Directions of ESG investments;
2. Problems and features of ESG investments;
3. Incentives of ESG investments;
4. Reduction of total greenhouse gas emissions;
5. Prospects (goals) of the company within the implementation of ESG-agenda.

Table 1 Comparative analysis between Rosneft and Equinor

Criteria	Rosneft	Equinor
Directions of ESG investments	Governance (first place) Social (second place) Environment (third place)	Governance (first place) Environment (second place) Social (third place)
Problems and features of ESG investments	– Forced measure – The influence of political and economic factors, including sanctions – Lack of standardized criteria and indicators (which may mislead investors) – Lack of public interest in the ESG concept	– Conscious measure – Development of renewable energy sources and green technologies – Lack of standardized criteria and indicators (which can mislead investors) – Public interest in the ESG agenda
Incentives of ESG investments	– The creation of an ecological image as part of access to foreign markets – Sales of products abroad – Implementation of projects abroad (access to the development of Arctic resources) – Lower costs for sales abroad; – Increase investment attractiveness	– Increasing confidence on the part of state authorities – Increasing investment attractiveness – Gaining access to participation in state programs and tenders of the oil sector – Enhancing an oil company's competitiveness – Obtaining tax benefits and preferences – Easier procedure for obtaining/extending licenses for oil production; – Reduced number of inspections by the state
Reduction of total greenhouse gas emissions	Sharp decline in 2021; the overall trend is mixed	A downward trend
Prospects (goals)	– Reduction of methane emissions to below 0.25% by 2035 – The company plans to achieve net carbon neutrality in emissions by 2050 – Increase in the level of ESG investments – Reduction of greenhouse gas emissions	– Between 2021 and 2026, the company will invest about $23 billion in renewable energy and green technology; the share of capital expenditures in the same area will increase to 50% by 2030 – The company plans to achieve net carbon neutrality in emissions by 2050 – Increase in ESG investments – Reduction of greenhouse gas emissions

Source Developed and compiled by the author

4 Conclusion

1. It is necessary to clearly define the criteria (standards) for oil companies as part of their compliance with the ESG agenda. Moreover, it is necessary to change, in the interests of the state and investors, the methods of control over oil companies so that all factors of the company's activities are considered.
2. After analyzing the activity of Russian oil companies in implementing the ESG concept, it is possible to conclude that a company is often evaluated not for its real contribution but for the company's possession of a number of formal documents on these aspects.
3. For foreign oil companies, there are more incentives to make ESG investments; it is a conscious measure, and there is a public interest in the ESG agenda.
4. There are much fewer incentives for Russian oil companies, which leads to the fact that ESG investments are a forced measure for them.
5. Rosneft occupies a leading position among Russian oil and gas companies in the field of ESG. In 2021, Rosneft made the largest amount of investment to improve its corporate governance (Governance).
6. The greatest results in increasing ESG investments have been achieved by those Russian oil companies that actively cooperate with foreign oil companies.
7. Equinor's investments in the G category (Governance) have been increasing for four consecutive years. The level of Equinor's investments in category S (Social) also has a positive trend.
8. In general, foreign oil companies make more ESG investments. According to statistics on reducing total greenhouse gas emissions, foreign companies show a downward trend, while the tendency for Russian companies is mixed.

Acknowledgements The author is grateful to the organizers for the opportunity to participate in the conference, present the research results, and hear the opinions of other conference participants on such an important topic. It is especially necessary to note the professionalism of preparation and holding this event, timely informing of the participants. The author would also like to thank the reviewers who read the research results and gave valuable comments.

References

1. Ministry of Energy of the Russian Federation (2020) Energy strategy of the Russian Federation for the period up to 2035 (Approved by Order of the Government of the Russian Federation on June 9, 2020 No. 1523-r). Moscow, Russia. Retrieved from https://minenergo.gov.ru/node/1026. Accessed 10 Sept 2022
2. Avetisyan I (2021) ESG conquers: Rosneft and Lukoil entered the world's largest companies rating. TEKFACE. Retrieved from https://tekface.ru/2021/09/17/esg-pokoryaet-v-rejting-krupnejshih-kompa/. Accessed 28 Sept 2022
3. Equinor ASA (2021) Equinor sustainability data hub. Retrieved from https://sustainability.equinor.com/. Accessed 5 Oct 2022

4. Equinor ASA (2021) Sustainability report, 2021. Retrieved from https://cdn.sanity.io/files/h61 q9gi9/global/d44ff2e9498e7d9cee9e88c4f01e6c4135c7a2f8.pdf?sustainaiblity-report-2021-equinor.pdf. Accessed 5 Oct 2022
5. Hill J (2020) Environmental, social, and governance (ESG) investing: a balanced analysis of the theory and practice of a sustainable portfolio, 1st edn. Academic Press, London
6. Kurnosova TI (2022) Domestic and foreign experience of using ESG-principles in designing oil and gas business development strategy. Econ Bus Law J 12(1):387–410. https://doi.org/10. 18334/epp.12.1.114058
7. ON&T Magazine (2020) Equinor driving renewable investment among oil majors. Retrieved from https://www.oceannews.com/news/energy/equinor-driving-renewable-invest ment-among-oil-majors. Accessed 25 Sept 2022
8. Petrov VO, Starikov IV, Furshchik MA (2021) ESG investments: foreign experience and prospects for Russia. Budget J 11:2–4
9. PJSC Rosneft. (2018) Sustainability report, 2018. Retrieved from https://www.rosneft.ru/upl oad/site1/document_file/Rosneft_CSR18_RU_Book.pdf. Accessed 2 Oct 2022
10. PJSC Rosneft (2019) Sustainability report, 2019. Retrieved from https://www.rosneft.ru/upl oad/site1/document_file/Rosneft_CSR2019_RUS.pdf. Accessed 2 Oct 2022
11. PJSC Rosneft (2020) Sustainability report, 2020. Retrieved from https://www.rosneft.ru/upl oad/site1/document_file/Rosneft_CSR2020_RUS.pdf. Accessed 2 Oct 2022
12. PJSC Rosneft (2021) Sustainability report, 2021. Retrieved from https://www.rosneft.ru/upl oad/site1/document_file/Rosneft_CSR2021_RUS.pdf. Accessed 2 Oct 2022
13. RAEX (2021) ESG ranking of Russian companies. Retrieved from https://www.raexpert.eu/ media/uploads/rankings/ESG_Ranking_15.12.2021_qjkQAoS.pdf. Accessed 26 Sept 2022
14. RAEX (2022) ESG ranking of Russian companies. Retrieved from https://www.raexpert.eu/ media/uploads/rankings/ESG_Ranking_03.10.2022_eng.pdf. Accessed 9 Oct 2022
15. Sheveleva AV, Tyaglov SG, Khaiter P (2021) Digital transformation strategies of oil and gas companies: preparing for the fourth industrial revolution. In: Konina N (ed) Digital strategies in a global market: navigating the fourth industrial revolution. Palgrave Macmillan, Cham, pp 157–171. https://doi.org/10.1007/978-3-030-58267-8_12
16. Silvola S, Landau T (2021) Sustainable investing: beating the market with ESG. Palgrave Macmillan, Cham. https://doi.org/10.1007/978-3-030-71489-5
17. Swedroe L, Adams S (2022) Your essential guide to sustainable investing: how to live your values and achieve your financial goals with ESG, SRI, and impact investing. Harriman House, Hampshire
18. Tyaglov S, Sheveleva A, Guseva T (2019) Justification of the need and feasibility of switching to renewable energy sources for the implementation of sustainable development principles. IOP Conf Ser: Earth Environ Sci 317:012004. https://doi.org/10.1088/1755-1315/317/1/012004
19. Tyaglov SG, Sheveleva AV, Rodionova ND, Guseva TB (2021) Contribution of Russian oil and gas companies to the implementation of the sustainable development goal of combating climate change. IOP Conf Ser: Earth Environ Sci 666:022007. https://doi.org/10.1088/1755-1315/666/2/022007
20. Zavyalova EB, Krotova TG (2021) Methods for achieving SDG 1, poverty eradication. In: Popkova EG, Sergi BS (eds) Modern global economic system: evolutional development vs. revolutionary leap. Springer, Cham, pp 1987–2001. https://doi.org/10.1007/978-3-030-69415-9_218
21. Zavyalova EB, Popkova EG (eds) (2021) Industry 4.0: exploring the consequences of climate change. Palgrave Macmillan, Cham. https://doi.org/10.1007/978-3-030-75405-1
22. Zavyalova EB, Studenikin NV (2019) Green investment in Russia as a new economic stimulus. In: Sergi BS (ed) Modeling economic growth in contemporary Russia. Emerald Publishing Limited, Bingley, pp 273–296. https://doi.org/10.1108/978-1-78973-265-820191011

Experiences of Climate Risk Management Based on Smart Green Innovations in Regions and Countries

Analysis of the Potential for Clustering Basic Industries in a Depressed Region in the Context of the Development of Green Innovation

Marina V. Palkina⬤, **Anastasia A. Sozinova**⬤, **and Olesya A. Meteleva**⬤

Abstract One of the priority tasks in the Russian Federation is to create regional innovative growth centers that ensure active and sustainable innovation and socio-economic development of the region. It is better to form these growth centers in the regions based on cluster formations. Due to the integration, cooperative, and innovative activities of various enterprises, cluster formations can significantly increase the competitiveness and innovative potential of growth centers. Creating cluster formations as growth innovative centers of regional economies is especially important in regions in which now, with the active application of the principles of green economy and deployment, the processes of implementation of green innovations are acquiring. In other words, regional clusters can become growth centers for their regions based on introducing all types of innovations (product, process, organizational, and marketing), leading to a reduction in the consumption of natural resources and environmental pollution. The research aims to assess the clustering potential of basic industries of a region and determine measures for developing a cluster approach based on green innovation. The authors used the official statistical data of the Federal State Statistics Service of the Russian Federation (Rosstat) for 2017–2021 to assess the clustering potential. The research results can be used by state and municipal authorities in the development and improvement of regional cluster policy, the formation of socio-economic development strategies, and the elaboration of measures for the sustainable development of their regions' economies.

Keywords Cluster · Industry · Clustering potential · Kirov region · "Green" innovations

JEL Classification R11 · R58 · Z18

M. V. Palkina · A. A. Sozinova (✉) · O. A. Meteleva
Vyatka State University, Kirov, Russia
e-mail: aa_sozinova@vyatsu.ru

M. V. Palkina
e-mail: palmavik@ya.ru

1   Introduction

In today's conditions, one of the priority tasks of Russian authorities at all government levels is to move to a post-industrial organization type at the federal, regional, and municipal levels. A cluster approach can solve this problem because it makes it possible to create the locomotives of economic growth [1]—clusters—within the boundaries of certain territories. This is confirmed by the level of productivity within clusters, which is higher than in the region or the country in general [1]. Analyzing the development potential of the cluster approach in these regions can help identify the preferable areas of clustering and, on this basis, prepare proposals for cluster policy in the context of the transition to a green economy and green innovation.

The research aims to assess the clustering potential of the basic industries of a depressed region. To achieve this purpose, we need to solve the following tasks:

- To clarify the content of the concept "cluster," its main characteristics;
- To identify and study methods of assessing the potential for clustering industries at the regional level;
- To assess the clustering potential of the basic industries of a specific region;
- To state the preferable directions of clustering in a specific region based on the development of green innovations.

2   Methodology

The authors used the following methods: structural analysis, system analysis, logical analysis, retrospective assessment, expert assessment, and graphical and tabular methods for visualizing the results of the study. The research object was the Kirov Region [2, 3].

The information base of this research includes regulatory legal acts and methodological documents on the clustering of the Russian Federation at various levels, official data from the Federal State Statistics Service of the Russian Federation (Rosstat), scientific articles, and official Internet sites.

Currently, foreign and Russian scientific literature provides various methods for assessing the potential for clustering industries at the regional level. Such authors as Battalova [4], Bogdanova and Gutman [5], Kapelyuk and Meikshan [6], Kovrov [7], Kulagina et al. [8], Pankratov et al. [9], Popkova et al. [10], Rastvortseva et al. [11], Sozinova and Meteleva [12], Sozinova et al. [13, 14], Volkov [15], deal with this issue.

Identification and systematization of methods for assessing the clustering potential made it possible to state that there is currently no generally recognized methodology. As a rule, the methods are based on determining the following main coefficients: localization coefficient of the cluster activity, coefficient of per capita production of the cluster, and cluster activity specialization coefficient.

The localization coefficient of cluster activity (Clca) is the ratio of the share of an industry (type of economic activity) in the production structure of a region to the share of the same industry (type of economic activity) in the country [16, 17]. We can calculate it according to the following formula:

$$Clca = (Vregion/GRPregion)/(Vrf/GRPrf) \tag{1}$$

where:

Vregion—the volume of goods of own production, own works, and services in the region; GRP region—the gross regional product of the region;
Vrf—the volume of goods of own production, own works, and services by types of economic activity in the Russian Federation;
GRPrf—the gross regional product of the Russian Federation [4].

The coefficient shows how many times the concentration of a given industry (type of economic activity) in a given region is more or less than in the whole country.

The coefficient of per capita output (Coc) is the ratio of the share of the industry (type of economic activity) of the region in the corresponding structure of the industry (type of economic activity) in the country to the share of the region's population in the country's population [16, 17]. We can calculate it according to the following formula:

$$Coc = (Vregion/Vrf)/(Pregion/Prf) \tag{2}$$

where:

Vregion—the volume of goods of own production, own works, and services in the region;
Vrf—the volume of goods of own production, own works, and services by types of economic activity in the Russian Federation;
Pregion—population of the region;
Prf—population of the country [16, 17].

The coefficient of specialization of the cluster activity (Csca) is the ratio of the share of the industry (type of economic activity) of the region in the corresponding structure of the industry (type of economic activity) to the share of the region's GRP in the GDP of the Russian Federation [4, 7]. We can calculate it according to the following formula:

$$Csca = (Vregion/Vrf)/(GRPregion/GDPrf) \tag{3}$$

where:

Vregion—the volume of goods of own production, own works, and services in the region;

Vrf—the volume of goods of own production, own works, and services by types of economic activity in the Russian Federation;

GRPregion—the gross regional product of the region;

GDPrf—country's gross domestic product [16, 17].

If the inequalities Clca > 1, Coc > 1, and Csca > 1 are satisfied, it is concluded that there are conditions for creating an industry cluster on the territory.

3   Results

The term economic cluster was first mentioned in the 1990s in "Competition" by M. Porter [16]. According to M. Porter, clusters are "geographically concentrated groups of interconnected companies, specialized suppliers, service providers, firms in related industries, organizations related to their activities (e.g., universities, standards agencies, and trade associations) in certain areas competing but working together" [16].

Nowadays, in the scientific community, there are many works in the field of clustering. Ketels, Schmitz, Swann, and Prevezer; Rosenfeld, Altenburg, and Meyer-Stamer; Roelandt and den Hertog; Simmie et al.; Cortright; and Beloglazova [18] define cluster. Having analyzed interpretations of the concept of a cluster by these authors, we can conclude that they are diverse, and there is no single and generally accepted definition among scientists and specialists.

Having studied regulatory legal acts and methodological documents on clustering in the Russian Federation, we can distinguish several definitions of a cluster currently used to create and develop clusters on the territory of Russia. An industrial cluster is "a set of business entities in the industrial sector connected by relations in this area due to territorial proximity and functional dependence and located on the territory of one subject of the Russian Federation or on the territories of several subjects of the Russian Federation" [19]. An international medical cluster is "a set of the infrastructure of the territory of the international medical cluster, project participants, and mechanisms of interaction between project participants" [20]. The infrastructure of the territory of the international medical cluster is "the totality of the territory international medical cluster and buildings, structures, and other facilities located there, including communal infrastructure facilities" [21]. A territorial cluster is "research and educational organizations connected by relations of territorial proximity and functional dependence in the sphere of production and sale of goods and services" [21].

Considering these interpretations of the cluster contained in scientific papers and methodological and regulatory documents, we can distinguish the following key characteristics of the cluster:

• A cluster is a group of enterprises and organizations of various profiles of activity (production, service, educational, research, etc.);

- Enterprises and organizations included in the cluster have common goals and development strategies and conduct joint activities for the production and sale of selected goods or services;
- Enterprises and organizations of the cluster are functionally dependent on each other and interact with each other within the same value chain;
- Enterprises and organizations that make up the cluster are geographically close.

To determine the most promising potential clusters in the basic industries of the Kirov Region, we calculate the localization coefficient for 2017–2021.

The Kirov Region has a high level of localization in only one type of economic activity "Electricity, gas, and steam; air conditioning." The dynamics of the values of the localization coefficient for this type of economic activity are negative. The values of the coefficient decrease throughout the analyzed period. If the localization coefficient for this type of activity continues to decrease, there will be no appropriate conditions and the possibility of creating a cluster in this industry in the short term. For other types of economic activity during 2017–2021, the level of localization is low (values do not exceed threshold 1). In these sectors, there are no conditions and the possibility of creating a cluster.

The results of calculating the coefficient of per capita production make it possible to conclude that the relative productivity is low and that there are no conditions for creating cluster formations for all major types of economic activity in the Kirov Region.

At the next stage of assessing the potential for clustering industries in the Kirov Region, we calculated the coefficient of specialization of the cluster activity.

The most promising type of economic activity for the formation of clusters in the Kirov Region is "Electricity, gas, and steam; air conditioning." The value of the coefficient of specialization of the cluster activity for this type of activity throughout the analyzed period is high (more than 1).

At the final stage of assessing the clustering potential, we calculated an integral indicator. The integral indicator is calculated by the formula [16, 17]:

$$C_{\text{integral}} = \sqrt[3]{Clca * Coc * Csca} \qquad (4)$$

where:

"Clca, Coc, and Csca are the values of the localization coefficient, the coefficient of per capita production, and the coefficient of specialization, respectively" [4, 7].

If the values of the integral indicator of the clustering potential are higher than 1, the considered type of economic activity is promising for cluster formations. The higher the values of the integral indicator, the higher the potential for the formation of cluster formations in the region. It is also necessary to consider the dynamics of the values of this indicator over time. An increase in the values of the integral indicator of the clustering potential in dynamics will indicate favorable prospects for developing cluster formations in the region and vice versa.

The values of the integral indicator were above one only for one type of economic activity—"Electricity, gas, and steam; air conditioning." The development of cluster formations in this industry is the most preferable for the region. For other types of economic activity, the value of the integral indicator is consistently low, which indicates a low potential for their clustering and the lack of opportunities and conditions for forming clusters in these industries in the Kirov Region.

In the context of the transition to a green economy, the leading principle of conducting the activities of all participants of the priority to create cluster formations in the field of "Electricity, gas, and steam supply; air conditioning" should be the principle of reducing the negative impact on the environment. The implementation of this principle is seen, on the one hand, based on the best available technologies [16] and, on the other hand, through the active implementation of green innovations.

Since 2015, the share of organizations in the Kirov Region that carried out innovations that improve environmental safety in five of the six areas of environmental improvement has been decreasing every year. This characterizes the negative situation with the implementation of green innovations and with the application of green economy principles in the region. In this regard, to increase the activity of applying the principles of the green economy in regional clusters in the field of "Electricity, gas, and steam supply; air conditioning" and in the economy of the Kirov Region, the following is advisable:

- To improve the regulatory and legal framework of the green economy at the regional level (e.g., by introducing incentives, tax benefits, etc. for participants of cluster formations to implement in their activities green innovation);
- To increase the investment appeal of green innovations in the region.

4 Conclusion

The analysis of the basic industries of the Kirov Region has shown that the prerequisites for cluster formation exist only in one industry—"Electricity, gas, and steam; air conditioning." For the successful formation and functioning of clusters in other industries, it is necessary to create the necessary conditions. It is advisable to consider this when preparing cluster policy and strategic planning documents for the region.

The paper proposes several measures to increase organizations' activity in the Kirov Region to implement innovations that increase the environmental safety of their production. The implementation of the proposed measures will make it possible to develop cluster formations and the regional economy based on green innovations, lead to the rational use of limited natural resources, improve the quality of life, and increase the welfare of the population, the region, and the country.

Data Availability Data on the dynamics of the localization coefficient, production coefficient per capita, coefficient of specialization of the main types of economic activities in the Kirov Region, 2017–2021, and the dynamics of the integral indicator of the cluster potential of the Kirov Region (2016–2020) described in the following section of the research are available at https://figshare.com/ with the identifier https://doi.org/10.6084/m9.figshare.21864587.

References

1. Komarova GB, Azizova SG (2022) Methodological approaches to the formation of cluster policy and clusters. Econ: Yesterday Today Tomorrow 12(2A):187–196
2. Palkina M, Kislitsyna V, Chernyshev K (2021) Analysis of the relationship of investment and demographic factors in the development of depressed regions. J Urban Regional Anal 13(1):113–124. https://doi.org/10.37043/JURA.2021.13.1.7
3. Savelyeva NK, Saidakova VA (2022) The level of economic security as an indicator of the region depression. In: Bogoviz AV, Popkova EG (eds) Digital technologies and institutions for sustainable development. Springer, Cham, pp 155–160. https://doi.org/10.1007/978-3-031-04289-8_26
4. Battalova AA (2013) Assessment of the clustering potential of the industry. Bull Eurasian Sci 6(19):8
5. Bogdanova TA, Gutman SS (2016) Methodological bases for realizing the potential of cluster development of the region. Econ Sci 5(138):57–63
6. Kapelyuk ZA, Meikshan YuV (2021) Assessment of the clustering potential of regional organizations of consumer cooperation in the Siberian Federal District. Bull Altai Acad Econ Law 12–1:105–113. https://doi.org/10.17513/vaael.1971
7. Kovrov GS (2014) Methodological aspects of clustering of basic industrial sectors in the regional economy. Problems Modern Econ 4(52):274–279
8. Kulagina NA, Lysenko AN, Noskin SA (2020) Assessment of regional conditions for development of the digital economy cluster. Bus Educ Law 3(52):76–80. https://doi.org/10.25683/VOLBI.2020.52.347
9. Pankratov AA, Musaev RA, Badina SV (2021) Approaches to identifying, measuring and predicting cluster effects. Problemy prognozirovaniya [Problems of Forecasting] 3(186):126–135. https://doi.org/10.47711/0868-6351-186-126-134
10. Popkova EG, Tyurina YuG, Sozinova AA, Bychkova LV, Zemskova OM, Serebryakova MF, Lazareva NV et al (2017) Clustering as a growth point of modern Russian business. In: Popkova EG, Sukhova VE, Rogachev AF, Tyurina YG, Boris OA, Parakhina VN et al (eds) Integration and clustering for sustainable economic growth. Springer, Cham, pp 55–63. https://doi.org/10.1007/978-3-319-45462-7_7
11. Rastvortseva SN, Snitko LT, Cherepovskaya NA (2013) Methodology for assessing the economic potential for the development of clusters in the region. Digest Finance 9(225):18–29
12. Sozinova AA, Meteleva OA (2022) Cluster development as a factor of sustainable economic development: Scientific analytics and management prospects. In: Popkova EG (eds) Imitation market modeling in digital economy: game theoretic approaches. Springer, Cham, pp 723–733. https://doi.org/10.1007/978-3-030-93244-2_78
13. Sozinova AA, Novikov SV, Kocnikov SN, Nemchenko GI, Alenina EE (2016) Peculiarities of isolated clusters operation. Int J Econ Financ Issues 6(S8):19–23
14. Sozinova AA, Okhimenko OI, Goloshchapova LV, Kolpak EP, Golovanova NB, Tikhomirov EA (2017) Industrial and innovation clusters: development in Russia. Int J Appl Bus Econ Res 15(11):111–118
15. Volkova YuA (2018) Assessing the prospects for cluster development in regions: methods and empirical study results: evidence from the Republic of Belarus. Econ Anal: Theor Pract 17(1):30–47. https://doi.org/10.24891/ea.17.1.30
16. Porter ME (1990) The competitive advantage of nations. Free Press, New York
17. Sakharova SM, Pavlova AV (2022) Integrated development of the territory of the Arctic zone of the Russian Federation based on the cluster approach. Bull Volga State Univ Service. Ser: Econ 18(2):35–40
18. Beloglazova SA (2019) Cluster form of economic organization: determination of the potential and directions of development in the regions of Russia (Synopsis of Dissertation of Candidate of Economics). Volgograd State University, Volgograd

19. Russian Federation (2014) Federal law "On industrial policy in the Russian Federation" (December 31, 2014 No. 488-FZ, as amended on May 1, 2022). Moscow, Russia: Legal Reference System "Consultant Plus." Retrieved from https://www.consultant.ru/document/cons_doc_LAW_173119/. Accessed 7 Oct 2022
20. Russian Federation (2015) Federal law "On the international medical cluster and amendments to certain legislative acts of the Russian Federation" (June 29, 2015 No. 160-FZ, as amended on July 26, 2019). Moscow, Russia: Legal Reference System "Consultant Plus." Retrieved from https://www.consultant.ru/document/cons_doc_LAW_181842/. Accessed 7 Oct 2022
21. Ministry of Economic Development of the Russian Federation (2008) Methodological recommendations on the implementation of cluster policy in the subjects of the Russian Federation (approved by the letter of the Ministry of Economic Development of the Russian Federation of December 26, 2008 No. 20615-ak/d19). Moscow, Russia: Legal Reference System "Consultant plus." Retrieved from https://spbcluster.ru/wp-content/uploads/2020/01/Metodicheskie-rekomendatsii-po-realizatsii-klasternoj-politiki.pdf. Accessed 7 Oct 2022

Developing the Mechanism of Foreign Economic Relations of the EU Countries in the Context of Implementing Priority Areas to Develop a Circular Economy

Diana M. Madiyarova⬤, Tatiana S. Malakhova⬤, Krasimir Shishmanov⬤, and Aigul T. Ageleuova⬤

Abstract The paper examines the current economic state of the EU countries and analyzes the key strategies and programs, specifically, the New Industrial Strategy for Europe and the new Circular Economy Action Plan implemented by the countries of the integration group. The theoretical and methodological basis of the research includes the historical and logical, dialectical principles and contradictions, and the method of scientific abstraction. The process-system approach, used in an in-depth analysis of the key economic indicators of the union members, was critical in justifying the need to develop a mechanism for foreign economic relations between countries in the context of implementing key strategies and programs for the development of Europe. Thus, a detailed analysis of the key strategies implemented in the EU countries is conducted, their timeliness and applicability are assessed, and the positions of leading Russian and international scholars on the issues of the transition of the EU countries to a circular economy and a new world economic context are studied. Thus, the research pays special attention to calculating economic indicators, particularly the index of industrial production, to identify changes in the scale of production in the EU countries. Besides that, the forecast data for this indicator is calculated up to the year 2025. The research then estimates oil and gas production volumes in some EU countries to identify a general trend in using these resources in the integration group. Also, the research defines the share of expenditures on environmental protection in the EU. The research presents the main elements and features of the new Circular Economy Action Plan, identifies the key areas for developing a

D. M. Madiyarova (✉)
RUDN University, Moscow, Russia
e-mail: mdm-diana@mail.ru

T. S. Malakhova
Kuban State University, Krasnodar, Russia

K. Shishmanov
D. A. Tsenov Academy of Economics, Svishtov, Bulgaria
e-mail: k.shishimanov@uni-svishtov.bg

A. T. Ageleuova
Kazakh Academy of Sport and Tourism, Almaty, Kazakhstan

circular production process, and describes the main opportunities that the EU might receive in case of its implementation. Based on this plan, the research developed a mechanism for the relationships between key economic bodies of the EU, which focus on the production and consumption of environmentally friendly products and the transition to innovation and digitalization.

Keywords Integration group · European Union · International economic relations · Risks and threats · Circular economy · Sustainability · Environment

JEL Classification F02 · F15 · F63 · F64

1 Introduction

In the current context, the foreign economic relations of the EU countries are relatively complex and contradictory. External and internal factors significantly affect their relationship within the integration group [9, pp. 38–41]. Thus, it is essential to develop a mechanism for foreign economic relations of the EU countries that help overcome future challenges and threats. Under current conditions, countries of the world economy are shifting to a new world economic and technological context [7]. Therefore, new industries and areas of production are actively developing. When developing the mechanism, it is necessary to consistently describe the range of economic bodies that will directly or indirectly participate in the transition to a circular economy [10, pp. 195–202]. The EU pays particular attention to the New Industrial Strategy for Europe [2], the new Circular Economy Action Plan for a cleaner and competitive Europe [1], and other vital initiatives. These documents shape the trajectory of further development of relations between various economic bodies of the EU. The documents specify that a circular economy can strengthen the industrial base of the EU and create a favorable business environment for the development of small and medium-sized enterprises (SMEs). A circular economy is an integral part of the broader transformation of industry towards climate neutrality and the long-term competitiveness of partner countries. Thus, it is vital to analyze certain individual indicators (particularly the industrial production index) and identify changes in the scale of production in the EU countries. An essential aspect of the research is the assessment of oil and gas production and environmental protection in individual EU countries. As a general matter, considering today's context, we highlight that various research schools and organizations conduct studies in this area. Some scholars report positive consequences for countries when transitioning to a circular economy; others see a threat that might transform traditional economic relationships between partner countries and lead to even greater dependence of developing countries on the developed ones (in terms of new technologies and management approaches). Therefore, we consider the positions of individual scholars and research schools that analyze the features of developing a circular economy in the EU and its impact on the partner countries of the union, which have different models of economic growth, socio-cultural characteristics, and scientific and technical potential.

2 Materials and Methods

Nowadays, the scholarly community is actively exploring the theoretical, methodological, and practical foundations of developing a circular economy. The countries of the European region actively analyze this research area; it is also of great interest to the global research space. Thus, Korhonen, Honkasalo, and Seppälä highlight that although the circular economy is a popular concept promoted by the EU in the current context, it still preserves a superficial and unstructured nature [8, pp. 37–46]. Nevertheless, in their studies, Velenturf and Purnell mention that the circular economy will allow using natural resources more efficiently. Besides, it might bring economic benefits while reducing environmental pressure [13, pp. 1437–1457]. However, some scholars (e.g., Pansera, Genovese, and Ripa) argue that the cyclical/circular economy studies often ignore socio-cultural and political aspects that include a gradual transition to a more sustainable future, joint creativity and work, social justice, and other essential areas [11]. Then, when implementing the circular economy, the issues related to its implementation in low-income and middle-income EU countries become particularly important. Scholars around the world are actively investigating these issues. Wright, Godfrey, Armiento, Haywood, Inglesi-Lotz, Lyne, and Schwerdtle report that applying a circular economy concept in low-income and middle-income countries might provide them opportunities to transit towards sustainable development, resource efficiency, and a low-carbon economy. Nowadays, the EU countries are particularly interested in reducing dependence on natural resources and have adopted this concept to achieve its goals. However, there are problems due to the relative disinterest of the "Global South" countries in implementing the circular economy concept [14]. Moreover, the scholars study an essential question of whether the circular economy can be a tool to support sustainable development [12]. Thus, based on the presented perspectives and approaches, we analyze particular economic indicators related to the mining and manufacturing industries and assess the conditions and costs of environmental protection in the EU. We will specifically focus on a new Circular Economy Action Plan that the EU countries are implementing; we will develop a mechanism for interaction between the participants of the union for its implementation.

3 Results

We analyze some economic indicators, particularly the industrial production index (IPI). The IPI is a relative indicator that characterizes the change in the production scale in the compared periods. In general, an increase in the IPI indicator shows the rise of the national economy, which entails an influx of investments and an increase in the value of shares of leading manufacturing companies. The formula for calculating the index is as follows:

$$\text{IPI} = \frac{Q_{cur}}{Q_{base}} \cdot 100\%,$$ (1)

where:

Q_{cur} Production in the current period;
Q_{base} Production in the base period.

Table 1 presents data on the industrial production index in the selected EU countries for 2017–2021 and the calculation of the forecast for this indicator until 2025. The dynamics of the IPI in the EU countries are rather unstable over the analyzed period. In Austria, the IPI was 105.0% in 2017, 104.8% in 2018, 100.5% in 2019, 102.7% in 2020, and 101.9% in 2021. In 2021, the IPI decreased by 3.1% compared to 2017. In Belgium, the IPI was 102.9% in 2017, 101.1% in 2018, 104.8% in 2019, 105.0% in 2020, and 104.5% in 2021. In 2021, the industrial production index increased by 1.6% compared to 2017. The calculations of forecast data show an increase in the IPI until 2025.

The IPI might be 105.8% in 2022, 107.0% in 2023, 107.0% in 2024, and 107.8% in 2025. In Bulgaria, on the contrary, a decrease in the industrial production index is observed over the entire analyzed period. It was 103.4% in 2017, 100.3% in 2018, 100.6% in 2019, 99.9% in 2020, and 98.8% in 2021. In 2025, the IPI might decrease by 7.9% compared to 2017. In Hungary, the industrial production index was 105.3% in 2017, 103.8% in 2018, 105.3% in 2019, 104.3% in 2020, and 106.0% in 2021.

In Germany, the industrial production index did not exceed 103.0% for the entire analyzed period. The indicator was 102.9% in 2017, 101.2% in 2018, 95.8% in 2019, 97.2% in 2020, and 95.0% in 2021. As a general matter, the calculations show a further decline in the industrial production index until 2025. In Greece, the IPI ranged from 101.0% to 104.1% for the analyzed period. In Denmark, this indicator is fairly stable. The forecast data show that the IPI might be 103.0%. Simultaneously, in 2025, the analyzed indicator might increase by 1.0% compared to 2017. Ireland has one of the lowest IPI rates for the entire analyzed period. The calculation of the forecast shows a further decrease in the IPI. It might be 84.0% in 2022, 79.5% in 2023, 73.5% in 2024, and 72.9% in 2025. In 2025, the industrial production index might decrease by 14.5% compared to 2021. As for Spain and Italy, there is an almost similar trend in the analyzed indicator. In 2020, the IPI in Spain and Italy decreased by 3.3% and 4.1%, respectively, compared to 2017. A further decrease in this indicator, to 96.3% and 93.6%, is predicted in Spain and Italy by 2025. In Latvia, the IPI was 108.4% in 2017, 101.5% in 2018, 101.0% in 2019, 101.3% in 2020, and 98.8% in 2021. The calculations show a decrease in this indicator until 2025. It might be 96.4% in 2022, 96.0% in 2023, 94.2% in 2024, and 92.3% in 2025. Compared to Latvia, the industrial production index is higher in Lithuania for the analyzed period. The exception was 2017, when the IPI in Lithuania was 1.5% lower than in Latvia. This indicator

Table 1 The assessment of the industrial production index (IPI) of individual EU countries for 2017–2021 and the calculation of the forecast (%) of this indicator until 2025

Country	2017	2018	2019	2020	2021	IPI forecast			
						2022	2023	2024	2025
Austria	105.0	104.8	100.5	102.7	101.9	100.5	100.0	100.1	98.9
Belgium	102.9	101.1	104.8	105.0	104.5	105.8	107.0	107.0	107.8
Bulgaria	103.4	100.3	100.6	99.9	98.8	97.7	97.4	96.3	95.5
Hungary	105.3	103.8	105.3	104.3	106.0	105.5	106.2	106.4	107.0
Germany	102.9	101.2	95.8	97.2	95.0	92.5	90.9	90.0	87.5
Greece	104.1	101.8	101.9	102.6	101.9	101.4	101.7	101.4	101.0
Denmark	102.0	102.0	102.7	103.0	102.0	103.0	103.0	103.0	103.0
Ireland	97.8	95.0	102.8	84.4	87.4	84.0	79.5	73.5	72.9
Spain	103.2	100.3	100.7	99.9	99.2	98.2	97.9	97.0	96.3
Italy	103.7	100.6	98.9	99.6	98.0	96.4	95.9	95.0	93.6
Latvia	108.4	101.5	101.0	101.3	98.8	96.4	96.0	94.2	92.3
Lithuania	106.9	104.8	103.4	104.5	104.4	103.2	103.3	103.3	102.7
Luxembourg	103.7	98.9	96.4	96.8	95.0	92.3	91.5	90.2	88.1
Netherlands	101.1	100.7	99.2	101.2	100.0	99.9	99.9	100.1	99.6
Poland	106.6	105.9	104.2	105.4	105.8	105.0	105.2	105.6	105.3
Portugal	103.5	100.1	97.6	97.8	95.7	93.6	92.5	91.1	89.2
Romania	107.9	103.5	97.7	100.1	97.6	94.1	92.9	91.8	89.0
Slovakia	103.3	104.3	100.5	100.0	99.0	97.5	95.7	95.0	93.4
Slovenia	107.7	104.9	103.3	103.7	102.0	100.5	99.8	98.8	97.4
Finland	103.2	103.5	101.9	103.9	102.7	102.9	102.8	103.1	102.7
France	102.1	100.6	100.2	100.1	99.9	99.1	99.0	98.6	98.2
Czech Republic	106.5	103.0	99.8	100.6	99.0	96.5	95.6	94.6	92.6
Sweden	104.3	102.4	101.5	102.0	101.0	100.0	100.0	100.0	99.0
Estonia	104.0	104.1	97.9	100.7	98.6	96.8	95.4	95.2	93.1

Source Compiled and calculated by the authors based on [6, pp. 157–305]

might be 103.2% in 2022, 103.3% in 2023, 103.3% in 2024, and 102.7% in 2025. As for Luxembourg, there was a sharp decrease in the analyzed indicator by 4.8% in 2018 compared to 2017. In 2019, This indicator was 96.4% in 2019 and 96.8% in 2020. In 2025, this indicator might decrease by 6.9% compared to 2021. Similar to Luxembourg, the Netherlands shows a similar trend. The forecast data shows a gradual decrease in this indicator. The IPI might reach 99.6% in 2025, which is 0.4% lower than in 2021. Current and forecast indicators are high for Poland. The industrial production index ranges from 104.2% to 106.6%. In Portugal, there was a sharp decrease in this indicator by 5.9% in 2019 compared to 2017. The calculations show a decrease in the IPI until 2025. In Romania, the IPI was 107.9% in 2017, 103.5% in 2018, 97.7% in 2019, 100.1% in 2020, and 97.6% in 2021. There is a gradual decrease in the analyzed indicator in Slovakia and Slovenia. In Slovakia, the IPI was 103.3% in 2017, 104.3% in 2018, 100.5% in 2019, 100.0% in 2020, and 99.0% in 2021. In Slovenia, the industrial production index was 107.7% in 2017, 104.9% in 2018, 103.3% in 2019, 103.7% in 2020, and 102.0% in 2021. Finland shows a high IPI for the entire period. However, the forecasts show the unstable state of this indicator: 102.9% in 2022, 102.8% in 2023, 103.1% in 2024, and 102.7% in 2025. In France, the IPI was 102.1% in 2017, 100.6% in 2018, 100.2% in 2019, 100.1% in 2020, and 99.9% in 2021. The calculations show a decrease in this indicator: 99.1% in 2022, 99.0% in 2023, 98,6% in 2024, 98.2% in 2025. In the Czech Republic, the industrial production index decreased by 6.7% in 2019 compared to 2017. The IPI might be 96.5% in 2022, 95.6% in 2023, 94.6 in 2024, and 92.6% in 2025. In Sweden, there is a similar trend as in the Czech Republic. The IPI decreased by 2.8% in 2019 compared to 2017. The IPI forecast values range from 99.0% to 100.0%. In Estonia, the IPI was 104.0% in 2017, 104.1% in 2018, 97.9% in 2019, 100.7% in 2020, and 98.6% in 2021. Thus, the IPI decreased by 5.4% in 2021 compared to 2017. A decrease in this indicator is predicted until 2025. Overall, based on the calculations of forecast data, there is a decrease in the industrial production index by 2025 across the EU countries. This trend might be connected to various internal and external factors. Therefore, it is necessary to pay special attention to the new Circular Economy Action Plan for a cleaner and competitive Europe [1]. For instance, in this plan, the central issues encompass developing environmentally friendly products, circularity in production processes, creating a market for secondary raw materials in the European Union, addressing the problems of waste export from the European Union, driving the transition to innovation and digitalization, and other significant matters. Precisely, particular attention in the new Circular Economy Action Plan is given to the issue of reducing the dependence of Europe on raw materials. The EU countries are focused on continuing the development of the circular economy at the global level. Besides, they intend to apply the experience of foreign countries and increase financial resources to implement sustainable development goals (SDGs) by 2030. Furthermore, the programs and plans specify that the use of various natural resources must be within certain limits so that they do not significantly impact the environment. Natural gas production continues to increase in some EU countries. Thus, in Austria, natural gas production increased by 0.1 billion cubic meters in 2017 compared to 2016. In Denmark and Ireland, there was an increase in natural gas production by 0.3 and 0.4 billion cubic meters, respectively, over the same period. A sharp rise in natural gas production was observed in Romania in 2017 (11.2 billion) compared to 2016 (9.8 billion cubic meters). Nonetheless, a decrease in natural gas production over the specified period was observed in Germany, Italy, the Netherlands, Poland, and Croatia by 1.1, 0.3, 6.8, 0.1, and 0.1 billion cubic meters, respectively. As for oil production, there was a reduction in almost all EU countries during the analyzed period, except for Italy and Croatia. In 2016, oil production in Italy amounted to 4.1 million tons and 4.5 million tons in 2017. In 2017, oil production increased by 0.4 million tons compared to 2016. In Croatia, in 2016, oil production amounted to 0.7 million tons, while, in 2017, it was 0.8 million tons. Oil production for the analyzed period increased by 0.1 million tons [5, pp. 122–123]. Nowadays,

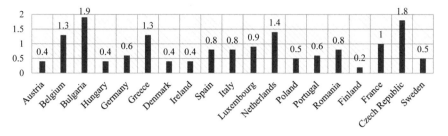

Fig. 1 The share of expenditures on environmental protection in selected countries of the European Union for 2018, % of GDP. *Source* Compiled by the authors based on the materials [4, pp. 154–295]

the EU countries pay much attention to environmental protection and the development of alternative forms of energy. The costs of environmental protection include the total amount of current expenditures and the capital investments of the countries, enterprises, and citizens focused on protecting and improving the quality of the environment. It is essential to highlight that the expenditures of the EU partner countries on national environmental protection in 2020 amounted to 273 billion euros. Thus, Eurostat estimates that spending on environmental protection increases by more than 2% annually. Overall, it has increased by 40% since 2006. Also, Eurostat reckons that spending on environmental protection in the percentage of the GDP has remained relatively stable over the past fifteen years (ranging from 1.8% to 2.0% of GDP) [3]. Thus, Fig. 1 shows the share of spending on environmental protection in selected EU countries. This research analyses this indicator in the dynamics. In Austria, the share of expenditures on environmental protection in 2018 was 0.4%. This figure was at the same level in 2016 and 2017. In Belgium, there was an increase in this indicator by 0.4% in 2018 compared to 2016. In Bulgaria, the analyzed indicator was 2.4% in 2016 and 1.9% in 2018.

In 2018, the indicator decreased by 0.5% compared to 2016. In Hungary, there was a similar decrease in the share of environmental spending from 1.2% in 2016 to 0.4% in 2018. In Germany, the indicator remained at 0.6% within the specified period. In Luxembourg and the Netherlands, there were also no significant changes in this indicator. In 2016 and 2018, the share of expenditures on environmental protection was 0.9 and 1.4%, respectively. In Poland, this indicator decreased by 0.1% in 2018 compared to 2016. On the contrary, in Portugal, there was an increase in the share of expenditures on environmental protection by 0.2% in 2018 compared to 2016. In France, this indicator was stable at 1.0% throughout the analyzed period. It is essential to mention that the New Industrial Strategy for Europe and the new Circular Economy Action Plan for a cleaner and more competitive Europe pay special attention to environmental protection and climate neutrality. Thus, it is necessary to present the mechanism of foreign economic relations of the key economic bodies of the EU and identify trajectories to develop the circular economy of this integration group.

4 Discussion

Figure 2 presents the mechanism of relations between countries and other EU bodies in the context of implementing a circular economy. The mechanism focuses on key bodies, namely active participants, implementing the new Circular Economy Action Plan. Large, medium, and small enterprises in the EU will continue developing environmentally friendly products, focusing on the circularity in production processes, developing sustainable products, and striving to increase the competitiveness of their products in the European and the world market. Nowadays, global competitors are forming their strategies and programs to expand new markets.

The EU market is desirable for the countries of the Asian region, particularly for China and India. Therefore, these bodies play a crucial role in the presented mechanism since they will need to transform their activities to new trends and realities in the context of developing a circular economy. It is important to note that this initiative comes from the leadership of the EU partner countries. The scholars identify that not all EU countries are ready for this transition. It is primarily because the Central and Eastern European countries (CEECs) and the countries of the Southern region have a less differentiated sectoral structure of the economy than the Western European countries. Besides, they have their own socio-cultural characteristics and economic development models that differ from Western European countries. The enterprises of the CEECs and the Southern region countries cannot compete in the European market, which also slows down the transition of these countries to a new technological context. Profound transformations in the structures of the economies of the EU countries might lead to problems and contradictions in the relationship between the union participants. Nevertheless, within the framework of a new Circular Economy Action Plan, there has been an increase in funding for relevant industries and sectors of the economy. Undoubtedly, in the presented mechanism, the banking structures of the EU countries are an essential element because they actively invest in new developments and technologies. Attracting public and private investments will help achieve sustainable development goals (SDGs) by 2030. Thus, financial institutions will develop new banking products, which will be distributed in the European monetary and financial market and on a global scale. If countries act as guarantors for implementing large-scale and high-risk projects when transitioning to a circular economy, banking structures and large enterprises will be involved in their development with great interest.

Civil society organizations, consumers of goods, and other economic bodies are also essential parts of this mechanism. If businesses redesign their operations when transitioning to a circular economy, consumers will need to transform their perceptions of how they use goods and services. A new Circular Economy Action Plan focuses on increasing product durability, reusability, repairability, and high-quality recycling. Therefore, manufacturers and consumers need to reshape their perception of using durable goods. Thus, there is no doubt that today, the indicated topic, related to the development of a circular economy, is relevant and debatable in the EU and other countries and regions of the world. Overall, countries need to apply the already

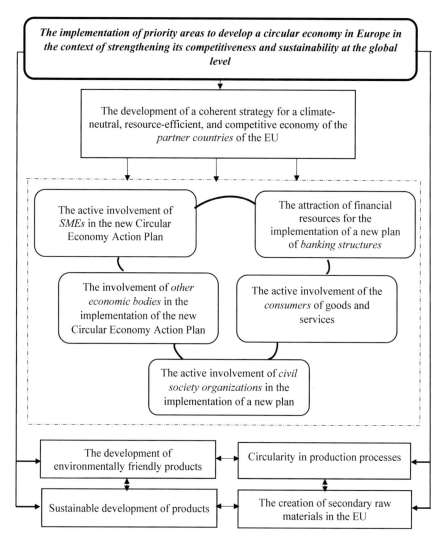

Fig. 2 The development of the mechanism of relations between countries and other bodies of the EU in the context of implementing a circular economy. *Source* Compiled by the authors

existing experience and capacity in implementing the outlined plans to develop a circular economy and, simultaneously, consider the modern challenges and threats in the global economy today.

5 Conclusion

First, the paper examined the perspectives and approaches of scholars in the field of developing a circular economy. As this research pinpoints, this field has a theoretical, methodological, and practical scope for studies in today's context. Generally, in the current context, the EU countries strive to implement the main areas of the New Industrial Strategy for Europe and a new Circular Economy Action Plan. Thus, the research emphasizes that not all EU countries strive to implement a circular economy. This transition might lead to the transformation of already established foreign economic relations with partner countries. Second, the industrial production index in the EU countries was analyzed, and the forecast of this indicator was calculated until 2025. The dynamics of the IPI in the countries of the integration group are relatively unstable for the analyzed period. In general, based on the calculations of forecast data, there is a decrease in the IPI in the EU countries by 2025. It might be due to the influence of internal and external factors. Thus, today, the EU countries strive to implement the main elements of the new Circular Economy Action Plan for a cleaner and more competitive Europe. Thus, the research evaluated the volumes of oil and gas production in selected EU countries and analyzed the expenditures of the union member countries on environmental protection. Third, based on the new Circular Economy Action Plan for a cleaner and more competitive Europe, the research developed a mechanism for relationships between countries and other bodies of the EU to implement a circular economy. The central bodies when implementing this mechanism are SMEs, banking structures, civil society organizations, consumers, and other essential bodies.

References

1. European Commission (2020) A new circular economy action Plan for a cleaner and more competitive Europe (COM(2020) 98 final). Brussels, Belgium. Retrieved from https://eur-lex. europa.eu/legal-content/EN/TXT/?qid=1583933814386&uri=COM:2020:98:FIN. Accessed 11 Mar 2020
2. European Commission (2020) A new industrial strategy for Europe (COM(2020) 102 final). Brussels, Belgium. Retrieved from https://eur-lex.europa.eu/legal-content/EN/TXT/? uri=CELEX:52020DC0102. Accessed 10 Mar 2020
3. Eurostat (2021) Environmental protection spending continues to increase. Retrieved from https://ec.europa.eu/eurostat/web/products-eurostat-news/-/ddn-20210707-1. Accessed 7 July 2021
4. Federal State Statistics Service of the Russian Federation (2018) Russia and world countries, 2018: statistical collection. Rosstat, Moscow, Russia
5. Federal State Statistics Service of the Russian Federation (2019) Russia and the European Union member states, 2019: statistical collection. Rosstat, Moscow, Russia
6. Federal State Statistics Service of the Russian Federation (2020) Russia and world countries, 2020: statistical collection. Rosstat, Moscow, Russia
7. Glazyev SY (2018) Leap into the future. Russia in the new technological and world economic structures. Litres, Moscow, Russia

8. Korhonen J, Honkasalo A, Seppälä J (2018) Circular economy: the concept and its limitations. Ecol Econ 143:37–46. https://doi.org/10.1016/j.ecolecon.2017.06.041
9. Malakhova TS, Kolesnikov NP (2019) Trends and contradictions of the global economy crisis and transformation of the world financial institutions. Eur J Econom Manage Sci 1:38–41
10. Malakhova TS, Dubinina MA, Maksaev AA, Fomin RV (2019) Foreign trade and marketing processes in the context of sustainable development. Int J Econ Bus Adm 7(SI2):195–202. https://doi.org/10.35808/ijeba/384
11. Pansera M, Genovese A, Ripa M (2021) Politicising circular economy: what can we learn from responsible innovation? J Responsible Innovation 8(3):471–477. https://doi.org/10.1080/232 99460.2021.1923315
12. Skvarciany V, Lapinskaite I, Volskyte G (2021) Circular economy as assistance for sustainable development in OECD countries. Oeconomia Copernicana 12(1):11–34. https://doi.org/10. 24136/oc.2021.001
13. Velenturf APM, Purnell P (2021) Principles for a sustainable circular economy. Sustain Prod Consumption 27:1437–1457. https://doi.org/10.1016/j.spc.2021.02.018
14. Wright C, Godfrey L, Armiento G, Haywood L, Inglesi-Lotz R, Lyne K Schwerdtle P (2019) Circular economy and environmental health in low- and middle-income countries. Globalization Health, 15:65. https://doi.org/10.1186/s12992-019-0501-y

The Role of Digitalization of Educational in the Sustainable Development of the Regions of Kyrgyzstan

Chinara R. Kulueva[ID], **Baktygul T. Temirova**[ID], **Bakty T. Marzabayeva**[ID], **Saikal S. Ibraimova**[ID], **and Urmatkan O. Amatova**[ID]

Abstract This research focuses on the problems of development digitalization of the educational system in the regions of the Kyrgyz Republic. The authors consider the policy of the Kyrgyz Republic in promoting the strategy for the development of digitalization in the regions of Kyrgyzstan, as well as the level of implementation of information and communication technologies in the territories of the country necessary for the development and improvement of the digitalization of the economy, in particular, education. The authors determine the role of development and improvement of the theory of digitalization, digital economy, digitalization of the national educational system, and regional development in the context of the adopted relevant programs, concepts, and territorial tasks focused on stabilization and creation of conditions for achieving positive dynamics in improving the level and quality of educational services provided through the introduction of advanced information and digital technologies for the development of regions and increasing the level of their training, acquiring competencies, and improving the level of digital literacy in the life of the population. Moreover, the authors clarify the definitions that enrich the terminology about digital education that has occupied a worthy place in economic science in recent years. Methods of observation and theoretical materials made it possible to reliably determine the role and significance of the digitalization of education in the sustainable development of Kyrgyzstan. Recently, the category of "digital education" has become the subject of research by many scientists studying its categories, which are enriched, interacted with, and revealed by modern scientific schools as a popular and directive direction. Much attention is paid to the importance of digitalization of education, training of competent specialists, fighting against negative manifestations in society, and raising the level of digital literacy. In the regions that closely

C. R. Kulueva (✉) · B. T. Temirova · B. T. Marzabayeva · S. S. Ibraimova · U. O. Amatova
Osh State University, Osh, Kyrgyzstan
e-mail: ch.kulueva@mail.ru

B. T. Marzabayeva
e-mail: bakty.2011.74@mail.ru

S. S. Ibraimova
e-mail: sibraimova@oshsu.kg

© The Author(s), under exclusive license to Springer Nature Switzerland AG 2023
E. G. Popkova (ed.), *Smart Green Innovations in Industry 4.0 for Climate Change Risk Management*, Environmental Footprints and Eco-design of Products and Processes, https://doi.org/10.1007/978-3-031-28457-1_37

interact based on the principle of "tripartism," the authors considered the necessity of creating new educational platforms between the state, the private sector, and the population aimed at improving the quality of educational services, expanding the radius of integration of educational programs, and creating new employment opportunities for graduates and employment that affect the socio-economic development of the territories.

Keywords Competencies · Education · New technologies · Educational technologies · Regions · Digitalization · Digital literacy

JEL Classification H70 · I20 · I29 · L86 · O15 · O32 · O38 · R19 · R59 · Z18

1 Introduction

Today's challenges related to the development of IT contribute to all aspects of everyday changeable processes of the dynamics of human life. Today's advances in technology may be obsolete tomorrow. It is also necessary to consider that the producers of such high-tech and dynamic technologies are essentially people who, deciding to simplify their life and satisfy their growing needs, knowing that resources are limited, takes a bold step into the human sphere of self-determination and development. In this case, each area of human activity must intensify its actions to receive appropriate benefits from innovations and correctly use the conditions created for citizens through information technologies. It especially concerns the issues of increasing digital literacy of people, particularly their digital education. Digital education or digital literacy can become the basis for solving many issues (problems) in the regions of Kyrgyzstan [1]. In this case, the government, represented by the relevant structure of the Ministry of Education and Science of the Kyrgyz Republic, and the country's population should be ready to accept a policy of widespread increase in digital literacy. In this case, we are talking about the media and information literacy of the population, which should carry the purely positive aspects of our comprehensive life, where special attention is paid to the modernization of educational standards at the school and university levels, considering the requirements of market transformations. It is necessary to agree that such work requires much money, time, and appropriate policy [2].

As life practice shows, advanced technological products, media, and telecommunication services provide the population of Kyrgyzstan with ample opportunities to communicate at a distance, exchange information, and gain access to voluminous world information products, which subsequently express their behavior, action, self-determination, and human development. Therefore, the current problem of human development should focus on choosing the information that pursues human development and mutual good-neighborly human relations, confused by aggressive, anti-human, and sometimes anti-tolerant products of information services disguised as educational communications. Therefore, the role of today's educational system in assessing the relevance and reliability of information is invaluable, where an open discussion of the ongoing processes of globalization should take place. The carriers of the educational system should be competent in all areas of human life, be ready for an open and, sometimes, aggressive environment of society, and be able to conduct good-neighborly, tolerant, and communicative relations, where the role of digitalization is invaluable if used in the above directions. We also note the role of teachers as the main subjects of transformations [3] and carriers of digital knowledge, which becomes a guide to the digital world.

Accordingly, the digital environment puts forward its own requirements for teachers. It requires completely different approaches in the learning process and different forms of work with students and listeners to have perfect digital literacy. It is necessary to have the ability to create and use digital communications and know how to and be able to communicate the required skills of communication, programming, searching, and exchanging information.

The real situation associated with the COVID-19 pandemic has accelerated the adoption of an innovative decision in the field of education in Kyrgyzstan (as in other countries)—the decision on the need to introduce and adapt digitalization as the safest and most acceptable mechanism for implementation of lifelong learning at all its levels.

2 Materials and Methods

Recently, the term digitalization has become more and more deeply embedded in human life. As our dynamic live shows, this irreversible process has become the basis of ongoing research in this direction. Many scientists try to get closer to the concept, processes, results, and phenomena associated with digitalization. They attempt to enrich a completely new direction, including new concepts and terminologies in the lexicon. Simultaneously, only a few manage to interpret the term digitalization and reveal its essence. According to Molchanova, "digitalization is not just the introduction of digital technologies in different areas of life to improve its quality, but also fundamental changes in thinking stereotypes and working methods" [4].

Abramchik [5], Abrosimova [6], Gancharik [7], Zenkov [8], Schwab [9], and others hold the view of digital transformation at all levels of government, as well as the educational system of the country aimed at student-centered learning and their compliance with the formed competencies.

With the transition of many countries to the priorities of the knowledge economy, the digitalization of education is predetermined by new knowledge and advanced technologies that occupy key positions and ensure the country's competitiveness. Life practice shows an increase in the importance of universities that conduct research and development. Nowadays, they act as a source, the most important resource for high-tech production.

It is impossible to ignore the relationship between the state (represented by the education sector) and the private sector, whose modern relations should move to the digital level. Accordingly, new forms of regulation of digital transformation in the educational system are required. The traditional form of training specialists does not meet current requirements. A new approach is required—coordinated relations between the field of education, science, and the manufacturing sector, where the role of introducing digital relations, especially in the higher professional education sector, is defined as one of the effective options for the development of higher education. The digitalization of the life of educational institutions in the current socio-economic conditions contributes to the establishment of "new specific rules" and objective trends in the functioning of the university, the main of which are as follows:

- Achieving transparency in the market of educational services;
- Accessibility to the necessary information from stakeholders: students, employers, entrepreneurs (and other representatives of the private sector), parents, etc.

Digital education in educational institutions provides an opportunity for them to develop entrepreneurial activities within their own walls (previously, self-financing between an enterprise and a university functioned under the USSR) and create an educational institution of an entrepreneurial type.

Clark pioneered the scientific rationale for the "entrepreneurial university." He tried to reveal its economic nature and coined the term "entrepreneurial university" [10]. Simultaneously, the scientists identified the significant features of digital education in today's educational institutions and proposed the corresponding areas of their activity: digitalization of management activities, digital links with the private sector, digital financial control, digital stimulation, and digital entrepreneurial culture [10].

3 Results and Discussion

The digitalization of the regions of Kyrgyzstan started on October 31, 2018, based on the National Development Strategy of Kyrgyzstan for 2018–2040, approved by the Decree of the President of the Kyrgyz Republic, where the country's digital ambitions were reflected in the Taza Koom National Digital Transformation Program [11] aimed at developing the infrastructure of each settlement, up to mountainous and high-mountainous territories, so that residents of different ages (in particular, teachers, children, farmers, pensioners, youth, etc.) feel the benefits of the introduction of digitalization. Moreover, a significant role is assigned to the priorities of digital education, that is, the provision of new technologies, innovations, and the Internet to each aiyl okmotu (village councils), where the subjects of the educational system are located—schools, kindergartens, vocational schools, etc.

More than 65% of the country's population lives in 484 local self-government bodies (LSGs) and 453 aiyl aimaks (village councils). The largest number is located in the Chui Region (23%); the smallest number is located in the Batken Region (about

7%). Of the total number, 17% of aiyl aimaks (79 village councils) function as self-sufficient territorial units; the rest of them (83%) survived on subsidized resources [12].

According to the analysis, 137 ayil okmotus (village councils) (30.2% of the total number of ayil okmotus) live and work outside of information technology [12]. Not to mention educational institutions, where classes in the 2019–2020 pandemic years have become a big problem for the villages and families with an average of 4–5 school-age children. The first reason for this problem is the lack or absence of free budgetary funds. The second reason is the underdevelopment of the innovation sector.

Despite the above problems, the school Internet connectivity has increased from 2.4 to 99%. The transparency of the process of admission of applicants to universities based on ORT data (general republican testing) is ensured; the system for reading barcodes of ORT certificates by each university is automated. Based on European standards (ESG-2015), Kyrgyzstan formed a regulatory legal framework to introduce procedures for independent accreditation of the quality of vocational education, recognition of diplomas of the Republic abroad, and integration of the education system into the global educational space [13].

The concept of "Digital Kyrgyzstan 2019–2023" [14] determined the foundations of digitalization of the education system. The essential tasks in this area are as follows:

- Introduction of digital technologies into the national education system;
- Development and improvement of IT education;
- Training of highly qualified specialists in the field of IT education;
- Training the population in digital skills, especially representatives of socially vulnerable groups;
- Development and improvement of digital content adapted to the state language.

The COVID-19 pandemic has accelerated the process of introducing and adapting digitalization to the educational sector. Since April 8, 2020, distance learning has taken place in the following formats: 1/3, 2/3, and 3/3. In the 1/3 format, 1400 video lessons were prepared, including sign language translation. These lessons were broadcast on the TV channels of the Kyrgyz TV and Radio Company "Balastan," ElTR "Bilim Ilim," "Channel 5," "Pyramid," and "Sanat" [2]. In the 2/3 format, distance professional development of teachers was organized; hotlines were opened to provide students with psychological assistance and methodological support to teachers. In the 3/3 format, work has been intensified on posting video lessons and presentation lessons on the portal of the Agency for Primary Vocational Education [13]; in secondary vocational education and universities, all contents were placed on the taught disciplines, namely teaching materials, lecture texts, tasks of the student independent work, programs, and tasks for intermediate and boundary control [12].

An online marathon called "EdTech—a new educational reality" was launched across the country, and a series of webinars were developed. This work is still ongoing in almost all educational institutions, from secondary to higher education, together with the Hi-Tech Park.

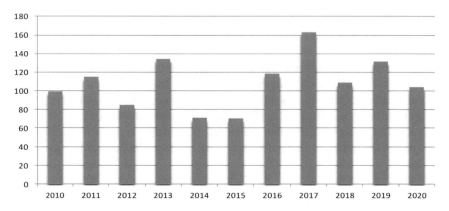

Fig. 1 Dynamics of growth of Internet access points in the educational sector of the Kyrgyz Republic for 2010–2020 (in %). *Source* Developed and compiled by the authors

As can be seen from Fig. 1, the national educational system of Kyrgyzstan is marked with stable dynamics of increasing the number of access points to the Internet with temporary deviations in 2014–2016 (the decrease amounted to 39.9%, 51.1%, and 39.3%, respectively) due to the lack of funds for the introduction of information and communication technologies in the educational sector. The situation changed for the better only in mid-2017 when many subjects of education supported the state policy on the transition to the digital economy and funds for the digitalization of education were found.

To this day, almost all universities in the country are connected to the system of interdepartmental electronic interaction (SIEI) "Tunduk." As part of the "Digital Kyrgyzstan 2019–2023," the process of digitalization of electronic licensing and accreditation continues. Osh State University (OshSU) is among the leaders. This university is one of the major universities in Central Asia. Osh State University intensified work on transitioning to a new advanced digital level [15]. On September 17–18, 2020, the Digital University conference was held within the walls of Osh State University. It was attended by the head of state and the rectors of about 40 leading universities in the country. At the conference, the task was set to completely transfer higher education to digital technologies within the framework of the platform "Electronic University," initiated by the staff of Osh State University.

The changes that took place in the country's life in the early 1990s, when the state was left without an ideology, any moral, and universal values, led to a decrease in the moral guidelines of some of the youth, who lost the opportunity to achieve their goals and understand by what means to do so. We consider it necessary to consolidate society on the joint responsibility of citizens to preserve the unity and cultural diversity of the Kyrgyz Republic [16]. In solving such problems, the role of the educational system for pupils and students is invaluable, especially in matters of respect for history, knowledge and veneration of the spiritual and cultural values of the people, and the formation of humanistic ideas of the Kyrgyz epic Manas, which promotes beliefs in the spirit of peace and non-violence [17].

Particular attention should be paid to an effective, balanced, and well-resourced state language policy as an essential part of the patriotic and spiritual education of pupils and students. We consider it necessary to digitalize the ideas of the epic "Manas" to be accessible to a wider audience of readers, not only for the title nation but also for representatives of other nations and nationalities as the basis for stabilizing society, as well as one of the ways to get out of the difficult state of the social system through knowledge, initiative, and labor reflected in the state ideology.

An important role plays the problem of digitalization in matters of combating human trafficking in the Kyrgyz Republic. It notes the importance of acquiring a competent attitude to one's social status in society through awareness, education, and upbringing [18]. Otherwise, there are enough well-established digital businesses in society that target potential victims from socially vulnerable strata of society with a low socio-economic status. The duality of digital technologies lies, on the one hand, in its development and use for criminal purposes in the organization of human trafficking and, on the other hand, in the use of digital by practitioners in combating this phenomenon in their response.

In Kyrgyzstan, as in other countries, labor resources are an invaluable resource and the main development factor. Despite the absence of problems with natural population growth, the Kyrgyz Republic faces a lack of highly qualified technical [19] and informational personnel. The reasons lie in the difficulties in finding employment and low wages, which have become the basis for a certain part of able-bodied creative-minded youth to leave Kyrgyzstan who successfully work in famous companies worldwide. Kyrgyzstan provides young people with the same knowledge for relatively minimal contract amounts. Having received higher education, the more advanced part implements all knowledge, skills, and abilities in another country more favorably for him or her and their family [20]. It turns out a paradox—Kyrgyzstan bears the costly part for training and acquiring competencies, and another country takes the "cream," the profitable part, providing the conditions for a trained worker to work only for good. This picture is one of the disturbing trends and still unresolved problems. Kyrgyzstan is far from condemning other countries. The reason is the citizens of the country. It is necessary to develop influence mechanisms to stimulate and support the work of the country's talented, gifted, and competitive youth.

4 Conclusions

Considering the issues of digitalization of the educational sector in the country from different latitudes and horizons, the authors are increasingly convinced of its necessity. High-speed globalization, ongoing global upheavals associated with the COVID-19 pandemic, human life with all its aspects, a constant struggle for resources and survival, integration of educational processes, information wars, corruption, etc., require an appropriate assessment, where the role of the digital order if properly used, is invaluable in preventing the "bad" and perfecting the "good." In the case of Kyrgyzstan, the authors believe that digitalization is a potential opportunity to improve the education system, through which it is possible to establish transparent business relations between educational institutions, consumers, and private business structures, which can positively affect the development of Kyrgyzstan and directly affect the socio-economic development of the territories.

In this research, the authors have taken out only the positive aspects of the digitalization of education and its huge potential to promote the socio-economic development of the regions of Kyrgystan, thereby activating the local market for educational services.

References

1. Kulueva CR, Ubaidullayev MB, Ismanaliev KI, Kuznetsov VP, Romanovskaya EV (2020) Digitalization of Kyrgyz society: challenges and prospects. In: Popkova E, Sergi B (eds) The 21st century from the positions of modern science: intellectual, digital and innovative aspects. Springer, Cham, Switzerland, pp 229–236. https://doi.org/10.1007/978-3-030-32015-7_26
2. Ministry of Education and Science of the Kyrgyz Republic and the Kyrgyz Academy of Education and Public Foundation "Media Support Center" (2020) Methodological guide for the introduction of digital education in the educational system of the Kyrgyz Republic. Bishkek, Kyrgyzstan. Retrieved from https://edu.gov.kg/media/files/7a31e659-fd0a-46e9-9390-a63689 5efb4c.pdf. Accessed 5 Sept 2022
3. Wilson C, Grizzle A, Tuazon R, Akyempong K, Cheung CK (2011) Media and information literacy of teachers. UNESCO, Paris, France. Retrieved from https://unesdoc.unesco.org/ark:/ 48223/pf0000192971_eng. Accessed 5 Sept 2022
4. Molchanova EV (2019) On the pros and cons of the digitalization of modern education. problems of modern pedagogical education, pp 64–4, pp 133–135. Retrieved from https://cyberl eninka.ru/article/n/o-plyusah-i-minusah-tsifrovizatsii-sovremennogo-obrazovaniya. Accessed 5 Sept 2022
5. Abramchik EL (2017) New possibilities of information technologies. Kiravanne y Adukatsyi 9:69–72
6. Abrosimova MA (2017) Information technologies in state and municipal management: textbook. Knorus, Moscow, Russia
7. Gancharik LP (2019) Open education system in management training in the digital economy. Open Educ 23(2):23–30. https://doi.org/10.21686/1818-4243-2019-2-23-30
8. Zenkov AR (2020) Digitalization of education: directions, opportunities, risks. In: Proceedings of Voronezh State University. Series: problems of higher education, vol 1, pp 52–55. Retrieved from http://www.vestnik.vsu.ru/pdf/educ/2020/01/2020-01-11.pdf. Accessed 5 Sept 2022
9. Schwab K (2016) The fourth industrial revolution. World Economic Forum, Geneva, Switzerland. Retrieved from https://law.unimelb.edu.au/__data/assets/pdf_file/0005/3385454/ Schwab-The_Fourth_Industrial_Revolution_Klaus_S.pdf. Accessed 5 Sept 2022
10. Clark BR (1998) "Creating Entrepreneurial Universities": organizational pathways of transformation. International Association of Universities Press and Pergamon-Elsevier Science, London, UK; New York, NY
11. President of the Kyrgyz Republic (2018) Decree "On the national development strategy of the Kyrgyz Republic for 2018–2040" (October 31, 2018 No. 221). Bishkek, Kyrgyzstan. Retrieved from https://mfa.gov.kg/uploads/content/1036/3ccf962c-a0fc-3e32-b2f0-5580bfc79 401.pdf. Accessed 5 Sept 2022

12. National Statistical Committee of the Kyrgyz Republic (2020) Analytical review: assessing the level of digital development in the Kyrgyz Republic. Institute of statistical research, Bishkek, Kyrgyzstan. Retrieved from http://www.stat.kg/media/files/82744364-3ebf-465e-a343-848cbb bf68b4.doc. Accessed 5 Sept 2022

13. Ministry of Education and Science of the Kyrgyz Republic (2020) The national online seminar on modern educational technologies Edutech KG. Retrieved from https://erasmusplus.kg/en/edutech2020/. Accessed 5 Sept 2022

14. Security Council of the Kyrgyz Republic (2018) The concept of digital transformation "Digital Kyrgyzstan 2019–2023" (approved on December 14, 2018 No. 2). Bishkek, Kyrgyzstan. Retrieved from https://www.gov.kg/storage/2020/12/files/program/12/kontsepts iya_tsifrovoy_transformatsii_tsifrovoy_kyrgyzstan_2019_2023.doc. Accessed 5 Sept 2022

15. Osh State University (nd) Official website. Retrieved from https://www.oshsu.kg. Accessed 5 Sept 2022

16. Ibraimova SS (2010) Culture as the basis of interethnic conflicts in the context of globalization. Soc Human Sci 5–6:16–19

17. Temirova BT (2019) The relationship of the trilogy "Manas" and Kyrgyz literature (Synopsis of Dissertation of Doctor of Philology). Bishkek, Kyrgyzstan. https://www.sovet.aitmatov.tk. Accessed 5 Sept 2022

18. Tashybaev AK, Kydyrmysheva (Marzarbaeva) BT (2021) Legislation and state initiatives in the field of combating human trafficking in the Kyrgyz Republic. Modern Sci 4–1, 339–345

19. Amatova UO (2016) International business and its development in Kyrgyzstan. Young Scientist 26(130):241–246

20. Kulueva CR, Suranchiev BT (2020) Problems and prospects of youth labor resources of the regions of Kyrgyzstan. Sci New Technol Innovations Kyrgyzstan 4:165–171. Retrieved from http://science-journal.kg/media/Papers/nntiik/2020/4/165-171.pdf. Accessed 5 Sept 2022

Climate Adaptation of Transport in the Eurasian Economic Union

Elena A. Barmina⊙, **Aleksandr L. Nosov**⊙, **Vladimir N. Volnenko**⊙, **Evgeny R. Ismagilov**⊙, and **Suj Bjao**⊙

Abstract The paper analyzes the consequences of climate influence on the transport complex in the EAEU and the Russian Federation. The authors determined the composition of the transport complex in relation to multidirectional climatic threats. The authors formed packages for adapting the institutional environment of the transport system and put forward technical proposals for climate adaptation of transport, using the example of the internal space of the Russian Federation, considering the inclusion of logistics flows to the EAEU countries. It is noted that the joint implementation of the proposed activities will effectively resist negative climate changes.

Keywords Transport · Climate change · Transport security · Proposals for adaptation · Institutional environment · Technical level

JEL Classification F15 · N7 · N70 · O19

1 Introduction

Transport is one of the most climate-dependent sectors of the economy. At the beginning of 2022, numerous problems related to climate and its impact on global transport logistics have aggravated worldwide. Simultaneously, ongoing events indicate the suspension of the development of national transport climate adaptation projects due to the COVID-19 pandemic and the diversion of major resources to combat it [1].

E. A. Barmina (✉)
Vyatka State University, Kirov, Russia
e-mail: ea_barmina@vyatsu.ru

A. L. Nosov
Vyatka State Agrotechnological University, Kirov, Russia

V. N. Volnenko · E. R. Ismagilov · S. Bjao
Belgorod State University, Belgorod, Russia

A literature review regarding applied instruments of technical regulation determined the following. In 2013, a group of experts from the United Nations Economic Commission for Europe produced a 270-page report "Climate change impacts for international transport networks and adaptation to them". The countries of Europe, America, and Japan were analyzed for the negative impact of climate change on transport systems. This report did not include Russia. Over the next decade, global climate threats have increased significantly [2].

In 2020, the Climate Center of Rosgydromet presented the results of research and development in the field of scientific and methodological substantiation of sectoral and regional adaptation strategies to current and expected climate changes [3]. The report notes the dynamics of the increase in the average temperature in Russia and the growing number of dangerous phenomena that cause physical and economic damage. Deformation of rail tracks, negative phenomena in port infrastructure, damage to pipeline systems, etc. are noted in the field of transport destruction of road surfaces.

Due to the critical importance of the topic, in February 2020, the Inland Transport Committee of the Economic Commission for Europe mandated the Expert Group established in 2019 to analyze information and facilitate the implementation of assessment of projects on climate change impact on transport operations and infrastructure [4]. The Chair of the Expert Group is to present the final report at the session scheduled for September 2025. Participation in the work of the Expert Group is open to all interested members of the United Nations and experts.

The costs of preventing the consequences of climate change are currently estimated at 1% of the world GDP [5]. For example, considering the need of transport in 2010–2030, costs of infrastructure development in the EU are estimated at more than 1.5 trillion euros, adaptation costs of vulnerable bridges have been estimated as $140–$250 billion through the twenty-first century [6]. Together, sustainable development of transport and logistics systems is possible through production management, quality management [7–9], and the introduction of new technologies. For example, infrastructure provision and organization of production based on the Internet of Things can serve as a tool that determines adaptation mechanisms to climate change [10–13].

2   Materials and Method

Research methods within the framework of the EAEU and the Russian Federation include the monitoring of identified threats and projects to counter negative climate change in the transport infrastructure.

The mechanisms of adaptation to the consequences of climate change in the field of transport management were analyzed. According to certain critical impact factors presented by the United Nations Economic Commission for Europe, adaptation mechanisms within the EAEU and the Russian Federation include monitoring of identified threats and development of projects to counter negative climate change in the transport infrastructure [1].

The Federal law "On transport security" (No. 16-FZ) adopted in Russia characterizes transport infrastructure facilities and vehicles in detail. However, it is limited to their protection from acts of unlawful interference and leaves human-made phenomena caused by climate impact without consideration [2, 14].

At the end of 2021, 79% of countries initiated adaptation planning mechanisms at the national level (7% more than in 2020), which implies a single mechanism for systematizing problems and solutions, starting from the level of strategic planning, development of policy and related laws and projects, implemented at the level of powers and their administrative units [15].

As for the Russian Federation, aspects of adaptation mechanisms to climate change are defined by the national action plan of the first stage for the period up to 2022. They contain key decisions that can be combined in three areas:

- Scientific and technological support for forecasting in the field of climate change;
- Development of the regulatory environment for management decisions in the field of state and municipal government;
- The complexity of planning, including monitoring the effectiveness of adaptation measures and their adjustment (if necessary) [2].

Actualization of development strategies for types of economic activity and sectors of the economy is determined as the key task. In terms of sectoral strategic decisions in transportation, climate change adaptation mechanisms have not found their way into the past period. Moreover, at the end of 2019 and 2020, a significant number of indicators of the transport strategy of the Russian Federation did not achieve the target values [16].

On November 21, 2021, the updated version of the Transport Strategy of the Russian Federation until 2030 [17] was approved with a forecast for the period until 2035. In the content part of the studied issue, the following provisions of the strategy deserve attention:

- The use of geographical advantages and the realization of the country's transit potential through integration into world transport chains as part of the Eurasian Economic Union;
- Increasing the transport accessibility of territories, including geostrategic, remote, and hard-to-reach territories.

As a confirmation of the importance and relevance of the considered issues at the country level, on February 17, 2022, the round table "Problems of adaptation to local climate changes as a factor in formation of climate resilience of the EAEU member countries" was held at the Scientific Center for Eurasian Integration. One of the key topics was the importance of creating supranational regulation systems in the field of ensuring climate resilience, which will then be integrated into the EAEU regulation system.

The concept of managing the sustainable development of transport companies is an ambiguous tool for strategic development in current conditions.

Given mechanisms for solving environmental problems and ensuring resilience to climate change as a tool, particular attention should be paid to the ESG concept, which determines the ratings of leading foreign companies in international markets. The ESG concept as sustainable development of commercial companies has recently gained great popularity as a tool for entering international markets and, accordingly, access to funds that support investments and innovations in green technologies and environmental safety. In 2004, UN Secretary-General Kofi Annan presented the report "The Caring Wins," which formulated the basic principles of the concept of sustainable development. According to the abbreviation, the essence of the concept, which is actively supported by corporate and state funds (US and EU), is as follows:

- Environmental—environmental protection;
- Social—social development;
- Governance—corporate governance.

Thus, each concept corresponds to several priorities (strategic directions) with certain KPIs specific to each industry. Based on ESG principles, independent expert organizations determine the ESG index of companies' attractiveness, which, in turn, determines the competitive advantages for foreign investors. About 800 companies in the USA have received the ESG rating in recent years; for comparison, Russia currently has no more than 135 companies [18].

According to 2021 data, the independent agency Raex-Europe published the ESG rating of Russian companies. It included 145 organizations; the top 5 included Russian Railways and Gazprom [18]. S7 Airlines and the road construction company Avtoban are among the companies that actively declare the concept of sustainable development in the transport sector.

S7 Airlines, which is one of the leaders in air transportation in the Russian Federation, also actively declares the ERG principles reflected in the published concept (according to the S7 Group ESG report for 2020) [19].

Measures are defined in terms of adaptation solutions to climate change; they are briefly systematized in Table 1.

Analyzing the experience of Avtoban and S7 Airlines, we can conclude that the events are mostly local, tactical in nature, and do not make it possible to fully implement the concept, according to the existing completeness of the western requirements for a number of objective reasons, the main of which are limited resources and lack of required investments.

Today, in connection with the current geopolitical situation, many experts see the concept of sustainable development as a political tool for restricting the entry of "undesirable" companies into the international market, which, in fact, is no longer related to environmental problems. These risks and the actual unavailability of western investment funds do not mean that this concept should be neglected. Guided by long-term solutions, it is apparent that it is necessary to include a mechanism for state regulation of the conceptual provisions of sustainable development as a tool for the development of the non-state sector of enterprises that form the industry (ESG technologies in Russia and the world) [18]. It should also be understood that such tools-requirements of the ESG concept as "carbon neutrality" and "carbon tax" are impossible in the conditions of functioning of a number of enterprises, especially in the transport industry of an industrial country, part of the transport routes of which

Table 1 A set of solutions of S7 Airlines in terms of adaptation mechanisms to climate change as part of the ESG concept

	Activity priorities	
	Basic principles of ERG	S7 airlines
Environmental	Resource-saving technologies	Fuel consumption is reduced by 5%;
	Responsible consumption	Reduction of energy consumption by 20% by installing heat regulators on radiators in all offices
		Reduction of consumption of cold and hot water in offices by 30% due to the installation of sensor mixers
		Reduction of paper use by 20 tons per year (equivalent to saving more than 200 trees) through the introduction of digital technologies, EFB, CCP, and SMI into the workflow, as well as accumulation of waste paper
	Reducing the negative impact on the environment	Reduction of the amount of de-icing liquid consumed at operating airports by 5–10% (300–400 thousand liters)
	Reducing carbon footprint	Ecological project "We are Siberia": more than 400 thousand trees were planted
Governance	Development of the aviation industry through innovation	Maintenance center at Pulkovo Airport as part of the investment agreement with the Government of St. Petersburg. The innovation center based on S7 TechLab was organized
	Strategic partnerships for sustainable development	S7 Airlines is the only Russian member of the Oneworld global aviation alliance

Source Compiled by the authors based on [19]

falls on the regions of the Far North. Thus, this concept requires a revision of key decisions based on priorities of the national economy and strategically important priorities of international cooperation. More specifically, key decisions are required in terms of governance in the form of formation and specification of requirements for transport companies, including in relation to possible adaptation to climate change, which does not depend on political decisions.

The active development of the state ESG paradigm in the Republic of Belarus at the state level and the fairly large interest of commercial enterprises in open forms of ESG reporting and the creation of its own rating in the Republic of Kazakhstan should be noted after analyzing the process of implementing the ESG agenda in other countries of the Commonwealth. The experience of Belarus deserves interest. The BIK Ratings rating agency prepared a report on the results of 2021 "Green financing, ESG and their development prospects in Belarus," which contains an overview of the ESG rating methodology being developed at the national level [20]. The review also contains a list of 28 Belarusian companies based on the concept of sustainable development, which unites all three areas of ESG, including financial institutions and retailers. However, there is not a single transport company [21].

In 2022, in the Republic of Kazakhstan, the company PwC Kazakhstan published the rating of "50 best companies of information disclosure in the field of sustainable development," the leader of which is JSC NC "KazMunayGas," a Kazakh operator for exploration, production, and transportation of hydrocarbons. This is also the only company on the list related to the transport industry [22].

As for joint decisions within the framework of the commonwealths of states, this task is complicated by the heterogeneity and the presence of geo-economic features of the EAEU members. The members of the EAEU have a significant difference in the volumes, lengths, and types of transport networks, as well as volumes of cargo flows, which makes it difficult to form general principles of transport policy and implement climate change adaptation mechanisms (Fig. 1).

The "road map" for implementation of the transport policy of the EAEU countries, agreed in February 2021, contains measures that develop provisions of the Treaty on the Eurasian Economic Union relating to transport and infrastructure. It considers the specifics of the development of road, rail, and water transport as it ensures the implementation of tasks and priorities of the transport policy. Nevertheless, it does not place due emphasis on the problems of climate change, which is a strategically important task.

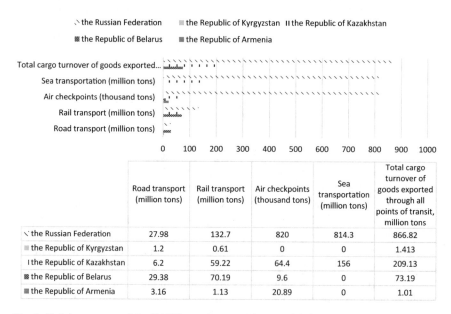

	Road transport (million tons)	Rail transport (million tons)	Air checkpoints (thousand tons)	Sea transportation (million tons)	Total cargo turnover of goods exported through all points of transit, million tons
the Russian Federation	27.98	132.7	820	814.3	866.82
the Republic of Kyrgyzstan	1.2	0.61	0	0	1.413
the Republic of Kazakhstan	6.2	59.22	64.4	156	209.13
the Republic of Belarus	29.38	70.19	9.6	0	73.19
the Republic of Armenia	3.16	1.13	20.89	0	1.01

Fig. 1 Freight turnover of the EAEU member countries, considering the types of checkpoints. *Source* Compiled by the authors based on [23–25]

3 Results

Let us develop a list of proposals for areas of activity in the field of climate adaptation of transport.

In general, adaptation measures in the transport sector should be aimed at reducing vulnerability and increasing the resilience of systems to climate factors. This implies not only the physical strength and durability of the infrastructure to withstand adverse impacts but also the ability to quickly restore at minimal costs.

At the heart of transport climate adaptation is a system for monitoring climate change manifestations and extreme climate events, which can have a wide variety of impacts on the transport infrastructure and transport services. By improving and expanding climatological support, it is possible to significantly increase the efficiency of the design and operation of the transport infrastructure [16].

The institutional group of proposals in terms of realizing the transport potential of the EAEU includes the following:

- Proposals on amendments to the requirements for project documentation. The potential impact of changing climate must be considered when planning, designing, constructing, and operating the transport infrastructure;
- Proposals on information and analytical support of the decision-making process on the relevant range of issues by the state authorities of the EAEU participants and the Russian Federation, local governments, and business entities, including the development of the legal and regulatory framework and organization of state regulation in the field of adaptation to climate changes;
- Proposals on the development of economic mechanisms related to the implementation of measures aimed at adaptation to climate changes;
- Proposals on monitoring new domestic and foreign technologies and experience in obtaining, collecting, transmitting, storing, and presenting climate and related information and information products necessary for solving regulatory, scientific, technical, economic, military-strategic, and other tasks when adapting to climate changes with a view to subsequent implementation in operational work;
- Proposals on the development of training and advanced training programs for specialists of the relevant specialization;
- Proposals on informing society about ongoing and expected climate changes, their causes and consequences, and possibilities of adapting to them. The climate provision of transport industries differs not only in the specifics of the information provided for each type of transport but also in their scale, which is due to the length of transport routes, measured by many thousands of kilometers;
- Proposals to promote the fulfillment of international obligations, including by the Russian Federation, in the field of monitoring, assessing, forecasting climate changes and their consequences, and adapting to them, including obligations under the UNFCCC, synchronization of management decisions regarding the ESG concept, and determination of common methods and ratings as part of agenda of sustainable development across the Commonwealth states.

The set of proposals on the example of the Russian Federation includes the following:

- Proposals on adjusting the strategy for the development of the transport industry, considering the need for its adaptation to climate changes and variability of climate;
- Proposals for land (road and rail) transport are associated with an increase in the number of hazardous phenomena, such as fog, heavy rains, snow avalanches, dangerous snowfalls and blizzards, and sandstorms;
- Proposals on winter maintenance of roads, safety, and uninterrupted traffic on roads in difficult climatic conditions. In winter, the danger on the road is due to its slipperiness associated with black ice and snow run-up. Frequent temperature fluctuations in winter also contribute to the destruction of road surfaces;
- Proposals on summer road maintenance. Rising temperatures and long periods of hot weather negatively affect roads, causing softening of asphalt pavement;
- Proposals on railway transport at high and low temperatures. Extended periods of extreme high and low temperatures can cause deformation of railway rails, resulting in reduced travel speeds and accidents on the railways;
- Proposals on the transport infrastructure in the north of the Russian Federation. Melting permafrost will damage the infrastructure. It causes subsidence of the subgrade and cryogenic swelling, which affects the strength and durability of roads and leads to deformation of the road surface;
- Proposals on responding to the expected increase in water runoff and river floods in certain regions of Russia. These phenomena will have the most catastrophic consequences precisely for transport networks because the main roads and railways run along the flood plains or cross them;
- Proposals on considering the effect of heat on inland water transport. Inland waterways can be severely affected by lower water levels during heat waves. This may lead to fewer shipping routes, shorter navigation periods, reduced cargo capacity, increased fuel costs, and more frequent groundings of ships;
- Proposals on changing timing and processes of freezing and opening of rivers and reservoirs. Due to the ongoing warming of the climate, a reduction in the period of freeze-up on Siberian rivers with a simultaneous decrease in the maximum thickness of ice should be expected. Significant changes are also possible in the timing and processes of freezing and opening of rivers and reservoirs;
- Proposals on the port infrastructure. Changes in the sea level and increased destructive power of storms will damage the port and cargo infrastructure due to flooding and increased costs for the construction and maintenance of ports. The permafrost retreat and the reduction of the Arctic ice area may also lead to disruption of the seaport infrastructure. A positive aspect of this process for Russia is an increase in the period of navigation along the Northern Sea Route, as well as a reduction in fuel costs;
- Proposals on air transport. Rising air temperatures could adversely affect various airport infrastructures, especially runways, while airports in northern regions could benefit from reduced snow and ice clearance costs. Frequent extreme

temperature rises can create operational problems such as increased power consumption by aircraft on the ground;

- Proposals for strengthening the role of science in adaptation to climate change. Proper proactive adaptation to climate changes will not work without deep, technological, supplied by workers, fundamental, and applied research. Simultaneously, the most important component of the scientific substantiation of adaptation measures is economic assessments that require methodological refinements;
- Proposals on pipeline transport. Dangerous consequences of climate changes for the oil and gas industry in the Arctic are related to the permafrost retreat;
- Proposals on priority projects of the land transport infrastructure. Among the priority projects for the development of the transport infrastructure in the Arctic, the Northern Latitudinal Railway can be singled out—the construction of a railway line (including a bridge across the Ob River), which will connect the existing sections of the Northern and Sverdlovsk railways;
- Systematization and integration of the listed proposals into a sectoral plan for adaptation to climate changes in the field of transport.

The earlier estimate of the cost of overcoming climate change in 1% of global GDP [5], which for the Russian Federation will amount to $14 billion or more than 1 trillion rubles (1% of $1.4 trillion of Russia's GDP). From the economic point of view, comparing the costs of maintaining the transport infrastructure of the Russian Federation, considering climate risks, pays off with the capacity of the freight transportation market in Russia. Thus, all invested funds are returned to the economy.

4 Conclusion

Taking the foregoing into consideration, it can be noted that the effective counteraction of the transport complex to climate change as part of the EAEU and the Russian Federation is implemented at the institutional and technical levels of regulation, with a significant degree of interest in cooperation in solving existing problems.

References

1. UNEP (2021) Step up climate change adaptation efforts or face huge disruption: UN report. Retrieved from https://www.unep.org/news-and-stories/press-release/step-climate-change-adaptation-efforts-or-face-huge-disruption-un. Accessed 24 Jan 2022
2. Climate Center of Rosgydromet (2020) Report on the scientific and methodological foundations for developing climate change adaptation strategies in the Russian Federation (within the competence of Rosgydromet). St. Petersburg, Amirit, Saratov, Russia. Retrieved from http://cc.voeikovmgo.ru/images/dokumenty/2020/dokladRGM.pdf. Accessed 24 Jan 2022

3. Federal Service for Hydrometeorology and Environmental Monitoring (Rosgydromet) (2021) Climate change: fact sheet, No. 88. Retrieved from https://rgmo.net/tmp/Izmenenie_kli mata_N88_DecJan_2020.pdf. Accessed 24 Jan 2022

4. United Nations Economic Commission for Europe (UNECE) (2020) A new 2020–2025 mandate and the terms of reference for the Group of Experts on assessment of climate change impacts and adaptation for inland transport. Retrieved from https://unece.org/DAM/trans/doc/ 2020/itc/ECE-TRANS-2020-6e.pdf. Accessed 22 Jan 2022

5. Medvedkov AA (2018) Adaptation to climate changes: the global environmental and economic trend and its significance for Russia. Bulletin of the Moscow State Regional University. Series: Natural Sciences, vol 4, pp 11–19. https://doi.org/10.18384/2310-7189-2018-4-11-19

6. United Nations Economic Commission for Europe (UNECE) (2013) Climate change impacts for international transport networks and adaptation to them. UNECE, New York, NY; Geneva, Switzerland. Retrieved from https://unece.org/DAM/trans/main/wp5/publications/climate_c hange_2014r.pdf. Accessed 27 Jan 2022

7. Sozinova AA, Lysova EA (2021) The marketing approach to managing the quality of company's products based on industrial and manufacturing engineering in the conditions of transnational capital transformation. Int J Qual Res 15(4):1089–1106. https://doi.org/10.24874/IJQ R15.04-05

8. Sozinova AA, Saveleva NK (2022) Marketing quality management in industry 4.0 in transborder markets. Int J Qual Res 16(3):955–968. https://doi.org/10.24874/IJQR16.03-20

9. Sozinova AA, Sofiina EV, Safargaliyev MF, Varlamov AV (2021) Pandemic as a new factor in sustainable economic development in 2020: scientific analytics and management prospects. In: Popkova EG, Sergi BS (eds) Modern global economic system: evolutional development vs. revolutionary leap. Springer, Cham, Switzerland, pp 756–763. https://doi.org/10.1007/978-3-030-69415-9_86

10. Bogoviz AV, Kurilova AA, Kozhanova TE, Sozinova AA (2021) Artificial intelligence as the core of production of the future: machine learning and intellectual decision supports. In: Ram M (ed) Advances in mathematics for industry 4.0. Academic Press, Chennai, India, pp 235–256. https://doi.org/10.1016/B978-0-12-818906-1.00010-3

11. Bogoviz AV, Kurilova AA, Kozhanova TE, Savelyeva NK, Melikhova LA (2021) Infrastructural provision and organization of production on the basis of the internet of things. In: Ram M (ed) Advances in mathematics for industry 4.0. Academic Press, Chennai, India, pp 211–231. https:// doi.org/10.1016/B978-0-12-818906-1.00009-7

12. Popkova EG, Saveleva NK, Sozinova AA (2021a) A new quality of economic growth in "smart" economy: advantages for developing countries. In: Popkova EG, Sergi BS (eds) "Smart technologies" for society, state and economy. Springer, Cham, Switzerland, pp 426–433. https:// doi.org/10.1007/978-3-030-59126-7_48

13. Popkova EG, Savelyeva NK, Sozinova AA (2021b) Smart technologies in entrepreneurship: launching a new business cycle or a countercyclical instrument for regulating the economic situation. In: Popkova EG, Sergi BS (eds) "Smart technologies" for society, state and economy. Springer, Cham, Switzerland, pp. 1722–1730. https://doi.org/10.1007/978-3-030-59126-7_188

14. Russian Federation (2007) Federal law "On transport security" (February 9, 2007 No. 16-FZ, amended on November 6, 2021). Moscow, Russia. Retrieved from http://www.consultant.ru/ document/cons_doc_LAW_66069/. Accessed 24 Jan 2022

15. UNEP (2021) Adaptation Gap Report 2020. Retrieved from https://www.unep.org/ru/resour ces/doklad-o-raznice-mer-adaptacii-k-izmeneniyu-klimata-2020-goda. Accessed 24 Jan 2022

16. Government of the Russian Federation (2019) Decree "On the approval of the national action plan for the first stage of adaptation to climate changes for the period up to 2022" (25 Dec 2019 No. 3183-r). Moscow, Russia. Retrieved from https://www.garant.ru/products/ipo/prime/doc/ 73266443/. Accessed 27 Jan 2022

17. Ministry of Transport of the Russian Federation (2008) Transport strategy of the Russian federation until 2030 (approved by the decree of the Government of the Russian Federation on November 22, 2008 No. 1734-r). Moscow, Russia. Retrieved from https://mintrans.gov.ru/doc uments/3/1009. Accessed 22 Jan 2022

18. itNews (2021) ESG technologies in Russia and the world: what is it, why and how actively is it being implemented? Habr. Retrieved from https://habr.com/ru/post/597177/. Accessed 30 Jan 2022
19. S7 Group (2020) ESG report of S7 Group for 2020. Retrieved from https://www.s7.ru/ru/about/sustainability/img/s7-ESG_2020.pdf?v4. Accessed 8 Mar 2022
20. BIK Ratings (2021) Green financing, ESG and their development prospects in Belarus: a report. Retrieved from https://bikratings.by/wp-content/uploads/2022/01/esg-i-zelyonoe-finansirovanie.pdf. Accessed 22 Jan 2022
21. Online Magazine "marketing.by" (2022) Belarus develops ESG rating for companies. Retrieved from https://marketing.by/novosti-rynka/v-belarusi-razrabatyvayut-esg-reyting-dlya-kompaniy/. Accessed 10 Mar 2022
22. Kapital.kz (2021) Top 50 Kazakhstani companies in regards to ESG information disclosure. Retrieved from https://kapital.kz/finance/101350/top-50-kazakhstanskikh-kompaniy-po-raskrytiyu-esg-informatsii.html. Accessed 21 Dec 2021
23. Analytical Center under the Government of the Russian Federation (2019) Freight transport in Russia: a review of current statistics. Bulletin No. 53. Retrieved from https://ac.gov.ru/archive/files/publication/a/24196.pdf. Accessed 4 Feb 2022
24. Eurasian Economic Commission (2021) Analysis of the state, dynamics and trends in the development of customs infrastructure in places where goods are moved across the customs border of the countries of the Eurasian Economic Union 2020: analytical review. Retrieved from http://www.eurasiancommission.org/ru/act/tam_sotr/dep_tamoj_infr/SiteAssets/CIDD3_DevCI/CIDD4_analysis_DCI_2021.pdf. Accessed 28 Feb 2022
25. National Statistical Committee of the Republic of Belarus (2020) Belarus in figures: statistical reference book, 2020. Minsk, Belarus. Retrieved from https://istmat.org/files/uploads/62683/belarus_v_cifrah_2020.pdf. Accessed 17 Feb 2022

Shaping a Green Public Procurement System in the Russian Federation

Natalya A. Guz⬤, **Marina V. Dubrova**⬤, **Tatyana Yu. Kiseleva**⬤,
Larisa N. Sorokina⬤, **and Nadezhda N. Zhilina**⬤

Abstract The paper aims to develop proposals for a green public procurement system in the Russian Federation. The indicated goal cannot be achieved without solving a list of tasks, in particular: (1) to explore the theoretical basis for the formation of a green public procurement system in the Russian Federation; (2) to analyze international experience with green public procurement; (3) to propose measures for the formation of a green public procurement system in the Russian Federation. The authors use the general scientific method of dialectic knowledge, system-structural, comparative-legal, and formal-logical methods, and special scientific methods. The research object is devoted to "green" public procurement, which is one of the innovative topics of improving the contracting system in Russia. A comprehensive analysis of current problems of forming a green public procurement system in the Russian Federation is presented for the first time; the theoretical content of the green public procurement system in the Russian Federation is disclosed in detail; ideas for forming a green public procurement system in the Russian Federation are proposed. The novelty of the research lies in the fact that it presents the first comprehensive analysis of contemporary problems of forming a green public procurement system in the Russian Federation and proposes ideas for forming a green public procurement system in the Russian Federation. These and other components of the scientific

N. A. Guz (✉) · M. V. Dubrova · T. Yu. Kiseleva
Financial University Under the Government of the Russian Federation, Moscow, Russia
e-mail: NAGuz@fa.ru

M. V. Dubrova
e-mail: MVDubrova@fa.ru

T. Yu. Kiseleva
e-mail: TKiseleva@fa.ru

L. N. Sorokina
RUDN University, Moscow, Russia
e-mail: lukshalar@mail.ru

N. N. Zhilina
Kazan Cooperative Institute (Branch) of the Russian University of Cooperation, Kazan, Russia
e-mail: znadnik@inbox.ru

novelty of this research allow us to formulate a conclusion about the theoretical and practical significance of scientific research.

Keywords Green public procurement system · Contracting system · Environmental criterion · Eco-labelling · Sustainable procurement

JEL Classification E64 · G18 · H11 · H23 · H57 · H72 · Q56 · Q57 · O13 · M38 · L15 · L66 · L68

1 Introduction

In the legal and economic literature, most authors refer to public procurement in a very simplified way, as the supply of goods, works, and services to meet federal needs [1]. The public procurement mechanism is one of the government's instruments for solving a wide range of social problems, mainly economic ones. According to the Federal State Statistics Service of the Russian Federation (Rosstat), the number of participants in public procurement and the number of contracts increase every year:

- 3,610,125 contracts worth 6.96 trillion rubles were signed in 2018;
- 3,645,413 contracts worth 8.28 trillion rubles were signed in 2019;
- 3,415,726 contracts worth 8.97 trillion rubles were signed in 2020;
- 3,321,207 contracts worth 9.47 trillion rubles were signed in 2021 [2].

Nowadays, the question of solving environmental problems has become a hotly debated issue. Green public procurement could be one of the instruments for overcoming environmental issues. This phenomenon is not yet widespread in the Russian legal framework. However, it has recently been actively discussed. The subject of scientific research is the social relations that develop during public procurement in the Russian Federation. The research aims to develop proposals for a green public procurement system in the Russian Federation. The indicated goal cannot be achieved without solving a list of the following tasks:

1. Exploring the theoretical basis for the formation of a green public procurement system in the Russian Federation;
2. Analyzing international experience with green public procurement;
3. Proposing measures for the formation of a green public procurement system in the Russian Federation.

2 Materials and Method

The authors use the general scientific method of dialectic knowledge, system-structural, comparative-legal, and formal-logical methods, and special scientific methods. The degree of development of scientific research is determined by the study of scientific works by Kondrat [1], Anchishkina et al. [3], and others.

The scientific novelty of this research lies in the fact that it presents the first comprehensive analysis of contemporary problems of forming a green public procurement system in the Russian Federation, reveals in detail the theoretical content of the green public procurement system in the Russian Federation, and proposes ideas for forming a green public procurement system in the Russian Federation. These and other components of scientific novelty allow us to formulate a conclusion about the theoretical and practical significance of this scientific research.

Before turning to the definition of green public procurement, it is necessary to analyze the relationship between sustainable procurement, green procurement, and green public procurement. The authors conclude that the sign of sustainability implies action not only on the scale of one country but also in a global sense—the world community, the environment, and the global economy [4]. Green procurement refers to public procurement that applies environmental criteria and requirements, such as requirements for the absence of harmful substances in the composition of products [5].

This definition contains an attribute of belonging to a certain state. Additionally, environmental procurement and green procurement act as synonyms. The World Wildlife Fund (WWF) defines green (environmentally responsible) procurement as part of a more general concept of responsible procurement that involves compliance with ethical, economic, and social standards [3]. Therefore, the authors believe that environmental or green public procurement is part of sustainable procurement. The content of the concept of green public procurement from the innovative side is also interesting, which implies rational (optimizing the spending of public funds through market opportunities to significantly increase environmental) and social benefits at the local and global levels [3]. Another attribute of green public procurement is the reduced environmental impact throughout their life cycle, compared to goods, services, and works with the same basic function that would otherwise be purchased [5].

The main purpose of green public procurement is to reduce the environmental impact. Additional purposes can include other social, economic, and political benefits, such as raising awareness of the impact of products and services on the environment or developing environmentally friendly green technologies [5]. It is important to note that the process of green public procurement should consider various consequences from:

- The way procurement is conducted, the use of non-renewable resources;
- Methods of production and manufacturing, logistics, and service delivery;
- The production and use of products, their reuse, and recycling;
- The ability of suppliers to anticipate and mitigate negative effects at the stages of supply chains [5].

Thus, the authors analyzed the relationship between the concepts of sustainable procurement, environmental procurement, and green public procurement and identified some features of "green" public procurement. Currently, the regulatory and legal framework for the contractual sphere of procurement of goods, works, and services for state and municipal needs is undergoing reform. Various optimization packages are being adopted that are aimed at reducing, unifying, and making auction and bidding procedures transparent. However, there is no separate area where green public procurement could be implemented yet.

Chronologically, government agencies and institutions support and recognize the need to form a system of green public procurement. In 2010, following a meeting of the Presidium of the State Council, an instruction was given to develop environmental parameters to be included in procurement documentation and preferences for participants in the auction and competitive procedures, which have a document of voluntary environmental certification. In 2013, the Government of the Russian Federation issued two instructions to ensure that environmental requirements are considered. In 2018, the Federation Council Committee on Agrarian and Food Policy and Environmental Management submitted for consideration a bill containing proposals to encourage customers working under the Federal law 44-FZ to give preference to goods with environmental properties when holding auctions and contests—energy-efficient equipment, office furniture made of environmentally friendly materials, paper from recycled materials, and electricity generated from renewable sources.

Nowadays, all initiatives have not yet been implemented. We believe this is due to the specifics of green public procurement—a deep elaboration of the functioning mechanism is needed. Difficulties arise with terminology (environmental criteria and environmental properties), methodology (eco-labeling, environmental certification, and environmental declaration), incentives for customers (choice of dispositive or imperative methods to determine the share of green public procurement), and preferences for participants of the auction and competitive procedures.

Currently, to evaluate the bids of procurement participants, the customer may use a separate group—environmental characteristics of the procurement object—along with qualitative and functional criteria. We believe that elaborating on these characteristics (at the initial stage in the form of explanations of state bodies and institutions) would allow customers to form an idea of their application. Additionally, environmental characteristics of procurement objects can be applied in electronic bidding for the procurement of goods, works, and services as a criterion for evaluating the bids of participants [6]. Analyzing the federal level of lawmaking and law enforcement, it is necessary to note an interesting practice of the subject of the Russian Federation—in 2010, Moscow introduced the following environmental requirements for quality and technical characteristics to the object of the procurement:

- On the level of emissions into the atmospheric air for motor fuel, cars, trucks, buses, lawnmowers, saws, etc.;
- On energy efficiency for electric and electronic equipment, lighting devices, and air conditioners;
- On the use of secondary raw materials for paper, etc. [7].

However, the practice is not widespread among other subjects of the Russian Federation; it also did not find a response at the federal level of the Russian Federation in the form of regulatory legal acts of state bodies and institutions. Therefore, the formation of the system of green public procurement should begin with normative legal acts of state bodies and institutions and then spread to the regional practice of the subjects of Russia. Additionally, enshrining basic provisions (e.g., terminology) at the legislative level does not mean the obligation and strict form of their application in describing the object of procurement. The legislation provides for the possibility to apply to the customer additional requirements, conventions, or justification of the need to use other indicators, requirements, conventions, and terminology [6]. Provisions on the regulation of environmental product labeling are also reflected in the current legislation norms in the form of technical regulations and GOSTs [8].

According to official statistics, in 2021, the predominant type of power plants in Russia were thermal power plants. They generated 67.7% of electricity, and most of them (about 50% of the total) used natural gas or, less well-known, associated petroleum gas as fuel. This is quite logical because the USSR and, subsequently, the Russian Federation were among the leaders in natural gas production almost throughout history, and the vast majority of thermal power plants were built during the Soviet Union when they received their basic equipment, and only some were upgraded in modern Russia. It was a fairly simple way of producing electricity with acceptable environmental sacrifices. Nevertheless, times are changing, and requirements and standards are also changing with them. In today's environment, society is extremely concerned about the global environmental crisis.

To summarize the above, it is necessary to draw the following conclusion. In the Russian Federation, there are prerequisites for the formation of a system of green public procurement, which is reflected in the basic Federal law [6] and in related legislation. There is a precedent for the mechanism of functioning of the green public procurement system at the regional level. We believe that a detailed elaboration of the methodology and legal and regulatory provisions is needed in the future.

3 Results

In 2021, the world faced an energy crisis. Europe, China, and India were the hardest hit. Europe suffered from a sharp rise in the price of natural gas, most of which was imported from Russia. In China, the development of industries and the rise in living standards outstripped the pace of electricity development. Thus, there was insufficient energy to supply all buildings and production facilities. On the other hand, India faced depletion of its coal reserves, while most of its power plants ran on coal. Other regions, such as Southeast Asia and North America, were also affected. Such a crisis has resulted in higher prices for goods and services, production stoppages, and logistical collapses. This translates into huge economic losses. The model of the formation of a green public procurement system in foreign countries comes down to the introduction of a list of environmental goods, works, and services (EU, China, and USA) and their further application (mandatory and non-mandatory options).

One of the key steps for forming the state green procurement system in Russia is the creation and adoption of a regulatory framework. Difficulties arising during the creation of regulatory provisions include disclosure of special procurement terminology (e.g., environmental criterion). In a general sense, the content of the environmental criterion of the object of procurement can include the property of a good, work, or service in the full production cycle or further use to cause the least harm to the environment. The assessment of the parameters of environmental friendliness of purchased products is aimed at compliance with the legality and effectiveness of the functioning of the contracting system, as well as compliance with the objectives of sustainable economic development in Russia. To form a system of state green procurement in Russia, foreign experience in developing the environmental parameters of purchased products, presented by the European Commission of the European Union, is of interest. The European Commission currently presents environmental criteria for state green procurement for 22 economic sectors [9].

The official website of the European Commission provides a scheme of the procedure for developing and revising environmental criteria and its description. The procedure for developing and revising environmental criteria based on duration can be standardized and shortened. The authors propose to consider the first variant of the procedure [9]. As can be seen, the stage of terminology development is rather complicated and laborious. However, it is necessary for the effective application of the legislation. The experience of the Republic of Belarus is interesting because it can be considered not only from the position of similarity of legislation but also as a member of the Eurasian Economic Community.

If we talk about compliance with the requirements for green procurement, we believe they can be considered in two aspects:

1. From the point of view of the process of organization and implementation of public procurement procedures;
2. From the perspective of the architecture of public procurement procedures and the content of the conditions for acquiring goods (works and services) [10].

The action of the legislation of the contractual system of the Republic of Belarus for green public procurement is implemented in the form of requirements (Fig. 1).

The requirements for the subject imply the presence of environmental criteria enshrined at the regulatory level, for example, environmental criteria for synthetic detergents [11], environmental criteria for refrigeration appliances [12], environmental criteria for wallpapers [13], or environmental criteria for televisions [14].

Requirements to participants imply the presence or absence of various legal states of the participant of procurement, for example, the absence of bankruptcy proceedings.

Requirements to the criteria and evaluation method imply the application of green public procurement terminology in the auction or competitive procedures—eco-labeling, environmental declaration, environmental certification, etc. [15, 16].

Requirements to the terms of performance of contracts concluded based on the results of procurement procedures imply various terms of performance—delivery of goods, the performance of works, or the rendering of services. Individual conditions are understood as, for example, warranty obligations or gratuitous elimination

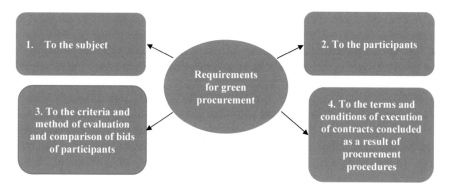

Fig. 1 List of requirements for green public procurement in the Republic of Belarus. *Source* Compiled by the authors based on UN Development Programme in Belarus [10]

of defects [17]. To summarize the above, the authors believe that to form a system of green public procurement in the Russian Federation, the experience of developing environmental criteria of the EU, as well as the requirements for green public procurement in the Republic of Belarus, are of particular interest. The formation of a green public procurement system in the Russian Federation should comprise several stages.

The first step is to define a list of environmental goods, works, and services. At the federal level, it is necessary to establish the degree of its compulsory nature (the share of certain purchases for the customer) and the system of preferences for the participant in the procurement.

The second step is to work out the terminology. In this case, it is necessary to define the range of terms, study related legislation, and use foreign experience.

The third step is to analyze the list of environmental goods, works, and services. Statistical and sociological studies can be carried out. Studies of foreign authors on the issues of environmental protection can also be considered. Additionally, it is necessary to use foreign experience in terms of procedure and discussion. It is important to elaborate on the creation of registers of goods (participants) and the procedure for certifying the environmental attribute.

The fourth step is to develop a mechanism for a green public procurement system:

1. When describing the subject matter of the procurement, the client has defined a product (work or service) included in the environmental list (synthetic detergent). The winner among the bidders shall be the one who provides a supporting document (eco-label, certificate, or declaration) as part of the bids.
2. When describing the subject matter of the procurement, the client has defined a product (work or service) included in the environmental list (office furniture with ecological properties). The winner among the bidders shall be the one who obtains the highest number of points, with the supporting document (eco-label, certificate, or declaration) being evaluated by additional points.

Additionally, an incentive mechanism should be developed for other stages of the auction and tender procedures (shortened contract terms, amounts of bid security, and contract or guarantee obligations), for example, the use of a preference model for the procurement of goods, works, and services by small businesses or other specific categories.

With the model already developed in the fifth stage, it is advisable to conduct a legal experiment. It is possible to test the system in a separate pilot project on the example of one or more regions. Provisions on experimental legal regimes should be considered within the current legislation. This stage involves the adoption of a legal and regulatory document (roadmap, development strategy, etc.) and its implementation.

The sixth step is to analyze the results of the implemented model and adjust or abandon the chosen strategy. If the results are successful, the decision should be made to extend the practices to other regions.

The seventh stage envisages legislative provisions and implementation of the model by the regions.

Federal law "On agricultural products, raw materials, and foodstuffs with improved characteristics" (June 11, 2021 No 159-FZ) establishes the legal basis for the implementation of agricultural and other activities related to the production, storage, transportation, and sale of agricultural products, foodstuffs, and industrial and other products with improved characteristics [18]. The adoption of this law implies the regulation of environmentally friendly or green products. In addition to definitions of basic concepts, there is also a provision for labeling agricultural products, foodstuffs, and industrial and other products with improved characteristics. According to the provisions of the rider, the producer of improved agricultural products, improved food, and industrial and other products has the right to apply the marking in the form of a graphic image (conformity mark) of improved agricultural products, food, and industrial and other products of a single sample from the date of entry of information about the company, the types of products produced, and other information in the unified state register of producers of agricultural products, food, and industrial and other products with improved characteristics for a period not exceeding the period of validity of the certificate of conformity of improved agricultural products, foodstuffs, and industrial and other products [19]. Additionally, GOST R 58,661-2019 "National standard of the Russian Federation. Products and food with improved characteristics. Conformity assessment" establishes the requirements for conformity assessment of products and food with improved characteristics [8].

4   Conclusion

Ecological or green public procurement refers to public procurement that applies environmental criteria and requirements, such as the requirement that products are free from hazardous substances in their composition. The main purpose of green public procurement is to reduce the negative impact on the environment. Other social, economic, and political benefits can be added. There is currently an opportunity for green public procurement, using the supply of food as an example. With further development of the newly adopted legislation, provisions of the federal legislation on agricultural products, foodstuffs, and industrial and other products with improved characteristics, as well as national standards, may be used by the purchaser to describe the object of procurement. The provisions on labeling of agricultural products, foodstuffs, and industrial and other products with improved characteristics, as well as the availability of information about the manufacturer in the unified register, may be used as a requirement for the criteria and method of evaluation. Such actions could significantly improve the overall economic security of the country, in fact, completely eliminating the possibility of an energy crisis in Russia while creating a large reserve of resources for the construction of knowledge-intensive and energy-intensive industries in the country or even full-cycle production complexes. In turn, this will increase the country's GDP, attract investment in the real sector, and strengthen the national currency.

References

1. Kondrat EN (2014) Financial crime in Russia. Threats to financial security and ways of counteraction. Justitsinform, Moscow, Russia
2. Unified Information System in the Field of Procurement (nd) Statistics. Retrieved from https://zakupki.gov.ru/epz/main/public/home.html#statAnchor. Accessed 10 Nov 2022
3. Anchishkina OV, Gracheva YA, Ismailov RA, Kuznetsova EM, Ptichnikov AV, Khmeleva EN (2020) Public green procurement: Experience of legal regulation and proposals for implementation in Russia. World Wildlife Fund (WWF), Moscow, Russia
4. UNEP (2012) Sustainable public procurement implementation guidelines. UNEP, Paris, France. Retrieved from https://www.oneplanetnetwork.org/sites/default/files/sustainable_public_procurement_implementation_guidelines.pdf. Accessed 10 Nov 2022
5. Kontturi C, Lankiniemi S, Yulirusi H, Kuznetsova E, Shadrina E (2021) Guidelines on the inclusion of environmental criteria in public procurement. Ecological Union, HSE University, and SUAI, St. Petersburg, Russia. Retrieved from https://ecounion.ru/wp-content/uploads/2021/04/rukovodstvo-po-vklyucheniyu-ehkologicheskih-kriteriev-v-gosudarstvennye-zakupki-2021.pdf. Accessed 10 Nov 2022
6. Russian Federation (2013) Federal Law "On the contract system in the sphere of procurement of goods, works and services for state and municipal needs" (April 5, 2013 No. 44-FZ). Moscow, Russia. Retrieved from http://www.consultant.ru/document/cons_doc_LAW_144624/. Accessed 10 Nov 2022
7. Moscow City Government (2010) Decree "On environmental requirements for quality and technical characteristics of products purchased under Moscow City Government order and directions for improvement of environmental certification and audit system" (April 20, 2010 No. 332-PP). Moscow, Russia. Retrieved from https://www.garant.ru/products/ipo/prime/doc/294521/. Accessed 10 Nov 2022
8. Federal Agency on Technical Regulating and Metrology (2019) National standard of the Russian Federation GOST P 58661-2019: products and food with improved characteristics. Conformity assessment (approved November 29, 2019 No. 1324-st). Rosstandart, Moscow, Russia. Retrieved from https://docs.cntd.ru/document/1200169972. Accessed 10 Nov 2022

9. European Commission of the European Union (nd) Section "Green Public Procurement." Retrieved from https://ec.europa.eu/environment/gpp/gpp_criteria_procedure.htm. Accessed 10 Nov 2022

10. GEF, UNDP, & Ministry of Natural Resources and Environmental Protection of the Republic of Belarus (2019) Methodological recommendations on organizing and conducting procurement of goods (works, services) using the principles of green procurement (public procurement, procurement at own expense and procurement of goods (works, services) during construction of facilities). Minsk, Belarus. Retrieved from https://minpriroda.gov.by/uploads/files/1323325-5071403-5071416.pdf. Accessed 10 Nov 2022

11. State Committee for Standardization of the Republic of Belarus (2007) STB 1733-2007 "Environment protection and environmental management. Environmental criteria for synthetic detergents" (Approved by Decree of the Gosstandart of the Republic of Belarus on February 23, 2007 No. 9). Minsk, Belarus: Gosstandart. Retrieved from https://tnpa.by/#!/DocumentCard/190926/288188. Accessed 10 Nov 2022

12. State Committee for Standardization of the Republic of Belarus (2007) STB 1742–2007 "Environment protection and environmental management. Environmental criteria for refrigeration appliances" (Approved by Decree of the Gosstandart of the Republic of Belarus on April 12, 2007 No. 23). Minsk, Belarus: Gosstandart. Retrieved from https://tnpa.by/#!/DocumentCard/192699/289990. Accessed 10 Nov 2022

13. State Committee for Standardization of the Republic of Belarus (2007) STB 1755-2007 "Environmental protection and nature management. Environmental criteria for wallpapers" (Approved by Decree of the Gosstandart of the Republic of Belarus on May 30, 2007 No. 32). Minsk, Belarus: Gosstandart. Retrieved from https://tnpa.by/#!/DocumentCard/194191/291493. Accessed 10 Nov 2022

14. State Committee for Standardization of the Republic of Belarus (2007) STB 1805-2007 "Environment protection and environmental management. Environmental criteria for televisions" (Approved by Decree of the Gosstandart of the Republic of Belarus on November 30, 2007 No. 62). Minsk, Belarus: Gosstandart. Retrieved from https://tnpa.by/#!/DocumentCard/201696/299097. Accessed 10 Nov 2022

15. State Committee for Standardization of the Republic of Belarus (2003) STB ISO 14024-2003 "Environmental labels and declarations. Environmental labeling type I." (Approved by Decree of the Gosstandart of the Republic of Belarus on March 19, 2003 No. 15). Minsk, Belarus: Gosstandart. Retrieved from https://tnpa.by/#!/DocumentCard/144984/166679. Accessed 10 November 2022

16. State Committee for Standardization of the Republic of Belarus (2002) STB ISO 14021-2002 "Ecological labels and declarations. Self-declared environmental declarations (Type II environmental labeling)" (Approved by Decree of the Gosstandart of the Republic of Belarus on September 27, 2002 No. 48). Gosstandart, Minsk, Belarus

17. State Committee for Standardization of the Republic of Belarus (2007) STB 1803–2007 "Environmental services" (Approved by Decree of the Gosstandart of the Republic of Belarus on November 30, 2007 No. 62). Gosstandart, Minsk, Belarus. Retrieved from https://tnpa.by/#!/DocumentCard/201691/299092. Accessed 10 Nov 2022

18. Russian Federation (2021) Federal Law "On agricultural products, raw materials, and foodstuffs with improved characteristics" (June 11, 2021 No. 159-FZ). Moscow, Russia. Retrieved from https://www.garant.ru/products/ipo/prime/doc/400788577/. Accessed 10 Nov 2022

19. Ministry of Agriculture of the Russian Federation (2022) Order "On approval of the procedure for marking in the form of a graphic image (conformity mark) of improved agricultural products, foodstuffs, and industrial and other products of a single design, as well as the said image and requirements for such marking" (February 11, 2022 No. 70). Moscow, Russia. Retrieved from https://rg.ru/documents/2022/02/28/minselhoz-prikaz70-site-dok.html. Accessed 10 Nov 2022

Nanotechnologization as Essential Component of Circular Economy: Chinese Versus Russian Experience

Alexander I. Voinov, Evgeny P. Torkanovskiy, and Vladimir S. Osipov

Abstract The authors investigate the relations between circular economy and nanotechnologies. For this purpose, they compare the approach of two countries (Russia and China) to circular economy and nanotechnologies, along with an investigation of international peers (the EU, the USA, and other countries). The research analyzed the patent activity in nanotechnologies using the methods of statistical analysis to make evident the trend for nanotechnologization and its implications for the national economy. The authors find that nanotechnologies solve urgent economic issues on circular economy principles, making it possible to ameliorate efficiency and performance, increase longevity, reduce waste, and ensure the recirculation of waste. Thus, the success of the circular economy is dependent on an integrated approach involving technology amelioration, particularly nanotechnologies. The research links the economic cost reductions achieved through nanotechnologization to the success of the circular economy. It also contributes to the literature on nanotechnologies offering analysis of patent activity as a proxy of nanotechnologization. The findings are relevant for policy-makers who may use nanotechnologization not only per se but as the vital component of circular economy strategy and for entrepreneurs—to introduce nanotechnologies in their circular enterprises.

A. I. Voinov (✉) · V. S. Osipov
MGIMO University, Moscow, Russia
e-mail: a.voinov@odin.mgimo.ru

V. S. Osipov
e-mail: vs.ossipov@inno.mgimo.ru

A. I. Voinov
International Institute of Energy Policy and Innovation Management, Odintsovo Branch of the MGIMO MFA of Russia, Odintsovo, Russia

Russian State Academy of Intellectual Property, Moscow, Russia

E. P. Torkanovskiy
Institute of Economics of the Russian Academy of Sciences, Moscow, Russia
e-mail: tor@gromanz.ru

V. S. Osipov
Lomonosov Moscow State University, Moscow, Russia

© The Author(s), under exclusive license to Springer Nature Switzerland AG 2023
E. G. Popkova (ed.), *Smart Green Innovations in Industry 4.0 for Climate Change Risk Management*, Environmental Footprints and Eco-design of Products and Processes, https://doi.org/10.1007/978-3-031-28457-1_40

Keywords Circular economy · Nanotechnology · Patents · China · Russia

JEL Classification Q32 · Q56 · O32

1 Introduction

The circular economy is a vibrant and evolving concept. The notion of a "circular economy" is modeled on self-sustaining or self-relying ecosystems and grounded in their complex, self-organizing, and circular flows of energy and matter. By cascading (passing along) waste energy and processing waste nutrients for reuse in the cycle, this closed-loop, the complex system reduces new resource inputs while eliminating waste, pollution, and emission outputs [1]. The opposite of the circular economy is the extractive or linear economy that developed along with the industrial revolution and transformed resources discarding the remnants after utilization into waste. The circular economy has its origins in industrial strategies for resource efficiency and waste prevention to reduce the risks and costs associated with waste and energy leakage [2]. There are a number of definitions of circular economy and significant literature dedicated to determining the correct definition (e.g., [3]). The authors of this paper rely on the definition suggested by Ellen MacArthur Foundation, which defines the circular economy as an economy that is restorative and regenerative by design, which means designing out waste and pollution, keeping products and materials in use, and regenerating natural systems.

The concept of a circular economy has been gaining momentum since its introduction in the late 1970s. Natural resources underpin the foundation of human activity. Individuals and organizations consume vast amounts of natural resources as a matter of routine without much cognizance of their continued availability in the future or the true cost of a depleting natural resource. Over the past decades of industrial activity, organizations, communities, and nations have acted to protect their interests by investing in and securing their supplies of natural resources that support economic growth [4]. An industrial complex, now variously termed as "extractive industries," supplies crucial non-renewable natural resources such as oil and coal for energy or iron and aluminum for construction [5]. Our societal reliance on the consumption of natural resources grows unabated such that the discussion of the sustainability of natural resources has taken primacy in policy and executive concerns. This largely explains the growing acceptance of the circular economy concept around the globe. In China, it is promoted by the national government as a key strategy. China's pioneering role in developing the circular economy concept may be partly explained by its role as the world's workshop and the necessity to use limited resources most efficiently. Its government has formulated a number of circular economy policies to facilitate circularity implementation. These policies focused on key applications such as composting, recycling, and eco-industrial parks. In our opinion, the Chinese experience forms a demonstration and a starting point for any sound strategy to develop the national circular economy plan. The Chinese experience is also significant because it provides examples and viability tests for circular economy projects at different levels. The development of the circular economy is an integral part of China's economic development [6]. China's circular economy policies and ambitions encompass the 3R framework—recycle, reduce, and reuse. Being a national strategy in China, circularity aims to include an increase in production efficiency, prevention of waste disposal, sustainable consumption, and life cycle extension. Circularity policies include demonstration projects where leading practitioners exhibit best practices for different circularity strategies. The total number of circular economy pilot and demonstration projects in China exceeds 2300.

This paper is structured as follows. Section 2 discusses the relationship between the circular economy and advanced technologies. We develop our view of nanotechnologization as an essential part of circular economy development. Section 3 discusses our conclusions regarding the parallel development of nanotechnology and the circular economy. Section 4 provides conclusions.

2   Methodology

China's circular economy consists of the small, middle, and macro levels and the recycling industry [7]. The small circular cycle refers to programs at the company level: promoting environmentally friendly product design and clean production, reducing the usage of materials and energy inputs and external services, and decreasing pollutant emissions. The benefits for business from circularity are closely linked to ecological modernization. They include (not limited to) an increase in business efficiency due to reduced pollution and waste production, reduction in future financial liabilities (e.g., clean-up of contaminated land), increase in labor productivity due to better working conditions, growth in sales of ecologically friendly goods and services, and sharing of pollution prevention and abatement technologies. The middle circular cycle refers to programs developed not by a single company but by their clusters. Based on industrial ecology principles, industrial eco-parks promote a symbiotic and resource-saving relationship among companies by integrating materials, energy, and information. The macro circular cycle refers to initiatives at the level of municipality, region, or even nation. The current focus lies on establishing eco-cities, eco-municipalities, and provinces. The recycling industry takes care of waste treatment and recycling at all levels, also encouraging cyclical utilization of resources in corporate production and households.

China's circular economy strategies are in line with the Chinese national strategy [8]. Specifically, the national goals laid out in China's 2025 and 2049 national strategy indicate that the country intends to become a leading player in many high-tech sectors, including energy-saving technologies and high-precision machinery. Securing control over the global supply of raw materials needed for such technologies guarantees significant advantages in achieving these ambitious goals.

In Russia, the circular economy remains mainly at the level of waste management concept. The dominant point of view is that the main task of a circular economy is to cope with an exponential growth of production and consumption waste caused by outdated production technologies and social and environmental problems [9]. Russia has started an ambitious program of waste collection and management, including the construction of waste incinerating plants all over the country and the set-up of a single state company—waste operator. Currently, the amount of production and consumption waste in Russia is growing rapidly, outpacing growth in the production and consumption of natural resources, as well as the country's GDP growth. From 2010 to 2018, the amount of generated waste increased by 1.6 times and amounted to about 7.3 billion tons in 2018. The rates of waste re-utilization and decontamination remain low, amounting, according to different researchers, to just 3–10% depending on the calculation method. There seems to be a dangerously increasing trend in waste generation at the macro level in relation to GDP. Since 2010, the specific factor of total waste generation per unit of GDP has grown by about 20%, thus confirming the tendency of "antidecoupling" or "negative decoupling" at the macro level, that is, when natural resource consumption and waste production trends outstrip GDP. According to the Federal Service for Supervision of Natural Resources (Rosprirodnadzor), more than 90 billion tons of waste have been accumulated with a total landfill area of 40,000 square km, which equals the territory of the Netherlands. It is also worth noting that 80% of goods in Russia end up in a landfill in the first six months of existence.

An essential aspect of the circular economy is its innovative use of advanced technology, saving energy and resources. Our research shows that circularity is not achievable without significant investments in R&D, breakthrough technologies, and innovative solutions. Otherwise, the economic cost of resources produced in the circular economy is significantly higher than the cost of resources achieved in a linear economy, which makes downstream production from circular resources uncompetitive. Government regulation can only partially offset the increased cost of production and needs technological advancement to complement the economic process of production to achieve circularity.

R&D has played an important role in China acquiring its dominant role over many critical raw materials, especially rare earth elements (REEs). The Baotou Research Institute of Rare Earths has been the world's largest research center of REEs ever since its establishment in 1963. The amount invested in R&D by the Chinese government has steadily increased over the years. The National High Technology Research and Development Program—known as Program 863—was set up in 1986 and was aimed at securing and strengthening China's global position in high-tech sectors by increasing innovation capacities.

In this respect, nanotechnologization becomes a critical building block for achieving a circular economy and national strategy. One of China's most notable R&D programs is the 15-year Medium-to-Long-Term plan (MLP) in Development and Science Technology. The one for 2006–2020 placed significant emphasis on research and innovation, with the goal for China to become a technological leader in new scientific areas. In 2006, China had a 10% share of the top 10% highly cited publications in science and technology, compared to approximately 22% for the USA and 28% for the EU, the latter having been dominant in R&D programs until then. Moreover, China's R&D expenditure was less than half of that of the EU in 2010. By 2015–2016, the situation changed drastically. China overtook the EU in the level of R&D expenditure and leveled with the USA on the most cited publications. In 2017, they both had a share of about 19% of the total 10% of most cited publications. The EU secured a stable share of 26% [10].

Like circular economy processes, nanotechnology requires an integral approach, uniting different sectors and technologies instead of "one-point application" technology. Due to the interdisciplinary nature of nanotechnology, it requires extensive multi-sector collaborations to foster more efficient results.

As a general rule, nanotechnologization, which reduces the need for new resources, can create the technical potential for the solution of environmental issues at a large scale. As a result, it would be possible to continue to live well, but differently, using 75% less energy per capita than in the developed economies today. However, national governments and their advisors worldwide have given less priority to the demand side options. A plausible trajectory towards a zero-carbon future might be a 75% cut in developed economy energy demand per capita matched with the simultaneous growth of developing country energy use to the same level, all to be supplied by the expansion of the new energy options. Unfortunately, there are no strong stakeholder groups that want to promote reductions in material demand. Industry 4.0 is being pursued with vigor by incumbent industries (eager to reduce costs and increase the speed of innovation) and supportive governments (striving to create knowledge-intensive barriers to international competition).

Our view is that technology and technological changes will play an overarching role in social and economic transformation linked to the notion of the circular economy. Our thinking is based on the fact that previous social and economic transformations, such as the Western industrialization that occurred in the nineteenth century or the most recent industrial revolution, were spurred by drastic technological changes (like the steam engine and railroad or semiconductors, the Internet and Global Positioning System, the latter innovations directed initially for military uses and encouraged by the governments). Similarly, R&D activities will play a key role in ensuring circularity at every step of the product value chains.

For example, at the end of the nineteenth century, the problem of horse manure was omnipresent in all major cities around the globe [11]. However, Moscow was one of the few cities spared by this crisis due to innovative technology for the year-round growing of asparagus developed by an agricultural expert turned entrepreneur that required huge quantities of horse manure as soil substitute and insulation in Russian asparagus-unfriendly climate. Special teams gathered all horse manure from Moscow streets 24/7 as it was a valuable resource and not just a waste.

Similarly, breakthrough nanotech innovations that have been actively developed and applied throughout the world economy during the last decade in high-tech and traditional industrial sectors, especially in energy, currently play a significant role in closing the loop for the circular economy. There has been a steady tendency for nanotechnologization as a promising source of national sustainable economic growth in the contemporary world of open innovations.

The contemporary world shifted to industrial and economic structures 4.0, pursuing the utilization of scientific achievements in the nanotech sector. Nanotechnological innovations may resolve key problems of today's civilization on a new technological basis [12].

Russia has substantial innovations in the sphere of nanotechnology and nanomaterials as far as the energy sector is concerned. This confirms the fact that nanotechnologization cannot appear as a leap from ancient technologies to contemporary industry but is usually applied in the most sophisticated and developed sectors of the national economy [13]. Government programs and road maps set the basis for the development of the nano industry until 2030. Kurchatov Institute of the Russian Academy of Sciences forecasts that wide utilization of nanotechnologies may lead to a significant economic effect in the high-tech and traditional industrial branches in the nearest decade. We may assume that the share of the nanotechnology sector in the economy of a particular country can determine its future role and place in the system of international economic relations. The prospects of growth of the real sector of the world economy in today's new normality lie in the modernization of industry through nanotechnology to achieve circularity. In the era of open innovations in industry 4.0, such nanotechnological solutions become instrumental for sustainable development.

3 Results

We have made an extensive analysis to evaluate the role and progress of nanotechnologies in the world and in China and Russia in particular, making use of the statistics of the American patent and trademark office (USPTO), which remains the overarching authority for the registration and protection of new technologies. In the USPTO database, on a case-by-case basis, we have selected patents associated with nanotechnologies, which are usually at the intersection of different sciences.

For the period 2013–2020, USPTO shows that the USA accounted for 35,528 nanotechnology patents and 45,155 published patent applications (78.7% registered), the share of the Republic of Korea—6724 and 9197, respectively (73.1% registered), the share of Japan—5938 and 6688 (88.8% registered) (Fig. 1). In total, the three leading nanotech nations account for 48,190 patents and 61,040 published applications, of which 79.0% were registered.

Our research shows a fourfold growth of Chinese nanotechnology patents in 10 years, with confidence catching up with the leading trio, whereas Russia remains at the margins of the nanotechnology movement (Tables 1 and 2). The total number of Chinese nanotechnology patents in 2013–2020 equals 3962, with 5776 published patent applications that assure China's fourth place among world nanotechnology leaders. However, the rate of approbation is lower than for leaders at 69%.

Our research shows that national government and private R&D expenditure is gradually transformed into nanotech leadership. China is now prepared to use

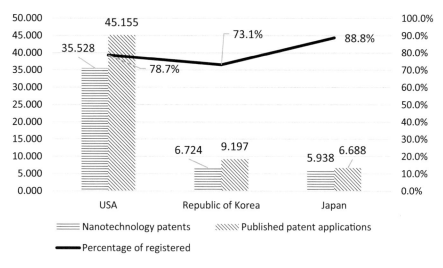

Fig. 1 Total amount of nanotechnology patents and published patent applications of the USA, Republic of Korea, and Japan in 2013–2020. *Source* Developed by the authors

Table 1 Chinese nanotechnology patent activity

	2013	2014	2015	2016	2017	2018	2019	2020	2021
Patents granted	270	357	393	416	524	520	746	736	824
Published patent applications	523	476	554	591	694	913	900	1125	1599

Source Developed by the authors

Table 2 Russian nanotechnology patent activity

	2013	2014	2015	2016	2017	2018	2019	2020	2021
Patents granted	4	11	8	12	7	1	4	7	9
Published patent applications	20	16	5	17	13	4	7	8	15

Source Developed by the authors

the technologies developed elsewhere to maintain the development of its circular economy. In the future, one of the Chinese export items may become circular economy technologies and associated equipment.

Russia envisages the development of scientific and technical innovations in nanotechnology and nanomaterials with a government nanotechnology industry plan for 2030. The plan is quite detailed, though it emphasizes already developed technologies or technologies near the completion stage. The Russian program for developing the hydrogen economy until 2030 also makes part of the nanotechnology industry plan. Similarly, government research initiated the NanoBioInfoCognito interdisciplinary alliance aiming to prepare artificial cognitive systems using nanotechnology components. However, such a detailed but patchwork approach barely works. According to USPTO statistics, from 2013 to 2020, Russia obtained only 54 international nanotechnology patents in the American market [14], with 90 patent applications (60.0% registered).

In our opinion, nanotechnologization in Russia follows the usual path of the linear economy—to respond to the growing demand, the governments and private sector seek ways to increase the offer. For example, global oil companies are ready to spend lavishly searching for new opportunities. These companies are in constant competition to increase oil extraction from available fields and search for new ones. Innovative production and mining technologies made it possible to extract up to one-third of the volume available underground [15]. Even a relatively small increase in the volume of oil extracted from already operating fields (if market participants extract 50% of hydrocarbons instead of 35%) can double the global extractable reserves of fossil energy, estimated today at 1.2 trillion barrels. We may forecast that it will take at least several decades of interdisciplinary research to reach this 50% mark. However, our view is that government strategy should focus more on the demand side of environmental issues.

4 Conclusion

As a replacement for the dominant economic modes of the industrialized world, the circular economy is espoused by environmentalists and increasingly by decision-makers as a crucial means of steering human society toward operating within the ecological limits of the planet. Thus, a circular economy may not only safeguard the integrity of the existing ecosystem essential for humanity's survival but also help rebuild natural capital. Society's political, economic, and environmental actors foresee the circular economy as a crucial pathway for decoupling finite resource consumption (and concomitant negative externalities) from economic growth.

The circular economy has the potential to help redress environmental degradation as well as economic instability that plagues many emerging economies on the new modern ESG basis. On their road to a low-carbon and equitable future, emerging economies can make all sectors of the national economy, especially the most advanced ones, work closely together to find synergistic opportunities to build closed-loop, circular systems. This can be achieved only on a new technological basis—using nanotechnologies on a wide scale. Circular entrepreneurs must be attentive to opportunities inherent in untapped or underused local resources. Such inter-sectoral and public–private cooperation can create sustainable resource access for remote and impoverished populations with multiple benefits. Chinese and Russian experience shows that it is not only the amount of money spent but the efficient combination of government regulation and entrepreneurship, as well as careful and integrated planning, that make these sustainable development goals achievable on the new technological platform. Redirecting our attention to collaborative demand

and supply problem solutions on a nanotechnology basis must be central to achieving the widely accepted goal of a more sustainable future.

References

1. Geissdoerfer M, Savaget P, Bocken NMP, Hultink EJ (2017) The Circular economy—a new sustainability paradigm. J Clean Prod 143:757–768. https://doi.org/10.1016/j.jclepro.2016. 12.048
2. Liu Q, Li H, Zuo X, Zhang F, Wang L (2009) A survey and analysis on public awareness and performance for promoting circular economy in China: a case study from Tianjin. J Clean Prod 17(2):265–270. https://doi.org/10.1016/j.jclepro.2008.06.003
3. Kirchherr J, Reike D, Hekkert M (2017) Conceptualizing the circular economy: an analysis of 114 definitions. Resour Conserv Recycl 127:221–232. https://doi.org/10.1016/j.resconrec. 2017.09.005
4. Torkanovskiy EP (2020) National competitiveness in the era of Industry 4.0 and new individual freedoms. World Futures 77(2):137–153. https://doi.org/10.1080/02604027.2020.1755213
5. Voinov AI, Lomakina OB (2022) Nanotechnologization—the determining factor of the VI technostructure of the world economy. In: Isachenko TM, Revenko LS (eds) A new paradigm for the development of international economic relations: challenges and prospects for Russia. MGIMO University, Moscow, Russia, pp 129–137
6. Yuan Z, Bi J, Moriguchi Y (2008) The circular economy: a new development strategy in China. J Ind Ecol 10(1–2):4–8. https://doi.org/10.1162/108819806775545321
7. Zhu J, Fan C, Shi H, Shi L (2019) Efforts for a circular economy in China: a comprehensive review of policies. J Ind Ecol 23(1):110–118. https://doi.org/10.1111/jiec.12754
8. M Pesce I Tamai D Guo A Critto D Brombal X Wang H, Cheng, Marcomini, A Marcomini (2020) Circular economy in China: translating principles into practice Sustainability 12(3):832. https://doi.org/10.3390/su12030832
9. Bobylev SN, Solovyeva SV (2020) Circular economy and its indicators for Russia. The World of New Econ 14(2):63–72. https://doi.org/10.26794/2220-6469-2020-14-2-63-72
10. Preziosi N, Fako P, Hristov H, Jonkers K, Goenaga X (eds) (2015) China, challenges and prospects from an industrial and innovation powerhouse. Publications Office of the European Union, Luxembourg. Retrieved from https://data.europa.eu/doi/10.2760/445820. Accessed 4 Nov 2022
11. Davies S (2004) The great horse-manure crisis of 1894: the problem solved itself. Foundation for Economic Education. Retrieved from https://fee.org/articles/the-great-horse-manure-crisis-of-1894/. Accessed 15 Oct 2022
12. Voinov AI (2014) Evolution of the concept of "nanotechnology." Invention 14(11):35–39
13. Voinov AI (2016) The creation of a national market of the results of intellectual activity in the field of nanotechnologies. Econ Law Issues 92:84–88
14. Voinov AI, Torkanovskiy EP, Shakirova AA (2021) Technology venturing in the USA: market development and practice of contemporary start-ups. Econ Sci 9(202):257–264. Retrieved from https://ecsn.ru/files/pdf/202109/202109_257.pdf. Accessed 4 Nov 2022
15. Salygin VI, Rybin MV, Voinov AI, Viktorov EI (2022) Venture in breakthrough technologies of the fuel and energy sector. Econ Sci 10(215):62–67. Retrieved from https://ecsn.ru/files/pdf/ 202210/202210_62.pdf. Accessed 4 Nov 2022

Noosphere Model as an Imperative for Sustainable Regional Development in View of the Green Economy Concept (Empirical Research Experience)

Tatiana N. Ivanova and **Nikita Y. Gulyaev**

Abstract The research examines the characteristics and features of the noosphere model of ecological security of the region as a key component of sustainable development of the green economy. The research includes a substantial empirical experience of a large-scale longitudinal sociological study devoted to the noosphere model of ecological security of the region. This research was written as part of the projects: on 12 areas of strategic development established by Presidential Decree "On national goals and strategic development objectives of the Russian Federation until 2024" (May 7, 2018 No. 204), Presidential Decree "On the strategy for the development of information society in the Russian Federation for 2017–2030" (May 09, 2017 No. 203), and Presidential Decree "On the strategy of environmental security of the Russian Federation until 2025" (April 19, 2017 No. 176). The study of the conceptual model of the green economy is the main stage of the research on the ecological noosphere region. It represents the main stage of structuring and analyzing the changes that contribute to the effective development of the Russian economy and provide a mobile interaction between nature and society. At the current stage, the level of socio-ecological knowledge contributes to the possibility of defining today's society as a socio-natural system, which regulates natural processes and covers the territorial forms of settlement, spheres of human life, socio-cultural values, social institutions, and activities of social communities. The authors focus on the research on the problems of the noosphere model as an imperative for sustainable development, as well as on the strategy of a prospective study of the concept of the green economy, which is contradictory in nature.

Keywords Noosphere model · Sustainable development · Region's concept of the green economy · City · Environmental safety

JEL Classification J6 · J60 · Z1 · Z13

T. N. Ivanova (✉) · N. Y. Gulyaev
Togliatti State University, Tolyatti, Russia
e-mail: IvanovaT2005@tltsu.ru

© The Author(s), under exclusive license to Springer Nature Switzerland AG 2023
E. G. Popkova (ed.), *Smart Green Innovations in Industry 4.0 for Climate
Change Risk Management*, Environmental Footprints and Eco-design of Products
and Processes, https://doi.org/10.1007/978-3-031-28457-1_41

401

1   Introduction

Work in this direction is planned as part of the consideration of socio-environmental and economic issues of the Samara-Togliatti agglomeration.

Environmental issues are currently pressing in the Samara Region.

This research aims to construct an innovative conceptual model of the social, ecological, and economic system of the Samara-Togliatti agglomeration. Moreover, the research aims to propose a research program of the social, ecological, and economic levels of the Samara-Togliatti agglomeration, the state of the regional and urban environment, and the range of problems faced by the population. Based on the assessment of everyday life, the residents of the Samara Region construct an image of their future and form their expectations. Thus, 23% of survey participants believe that knowledge about the environmental situation can serve as a means of increasing the population's social and political activity. For 16% of survey participants, information about ecology broadens their horizons and erudition. More than half of the environmental studies students receive information about the state of the environment from the media. According to 4% of the respondents, they receive information about the environmental situation from the participants in environmental movements.

2   Methodology

In the context of sustainable development, problems of the formation of environmental security factors have been considered by such social scientists as Akhmedov et al. [1], Babicheva [2], Bogolyubov [3], Danilov-Danilyan [4], Ignateva [5], Kubarev and Ignateva [6], Savon and Karibzhanova [7] and others.

Environmental security and ecology are analyzed in the New Ecological Paradigm as a branch of social ecology. It became widespread in the 1960s. The ecological paradigm is a sustainable system of norms, principles, orientation values, views, and various attitudes, defined in the "nature-society" system for our time. This paradigm is a specific form of the social paradigm.

3   Results

The survey provided data on assessing respondents' awareness of various aspects of the environmental situation and legal regulation of activities. According to the results obtained, 52% of the survey respondents believe that they are well informed about the rules of conduct in nature. Only 7% feel uninformed about these issues. Accordingly, 38% of the respondents had heard something about the rules of conduct in nature. Half of the respondents are insufficiently informed about other aspects of environmental control and the environmental situation.

The respondents are informed about environmental norms in the context of the impact of ecology on human health. Only 11% of the respondents are uninformed. One-third of the respondents are well-informed about the rules of household waste disposal, and the number of uninformed participants in the survey is half as much. A similar distribution ratio is observed between those who are well-informed about environmental legislation and those uninformed about it. A quarter of the respondents are well aware of the protected natural objects. The number of uninformed respondents is 16% and 21%, respectively. Only 11% of the respondents are informed about environmental monitoring results; two-fifths of survey participants know nothing about it. Survey participants expressed their opinions on the core of environmental culture. One-third of those surveyed believe that an environmental culture is necessary to improve the quality and standard of living. According to 28% of the respondents, environmental culture manifests itself in developing resource-saving technology. One-tenth of the respondents believe that globalization and awareness of the integrity of humanity determine the foundations of ecological culture.

Speaking of the state of regional and urban environments and the range of problems faced by the population, residents of the Samara Region assess their everyday life and, on this basis, form their expectations of the future and construct an image of their future and the future of the region.

The results of the first stage of the sociological study identified the most pressing problems of the region and the city. These problems are related to the economy, social sector, ecology, innovative scenarios of regional development, quality of life, health, culture, youth policy, and the development of small and medium-sized businesses. To conduct an in-depth analysis, the Department of Sociology of the Togliatti State University conducted an additional study using the qualitative method. In our opinion, the most appropriate method was to conduct research using foresight technology (design method).

During the research, interviews were conducted, focusing on a particular issue of regional and urban life. All stages were conducted according to the same script. During each interview, there were also additional questions that were most relevant to the studied area. Expressing an emotional attitude toward the health care system in the region and cities, respondents give a negative assessment, noting the low quality of services. This is how a man who works in an industrial plant expressed his opinion, "Health is the most important structure, but it is not treated this way. It seems to me that plumbers are better at their jobs than medical professionals are at their." The survey participants supported this opinion. Respondents noted the lack of proper attention to patients. Only one middle-aged woman noted the following, "I deal with medicine every day. I have young children who are often sick. Thus, I have to spend much time getting an appointment and waiting for my turn to see a doctor."

The respondents highlighted such a positive trend as "Electronic registrar." This service allows one to make an appointment with a doctor from home at a convenient time. However, the shortcomings were revealed here as well. One of the respondents noted, "Making an appointment using the electronic registrar is convenient.

Nevertheless, let us see at what it is turned into. A doctor needs at least 20 min to examine a patient, while the regulation sets a standard of 30 min. The time indicated in the electronic registrar between registration cards is 2–3 min. That is, you have an appointment for 19:07, and the next person has an appointment for 19:11."

It must be concluded that this service is a positive innovation. Nevertheless, it is not finalized either technically or organizationally. Another positive point is the improvement of medical facilities. The respondents noted that hospitals have become cleaner, repairs are being made, and new equipment is appearing. The so-called "birth certificate" is also a useful innovation. According to this certificate, young mothers and newborns receive decent treatment at medical facilities. Nevertheless, these services are valid only until the child turns one year old.

The main problems of medical care are identified. One of the respondents noted that there were no such queues ten years ago. According to this respondent, the reason is that medical personnel cannot perform their job properly or that people get sick more often. The audience noted noticeable results in the modernization of the health care system. New equipment has been brought to medical facilities. Nevertheless, such services as ECG, CT, scans, and MRIs remain too expensive for patients. Even if a patient has a referral from a doctor, it is still necessary to pay. The claim that Russia has free medical services is refuted in this case. Therefore, according to the respondents, local governments can do almost nothing to improve this area except for regular repairs and monitoring of the performance of medical personnel. This problem has a national level. As noted by the respondents, this environment is important for the region's future economy. The respondents listed the main problems they encountered during their activities. The most important difficulty in developing a business is taxation. Respondents also expressed dissatisfaction with the corrupt regime that now reigns throughout the country. Given the negative impact of corruption on the development of entrepreneurship and a competitive economy, the Samara Region has to fight it, which is not much of a possibility. According to some respondents, "The fight against corruption is an incomprehensible task for most business people due to their social connections and the established "better-to-pay" habit. Thus, most of them remain silent."

In addition to the discussed issues, respondents identified a long list of problems that hinder the normal development of small and medium-sized businesses. These problems must be solved at the national and municipal levels. That is why we asked what measures local governments and federal authorities can take to overcome the problems in small and medium-sized businesses. This is what the director of the web design studio noted, "Currently, Togliatti creates opportunities for the normal functioning of the business. Our company plans to move to the "Zhigulevskaya Valley" Technology Park, which is under construction and is to provide office, laboratory, and production facilities for rent, as well as exhibition space and conference and training rooms convenient for our type of activity. We received the right to preferential rent for all premises. That is all well, but the construction of this place has taken too long. We planned to move there in early 2018, and it is still unknown if we will move there in 2020." Another respondent noted, "Our city has a special economic zone, which implies a special and convenient mode of entrepreneurial activity. I am interested in this project, as my business is connected with the industry. We are residents of this project. As always, the authorities promise us incredible opportunities, and yet it all comes down to social connections, corruption, and convenient cooperation for the authorities."

Thus, the results of the sociological study show that most of the factors affecting the development of entrepreneurship are outside the decision-making power. However, the awareness of negative consequences seems to have reached the federal authorities, as evidenced by recent publications in the government press. Hopefully, appropriate changes in the system of state regulation will help in the future to avoid problems in doing business. After all, the whole picture of the socio-economic indicators of the region shows its enormous potential. The municipal authorities need clarity in the system of management, which will help realize the strong potential of economic development of the region, which will raise the quality of life of residents to a new level.

A variety of emissions has hopelessly spoiled the environmental situation of Russian megacities. The Samara Region is no exception. Respondents were asked to answer several questions:

- How do you assess the environmental situation?
- Are you satisfied with the conditions created to improve the environmental situation?

It should be noted that respondents consider the environmental situation in the region to be unsatisfactory, pointing to the large number of industrial enterprises that pollute the environment. This is how a resident of the Central District of Togliatti expressed his opinion, "Residents of the region have been concerned about these issues for years. For example, residents of Togliatti go to unsanctioned rallies and write to the city administration and the environmental committee. Nevertheless, the situation does not change. The higher authorities claim that this phenomenon is not a plant emission, which seems an absurd assertion to city residents because these "adverse weather conditions" occur regularly." Some respondents commented on solving problems related to the consequences of forest fires, "This spring, we learned purely by chance about the campaign "We Plant a Forest." We arrived and planted our lane of plants. However, the organization was terrible. It is not clear how the people were gathered together. Maybe that is why they did not participate in the campaign, due to this previous experience. I also saw an announcement that there was a clean-up of litter in the forest. I would certainly go with my family because I would like to teach my children not to litter. However, we did not know about this event." From the above, we conclude that the Samara Region lacks information support; people are unaware of the events in the region. Respondents expressed dissatisfaction with the number of green spaces in the region's cities.

The study showed that many people take part in clean-up days, considering it necessary for the region. The respondents also noted the importance of these activities. The respondents highlighted measures that could be effective on the part of local and federal authorities in overcoming problems in this area. Respondents suggested organizing a separate collection of household waste to facilitate further disposal.

Overall, the survey showed that the region does nothing to improve the environmental situation. Currently, no document (i.e., an environmental policy or a similar document) organizes environmental activities in the region. In our opinion, the legislation is largely to be blamed. The powers of local governments are mainly focused on collecting and disposing of household and industrial waste. Simultaneously, local governments are deprived of the right to control pollution sources by law. The implementation of these powers is still weak, their funding is clearly insufficient, and there is no material base.

According to the respondents, youth policy is a positive phenomenon in Russia and its regions. The government has started to develop some mechanisms to promote youth activism. However, insufficient funds are allocated for the implementation of projects. A member of the affairs committee said, "As for the structure of youth policy, I would like to say—there is not enough young personnel. More often than not, young people act in isolation from the main youth policy of the country. Basically, these are people who are close to social and political movements." Of the positive changes in this area, the respondents highlighted the work of public youth organizations and the actions of various foundations. In general, the respondents note that the work in this area is quite active and extensive. Nevertheless, the information part of the youth policy is underdeveloped. The group of respondents came to the general conclusion that the effective implementation of youth policy directly depends on forming a unified information space in the region. The respondents believe that the most effective measures on the part of the authorities for the development of youth policy are as follows:

- Expansion of the leisure spectrum and development of tourism;
- Development of family recreation;
- Activities for the development of culture and sports;
- Dialogue with young people (events at schools, universities, etc.).

Thus, the current society and the country have not yet fully overcome the consumer attitude toward young people, which has formed the dependent position of the younger generation. The task of society and the state is to provide full support to youth public associations, which direct the activity of young people in the direction of public interests and the interests of state development and should form an ideology of positive forms of youth and children's leisure and healthy lifestyles. The adopted directions and programs of youth work to ensure its effectiveness should be systemic, long-term, and stable.

The region and urban space begin to compete for human resources and talents, as well as for the right to take place among global leaders. Under these conditions, the importance of culture as a factor in the competitiveness of cities and regions increases manifold. The respondents expressed their emotional feelings about the level of culture in the region. Most respondents agreed that the level of culture in the region is quite high. Respondents note the regularity of various music festivals, exhibitions, and concerts, sufficient for people who are not very interested in culture and art. However, there was a different opinion, expressed by female students in the design department: "As someone who wants to develop in this area actively, it seems to me that the choice of events in the region is very small. The situation is even more complicated as far as contemporary art is concerned. One has to go to Moscow or other cities to satisfy their needs in the art. Even though there is now a program for the development of culture in the regions, the Samra Region remains very closed to it. We faced with this when a cultural alliance was formed. We tried to participate but failed to implement our projects due to the lack of suitable space and funding. To satisfy my musical preferences, I have to wait for several months because the bands I am interested in rarely play in our region. There are concerts, but very rarely. If we look at Moscow playbills, there is considerably more choice. If we talk about self-fulfillment in this area, all doors are closed in the cities of the Samara Region. I do not see any prospects for the development of a person engaged in the arts in the regions." After listening to this opinion, respondents concluded that the development of culture in the region depends not only on regional authorities but also on the level of education. The inculcation of an interest in culture and arts should be carried out from childhood. If this does not happen, no museums, libraries, or exhibitions will be able to raise the level of culture in the region.

According to the respondents, the main measures that can be taken by regional authorities to increase the level of culture are as follows:

- Create conditions for the recreation and leisure of citizens in parks and squares;
- Modernize the material and technical base and increase the collections of libraries and museums;
- Repair and equip educational institutions of culture and art in accordance with today's requirements;
- Create conditions for the revival, preservation, and development of folk traditions of the multinational region and the city, as well as improvement of the quality of the ecological situation in the region.

Thus, the social, ecological, and economic environment of the Samara Region is at the formation stage. However, the elements that have already been established affect residents of the region, educate them and form people who live in the region and lead an urban lifestyle. As in any other region, several problems must be solved at the national level.

In conclusion of this research, we can state that the Samara Region is highly urbanized and densely populated. It is a center with a developed industry and an average income level among the population. The outlook for the region's development and bringing it to a high level of development is determined by the Strategy for the socio-economic development of the Samara Region until 2030. The main areas of economic specialization of the region are the production of automobiles and automotive components, aerospace engineering, oil production and refining, non-ferrous metallurgy, chemistry, electricity, and agriculture. The implementation of this scientific research was based on the ecosystem approach. It is based on the criteria of sustainable development. The heuristic potential was realized by using the principles of sustainable development when studying the problems of improving the regional territory. The novelty of the implemented approach is that there is a pluralism of ideas about ecological culture, which are reflected in science and public consciousness. The fourth approach focuses on the active involvement of the region's social institutions in the formation and development of the region's institutional indicator. In the fifth approach, the authors focus on the structural and functional analysis of the ecosocialization of the region's population, considering the social, environmental, and economic specifics of the Samara Region. The novelty of the research findings consists in the fact that the correlation analysis of the results of the questionnaire provided an opportunity to identify the effectiveness of methods of solving environmental problems from the perspective of respondents. The analysis identified the following new trends. The results of the sociological study can be used in innovative regional social, environmental, and economic programs of the Samara Region within the following frameworks:

1. Restructuring and creation of conditions for regional development (2017–2020):

    (a) To develop important industrial clusters in the Samara Region—petrochemical, aerospace, and automotive in the context of their modernization.

2. Russia's leading manufacturing hub (2021–2025):

    (a) Provide technological capabilities and increase the technological level;
    (b) To accelerate the development of existing cluster areas, which will create an information base for the development of innovative directions, which will contribute to the diversification and sustainability of the economy of the Samara Region and ensure the stable development of the region in the long-term;

3. Industrial Innovation Center (2026–2030):

    (a) To develop industrial innovation and intensify enterprises of the region producing innovative products;
    (b) To strengthen state measures to develop the innovation infrastructure and support the development of activities.

Innovative technology for the formation of the environmental and social potential of the city and region includes the development of the three horizons (stages) of the Samara Region. The principle of social, environmental, and economic indicators is aimed at studying the issues of constructing the indicators of the social, environmental, and economic system of the region and stimulating environmental modernization in the socio-economic transformation of Russian society. According to the principle of ecological balance, it is necessary to reduce the negative impact on the environment. Scientific and statistical materials that reflect the specifics of ecology in the Samara region will be analyzed when creating the theoretical framework. Statistical materials on environmental problems in the region will be systematized.

4 Conclusion

The scientific value of the results is that the process of ecosocialization of the population is defined in the context of the methodological principles of building an innovative conceptual social, environmental, and economic model through the formation of the environmental culture of the population, which include the values of the sustainable development of the region. Thus, the novelty of the proposed innovative conceptual regional model lies in the fact that it implements ecosystem, environmental, economic, institutional, structural, and functional approaches to the analysis of ecosocialization of the population of the region, considering social, environmental, and economic specificity of the Samara Region.

According to these theories, the quality and life of the population are primarily determined by the possibilities of forming a positive environmental consciousness, which will meet a variety of needs of people: economic, social, cultural, recreational, etc.

The conducted sociological study shows that the innovative conceptual theory of social, ecological, and economic structuring of the Samara Region is a multilevel and complex system, which acts as a mobile system of positive future-oriented future aimed at positive ecosocialization of the population. In a turbulent environment, it is important to plan and forecast the ecological, societal, and economic foundations of an innovative region in the context of a step-by-step formation of the present and future.

According to this concept, socio-innovative indicators of the social, environmental, and economic system of the region are highlighted; the program of research and development of effective social, environmental, and economic development of the region is proposed; the model of ecosocialization process, considering the needs of the population of the Samara Region in different territories, is proposed; recommendations for organizations, institutions, and enterprises to improve the mechanism and ecosocialization of population and the territories are proposed.

Suggestions and recommendations oriented to further stages of research on theoretical problems are as follows:

1. The results of the sociological study indicate an ecosystem approach in the development of the territories of the Samara Region;
2. It is advisable to intensify and prolong sociological research on ecoregions, considering the application of the targeted scenario development of the region within the framework of the complex ecosystem, ecological, economic, institutional, and structural–functional approaches.

Thus, the well-being of the region's inhabitants depends mainly on the state of various spheres of territory. Nowadays, the correlation of regional problems with the factors of environmental safety of the territory and environmental safety of the region is an important, extensive, and vital action to qualitatively improve the region and living conditions [3].

The concept of the green economy is particularly relevant. It can become the basis for the design and formation of the noosphere model as an imperative for sustainable development of the region in the prospects of the green economy through the introduction of innovative technologies and a humane attitude of humanity to the environment.

References

1. Akhmedov RM, Ivanova YuA, Morodumov RN (2019) On the issue of environmental safety. Vestnik Econ Secur 4:108–110. https://doi.org/10.24411/2414-3995-2019-10224
2. Babicheva NE (2013) Integrated methodology of economic analysis of the development of organizations using the resource approach. Econ Anal: Theory Pract 12(1):10–18
3. Bogolyubov SA (2016) The laws of nature and the laws of society. State and Law 11:21–31
4. Danilov-Danilyan VI (2014) On the correlation of legal and economic aspects in nature protection. In: Tikhomirov YuA, Bogolyubov SA (eds) Law and ecology: proceedings of the international practical school for young legal scholars. Infra-M, Russia, Moscow, pp. 48–58
5. Ignateva MN (2014) The main provisions of the geo-eco-socio-economic approach to the development of natural resources. News of the Ural State Mining University 3(35):74–80. Retrieved from http://www.iuggu.ru/download/2014-3-Ignatieva.pdf. Accessed 15 Nov 2022
6. Kubarev MS, Ignateva MN (2018) Environmentally friendly nature management as one of the main conditions for sustainable development. News of the Ural State Mining University 1(49):94–100. https://doi.org/10.21440/2307-2091-2018-1-94-100
7. Savon DJ, Karibzhanova EL (2012) Ecologization of the industrial enterprises economic activity as the basis of the sustainable development of the region economy. Econ Humanitarian Stud Reg 4:113–121

Regional Aspects of the Digital Divide and Its Overcoming in the Sustainable Development Goals

Alexey V. Sysolyatin[ID]**, Anastasia A. Sozinova**[ID]**, Mikhail Yu. Kazakov**[ID]**, Victoria V. Kurennaya**[ID]**, and Sergey A. Fomichenko**[ID]

Abstract The paper considers the problem of digital gaps in the regional development of Russia. Digitalization of all areas of life is now becoming an essential requirement for the sustainable development of countries and regions that want to ensure stable economic growth and competitiveness in a dynamic and changing world. The quality of life of the territories of the subjects of the Russian Federation and the country's place in the international arena depends on the progress of the regions in the digitalization of social and economic life. Therefore, it is important to ensure the acceleration of the digital transition and its uniformity in the regional context and develop a mechanism to overcome the digital divide, giving new opportunities for the socio-economic development of the territories of the Russian Federation. The growth of the added value of the aggregate product in the economy is due to the significant relationship between digital development and the rate of economic growth of regions. The paper highlights the strategic directions of leveling the digital divide in the regional development of Russia.

Keywords Digital economy · Digital discontinuities · Regional development · Digital development · Digital divide · Regional economy

JEL Classification O14 · O33 · R11

A. V. Sysolyatin · A. A. Sozinova (✉)
Vyatka State University, Kirov, Russia
e-mail: aa_sozinova@vyatsu.ru

M. Yu. Kazakov · S. A. Fomichenko
Kuban State Technological University, Krasnodar, Russia

V. V. Kurennaya
Moscow Polytechnic University, Moscow, Russia

© The Author(s), under exclusive license to Springer Nature Switzerland AG 2023
E. G. Popkova (ed.), *Smart Green Innovations in Industry 4.0 for Climate Change Risk Management*, Environmental Footprints and Eco-design of Products and Processes, https://doi.org/10.1007/978-3-031-28457-1_42

1 Introduction

The problem of the digital divide has been the subject of academic research for more than 20 years. Its importance in the paradigm of regional sustainable development is of particular relevance as the national program "Digital economy of the Russian Federation" is being implemented on the country's territory. Digital transformation covers almost all areas of social life, increasing the opportunities to implement advanced technologies in business, increasing national security, and improving the quality of life [1, 2].

The process of digitalization in the country is gradually developing, but its pace does not allow us to classify Russia as a highly developed country. In terms of digital transformation, the Russian Federation is closer to the average European country. Public services are the leading sector for introducing advanced information and communication technologies (ICT); digital products are also actively used among the population [3].

The main problems of digitalization in Russia are as follows:

- The prevalence of foreign equipment and software, the insufficient innovative activity of businesses, and the low share of investment in R&D;
- Low level of interaction between science and production, weak implementation of the mechanism of innovation system activity;
- The need to develop digital competencies and skills in the training and retraining system.

The problem of overcoming the digital divide in the regions is also relevant among the obstacles to developing the digital economy.

This fact confirms the importance and relevance of developing a strategy to overcome the digital divide based on a systemic approach, including the development of skills and competencies in personnel training, stimulation of innovative activity and digital solutions in business and building a support mechanism from the state in the transition to the digital economy.

The research aims to identify the principles of sustainable development and intensification factors of regional economies based on the modernization of digital development and overcoming the digital divide.

The essence of the concept of sustainable development of economic entities in the broad sense is to ensure the stable growth of indicators in the environmental, social, and economic spheres in the long term [4].

2 Methodology

The methodological basis of this research includes the provisions of micro- and macroeconomics and the relationship between their elements based on the use of modern advances and innovative approaches to developing the digital economy to form and expansively reproduce the economic system on the principles of sustainable development.

The substantiation of the main provisions of the work concerning the process of overcoming the digital divide in the regional context to launch the mechanism of sustainable development was carried out using the methods of deduction and induction, systemic and integrated approaches. The empirical part of the work is based on the use of monographic methods of information analysis.

The problems of the digital divide have been studied in the works of such foreign authors as Trappel et al. [5–8]. The digital divide was first considered in the context of policy decisions on transitioning to an information society and the program to overcome the digital divide [7]. Subsequently, the scientific understanding of the digital divide has been constantly transformed. The initial focus on the digital divide was on the uneven access of citizens to the Internet and ICT [8]. This problem was presented mainly by considering the opposing theories of accessibility and the impossibility of using network infrastructure by particular segments of society and individuals [5]. The considered problem has a much broader set of diverse factors, including the specifics of economic development, the social organization and geographic location of the regions, and others. The above circumstances undoubtedly determine the access and nature of ownership and use of ICT in the territorial context [6].

The Russian experience of overcoming the digital divide in the regions is also of great interest, including the works of Alexandrova [9], Dyatlov and Selischeva [10]; Kataeva et al. [4], Polozhikhina [3], Popkova et al. [11], Rusanovskiy and Markov [12], Selishcheva and Asalkhanova [13] and other scholars studying the conditions and nature of the transition to an information economy.

The above authors revealed the essence and the role of the digital economy and the problems of digital inequality in sufficient detail. Nevertheless, it is necessary to complement the theoretical concepts, considering the complex nature and conditions of the digital breakthrough as a factor of sustainable development. It is also required to conduct an in-depth study of the mechanism of modernization and overcoming the digital divide at the level of the subjects of the Russian Federation.

3   Results

The digital economy is at the heart of the digital world. The digital economy is defined as activities based on data in digital form, processing large volumes of data through ICT, which can significantly improve the efficiency of production, logistics, and use of material and technological resources, as well as increase labor productivity [14].

The generalization of economic works on the problem of the "digital divide" revealed two approaches to the interpretation of its essence. For example, in a 2001 OECD report, the digital divide is represented by different opportunities for access to advanced ICT and the Internet [15]. Based on this definition, we can state that the first approach focuses on the form of the digital divide. The digital divide entails increasing inequality in economic and social development [10]. The second approach is distinguished by a deeper analysis of the causes and consequences of the digital divide, highlighting its levels. For example, P. Norris distinguishes "global inequality," "social inequality," and "democratic inequality." Global inequality lies in the different access to the Internet in developed and developing countries. Social inequality is manifested in the differentiation of the amount and level of information between segments of the population. Democratic inequality determines the level of implementation of democratic rights and freedoms through the Internet [16].

At the present stage, Russia cannot be attributed to the group of leaders in the digital transformation of the economy. The contribution of the digital economy to GDP is not significant; there is a significant delay in the adoption of advanced foreign technologies. The level of ICT development, which is a driver of the development of digital transformation, varies by region in the Russian Federation [12].

The digital divide "is the lack of access to advanced information and communication systems" [9].

On the digital map, regions of the Russian Federation differ significantly in their potential for Internet access. The development of the technical component makes a certain contribution to the informatization of regions. Nevertheless, it does not solve all problems, including socio-economic ones. The solution to this problem seems to be the development of digital culture, literacy of the population, and digital infrastructure.

In the summer of 2021, the Ministry of Digital Development, Communications, and Mass Media of the Russian Federation presented the results of the rating of "digital maturity" of Russian regions. Among the 85 subjects of the country, nine were classified as highly mature. These are mostly large, economically strong territories with large budgets—Moscow, St. Petersburg, Tatarstan, and others. Most (62 regions) are characterized by an average level of digital maturity. Finally, 14 regions (e.g., Dagestan, Ingushetia, Chukotka Autonomous District, and others) have a low level of maturity. There is a clear correlation between the level of the economic well-being of the region and its digital maturity. In turn, the development of the digital economy is a multiplier for improving socio-economic development and living standards. The Government of the Russian Federation plans to use the rating data to develop individual trajectories to achieve the targets by 2030 for each subject of the Russian Federation [17].

Opportunities for the digital transformation of regions are determined by the level of ICT spending in regional budgets. In 2021, ICT expenditures of the regions decreased by only 9% (to 205.1 billion rubles) compared to 2020). In 2022, significant growth of 19% is planned. In recent years, central regions have held the lead: in 2021, Moscow (76.3 billion rubles), St. Petersburg (21.5 billion rubles), and the Moscow Region (6.9 billion rubles) [18].

The digital divide between regions was estimated by the Moscow School of Management Skolkovo and EY (Ernst and Young) in 2020. As a result of the study, the index of the digital life of cities was calculated. The value of this index differs between leaders and outsiders by a factor of five. Among federal districts, the leaders are the Ural Federal District and the Central Federal District, followed by the North Caucasus Federal District [19].

Table 1 Number of subscribers to broadband Internet access per 100 people in 2021 by Russian regions

Subject of the Russian Federation	Number of broadband Internet subscribers per 100 people
Moscow	38.36
Republic of Karelia	34.82
Novosibirsk Region	31.56
Murmansk Region	31.26
Voronezh Region	29.44
Republic of Tuva	8.87
Republic of Adygeya	8.58
Chechen Republic	6.32
Republic of Dagestan	4.48
Republic of Ingushetia	2.09

Source Compiled by the authors based on the data from the Federal State Statistics Service of the Russian Federation [20]

The data on the number of subscribers to broadband Internet access in 2021 confirms the conclusion about the unevenness of digital development in the regional context (Table 1).

The number of fixed broadband Internet subscribers per 100 people in Moscow exceeds the minimum value in the Republic of Ingushetia by 18 times. However, the gap narrowed, as it was 36 times in 2017. While the distribution of leading regions by the considered indicator is fairly even across federal districts, the North Caucasus Federal District stands out sharply among the lagging regions-outsiders. The North Caucasus Federal District accounts for five regions out of 10 outsiders at once. The average number of subscribers to fixed broadband Internet access per 100 people in this district was 10.76, which is several times lower than in other federal districts.

The greatest success in digital development is noted in the area of "e-government" (website "Gosuslugi"). Let us analyze the proportion of the population who used the Internet to obtain state and municipal services in the context of federal districts in accordance with the ICT surveys of the Federal State Statistics Service of the Russian Federation (Rosstat) in 2017 and 2020 (Table 2).

The share of the population that used the Internet to obtain state and municipal services increased from 64 to 81% during the reviewed period. The digital gap between federal districts was estimated based on dispersion. Despite its reduction from 46% in 2017 to 27% in 2020, the differentiation between regions on this indicator is still significant.

In accordance with the Concept of Informatization of Regions of December 29, 2014 No. 2769-r, each subject of the Russian Federation has developed and implemented a program of informatization activities to level out the digital divide [21].

Table 2 The share of the population that used the Internet to obtain state and municipal services in the total number of people who received state and municipal services by federal districts, %

Federal district	2017	2020
Russian Federation	64.3	81.1
Central Federal District	71.3	88.5
Volga Federal District	67.2	79.7
Southern Federal District	64.3	82.0
Siberian Federal District	61.7	74.2
Ural Federal District	58.9	78.3
Northwestern Federal District	56.0	75.5
North Caucasus Federal District	53.0	74.5
Far Eastern Federal District	52.4	72.8
Scatter of values (variance) to estimate the digital divide	46	27

Source Calculated and compiled by the authors

The above facts prove that the digital gap between the regions appears to be a significant obstacle to the successful implementation of the national project to digitalize the economy of Russia.

Active digitalization of regional economic systems contributes to their qualitative transformation. It helps change the trajectory of development toward competitive advantage. To overcome the digital divide, it is necessary to develop action plans to reduce the level of digital differentiation in these regions. Among other things, great attention should be paid to the quality of the digital environment, accessibility of advanced digital services to citizens, training people in digital skills, providing them with access to elements of the digital infrastructure, and creating conditions for more intensive innovation in the implementation of digital solutions in the business. In our opinion, the interest of regional authorities in overcoming the digital divide is important. Digitalization is already stimulated by the inclusion of several indicators in the governors' performance rating, characterizing the level of development of the region's digital economy. Nevertheless, this requires financial and organizational support from the federal authorities, including through implementing the national project "Digital economy."

The main areas of action to overcome the digital divide in the regions of Russia are as follows:

- Disseminating and increasing the availability of information networks in the regions based on high-speed fiber-optic Internet and fifth-generation wireless mobile communication technology;
- Transfer of the organizations in the regions to digital platforms that simplify the organization of business;
- Creation of regional centers of digital competencies to improve digital literacy;
- Supplementing regional development programs with projects to develop digital technologies, products, and services;

- Continuation of the practice of e-government services and expansion of their application in the regions;
- Determination of the digital divide in the development of digital transformation of Russia based on the calculation of additional indicators [9].
- The use of smart technology, which can have a delayed effect in the form of economic growth and management of the cyclicality of the economy [11].

4 Conclusion

Inequality of the subjects of the Russian Federation in digital development is a considerable challenge in the implementation of digital transformation, holding back the economic development of the regions and the country. Therefore, reducing the digital divide in the regions should be a priority in the government's objectives [13].

Russian regions need to adopt digital technologies at an entrenched pace through the technical improvement of high-speed Internet and data transmission networks, expanding access for citizens and businesses, developing digital skills among the population, providing data protection and cyber-security technologies, and forcing organizations to switch to electronic document management.

Intensified implementation of ICT in the public life of regions and bridging the digital divide will improve the efficiency of regional economies and bring a new level of quality of life.

References

1. Gladkova AA, Garifullin VZ, Ragnedda M (2019) Model of three levels of the digital divide: current advantages and limitations (as exemplified by the Republic of Tatarstan). Vestnik Moskovskogo universiteta, vol 4. Zhurnalistika, Seriya 10. pp 41–72. https://doi.org/10.30547/vestnik.journ.4.2019.4172
2. Government of the Russian Federation (nd) Project "Digital economy." Retrieved from https://национальныепроекты.рф/projects/tsifrovaya-ekonomika. Accessed 7 Sept 2022
3. Polozhikhina MA (2018) The national models of the digital economy. Econ Soc Problems of Russia 1:111–154. Retrieved from https://cyberleninka.ru/article/n/natsionalnye-modeli-tsifrovoy-ekonomiki. Accessed 15 Sept 2022
4. Kataeva N, Sysolyatin A, Feoktistova O, Starkova D (2020) The concept of sustainable development environmental aspects and project approach. E3S Web of Conf 244:11027. https://doi.org/10.1051/e3sconf/202124411027
5. Compaine B (ed) (2001) The digital divide: facing a crisis or creating a myth? MIT Press. Cambridge, MA. https://doi.org/10.7551/mitpress/2419.001.0001
6. Ragnedda M, Muschert GW (eds) (2013) Routledge, New York, NY
7. Trappel J (ed) (2019) Digital media inequalities: policies against divides, distrust and discrimination. Nordicom, Gothenburg, Sweden
8. van Dijk J (2013) A theory of the digital divide. In: Ragnedda M, Muschert GW (eds) The digital divide. The Internet and social inequality in international perspective. Routledge, New York, NY, pp 28–51

9. Alexandrova TV (2019) Digital divide regions of Russia: causes, score, ways of overcoming. Econ Bus: Theo Pract 8:9–12. https://doi.org/10.24411/2411-0450-2019-11101
10. Dyatlov SA, Selischeva TA (2014) Regionally spatial characteristics and ways to bridge the digital divide in Russia. Econ Educ 2:48–52
11. Popkova EG, Savelyeva NK, Sozinova AA (2021) Smart technologies in entrepreneurship: launching a new business cycle or a countercyclical instrument for regulating the economic situation. In: Popkova EG, Sergi BS (eds) "Smart technologies" for society, state and economy. Springer, Cham, Switzerland, pp 1722–1730. https://doi.org/10.1007/978-3-030-59126-7_188
12. Rusanovskiy VA, Markov VA (2018) Digitalization as a driver for the advanced development of agglomerations in Russia. Sci Anal J "Sci Practice" Plekhanov Russian Univ Econ 10(4):17–27
13. Selishcheva TA, Asalkhanova SA (2019) Problems of digital inequality of Russia's regions. Probl Mod Econ 3(71):230–234
14. Presidential Executive Office (2017) Decree "On the Strategy for the development of information society in the Russian Federation for 2017–2030" (May 9, 2017 No. 203). Moscow, Russia. Retrieved from http://www.kremlin.ru/acts/bank/41919. Accessed 10 Sept 2022
15. OECD (2001) Understanding the digital divide. OECD Publishing, Paris, France. Retrieved from https://www.oecd.org/sti/1888451.pdf. Accessed 15 Sept 2022
16. Norris P (2001) Digital divide: civic engagement, information poverty and the Internet worldwide. Cambridge University Press. Cambridge, UK; New York, NY. Retrieved from https://archive.org/details/digitaldivideciv0000norr. Accessed 13 Sept 2022
17. Petrova V (2021) Governors were given digits. The rating of "digital maturity" of the regions was presented. Kommersant. Retrieved from https://www.kommersant.ru/doc/4938764. Accessed 12 Sept 2022
18. Rudycheva N (2022) Russian regions have planned for significant growth in ICT expenditures in 2022. CNews. Retrieved from https://www.cnews.ru/articles/2022-03-09_ikt-rashody_regionov_v_2022_godu_vyrastut. Accessed 12 Sept 2022
19. Moscow School of Management SKOLKOVO (2019) Report "Index 'Digital Russia'" Retrieved from https://sk.skolkovo.ru/storage/file_storage/00436d13-c75c-46cf-9e78-89375a6b4918/SKOLKOVO_Digital_Russia_Report_Full_2019-04_ru.pdf. Accessed 15 Sept 2022
20. Federal State Statistics Service of the Russian Federation (2021) Regions of Russia: socio-economic indicators—2021. Retrieved from http://www.gks.ru/bgd/regl/b18_14p/Main.htm. Accessed 16 Sept 2022
21. Government of the Russian Federation (2014) Order "On approval of the regional informatization concept" (December 29, 2014 No. 2769-p, as amended on October 18, 2018). Moscow, Russia. Retrieved from https://www.consultant.ru/document/cons_doc_LAW_173678/. Accessed 15 Sept 2022

Economic Evaluation of Ecosystem Services of the State Nature Park "Salkyn-Tor"

Damira K. Omuralieva[ID]**, Kubatbek Asan uulu**[ID]**, Nurila M. Ibraeva**[ID]**, Seil J. Zholdoshbekova**[ID]**, and Aida M. Bekturova**[ID]

Abstract The research focuses on the issues of assessing the ecosystem services of specially protected natural areas (SPNA). The authors evaluated land, water, and forest ecosystems of the State Nature Park "Salkyn-Tor" (SNP). The authors also estimated the cost of carbon absorption by forest resources in this reserve. The subject of economic evaluation of natural resources is primarily their consumer properties. When determining the objective value of resources, socially necessary costs for protecting and reproducing resources are also considered. Therefore, these methodological approaches are the theoretical basis for determining the objective price of all natural resources in their economic evaluation.

Keywords Natural resources · Specially protected natural areas · Objective price of natural resources · Water resources · Land resources · Biodiversity · Forest resources · Recreational services · Ecosystem services · Economic evaluation of ecosystem services

JEL Classification Q57

1 Relevance of the Research

At the present stage of developing relations between society and nature, there is an urgent need to determine the real economic value and cost of natural services and resources. Simultaneously, the economy currently faces the problem of correctly determining the value of nature. Specially protected natural areas (SPNA) play a special role in providing environmental services. Thus, the justification of the need and significance of economic evaluation of environmental services is becoming

D. K. Omuralieva (✉) · K. A. uulu · N. M. Ibraeva · S. J. Zholdoshbekova · A. M. Bekturova
Kyrgyz National Agrarian University named after K. I. Skryabin, Bishkek, Kyrgyzstan
e-mail: d-omuralieva@yandex.com

© The Author(s), under exclusive license to Springer Nature Switzerland AG 2023
E. G. Popkova (ed.), *Smart Green Innovations in Industry 4.0 for Climate Change Risk Management*, Environmental Footprints and Eco-design of Products and Processes, https://doi.org/10.1007/978-3-031-28457-1_43

increasingly important. In the Naryn oblast, the SPNA is represented by three institutions: the Karatal-Zhapyryk State Nature Reserve, the Naryn State Nature Reserve, and the State Nature Park "Salkyn-Tor."

2 Materials and Method

The methodological basis of this research includes the works of foreign and Kyrgyz researchers. The systematization of human benefits from natural systems and attempts to assess them in science began in the second half of the twentieth century and continues to this day. There are as many definitions of ecosystems in the literature as there are approaches to classifying them. A significant study of the value of ecosystems in the world has been the work led by Costanza et al. [1]. Daily [2] and De Groot et al. [3] have pioneered studies of the value of ecosystem services and natural capital on a global scale.

There is no fundamental research on the economic evaluation of ecosystem services in Kyrgyzstan. A breakthrough in this direction can be considered the most important research conducted with the support of international projects assisted by UNDP [4], the experience of Russia [5], as well as studies of Kyrgyz scientists Razhapbaev [6] and Sabyrbekov [7]. The next step in the studied direction is the results of the work of our research group on the assessment of ecosystem services of specially protected natural areas in the Naryn district, conducted within the framework of the grant project of the Ministry of Education and Science of the Kyrgyz Republic.

The methods typical of quantitative methods—random sampling and a questionnaire—were applied to form the sample and develop the research tools. This allows for characterizing the present study as a qualitative study with elements of quantitative methods. The principle of the presence of all nearby villages and the city of Naryn as the main stakeholders of ecosystem services was applied to select study areas. The survey involved tourists who arrived at the time of the study from other countries and regions. Empirical data was collected by gathering quantitative data in the sphere of ecosystem service recipients. The prepared questionnaires were used to collect data in this area. To collect environmental data, the authors used a qualitative method of semi-structured focused interviews to obtain detailed information on fewer questions. During the interviews, additional questions were asked about the situation. These questions were aimed at an in-depth study of the quality of ecosystem services. Questions were prepared in advance and clearly read out to respondents; the answers and other conscious reactions to them were noted and recorded. Questionnaires and interviews with the respondents lasted between 30 min and 1 h. In total, 10 people participated in the interviews, and 89 people took part in the questionnaire.

3 Results

In our opinion, the meaning of the economic valuation of ecosystem services is to determine their cost, that is, the importance for humans of the various benefits derived from nature. In another way, such an assessment can be called natural or physical capital.

Ecosystems, considered natural capital, have advantages over physical capital because they can recover if competently managed. Nevertheless, like physical capital, natural capital is subject to depletion due to reducing future production possibilities.

It is advisable to evaluate the results of protected areas in terms of how their activities allow for improving the environment and preserving and increasing national wealth. In other words, when evaluating the effectiveness of protected area management, one should consider the impact of protected areas on the natural resource potential and existing living conditions of the population of the metropolis. This allows us to form a generalized concept of efficiency as the ratio of all types of results achieved (economic, environmental, and social) with the costs that cause them.

We have proposed the following methods for assessing the ecosystem services of SPNA of the Kyrgyz Republic.

1. Land resources—direct market valuation, normative price;
2. Forest resources: wood resources and non-wood resources—direct market valuation;
3. Forests carbon absorption—indirect market valuation, value recovery method;
4. Recreational resources—transport cost method, direct market valuation;
5. Water resources—direct market valuation;
6. Normative price—direct market valuation, the method based on damages rates;
7. Animals and plants not included in the Red List—direct market valuation, corresponding to black market prices, the method based on damages rates.

"Salkyn-Tor" is located in the central part of the Middle Naryn valley, on the territory of the Naryn district. The length of the territory from north to south is about 15 km, and from east to west—30 km. The total area of the State National Nature Park (SNNP) is 10,419.0 ha. The regime of the protection zone is as follows.

- The protected zone—2472.8 ha;
- Tourist recreational zone—276.4 ha;
- The zone of limited economic activity—7669.8 ha.

For the first time, we determined the total value of the main ecosystem services provided by the nature park.

3.1 Evaluation of Recreational Services

The considered natural park is visited mainly by the residents of Naryn city and nearby villages located within a radius of up to 30 km. The estimated number of visitors was 13,013 people per year. With an estimated number of 13,013 people per year, the predicted income of the natural park for the year is 9,408,868 soms. According to our calculations, the potential capacity of the nature park is 1,902,278 soms.

The analysis of the income beneficiaries of ecosystem services is very valuable in the economic valuation of these services. This analysis shows the efficiency of the recreational area. For this calculation, the prices for visiting the area, renting the premises, and additional services have been determined.

1. Visiting the park area—260,260 soms (13,013 people × 20 soms);
2. Rent of premises—72,000 soms (2000 soms/month × 36 times per year);
3. Extra services—36,000 soms (150 soms per hour × 240 times per year).

Thus, the park receives 368,260 soms per year from the provision of recreational services.

The data from the tourist survey shows that the park's services are mainly used by local residents. Of the visitors, tourists who came from outside made up only 8%; the rest were locals. To the question, "How do you assess the park's infrastructure?"—26.9% responded "good," 53.8%—"satisfactory," 15.4%—"unsatisfactory," and 3.9%—did not answer. Most respondents expressed a wish to improve infrastructure: the opening of new attractions, cafes, and shops, including souvenir shops. To the question: "Are you satisfied with the quality of work of the park staff," most tourists (88.5%) gave a positive answer. An interesting fact of the research was the willingness of tourists and the local population to pay for the maintenance of the nature park in the absence or termination of funding. Of the tourists surveyed, 88.5% expressed their willingness to pay for the preservation of the natural park, which was pleasant for the research team.

It should be noted that 50% of the survey participants were willing to contribute monetary funds in cash for the park's preservation. Nevertheless, they did not indicate a specific amount. Most responded that they would provide financial support to the best of their ability. Moreover, 26.9% of tourists are ready to volunteer, and 30.8% are ready to provide campaigning and educational activities in favor of the park.

The opinion of tourists and visitors on their willingness to pay for the conservation of the state nature park is also fully supported by the local population. Of these, 96.8% are willing to pay. The local population is ready to support the park in case of lack of state funding in the following forms: volunteering—19.7%, campaigning and educational activities—54.8%, and support in monetary form and agitation work—32.3%.

The demographic profile of the surveyed tourists is as follows.

- 38.5% work in the civil service;
- 34.6% are temporarily unemployed;

- 19.2% are engaged in housework.

Of these, 30.8% have higher education, 30.7% have special secondary education, and 38.5% have secondary education.

Tourists noted the following reasons for preserving the natural park as a priority.

1. The protected area has a high aesthetic value (69.2%);
2. The protected zone is the national pride of Kyrgyzstan (66.4%);
3. Zone of endemic animals (65.4%);
4. Nature protection (61.5%).

The response of the local population to the same question was as follows.

1. The protected area has a high aesthetic value (41.9%);
2. The protected zone is the national pride of Kyrgyzstan (54.8%).;
3. Nature protection (48.4%).

Tourists were asked the question what measures should be taken to improve the protection of the natural park. The answers to this question were as follows.

- It is necessary to carry out environmental measures (42.3%);
- Awareness-raising activities should be conducted (34.6%);
- It is necessary to develop ecological tourism (26.9%).

Local residents gave the following answers to the same question.

- It is necessary to carry out environmental measures (64.5%);
- Awareness-raising activities should be conducted (58.1%);
- It is necessary to develop ecological tourism (48.4%).

Of the respondents (local population), 51.6% were women, and 48.4% were men. Of them, 45.2% have higher education, 25.8% have special secondary education, and 29% have secondary education. The social status of the respondents is as follows.

- 25.6%—housewives;
- 22.6%—entrepreneurs;
- 22.6%—pensioners;
- 19.4%—civil servants.

3.2 Determining the Value of Forest Resources

The forest is a large and complex ecosystem that supports high species diversity and provides many valuable services to consumers. The total area of the natural park covered with forests is 2316.5 ha. The forest cover index is only 22.2% in relation to the total area of the SNP "Salkyn Tor." The forests are unevenly distributed, concentrated mainly in separate tracts, scattered in various sizes along the slopes of the gorges; the northern slopes of the mountains are the most wooded. The total stock of forest resources is 130,074.8 m^3.

According to the chronicle of nature of the SNP "Salkyn-Tor," forest plantations are represented by thickets of Tien Shan spruce (82.6%) and creeping juniper (14%), shrub willow (1%), honeysuckle (0.9%), caragana (0.4%), and birch (0.3%). Based on the available chronicle data, namely, the growing stock (cubic meter per ha) and the total area of the natural park, the growing stock was calculated for its entire territory: Tian Shan fir—126,620.3 m^3, Larch tree—225.0 m^3, and Birch Tree—239.2 m^3. By determining the total growing stock of each kind, we can calculate the economic value of wood resources using direct market valuation. The total value of tree resources of the park was 1,453,643.9 thousand soms.

To assess the shrub species of tree resources of the SNP, we used the size of rates for fauna and flora, mummy-containing mineral raw materials, and mushrooms by legal entities and individuals from May 3, 2013 No. 224 "On approval of rates for calculating the amount of penalties for damage caused to objects of fauna and flora, mummy-containing mineral raw materials, and mushrooms by legal entities and individuals" [8].

According to our calculations, the volume of shrub species amounted to 2464.8 thousand soms. Thus, the total amount of wood resources of the park is 1,456,108.7 thousand soms.

3.3   Forests Carbon Absorption

In forests, biodiversity allows species to evolve and dynamically adapt to changing natural conditions (including climatic conditions), maintain the ability to the selection of trees and improvement of tree species (to meet human needs for goods and services and to change operational requirements for them), and maintain their functions in the ecosystem. Forests of any territory, including the SNP "Salkyn-Tor," absorb carbon dioxide and release oxygen in the process of photosynthesis; that is, they purify the atmospheric air. Such benefits provided by the forest ecosystem, outside of itself, can be used to calculate the indirect value of forest resources.

The calculation of the value of forest resources of the SNP "Salkyn-Tor" in terms of carbon dioxide absorption by forests was based on the average indicators of biological productivity of coniferous and deciduous stands, which can absorb 2025 t/ha of carbon dioxide or 5.5 t/ha of carbon during the growing season. During the growing season, the forest resources of the park, the area of which is 2316.5 ha with an average biological productivity of 5.5 tons per ha, can process carbon dioxide in the volume of 12,740.8 tons. To translate this volume into a valuation, we will use the cost of a carbon dioxide emission quota on the stock exchange. The Intercontinental Exchange currently lists carbon emission quotas for sale at 50.0 euros per metric ton. The exchange rate of the National Bank of the Kyrgyz Republic (as of August 22, 2022) for one euro was 80.73 soms, then the cost of the quota was 4037.0 soms. Referring to the above, we can calculate the economic assessment of carbon dioxide absorption by the forest resources of the SNP "Salkyn-Tor": 12,740.8 × 4037.0 = 51,434.6 thousand soms.

3.4   Determining the Value of the Park's Land Resources

In determining land value, we recommend using the standard and market prices. The area of lands of the SNP "Salkyn-Tor," transferred for unlimited use, is 10,419.3 ha. Of this land, 2069.1 ha (20%) is forest land, 1200.9 ha (11%) is land not covered with forest, 3951.5 ha (38%) is agricultural land (pasture), and 3226.5 ha (31%) is other land.

The value of forest land is considered above in determining the valuation of forest resources. Therefore, all categories of land will be evaluated, except forest land.

3.5   Determination of the Normative Price of Pastures

To determine the pasture lands of the SNP "Salkyn-Tor," we will be guided by the "Procedure for determining the cost evaluation (normative price) of agricultural land," approved by the Decree of the Government of the Kyrgyz Republic No. 33 (January 24, 2013) [9]. The principles of determining the standard price of pastures and hayfields are based on the factors of natural and climatic conditions in conjunction with the quality state of the land. According to our calculations, the total cost of the pastures is 3951.5 ha $\times$ 4300 = 45,197.3 thousand soms.

According to the Law of the Kyrgyz Republic "On specially protected natural areas," lands of specially protected areas are not sold or leased. Since there are currently no special regulations for assessing lands of specially protected natural areas, we will use the "Regulations on the procedure of granting land plots that are in state ownership" [10]. We assume that the lands of the nature park have been provided to legal entities and individuals. Due to the lack of government regulations, we will use data from the municipalities that have leased their land through auctions. On June 1, 2021, the starting price of one hectare of the state fund for redistribution of agricultural land in the municipalities was 4000 soms [11]. We will use this price. Next, we will estimate non-forested (1200.9 ha) and other lands (3226.5 ha), totaling 4427.4 ha. The cost of non-forested and other lands is 4427.4 $\times$ 4000 = 17,709.6 thousand soms. Thus, the total cost of all lands of the SNP "Salkyn-Tor" was 62,907.0 thousand soms.

3.6   Biodiversity Assessment

Currently, 67 bird species, 22 mammals, and 4 reptile species inhabit the park.

To calculate the total value of the wildlife of Nature Park, we will use the amount of tax for calculating the amount of damage caused to hunting resources. In the Kyrgyz Republic, the rates for calculating the number of claims for illegal seizure of game animals were approved in accordance with the Decree of the Government of the Kyrgyz Republic No. 224 (March 3, 2013) [10]. A special place in the animal world is occupied by animals included in the Red Book. The following species of these animals inhabit the SNP "Salkyn-Tor": leopard, saker falcon, marten, bear, and lynx.

The SNP "Salkyn-Tor" keeps species records of animals listed in the Red Book. To evaluate the animals listed in the Red Book, we will use the amount of tax for each individual (head, pelt, and one specimen) regardless of age, size, and weight (in soms), established for the illegal extraction (shooting, trapping) of exotic animals and plants. According to our calculations, the number of exotic animals included in the Red Book in the nature park was 2920.000 soms, while the total amount of wildlife in the SNP "Salkyn-Tor" was 10,780.100 soms.

3.7 Water Resources Assessment

The hydrogeographical network of the park's territory is formed by numerous rivers originating from glaciers and snowfields formed from snow avalanches of the Naryn-Too ridge. These include rivers and brooks: Teke-Sekirik, Alysh, Kur-Sai, Kurgak, Dubeli, and Kuu-Dongoch. To estimate the water resources of the investigated river, we use tariffs of 3 tiyins per 1 m^3 according to the Law of the Kyrgyz Republic "On establishing tariffs for irrigation water supply services for 1999," which is still in force [12]. According to this tariff, the cost of water resources in the park was 6560.8 thousand soms.

The total value of the ecosystem services of the nature park is shown in Table 1. To obtain a summary of the ecosystem services value, we summarize the cost of all kinds of services that we have calculated and report data on additional sources of funding. We included the following amounts from the financial reports of the park: the SNP receives 45.5 thousand soms annually from the sale of seedlings and firewood to the population; the SNP receives 45.1 thousand soms to the budget of the park from the lease of land to tenants.

The calculation result of the all ecosystem services amount show the total SNP "Salkyn Tor" sum is 1 588 251.0 soms.

4 Conclusion

The main scientific conclusions and results based on the conducted research are as follows:

1. The SNP "Salkyn-Tor" is one of the most popular recreational areas in the Naryn district. With an estimated number of 13,013 visitors per year, the projected annual income of the nature park is 9,408,868 soms. According to our calculations, the potential capacity of the nature park is 1,902,278 soms.

Table 1 Results of the economic evaluation of the ecosystem services of the Salkyn-Tor Nature Park

An ecosystem service	Belonging to a type of ecosystem service	Ecosystem value, thousand soms	Applicable evaluation method
Land resources	Providing	62,907.0	Market evaluation, normative evaluation
Absorption of carbon by forests	Regulating	51,434.6	Market evaluation
Tourism and recreation	Cultural	368.3	Market evaluation, transport costs, willingness to pay
Water resources	Providing	6560.8	Market evaluation, normative evaluation
Forest resources	Providing	1,456,108.8	Market evaluation
Biodiversity	Providing	10,780.1	Market evaluation, the amount of the damage tax
Land rentals	Providing	45.1	Contractual evaluation
Selling of seedlings	Providing	23.0	Market evaluation
Timber products (firewood)	Providing	22.5	Market evaluation
Total	1,588,251.0		x

Source Developed by the authors

2. Considered as natural capital, ecosystems have advantages over physical capital, since, if competently managed, they can recover. Nevertheless, like physical capital, natural capital is subject to depletion, reducing future production possibilities. The total amount of ecosystem services provided by the park is 1,588,251.0 thousand soms.
3. It is advisable to evaluate the outcomes of specially protected areas in terms of what their activities improve the environment and preserve and increase national wealth. An analysis of methodological approaches to the valuation of ecosystem services has shown that only a comprehensive consideration of all the functions of these services will provide a complete and adequate assessment.

References

1. Costanza R, D'Arge R, de Groot R, Farberk S, Grasso M, Hannon B, van den Belt M et al (1997) The value of the world's ecosystem services and natural capital. Nature 387:253–260. https://doi.org/10.1038/387253a0
2. Daily GC (ed) (1997). Island Press, Washington, DC

3. de Groot RS, Fisher B, Christie M, Aronson J, Braat L, Haines-Young R, Ring I, et al (2010) Integrating the ecological and economic dimensions in biodiversity and ecosystem service valuation. In: Kumar P (ed) The economics of ecosystems and biodiversity (TEEB): the ecological and economic foundations. Earthscan, London, UK; Routledge, Washington, DC, pp 9–40

4. Rodina EM, Sabyrbekov R, Choduraev TM (2018) Assessment of the ecological and economic potential of natural areas (assessment of ecosystem services): textbook for high schools. Bishkek, Kyrgyzstan

5. Bukvareva EN, Sviridov TV (eds) (2020) Ecosystem services in Russia: a prototype of the national report. In: Biodiversity and ecosystem services: principles of accounting in the Russian Federation, vol 2. Publishing House of the Biodiversity Conservation Center, Moscow, Russia. Retrieved from https://teeb.biodiversity.ru/publications/Ecosystem-Services-Russia_V2_web. pdf. Accessed 20 Aug 2022

6. Razhapbaev MK (2010) Guidelines for assessing the value of forest land in the Kyrgyz Republic. Bishkek, Kyrgyzstan

7. Sabyrbekov R (2017) Express assessment of ecosystem services of the state natural park "Chon-Kemin". Bishkek, Kyrgyzstan

8. Government of the Kyrgyz Republic (2017) Decree "On amendments to the Decree of the Government of the Kyrgyz Republic 'On approval of fees for calculating the size of penalties for damage caused to fauna and flora, mummy-containing mineral raw materials, and mushrooms by legal entities and individuals' dated 3 May 2013 No. 224" (18 Aug 2017 No. 501). Bishkek, Kyrgyzstan. Retrieved from http://cbd.minjust.gov.kg/act/view/ru-ru/100233. Accessed 20 Aug 2022

9. Government of the Kyrgyz Republic (2002) The procedure for determining the valuation (standard price) of agricultural land (4 Feb 2002 No. 47, amended by the Decree dated on 24 Jan 2013 No. 33). Bishkek, Kyrgyzstan. Retrieved from http://cbd.minjust.gov.kg/act/view/ru-ru/ 53146. Accessed 20 Aug 2022

10. Government of the Kyrgyz Republic (2019) Regulations on the procedure for providing state-owned land plots (9 Oct 2019 No. 535, amended by the Decree dated on 14 Apr 2021 No. 146). Bishkek, Kyrgyzstan. Retrieved from http://cbd.minjust.gov.kg/act/preview/ru-ru/157147/35? mode=tekst. Accessed 20 Aug 2022

11. Dobretsova N (2021) The political fate of women in local government. Municipality Mag 6(116):10

12. Kyrgyz Republic (1999) Law "On setting tariffs for irrigation water supply services for 1999" (24 Mar 1999 No. 32). Bishkek, Kyrgyzstan. Retrieved from http://cbd.minjust.gov.kg/act/ view/ru-ru/199?cl=ru-ru. Accessed 20 Aug 2022

The Main Stages of Building a Strategy for the Development of a Regional Socio-Economic System

Vasyli E. Krylov and **Zhanna A. Zakharova**

Abstract The research aims to develop an algorithm for constructing a strategy for the development of a regional socio-economic system and an analytic-geometric method that makes it possible to calculate the parameters of the system (characteristics of its internal environment, considering the influence of external factors) and visualize its development over time. At the beginning of the work, the authors introduced an integral characteristic of the internal environment of the system—the state. Next, the authors determined a graphical model of the development goal of the socio-economic system and introduced the characteristics of the goal. Next, the issues of determining the influence of environmental factors are discussed. Particular attention is paid to the analytical and geometric model of the development strategy of the socio-economic system. This model makes it possible to determine the degree of compliance of the development plan with the actual indicators and assess the possibility of achieving the goal. An important issue of strategizing is the assessment of the resource required by the system to achieve the targets. The corresponding formulas are given in the work. In conclusion, an algorithm for constructing a strategy for a regional socio-economic system is given. It is presented in block diagram form.

Keywords Socio-economic system · Regional socio-economic system development strategy · Internal environment · Internal environment factors · System state · External environment · External environment factors · System development goal

JEL Classification C10 · C53

V. E. Krylov (✉)
Vladimir branch of the Financial University, Vladimir, Russia
e-mail: wek_70@mail.ru

Z. A. Zakharova
Nizhny Novgorod State Technical University named after R.E. Alekseev, Nizhny Novgorod, Russia

1 Introduction

The most important task of managing the regional socio-economic system is the problem of developing a strategy for its development. The optimal strategy ensures the achievement of targets in the shortest time possible. Using the optimal strategy makes it possible to minimize own and attracted resources, as well as to find the points in time at which the use of resources will give the maximum economic effect.

In the Russian Federation, the regulatory and legal framework for the development of a regional strategy is Federal law "On strategic planning in the Russian Federation" (June 28, 2014 No. 172) [1] and Methodological recommendations for the development and adjustment of a strategy for the socio-economic development of a constituent entity of the Russian Federation and an action plan for its implementation approved by Order of the Ministry of Economic Development of the Russian Federation dated March 23, 2017 No. 132 [2].

Several regions of the Russian Federation have adopted development strategies. Examples of these strategies are the development strategy of the Vladimir Region 2030 [3], the development strategy of the Leningrad Region 2030 [4], the development strategy of the Nizhny Novgorod Region 2035 [5], and others.

The developed regional development strategies are based on the application of empirical methods. The subjective attitude of the developer of the strategy to the problem plays an important role in them. The regional development strategy pays little attention to the probabilistic and stochastic nature of the environmental factors in which the system develops. Additionally, there is no feedback and practically no opportunities for adjusting the strategy.

The above problems determined the need to obtain an algorithm for constructing a strategy for developing a regional socio-economic system based on the use of formal, mathematical methods.

This research aims to provide a brief summary of the essence of these methods and define the stages of building a strategy for developing a regional socio-economic system.

2 Methodology

In several works, the authors proposed a system of economic and mathematical methods that make it possible to build a strategy for developing socio-economic systems. The methods are based on space–time diagrams (STD)—geometric objects built in the Cartesian coordinate system tOs, where t is time, and s is the state of the system (the concept of state will be introduced below). The main approaches to the construction of the STD are presented by Krylov [6].

The internal environment of the system is modeled as a continuous function:

$$s = s(t), \tag{1}$$

in which the argument is the time, and the function's value is the state. Let us briefly describe the method for obtaining the form of function (1). This issue is considered in detail in our previous work [7].

Empirical data are collected at discrete times

$$t_1, t_2, \ldots t_n. \tag{2}$$

We adhere to the classical approach, according to which the internal environment of the socio-economic system (in particular, the regional system) is characterized by seven factors: goal, objectives, structure, technology, resources, budget, and human potential.

Each factor is assigned a numerical value

$$f_i = f_i(t_j), 1 \le f_i \le 10. \tag{3}$$

Values (3) of factors in the observed periods of time form the matrix of the internal environment are as follows:

$$F_{IE} = \left(f_i(t_j) \right) = \begin{pmatrix} f_1(t_1) & f_1(t_2) & \ldots & f_1(t_n) \\ f_2(t_1) & f_2(t_2) & \ldots & f_2(t_n) \\ \vdots & \vdots & \ldots & \vdots \\ f_7(t_1) & f_7(t_1) & \ldots & f_7(t_n) \end{pmatrix}, \tag{4}$$

In matrix (4), rows correspond to factors, and columns correspond to time points (2).

Each factor of the internal environment has a different effect on the system; this effect may change over time. We determine the level of influence f_i of the factor at a point in time t_j using the weight

$$p_i = p_i(t_j),$$

wherein

$$\sum_{i=1}^{7} p_i(t_j) = 1.$$

The weights form a matrix are:

$$P = \left(p_i(t_j) \right) = \begin{pmatrix} p_1(t_1) & p_1(t_2) & \ldots & p_1(t_n) \\ p_2(t_1) & p_2(t_2) & \ldots & p_2(t_n) \\ \vdots & \vdots & \ldots & \vdots \\ p_7(t_1) & p_7(t_1) & \ldots & p_7(t_n) \end{pmatrix}. \tag{5}$$

Multiplying matrix (4) by the matrix transposed to (5), we obtain a matrix—a row S of empirical values of the system states at time points (2):

$$S = (s_1 s_2 \ldots s_n) = F_{IE} \times P^t, \; s = s_j = s(t_j) = \sum_{i=1}^{5} f_i(t_j) \cdot p_i(t_j). \qquad (6)$$

Based on the empirical values (6) of the states, the methods of regression analysis (e.g., using the least squares method) determine the type of function (1).

An important stage in the formation of a strategy is modeling the development goal of a socio-economic system [8]. The target model is built in the Cartesian coordinate system tOs in the form of a figure T.

Let us list the parameters of the goal (in terms of space–time diagrams).

1. The time interval for achieving the goal:

$$\Delta t_T = t_T^{max} - t_T^{min}.$$

2. Interval of desired system states:

$$\Delta s_T = s_T^{max} - s_T^{min}.$$

3. Equation of the boundary of the region T. It has the following form:

$$f = f_T(t; s). \qquad (7)$$

The boundary of the region T can be represented as a combination of two functions $f(t)$ and $g(t)$, corresponding to the condition: $f(t) \geq g(t)$.

4. Coordinates of the geometric center (barycenter) of the region T:

$$t_c = \frac{1}{S_T} \int_a^b t \cdot (f(t) - g(t)) dt,$$

$$s_c = \frac{1}{S_T} \int_a^b \frac{f(t) + g(t)}{2} \cdot (f(t) - g(t)) dt$$

5. T figure area:

$$S(T) = \iint_T f_T(t; s) dt ds,$$

$$S_T = \int_a^b (f(t) - g(t)) dt$$

The development of the socio-economic system occurs in the external environment. The influence of the external environment is manifested in the change in the rate of the system's development. If the environment favors the development of the system, it accelerates its development and acts (in terms of SWOT analysis) as an opportunity. On the contrary, the external environment can hinder the development of the system and act as a threat. In this case, the rate of the system's development slows down to a complete stop.

The influence of the external environment on the system will be modeled in the form of influence coefficients. The coefficient of influence determines the degree of influence of the environmental factor on the system F_i, which is in the state s_l at the moment of time t_j. It can take values from $-\infty$ to $+\infty$,

$$-\infty < k_{F_i}^{t_j, s_l} < +\infty.$$

If $k_{F_i}^{t_j, s_l} < 0$, then the factor F_i is perceived as a threat. If $k_{F_i}^{t_j, s_l} > 0$, then F_i is a possibility. In the case $k_{F_i}^{t_j, s_l} = 0$, it is stated that the factor F_i is neutral and does not affect the system.

The set of values of the influence coefficients forms a matrix $F_i = \left(k_{F_i}^{t_j, s_l} \right)$.

$$F_{EE}^r = \sum_i F_i = \sum_i \left(k_{F_i}^{t_j, s_l} \right) = \left(\sum_i k_{F_i}^{t_j, s_l} \right), \tag{8}$$

equal to the sum of matrices for each environmental factor. The constructed matrix F is an analytical model of the external environment. Each of its elements represents the coefficient of influence of the external environment, corresponding to a certain state of the system at a given time.

Based on the available empirical data, for each possible state of the system (for simplicity, it is assumed that the system assumes its states at discrete times), we construct an exponential function:

$$f\left(k_{F_i}^{t_j, s_l} \right) = a^{k_{F_i}^{t_j, s_l}}. \tag{9}$$

Then, if v_0 is the initial rate of the system's development, then under the influence of environmental factors, it changes and becomes equal to

$$v_1 = f\left(k_{F_i}^{t_j, s_l} \right) \cdot v_0.$$

Note that the rate of development of the system can be defined as the derivative of function (1) with respect to time:

$$v = s'(t).$$

Matrix (8) [in the retrospective case built on empirical data and in the prospective built by predicting the coefficients of influence according to formula (9)] can be represented as a space–time diagram.

In the diagram, equal values of influence coefficients are united into zones highlighted in the same color. Black color corresponds to impassable zones, $f\left(k_{F_i}^{t_j,s_l}\right) = 0$.

The development strategy of the socio-economic system is modeled as a directed line, the graph of which is the graph of the function (1).

Note that during the initial planning of the strategy, the influence of environmental factors is not considered. When modeling a strategy, considering the influence of environmental factors, it is necessary to correct the initial line U.

A very important step in building a development strategy for the socio-economic system is to check the achievement of target indicators using the chosen strategy. Note that the existing methods for constructing a development strategy do not provide for such a check, which is necessary. After all, if it turns out that the goal has not been achieved during the implementation of the strategy, then it is necessary to attract an additional (preferably accurately calculated) resource necessary for the system to achieve the goal using an adjusted strategy.

Thus, in terms of space–time diagrams, to achieve the goal of development of the socio-economic system with the help of the chosen strategy, the following conditions must be met.

1. The time t_1 the goal must belong to the following interval:

$$t_T^{\min} \le t_1 \le t_T^{\max}.$$

2. The final state of the system s_1 must belong to the following interval of target indicators:

$$s_T^{\min} \le s_1 \le s_T^{\max}.$$

3. Equation

$$f_T(t) = s(t)$$

has a solution. In this case, the solution of the equation satisfies the condition: $t_1 \le t \le t_2, s_1 \le s \le s_2$. If $t = t_1, s = s_1$, then the system reaches the goal at its boundary.

4. It is advisable to model the final state of the system not as a point but as areas (zones) of uncertainty. Thus, under certain circumstances, the system will be able to achieve the targets with some probability. The probability p_T of this event can be calculated using the geometric definition of probabilities:

$$p_T = \frac{S(T \cap D)}{S(D)}.$$

$S(D)$ is the area of the zone of uncertainty and $S(T \cap D)$ is the area of the common part of the regions D and T.

Building a strategy is impossible without assessing the resource required for its implementation. In terms of spatio-temporal diagrams [9], the formula for the total resource R required to implement the development strategy of the regional socio-economic system is as follows:

$$R = R_U + R_{on}^+ + R_{on}^-.$$

where R_U is the volume of resources allocated for the implementation of the strategy. Its value is as follows:

$$R_U = \sum_i R_{U_i} = K_1 \cdot \sum_i k_{U_i} \cdot l_{U_i},$$

where the summation is carried out over all sections with constant values of the coefficients of the influence of the external environment; K_1 is the coefficient of dimension, and k_{U_i} is the value of the non-linear function

$$g = g\left(k^{t,s}\right) = k_U,$$

l_{U_i} is the length of the section of the model of the development strategy of the regional socio-economic system, corresponding to a constant value of the influence coefficient,

$$l_U = \int_{t_0}^{t_t} \sqrt{1 + (s'(t))^2}\,dt.$$

The amount of operational, additional resource R_{on}, consists of two parts: R_{on}^+ allocated to the system to adjust the strategy, and the resource R_{on}^- released in the presence of favorable circumstances. The volume of the operational resource is determined by the following formula:

$$R_{on} = \sum_j R_{on,t_j} = \sum_j K_2 \cdot \Delta k_v^{t_j} = K_2 \sum_j \Delta k_v^{t_j},$$

where K_2—the coefficient of dimension, depending only on the characteristics of the socio-economic system, and $\Delta k_v^{t_j}$—an indicator of changes in environmental conditions, determined by the formula:

$$f = f\left(k^{t,s}\right) = k_v.$$

3   Results

Let us now turn to the description of the main stages of building a strategy for the development of a regional socio-economic system. The algorithm's complexity lies in the fact that it is not linear. Some steps must be performed in strict sequence. Others are in parallel. The general scheme for building a strategy is shown in Fig. 1.

When designing a strategy, the following steps must be taken.

First, it is necessary to determine the values (2) of the time intervals in which the necessary empirical information will be collected.

Next, we model the internal environment of the system. Using the procedures described above, we set the type of function (1).

We are building a model of the development goal of the socio-economic system.

Finally, simulating the internal environment of the system, we establish the initial time t_0 and the initial state s_0 of the strategy implementation.

We analyze the external environment of the system. Based on the available empirical data, we determine the coefficients of influence [we form matrix (8)] and set the type of function (9). We build a space–time diagram—a retrospective graphical model of the external environment. Based on the data obtained, we obtain a promising model of the external environment. After building, we need to adjust the target. The adjustment consists in changing the function (7) if the initial target model is superimposed on insurmountable sections of the external environment.

Drawing the tangents to the figure T through the point s_0, we get the phase space F_{EE}^p—the figure on the space–time diagram, inside which the strategy will be built.

From the point s_0, we build a strategy U for the development of the socio-economic system. If necessary, we can build several options U_i for the strategy.

The next step is to check the conditions for achieving the goal. If the goal is achieved, then proceed to the next stage. Otherwise, it is necessary to either adjust the goal, change the strategy U, or consider another strategy variant U_i.

Finally, resources are estimated. If it is enough to implement the strategy option, then it is taken into work. Otherwise, it is necessary to adjust the strategy so that the volume of available resources is sufficient for its implementation.

The space–time diagram corresponding to the described algorithm is shown in Fig. 2.

4   Conclusion

Building a strategy for the development of a regional socio-economic system is an important and very urgent task. The system of methods proposed in the research offers a new, different from the already established, approach to its solution. Distinctive features of the proposed system of methods from the known ones are as follows.

1. A formal, mathematical approach to solving the problem;
2. Minimizing the human factor in developing a strategy;

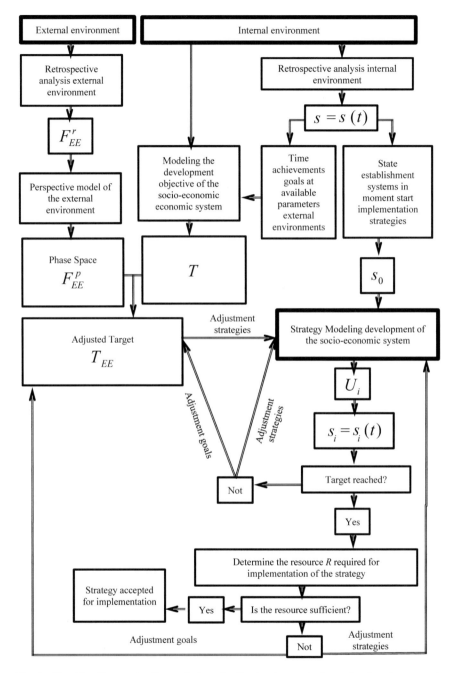

Fig. 1 Algorithm for constructing a strategy for the development of a regional socio-economic system. *Source* Developed by the authors

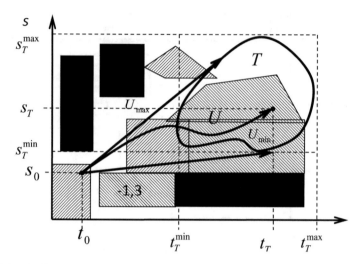

Fig. 2 Model of the development strategy of the regional socio-economic system. *Source* Developed by the author

3. The ability to vary the available resources and make operational adjustments to the strategy as it is implemented when receiving information about changes in the external and internal environment of the socio-economic system.

References

1. Russian Federation (2014) Federal law "On strategic planning in the Russian Federation" (28 Jun 2014 No. 172-FZ). Legal Reference System "Consultant Plus", Moscow, Russia. Retrieved from https://www.consultant.ru/document/cons_doc_LAW_164841/. Accessed 1 Jun 2022
2. Ministry of Economic Development of the Russian Federation (2017) Guidelines for the development and adjustment of the strategy for the socio-economic development of the subject of the Russian Federation and the action plan for its implementation (approved by order on 23 Mar 2017 No. 132). Legal Reference System "Garant", Moscow, Russia. Retrieved from https://www.garant.ru/products/ipo/prime/doc/71542236/. Accessed 1 Jun 2022
3. Government of the Vladimir Region (2009) Strategy for socio-economic development of the Vladimir Region until 2030 (approved by decree of Governor of the Vladimir Region on 2 Jun 2009 No. 10). Legal Reference System "Garant", Vladimir, Russia. Retrieved from https://base.garant.ru/19416240/53f89421bbdaf741eb2d1ecc4ddb4c33/. Accessed 1 Jun 2022
4. Legislative Assembly of the Leningrad Region (2019) On amending the regional law "On strategy for social and economic development of the Leningrad Region until 2030" (3 Dec 2019 No. 100-oz). Committee for Economic Development and Investment Activities of the Leningrad Region, St. Petersburg, Russia. Retrieved from https://econ.lenobl.ru/ru/budget/planning/concept2030/. Accessed 1 Jun 2022
5. Government of the Nizhny Novgorod Region (2018) Resolution "On approval of strategy for socio-economic development of the Nizhny Novgorod Region until 2035" (21 Dec 2018 No. 889, as amended on 31 Dec 2021). Electronic fund Legal and Regulatory Technical Documents,

Nizhny Novgorod, Russia. Retrieved from https://docs.cntd.ru/document/465587311. Accessed 1 Jun 2022

6. Krylov VE (2012) Space-time diagram in the strategic management of socio-economic systems. J Dyn Comp Syst—XXI Century 6(4):35–41

7. Krylov VE (2021) Application of STD—Methodology in SWOT—Analysis. In: Proceedings of the GCPMED 2020: 3rd international scientific conference "global challenges and prospects of the modern economic development", pp 931–938. https://doi.org/10.15405/epsbs.2021.04.02.111

8. Krylov VE (2020) Analytical approach to development goal modeling for the socio-economic system. In: Proceedings of the GCPMED 2019: 2nd international scientific conference "global challenges and prospects of the modern economic development". Samara State University of Economics, Samara, Russia, pp 1511–1518. https://doi.org/10.15405/epsbs.2020.03.217

9. Krylov VE (2021) Approaches to assessing the resource allocated for the implementation and adjustment of the development strategy of the socio-economic system. Econ Manag Control Syst 2(40):34–40

Digital Technologies of the Project "Moscow 'Smart City—2030'": The Transport Sector

Aleksandr A. Matenkov⬤, **Ruslan I. Grin**⬤, **Markha K. Muzaeva**⬤, **and Dali A. Tsuraeva**⬤

Abstract The research deals with the priority areas of digitalization in the transport sector in interpreting the strategy "Moscow 'Smart City—2030'." The research aims to study the priority areas of digitalization of transport flows of the metropolis and the potential impact of digitalization on the functioning of the territory. By applying the methods of content analysis and the regulatory-legal method in the research, the authors assessed the position of the city authorities on the most sought-after areas of innovation in the transport sector and determined the composition of socio-economic benefits of digitalization of the transport sector. The analysis of statistical indicators of the development of the transport sector of the Moscow urban agglomeration has confirmed the growing need to improve the efficiency of transport infrastructure in the broad sense, including an increase in the level of connectivity of the city districts and the level of sustainability of the transport system. The results show certain disproportions between the priority areas of transport development and the actual needs of the urban infrastructure, as well as the presence of significant legal constraints in implementing uncrewed transport concepts. It is demonstrated that there is a certain consensus between the municipal authorities and the population on the issue of assigning the transport sector among the priorities for implementing digital technology. The specifics of the metropolitan area (high concentration of capital and innovation activity) allow for considering Moscow as a model example of the introduction of innovative technologies. In this regard, it is necessary to optimize the legal restrictions on the introduction of innovations in the field of transport (on the model of a legal sandbox, Regulatory Sandbox).

Keywords Digitalization · Smart city · Transport · Moscow · Digital technologies · Innovations

JEL Classification D6 · E2 · E6 · H4 · O1 · O3 · P3 · R4

A. A. Matenkov (✉) · R. I. Grin · M. K. Muzaeva · D. A. Tsuraeva
MGIMO University, Moscow, Russia
e-mail: matenkov2012@yandex.ru

© The Author(s), under exclusive license to Springer Nature Switzerland AG 2023   441
E. G. Popkova (ed.), *Smart Green Innovations in Industry 4.0 for Climate Change Risk Management*, Environmental Footprints and Eco-design of Products and Processes, https://doi.org/10.1007/978-3-031-28457-1_45

1 Introduction

The impact of the development of Russian megacities on the state of the transport sector is manifested in the overall growth of the load on the infrastructure, along with the increasing need to accelerate the movement of people (including—as active participants in the labor market and consumers) and goods. For the Russian Federation, the problems of improving the efficiency of transport infrastructure management in large metropolitan areas play a particularly important role in the context of imbalances in territorial development and related consequences for the national economy. At the turn of 2020, Russia has 20 large urban agglomerations (according to the Government of the Russian Federation, about 40 urban agglomerations [1]), which account for over 30% of the population and about 40% of GDP [2]. In general, the trend of localization of the economically active population and capital in large megacities is a worldwide trend. PricewaterhouseCoopers estimates that by 2030, 24% of the world's population will live in agglomerations with a population of more than 1.5 million people. Additionally, PricewaterhouseCoopers states that the contribution of agglomerations to global GDP will increase from 38% in the late 2010s to 43% in 2030 (a quarter of the world's population is predicted to live in cities by 2030) [3].

The Moscow agglomeration is the largest Russian agglomeration. It unites the city of Moscow and the surrounding territories. It is highly probable that in the foreseeable future, Moscow and the Moscow agglomeration will continue playing the role of a center of attraction of resources of all forms with the corresponding consequences in the form of increased load on all elements of infrastructure, logistical, and operational risks. This argument logically leads to the recognition of the importance of theoretical and methodological issues of infrastructure development of agglomerations with a focus on improving their efficiency and carrying capacity. The introduction of digital technologies in the context of the project "Moscow 'Smart City—2030'" implemented by the Moscow authorities is considered a tool to solve this problem.

2 Methodology

To achieve the research goal, the work with sources and literature relied on general scientific methods of analysis and synthesis. Given that the practice of using information technology in the field of transport is studied based on program and strategic documents of the federal level and the level of the subject of the federation, the methods of legal analysis were in demand. The study of the practice of using information technologies in the sphere of transport in Moscow is based on the application of methods of content analysis and historical analysis to analytical materials and statistical collections.

3   Results

The Moscow agglomeration is among the largest in the world. Using the normative-legal approach to considering the essence of agglomerations, we will consider them as "a set of compactly located settlements and territories between them, connected with the joint use of infrastructure facilities and united by intensive economic, including labor, and social ties" [1]. Attempts are made (e.g., within the framework of the general plans of development of Moscow and the Moscow Region) to indirectly influence the dynamics of migration inflow to Moscow agglomeration ("pull away" part of the population). Nevertheless, as the statistics show, these attempts are not successful. The logical consequence of placing Moscow among the largest on a national and global scale is the high social, administrative, and financial potential, adjacent to the excessive load on the transport system. In a broad sense, infrastructure is understood as a set of certain institutions (organizations), norms and rules by which they function, and technical systems that support their activities. The purpose of infrastructure is to ensure the functioning of the agglomeration in a variety of areas of this functioning.

A characteristic feature of agglomerations as areas of population concentration is a progressive increase in the load on the elements of infrastructure: social, transport, household, and housing. This increases logistics costs and operational risks and requires improving the quality of management and design of transport and urban infrastructure using digital technology [4].

According to the estimates of international organizations, relatively recently, Moscow was among the world's top five megacities. Moscow was one of the world's five largest megacities with a high load on the transport system [5]. At the turn of the 2020s, the city managed to improve the efficiency of the transport infrastructure significantly. A significant role in this improvement was played by the consistent actions of the city authorities to integrate information technology into the transport infrastructure. The current stage of development of these efforts is included in the priorities of the project "Moscow 'Smart City—2030'".

First, it is necessary to illustrate the context of the practice of implementing information technology in the transport sector in Moscow. As the political, administrative, and economic center of the Russian Federation, Moscow is one of the relatively few constituent entities of the federation capable of covering its expenditures with budget revenues (with revenues exceeding 3 trillion rubles in 10 months of 2022). Moscow is the location of the head offices of major companies (including innovative and communications industry companies) and the attraction region for significant flows of incoming migration. Moscow is also characterized by a relatively high level of development of information and communications infrastructure. A 2021 survey conducted among 13.5 thousand respondents in a number of European countries, including the Russian Federation, showed a high level of digital technology penetration in everyday life (Cisco Broadband Index Survey) [6].

Until 2010, the transport situation in Moscow was close to critical: the road network had reached its maximum capacity, and Moscow had one of the worst traffic situations in the world. That is why, in 2011, the Government of Moscow City and leading Russian and international experts worked out the State Program for Moscow Transport Development until 2020. The plan focuses on analyzing a large amount of commuter data to reduce the road load through a strategic approach to modernization and new construction, as well as the launch of the Intelligent Transport System (ITS). Subsequently, program activities in the field of transport development were integrated into the project "Moscow 'Smart City—2030" [5].

High budget possibilities allow the city authorities to develop the transport network intensively, but the number of cars registered in the Moscow transport hub is outstripping even the record pace of road construction [5]. There remains an imbalance between the residential and working areas in terms of load and requirements to transport infrastructure: 59% of jobs are located within the Third Ring Road, but only 9% of the population lives there. There are about 3.7 million cars daily on Moscow roads; in conjunction with the Moscow region, this figure reaches 8.4 million cars, creating a huge load on the street and road network [7]. That is why finding and implementing innovative solutions for traffic management that meet the growing mobility needs of Muscovites and visitors to the capital remains one of the key objectives of the development of urban transport infrastructure.

Already in the 2010s, Moscow made significant progress in introducing digital technologies in the field of transport. The Intelligent Transport System (ITS) has been in operation in Moscow since 2011. Initially, it covered 30% of the city's territory, reaching 100% by 2017 [8]. ITS is a comprehensive monitoring system for managing traffic and public transport. In 2013, the Moscow Traffic Management Center launched the Control Center, which analyzes data from the equipment installed throughout the city—speed sensors, adaptive traffic lights and traffic safety cameras, monitored surveillance cameras, and GPS/GLONASS sensors on public transport [9]. Thus, we can judge about the formed complex of hardware and software integrated into the transport system of the city. The intelligent transport system of Moscow tracks 10,000 ground vehicles, more than 72,000 cabs, and 11,000 cars in the car-sharing network [8]. The Traffic Control Center is the largest in Europe. Every day, the Traffic Management Center receives more than 350 million packets of data from various locations, including the following:

- 80 million trips;
- 45 million speed measurements from sensors;
- More than 60 million records of vehicle telematics data in the Regional Navigation Information System (RNIS).

The results of the introduction of information technology in the sphere of transport in Moscow indicate objectively high social and economic usefulness, including the following:

- The average speed of private transport within the transport infrastructure of Moscow will increase in 2019–2020 to the level of 2010 by 20%;
- Punctuality and reliability of ground transportation services using dedicated infrastructure reached 97% in 2019–2020 (compared to 76% in 2010);

- 42% reduction in traffic fatalities (down to 2.9 deaths per 100,000 residents) compared to 2010;
- 20% increase in average speed from 2010 to 2019 [5].
- Reducing the number of traffic accidents by more than four times in 2019 compared to 2010;

Moscow is at the forefront of change, introducing the most advanced technology and the best national and international innovations. The introduction of the intelligent transport system has led to a seismic change in the traffic situation thanks to smart traffic lights, a network of cameras and sensors that analyze and regulate traffic flows, and other IT solutions that often remain hidden from the view of Moscow residents. With the development of the Internet of Things and computer modeling, it became possible to create a digital model of any physical object. A digital twin of Moscow is now being developed. A digital twin is a prototype of a real city, by which one can analyze the real situation on the roads, providing a reaction to possible changes and external influences [10]. This is an accurate reflection of the city in the digital realm, with information from various sensors, monitoring systems, and resource meters. A dynamic traffic model is already working in Moscow, enabling real-time assessment of the traffic situation, making short-term forecasts, and informing the residents 24/7. Muscovites receive targeted SMS and push notifications to their smartphones about changes in the operation of public transport and traffic situation. The information is based on a specific person's transportation behavior, profile, and situational triggers. Currently, more than 4.5 million Moscow residents have received the information.

One of the distinctive features of Moscow as a territory of the implementation of information technologies in the sphere of transport is also the achievement of consensus between the city authorities and the population on the issue of digitalization. As follows from the results of the survey of Moscow respondents on the priority areas of digitalization within the framework of the smart city system, transport is among the five areas in which the population is already actively using the achievements of digitalization (Fig. 1).

In the second half of the 2010s, the infrastructure for implementing information technology in the urban transportation system expanded. In 2017, the Transportation Security Management Center was opened. It receives data from all CCTV cameras in the metro and has access to cameras in the Moscow Metro. Currently, more than 7700 security officers are on duty at stations and metro entrances. There are emergency call points at all stations; security posts with specialized equipment to detect prohibited items and substances are installed at subway entrances. The comprehensive approach applied in 2017 made it possible to reduce the number of crimes committed in the subway by 35% compared to the previous year and the number of administrative offenses by 21% [8].

One of the integrated technological solutions in the transport sector in Moscow is the concept of Mobility as a Service or Vehicles as a Service (MaaS) [11]. In its essence, this concept means the abandonment of personal transport in favor of public or rental vehicles. Cabs and car-sharing platforms like Yandex.Drive, Citymobil, Delimobil, BelkaCar, and others, as well as electric scooter rental companies Urent,

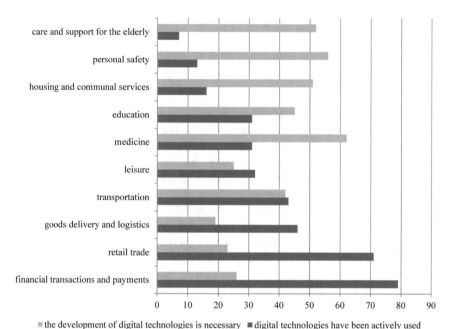

Fig. 1 The results of the survey of Moscow respondents on the priority areas of digitalization within the smart city system. *Source* Calculated and built by the authors

Whoosh, and Samokat Sharing, form the infrastructure support for this concept based on their own technological solutions. A further logical step for the city authorities was to create a unified mobile application based on the MaaS concept, uniting all types of public transport and providing the opportunity to create a multimodal route. All operators of cabs, car-sharing services, bicycle rentals, and scooters were invited to join the platform. In 2021, the platform "Moscow Transport" was launched, which partially provides access to the infrastructure based on MaaS. This application allows users to lay out a convenient route, pay once, choose a suitable mode of transport, and immediately get all accompanying information about the trip, restrictions, closures, and tariffs of different operators [1].

The development of uncrewed transport is among the ambitious tasks of developing Moscow's transport infrastructure. As part of the calculation of the Autonomous Vehicles Readiness Index (AVRI), KPMG surveys of experts in the Russian Federation on the perception of uncrewed technologies put the country in fourth place after India, Mexico, and the United Arab Emirates [12]. The Department of Information Technology of Moscow assessed public opinion, including the issues of uncrewed mobility, as part of the evaluation of the prospects of the strategic plan for the megalopolis "Concept of Moscow 2030." The results of the survey of more than five thousand Moscow respondents aged 18–65 years showed that the city population, in general, is actively using innovative technologies in their daily lives.

Simultaneously, 37% of respondents indicated that they would like to see uncrewed transport in the digital city of the future [11].

One of the few studies on the problems of smart cities and the development of uncrewed mobility in the Russian Federation is the work of E. N. Yadova and P. A. Levich, which reflects the results of a questionnaire survey of 2314 respondents by spontaneous sampling [9]. The study showed results that are natural for the sample: 89.4% of respondents were positive or rather positive about the prospects of introducing uncrewed cars into everyday life, and 78.9% showed a willingness to purchase or use such technology.

The results obtained by the researchers allow for illustrating one of the critical dependencies for promoting the smart city concept. It consists of the fact that respondents tend to give a more positive assessment of the implemented technologies in the presence of evidence-based benefits from the introduction of innovations. A controversial characteristic from the point of view of sample representativeness is the research of E. N. Yadova and P. A. Levich, which used the method of questioning by placing questionnaires on the Internet. It is apparent that respondents with access to network technologies (due to geographical, social, or economic factors) initially demonstrate a higher predisposition to assess the smart city concept positively. This indicates a higher quality of human capital, skills, and willingness to use digital capital in the population of the most urbanized regions of the country [13].

One of the significant limitations in the further development of information technology in the smart city project of Moscow is the ability of the municipal authorities to ensure the maintenance and further development of the infrastructure in the context of high dependence on imports of high-tech equipment and the imposed sanctions restrictions. As follows from the reporting materials of the Department of Transport of Moscow, as of 2020–2021, the functioning of the infrastructure relied on the following elements:

- 180,000 units of surveillance cameras;
- 2960 units of recording equipment;
- More than 40,000 units of traffic control devices;
- 3900 units of road transport scanners [5].

On the scale of the software and hardware complex (considering the need to maintain it in working order, modernization, and renewal), we can imagine the challenges facing the city authorities in terms of ensuring the infrastructure's functioning [14]. The level of personnel training with regard to the requirements of the digital economy and the lack of investment in the knowledge economy are considerable technological challenges, the overcoming of which will help remove the constraints on the implementation of these projects [15].

4 Conclusion

In the context of the arguments of the growing load on the transport infrastructure of Moscow, the priorities selected by the city authorities for the introduction of information technology in the transport sector seem relevant and justified. The promotion of uncrewed transport, intelligent transport system, and transport sharing are positioned by the municipal authorities as the key priorities for developing information technology in the transport sector. In a broader format, the goals of introducing digital technologies in the transport sector of Moscow are also aimed at reducing the need for physical transport channels (digital technologies as a tool for equating virtual presence with real presence, promoting logistics services and services, and abandoning the use of personal transport) and achieving the goals of improving the quality of life and protecting the environment. Moscow has a relatively favorable legal and institutional environment for implementing information technologies in the transport sector: a relatively high standard of living for the population, access to capital, and the ability of the city authorities to promptly make changes to the legal environment and finance capital-intensive projects.

5 Recommendations

The current development of the transport sector in Moscow is characterized by a high level of penetration of information technology, with the exception of uncrewed transport. Their further penetration and the ability to achieve the goals set out in Strategy "Moscow—2030" largely depend on the state of the regulatory environment (restrictions on uncrewed transport), the dynamics of investment activity, the speed of social change, and the availability of technology (digitalization of the transport sphere requires appropriate infrastructure development in conditions of high dependence on imports of knowledge-intensive products). As tools to remove these constraints, it is advisable to increase the mobility of city authorities in the field of legal regulation of innovations in the transport sector (on the model of the legal sandbox, Regulatory Sandbox), strengthen links of production and innovation centers (based on networking principles, active use of technology parks and business incubators), and create conditions for the development of public–private partnerships, including in the field of digital education.

References

1. Government of the Russian Federation (2022) Decree "On approval of the strategy of spatial development of the Russian Federation until 2025" (13 Feb 2019 No. 207-r, as amended on 30 Sep 2022). Moscow, Russia. Retrieved from https://docs.cntd.ru/document/552378463. Accessed 10 Dec 2022

2. Polidi T (2017) Liberation of cities: How agglomerations will help Russian economy. RBK Daily. Retrieved from https://www.rbc.ru/opinions/economics/11/10/2017/59dde2ce9a79475 a5f5e5df5. Accessed 10 Dec 2022

3. Demidova A, Gubernatorov E (2017) A quarter of the world's population got predicted life in giant cities by 2030. RBK Daily. Retrieved from https://www.rbc.ru/business/06/07/2017/595 df2c19a794776e863d1b3. Accessed 10 Dec 2022

4. Ivanitskaya NV, Baybulov AK, Safronchuk MV (2020) Modeling of the stress-strain state of a transport tunnel under load as a measure to reduce operational risks to transportation facilities. J Phys: Conf Ser 1703(1):012024. https://doi.org/10.1088/1742-6596/1703/1/012024

5. Ministry of Transport of the Russian Federation (2020) Building a transport system of the future: the traffic control center's performance report 2020. Retrieved from https://www.polisn etwork.eu/wp-content/uploads/2021/02/MTCC_EN.pdf. Accessed 10 Dec 2022

6. Bakhur V (2021) Cisco broadband index survey: Russians consider Internet access no less important than utilities. CNews. Retrieved from https://www.cnews.ru/news/line/2021-07-19_ issledovanie_cisco_broadband_index. Accessed 12 Dec 2022

7. Autonews (2021) Authorities named the total number of cars in Moscow. Retrieved from https:// www.autonews.ru/news/61c853cb9a794703b66ac3d4. Accessed 10 Dec 2022

8. Department of Transport of Moscow (2017) Digitalization of Moscow transport: department of transport of Moscow. Retrieved from https://report2010-2017.transport.mos.ru/pdf/ar/en/ mega-projects_digitalization.pdf. Accessed 10 Dec 2022

9. Yadova EN, Levich PA (2020) Analysis of preparedness to the modern (or up to date) technologies in conceptual frame of STS and RRI. Technologos 2:25–41. https://doi.org/10.15593/ perm.kipf/2020.2.03

10. RAI Amsterdam (2021) Three smart cities in traffic management: Perth, Moscow, Mexico City. Retrieved from https://www.intertraffic.com/news/traffic-management/three-smart-cities-in-traffic-management-perth-mexico-city-moscow/. Accessed 10 Dec 2022

11. Department of Information Technology of Moscow (2018) Concept of Moscow 2030. Retrieved from https://2030.mos.ru/netcat_files/userfiles/documents_2030/opros.pdf. Accessed 10 Dec 2022

12. KPMG (2020) Autonomous vehicles readiness index. Retrieved from https://home.kpmg/xx/ en/home/insights/2020/06/autonomous-vehicles-readiness-index.html. Accessed 12 Dec 2022

13. Safronchuk MV, Sergeeva MV (2021) The concept of economic growth through digital economy perspective. In: Popkova EG, Sergi BS (eds) Modern global economic system: evolutionary development vs. revolutionary leap. Springer, Cham, Switzerland, pp 1264–1271. https://doi.org/10.1007/978-3-030-69415-9_138

14. Ivanov OV, Shamanina EA (2021) PPP as a tool to achieve sustainable development goals and implement the concept of "Quality infrastructure investments". In: Zavyalova EB, Popkova EG (eds) Industry 4.0: exploring the consequences of climate change. Palgrave Macmillan, Cham, Switzerland, pp 309–322. https://doi.org/10.1007/978-3-030-75405-1_28

15. Safronchuk MV, Ivanitskaya NV, Baibulov AK (2022) Global labor market and challenges of digitalization. In: Popkova EG (ed) Imitation market modeling in digital economy: game theoretic approaches. Springer, Cham, Switzerland, pp 142–150. https://doi.org/10.1007/978-3-030-93244-2_17

The Impact of the COVID-19 Pandemic on the Economic Security of Russian Regions: Assessment of Resistance to the Spillover Effects of Epidemic Risks

Nadezhda V. Kapustina⬛, Yury V. Kuznetsov⬛, Nadezhda V. Pilipchuk⬛, Elena S. Materova⬛, and Ekaterina V. Lisova⬛

Abstract The research aims to identify the spillover effects of epidemic risks for the economic security of Russian regions and propose a system of measures for their leveling at the stage of post-pandemic recovery. The research substantiates the hypothesis that epidemic risks of economic security have spillover effects on the economy of Russian regions, increasing the impact of a combination of traditional risks and provoking the emergence of new risks. The authors proposed a system of indicators and a methodology for assessing the stability of regional economic systems to the spillover effects of epidemic risks based on the use of a statistical method for calculating integral indicators. The results of the calculations made it possible to rank the regions of Russia by the level of resistance to the spillover effects of epidemic risks to economic security. It is proved that the Southern Federal District has the greatest resistance to spillover effects. Ranking of regions by the indicator of resistance to spillover effects allowed the authors to propose a model of economic security management for two types of regional economic systems. For regions resistant to spillover effects, it is recommended to implement a set of measures aimed at strengthening internal potential. For regions unstable to the spillover effect, a sequence of actions is proposed based on leveling the negative impact of the spillover effects and the subsequent implementation of measures to strengthen internal potential.

N. V. Kapustina (✉)
Financial University Under the Government of the Russian Federation, Moscow, Russia
e-mail: economresearch@mail.ru

Y. V. Kuznetsov
St. Petersburg State University, St. Petersburg, Russia
e-mail: y.kuznetsov@spbu.ru

N. V. Pilipchuk
Tver State University, Tver, Russia

E. S. Materova
Samara State University of Economics, Samara, Russian Federation

E. V. Lisova
ANO HE "Institute of Business Career", Moscow, Russia

© The Author(s), under exclusive license to Springer Nature Switzerland AG 2023
E. G. Popkova (ed.), *Smart Green Innovations in Industry 4.0 for Climate Change Risk Management*, Environmental Footprints and Eco-design of Products and Processes, https://doi.org/10.1007/978-3-031-28457-1_46

Keywords Economic security · Region · Regional economic system · Pandemic · Epidemic risks · Spillover effects · Risk leveling · Post-crisis recovery

JEL Classification R11 · R13 · R58

1 Introduction

The development of regions is determined by the cumulative influence of factors that differ in nature and impact. Regional farms have learned to adapt to the adverse effects of financial and economic factors in the process of evolutionary development (to predict, analyze, evaluate, and level risks). Evidence of this is the overcoming of turbulence by the regions of Russia caused by financial crises, for example, the global financial crisis of 2008–2009 or the economic crisis in the Russian economy of 2014 [28, 29]. During the development of digitalization processes, it also became clear how technological factors can influence regional economic systems. To bridge the digital divide with other regions, public authorities connected all available tools, stimulating economic entities to expand digital infrastructure and use digital technologies [11, 15]. However, the world has many faces in terms of the risks that exist in it. Thus, 2020 was the beginning of the widespread impact of epidemic risks. Rapidly spreading in the world, the new coronavirus infection has endangered the life and health of the world's population and created prerequisites for a sharp weakening of the economic security of countries and regions [8, 10, 12]. The costs for regional economies are still difficult to estimate due to the high differentiation of territories and the current procedure for considering risk factors [20, 21].

2 Methodology

Various structural elements can be distinguished in the structure of the economic security of the region. The state of these elements is determined by the influence of financial, economic, political, social, and other risks [6, 9, 14, 17, 23, 27]. In the traditional sense, the economic security of the regions is free from the impact of epidemic risks. Until 2020, Russian regions did not experience such risks. However, the rapid spread of COVID-19 has affected the state of economic systems in all regions [6, 18]. Significant pressure on regional budgets was exerted by the emergency financing of the healthcare system caused by the need for prevention and medical care for infected people, which also affected the reduction of the economic security of the regions [16].

The authors propose two hypotheses:

- Hypothesis (1). Epidemic risks of economic security cause spillover effects in the economy of Russian regions, increasing the impact of a combination of traditional risks and provoking the emergence of new risks.

- Hypothesis (2). Assessment of the stability of regional economic systems to the spillover effects of epidemic risks makes it possible to differentiate the priorities of regions in managing economic security.

The research aims to identify the spillover effects of epidemic risks for the economic security of Russian regions and propose a system of measures for their leveling at the stage of post-pandemic recovery.

The research objectives are as follows:

1. To systematize the risks of regional economic security;
2. To analyze the spillover effects of epidemic risks to regional economic security;
3. To assess the resilience of regional economic systems to the spillover effects of epidemic risks and propose a model for economic security management in the region during post-pandemic recovery.

To calculate the integral indicator of the stability of the regional economic system to the spillover effects of epidemic risks of economic security, the authors carried out the procedure of normalization of each group of indicators by dividing the indicator by the reference. The significance of each indicator was determined using the expert survey method. The calculation of the final integral indicator was carried out by multiplying each normalized indicator by its weight and further summing it up to obtain an integral value. A sustainability indicator value equal to one is considered ideal, and a region with such an indicator value has the highest level of resistance to spillover effects and the highest level of economic security, and vice versa.

3 Results

During the pandemic, it is advisable to investigate the spillover effects [13] of epidemic risks to the economic security of regions, that is, the concomitant effects on the region's economy from the implementation of epidemic risks and the emergence of new ones, not directly related to them (Fig. 1).

Epidemic risks can manifest themselves as independent risks associated with the implementation of threats to the life and health of the population of the region. In this case, it is worth evaluating indicators such as morbidity and mortality, as well as the loss of the working capacity of the population for the period of the disease [16]. The spread of COVID-19 led to a sharp increase in the incidence of the population, while the impact of this infection had direct (in the form of signs of this disease) and indirect consequences (in the form of exacerbation of chronic diseases as a consequence of the transferred coronavirus). However, operational data reflecting the dynamics of population morbidity in the context of federal districts are not presented by state statistics, including in terms of establishing the consequences of coronavirus. Therefore, to assess the direct impact of epidemic risks associated with the spread of COVID-19 on economic security, it is advisable to conduct an assessment using population mortality indicators (Table 1).

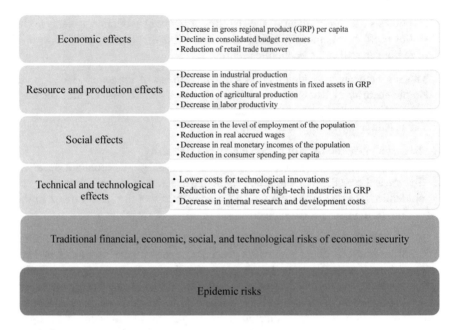

Fig. 1 Spillover effects and risks of economic security of regions during the pandemic. *Source* Compiled by the authors

Table 1 The total number of deaths from COVID-19 in Russia, 2020

Indicator	The number of deaths in Russian cities from COVID-19	The number of deaths in rural areas from COVID-19	Total number of deaths in 2020
The number of deaths from coronavirus infection caused by COVID-19	119,679	25,012	144,691

Source Compiled by the authors based on [3]

Simultaneously, epidemic risks are a factor of turbulence in the external environment of the region, which contributes to the aggravation of internal risks of regional development, leading to a decrease in economic activity and recession. In other words, epidemic risks have direct consequences for the population in the form of an increase in the number of cases and deaths and in the form of side effects for the region's economy, that is, spillover effects.

Traditionally, the economic security of a region is measured by the degree of fluctuations in the gross regional product (GRP) per capita, depending on external changes [7]. However, when conducting a comparative analysis of economic spillover effects on the economic security of regions during a pandemic, it is appropriate to apply other indicators of regional development, for example, consolidated budget revenues or retail trade turnover in dynamics [1]. To assess the resource-production spillover effects of epidemic risks to the economic security of regions, it is advisable to use indicators of industrial production and labor productivity in the context of industries and consider the volume of investment in fixed assets, which determines the future vector of development of the regional industry. The pandemic affects every member of society, their ability to perform labor functions, and their financial well-being. Therefore, a mandatory stage of analysis is the study of indicators reflecting the social side of society (employment, income, and expenses) [24]. During the period of temporary release of labor due to sanitary restrictions, digital technologies become tools for maintaining production processes. Innovative equipment of production and involvement in R&D can guarantee the region's enterprises sustainable development in the future and accelerated post-crisis recovery [26].

We present data on the assessment of the stability of regional economic systems to the spillover effects of epidemic risks of economic security in Table 2 which is available at repository.

In general, in Russia in 2020, there was a decrease in GRP per capita by − 1.1% compared to 2019; it spread to three of the eight federal districts. The largest reduction in GRP per capita occurred in the Ural Federal District (−12%) [3]. The North Caucasus Federal District has the best indicator of the GRP per capita ratio in 2020–2019. The COVID-19 pandemic and the fall in oil prices have led to a decrease in the own tax and non-tax revenues of regional budgets. Income from income tax experienced the most significant drop [5, 22]. Thus, in 2020, there was an increase in the volume of federal transfers (by 54%), which made it possible to compensate for the reduction in tax and non-tax revenues [4, 5]. The negative impact of the COVID-19 pandemic, which caused an economic spillover effect for economic security, was the reduction in retail trade turnover in Russia as a whole: in 2020, it decreased by 2.2% compared to 2019 in comparable prices [25]. The maximum reduction in retail trade in 2020 was achieved in the North Caucasus Federal District. The average decline in industrial production in Russia was 5.5%. In most regions of Russia in the pandemic year of 2020, there was also a decrease in the indicator with the maximum value in the Siberian Federal District (95.9% relative to 2019) [3]. By the end of 2020, agricultural production increased in all regions of the country except the Ural Federal District (98.8% compared to 2019). The expected reduction in investments in fixed assets was almost compensated [19]. The only region with a noticeable reduction in investment in 2020 was the Far Eastern Federal District (−6.2%) [2]. The negative impact of the COVID-19 pandemic was the reduction of employment. Such dynamics were typical for all regions of Russia, without exception. The maximum reduction occurred in the North Caucasus Federal District (95.8% compared to 2019).

In 2020, the real monetary incomes of the population decreased, including in the context of all federal districts; the maximum reduction was typical for the North Caucasus, Volga, and Ural districts (97.2% compared to 2019). Simultaneously, real accrued wages in 2020 increased in all federal districts, which can be explained by a 3.1% increase in inflation and the delayed effect of the pandemic. The reduction in employment of the population and their monetary incomes led to a decrease in the amount of consumer spending per capita in all regions of the country, with the maximum reduction in the Central Federal District (93.6% in 2020 compared to 2019). The technical and technological spillover effect for the economic security of regions can be estimated using the costs of innovative activities of organizations, the share of high-tech industries' products in GRP, as well as internal R&D costs.

The assessment of the stability of regional economic systems to the spillover effects of epidemic risks to the economic security of Russian regions allowed us to draw the following conclusions. The Southern Federal District ranks first among the regions of Russia in terms of economic security and resistance to the spillover effects of epidemic risks. The North Caucasus Federal District is in second place. The Central Federal District is in third place. The North Caucasus Federal District is the leader in terms of resistance to economic spillover effects. The Central Federal District has the best indicator of resistance to resource-production spillover effects. The maximum value of the integral indicator of resistance to social and technical, and technological spillover effects is observed in the Southern Federal District. Thus, the Southern Federal District demonstrated the greatest resistance to epidemic risks and the highest level of economic security during the COVID-19 pandemic.

The management of the economic security of the region in the conditions of a pandemic will be built in accordance with two key vectors:

- Strengthening the internal potential of the regional economic system and identifying new vectors of its development, which will form a reliable protection against external turbulence;
- Leveling the turbulence of the external environment by reducing the spillover effects of epidemic risks (Fig. 2).

The regions of Russia that have a high level of resistance to the spillover effects of epidemic risks (1–4 place in the rating of regions) and a high level of economic security should focus on the first vector of development—strengthening the internal potential of the regional economic system, including the following:

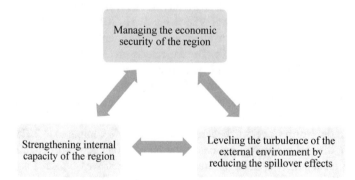

Fig. 2 The model of managing the economic security of the region at the stage of post-pandemic recovery. *Source* Compiled by the authors

1. To develop, test, and optimize organizational and economic mechanisms for ensuring import substitution in the critically import-dependent sectors of the real economy of the regions of Russia at the stage of post-crisis recovery; implement appropriate state programs to stimulate import substitution;
2. To launch educational programs to improve the skills of specialists of enterprises in the region and retraining of workers who dropped out of employment during the pandemic; develop online educational services for colleges and universities in the region;
3. To implement accelerated digitalization measures to equalize the possibilities of Internet access for the population and economic entities of the region and ensure uninterrupted operation in the event of new epidemic risks.

For regions with a low level of resistance to the spillover effects of epidemic risks, it is advisable to implement management efforts in the direction of strengthening the internal potential and leveling the turbulence of the external environment, including the following:

1. To prevent the likelihood of new epidemic risks, including equipping medical institutions with special equipment and digitalization of medical services at the regional level, increasing the number of specialized medical personnel, allowing them to respond promptly to the outbreak of new viruses;
2. To develop regional programs and grants for tax and financial incentives for entrepreneurial activity;
3. To implement measures of social protection of the population and support the population in a difficult life situation during the pandemic.

4 Conclusion

First, the authors identified the features of the manifestation of the spillover effects of epidemic risks to the economic security of the regions of Russia, determined economic, resource-production, social, and technical–technological spillover effects, and disclosed their content. Second, the author assessed the level of stability of regional economic systems to the spillover effects of epidemic risks of economic security. A method for calculating the integral index of resistance to spillover effects is proposed. The regions of Russia are ranked according to this indicator. It is concluded that the Southern Federal District has the greatest resistance to the spillover effects of epidemic risks to economic security. Third, the research proposes a model of economic security management of the region at the stage of post-pandemic recovery, which involves strengthening the internal potential of the regional economic system and leveling the turbulence of the external environment by reducing spillover effects. The research also defines a set of measures for implementing the proposed model.

Data Availability Data on the assessment of the stability of regional economic systems to the spillover effects of epidemic risks of economic security (Table 2) is available at https://figshare.com/search?q=10.6084%2Fm9.figshare.21623751.

References

1. Alikina EB (2020) Comparative analysis of economic security of regions. In: Rudenko MN, Subbotina YD (eds) Economic security: problems, prospects, development trends: proceedings of the VI international scientific and practical conference. Perm State University, Perm, Russia, pp 28–39
2. Federal State Statistics Service of the Russian Federation (2020) Report: socio-economic situation of Russia. Retrieved from https://rosstat.gov.ru/compendium/document/50801. Accessed 12 Nov 2022
3. Federal State Statistics Service of the Russian Federation (2021) Information for monitoring the socio-economic situation of the subjects of the Russian Federation. Retrieved from https://rosstat.gov.ru/folder/11109/document/13259. Accessed 12 Nov 2022
4. Federal Treasury of the Russian Federation (2021) Consolidated budgets of the subjects of the Russian Federation and budgets of territorial state extra-budgetary funds. Retrieved from https://roskazna.gov.ru/ispolnenie-byudzhetov/konsolidirovannye-byudzhety-subektov/. Accessed 12 Nov 2022
5. Foundation "Center for Strategic Research" (2021) Execution of consolidated budgets of the subjects of the Russian Federation. Retrieved from https://www.csr.ru/upload/iblock/623/gd1 sf3nv0o1m7rpls2yuhyvqkohsjffz.pdf. Accessed 12 Nov 2022
6. Fraymovich DY, Konovalova ME, Roshchektaeva UY, Karpunina EK, Avagyan GL (2022) Designing mechanisms for ensuring the economic security of regions: countering the challenges of instability. In: Popkova EG, Polukhin AA, Ragulina JV (eds) Towards an increased security: green innovations, intellectual property protection and information security. Springer, Cham, Switzerland, pp 569–581. https://doi.org/10.1007/978-3-030-93155-1_63
7. Greer B (1986) European economic security. In: Flanagan SJ, Hampson FO (eds) Securing Europe's future: a research volume from the center of science and international affairs, Harvard University. Routledge, London, UK, pp 221–241
8. Gukasyan ZO, Tavbulatova ZK, Aksenova ZA, Gasanova NM, Karpunina EK (2022) Strategies for adapting companies to the turbulence caused by the COVID-19 pandemic. In: Popkova EG (ed) Business 4.0 as a subject of the digital economy. Springer, Cham, Switzerland, pp 639–645. https://doi.org/10.1007/978-3-030-90324-4_102
9. Karpunina EK (2022) Current approaches to assessing the economic security of regions in the era of COVID-19. In: Gerasimov VI (ed) Proceedings of the XXI national scientific conference with international participation. Moscow, Russia, pp 1105–1108
10. Karpunina EK (ed) (2022) Modern approaches to ensuring the economic security of the state and regions in the era of uncertainty: monograph. Ruscience, Moscow, Russia
11. Karpunina EK, Lapushinskaya GK, Arutyunova AE, Lupacheva SV, Dubovitski AA (2020) Dialectics of sustainable development of digital economy ecosystem. In: Popkova E, Sergi B (eds) Scientific and technical revolution: yesterday, today and tomorrow. Springer, Cham, Switzerland, pp 486–496. https://doi.org/10.1007/978-3-030-47945-9_54
12. Karpunina EK, Moskovtceva LV, Zabelina OV, Zubareva NN, Tsykora AV (2022) Socio-economic impact of the COVID-19 pandemic on OECD countries. In: Popkova EG, Andronova IV (eds) Current problems of the world economy and international trade. Emerald Publishing Limited, Bingley, UK, pp 103–114. https://doi.org/10.1108/S0190-128120220000042011
13. Karpunina E, Gubernatorova N, Daudova A, Stash Z, Kargina L (2020) The spillover effects of the digital economy. In: Proceeding of the IBIMA 2020: 36th international business information management association conference. Granada, Spain, pp 942–954
14. Karpunina E, Nazarova I, Iljina L, Shvetsova I, Chernenko E (2022) The economic security threats of the region in terms of digitalization: assessment and development of leveling tools. In: Popkova EG, Sergi BS (eds) Geo-economy of the future: sustainable agriculture and alternative energy. Springer, Cham, Switzerland, pp 515–525. https://doi.org/10.1007/978-3-030-92303-7_55

15. Karpunina E, Shurchkova J, Borshchevskaya E, Konovalova M, Levchenko L (2019) Opportunities of advanced development of the digital economy ecosystem. In: Proceedings of the IBIMA 2019: 33rd international business information management association conference. Granada, Spain, pp 7454–7461
16. Karpunina E, Zabelina O, Galieva G, Melyakova E, Melnikova Y (2020) Epidemic threats and their impact on the economic security of the state. In: Proceeding of the IBIMA 2020: 35th international business information management association conference. Seville, Spain, pp 7671–7682
17. Korableva A, Karpov V (2019) Assessment of the regional financial system on the basis of economic security indicators. Russ J Soc Sci Humanit 2(36):158–165. https://doi.org/10.17238/issn1998-5320.2019.36.158
18. Korolyuk E, Rustamova I, Kuzmenko N, Khashir B, Karpunina E (2021) Diagnostics of regional economic security problems during the 2020 crisis. In: Proceeding of the IBIMA 2021: 37th international business information management association conference. Cordoba, Spain, pp 5248–5257
19. Kudrin AL, Mau VA (eds) (2021) Section 4. The real sector of the economy. In: Russian economy in 2020: trends and outlooks (Issue 42). Moscow, Russia: Gaidar Institute for Economic Policy. Retrieved from https://www.iep.ru/files/text/trends/2020/Book.pdf. Accessed 12 Nov 2022
20. Look C (2020) Lagarde primes ECB for more economic stimulus. Bloomberg. Retrieved from https://www.bloomberg.com/news/articles/2020-10-29/lagarde-primes-ecb-for-more-stimulus-as-virus-derails-recovery. Accessed 21 Nov 2022
21. Lovkova ES, Kashitsina TN, Kapustina NV, Rustamov NNO, Sultanova AV (2022) The problem of providing a highly effective flexible methodology in the management of regional marketing projects and its solution. In: Popkova EG (ed) Imitation market modeling in digital economy: game theoretic approaches. Springer, Cham, Switzerland, pp 73–79. https://doi.org/10.1007/978-3-030-93244-2_9
22. Ministry of Finance of the Russian Federation (2020) The main directions of budget, tax and customs tariff policy for 2021 and for the planning period of 2022 and 2023. Retrieved from https://minfin.gov.ru/common/upload/library/2020/10/main/ONBNiTTP_2021_2023.pdf. Accessed 12 Nov 2022
23. Mityakov SN, Mityakov ES, Romanova NA (2013) The economic security of the Volga Federal District regions. Econ Reg 3(35):81–91. https://doi.org/10.17059/2013-3-6
24. Nazarova I, Galieva G, Sazanova E, Chernenko E, Karpunina E (2022) Labor market and employment problems: analysis of long-term dynamics and prospects of development in Russian regions. In: Popkova EG (eds) Imitation market modeling in digital economy: game theoretic approaches. Springer, Cham, Switzerland, pp 711–722. https://doi.org/10.1007/978-3-030-93244-2_77
25. Oveshnikova LV, Sibirskaya EV (2021) COVID-19 and its destructive impact on the economic security of Russian regions. In: Gerasimov VI (ed) Russia: trends and prospects for development: proceedings of the XX national scientific conference. Institute of Scientific Information on Social Sciences of the Russian Academy of Sciences, Moscow, Russia, pp 1052–1055
26. Sadueva M, Kuzmina O, Kukina E, Shurupova A, Karpunina E (2020) Investment in R&D as a trigger for accelerated development of the digital economy ecosystem: a comparative analysis of OECD countries. In: Proceeding of the IBIMA 2020: 36th international business information management association conference. Granada, Spain, pp 7984–7995
27. Shubina N (2017) Conceptual approaches to the understanding of economic safety of region: essence, structure, factors and condition. Bull Ural Fed Univ Ser Econ Manage 16(2):288–307. https://doi.org/10.15826/vestnik.2017.16.2.015

28. Sukhadolets T, Stupnikova E, Fomenko N, Kapustina N, Kuznetsov Y (2021) Foreign direct investment (FDI), investment in construction and poverty in economic crises (Denmark, Italy, Germany, Romania, China, India and Russia). Economies 9(4):152. https://doi.org/10.3390/economies9040152

29. Tsypin P, Macheret D, Kapustina NV (2021) The problem of specific railway transport resources sharing. In: Gaol F, Filimonova N, Acharya C (ed) Impact of disruptive technologies on the sharing economy. IGI Global, Hershey, PA, pp 13–27. https://doi.org/10.4018/978-1-7998-0361-4.ch002

Assessment of the Level of Digitalization of Russian Regions Under Conditions of Socio-economic Uncertainty

Natalia M. Fomenko⬡, **Olga M. Markova**⬡, **Konstantin N. Ermolaev**⬡, **Julia V. Ioda**⬡, **and Tatyana S. Zhigunova**⬡

Abstract The paper aims to identify the peculiarities of the digital development of Russian regions in terms of socio-economic uncertainty and determine the vector of regional policy to ensure accelerated digitalization of territories. The authors analyzed the main indicators of digital development households, organizations, and public authorities of Russian regions in the pre-pandemic period compared to the period of an active course of the pandemic, as well as at the stage of military and political instability. The research identifies Russian regions where digital development processes accelerated during the period of social and economic uncertainty caused by the COVID-19 pandemic, as well as regions where digitalization limitations became evident. The analysis allowed the authors to conclude about the multi-directional influence of socio-economic uncertainty on the course of digital development processes in Russian regions. On the one hand, the transformation in consumer behavior and the new external conditions of socio-economic uncertainty during the pandemic led to an accelerated digitalization of regional economic systems. On the other hand, the socio-economic uncertainty of 2020 has caused dramatic changes in the implemented business models, formats of organizations, and the nature of employment, which manifested itself in the reduction of most indicators of digital development of organizations in 2020. For each selected group of Russian regions,

N. M. Fomenko (✉)
Plekhanov Russian University of Economics, Moscow, Russia
e-mail: economresearch@mail.ru; fnata77@mail.ru

O. M. Markova
Financial University under the Government of the Russian Federation, Moscow, Russia
e-mail: OMMarkova@fa.ru

K. N. Ermolaev
Samara State University of Economics, Samara, Russia

J. V. Ioda
Lipetsk Branch of the Financial University under the Government of the Russian Federation, Lipetsk, Russia

T. S. Zhigunova
St. Petersburg State University, Saint Petersburg, Russia

© The Author(s), under exclusive license to Springer Nature Switzerland AG 2023
E. G. Popkova (ed.), *Smart Green Innovations in Industry 4.0 for Climate Change Risk Management*, Environmental Footprints and Eco-design of Products and Processes, https://doi.org/10.1007/978-3-031-28457-1_47

the authors proposed state policy measures in the field of digital development, which is supportive and accelerated in nature.

Keywords Region · Digitalization · Digital development · Socio-economic uncertainty · Pandemic · Political-military conflict · Accelerated digitalization · Public policy

JEL Classification O33 · R11 · R58

1 Introduction

Recently, the processes of social development have been accompanied by increasing uncertainty and turbulence. This applies to the regular emergence of economic crises (global and local) and, as a consequence, the aggravation of social tensions (deterioration of the quality of life, increasing social differentiation, dissatisfaction with the implemented policy, the development of an unfavorable situation in the labor market, etc.). The year 2020 and 2021 showed the vulnerability of global and national economies to epidemic threats. In a short period, the well-being of society was undermined, and existing problems of a socio-economic nature only intensified. The year 2022 became even tenser in terms of the unfolding of military and political conflicts, as well as a number of government decisions to impose sanctions on the Russian economy by a number of developed EU countries and the USA. This was another event contributing to increased uncertainty and turbulence in the world, regardless of the involvement of national economies in the immediate conflict [5, 8, 14, 19].

Under conditions of increasing uncertainty and turbulence, the objective processes in economic systems take on a somewhat different character. Particularly, digitalization processes, previously differentiated in terms of countries and regions of the world, have reached a new stage in their development, pushing people, businesses, and government agencies to improve digital infrastructure and increase the level of access to the use of ICTs [2, 11].

As for the digital development of Russian regions, it should be noted that the pandemic and military and political tensions were factors that pushed regional authorities to accelerate digitalization as a prerequisite for the sustainable development of regional economic systems, improving the level and quality of life of the population and inclusion in global communications processes.

2 Methodology

Features of digital development of countries and regions are described in the works of Bychkova et al. [2], Karpunin et al. [9–11], and Molchan et al. [15]. The authors emphasize the problems of providing access to the Internet to the population and businesses, which hinder the penetration of digital technology in social processes. Another problem of the digital development of territories is the lack of motivation of enterprises in the regions to digitalize their own activities because of the need for additional investment in ICT. According to researchers, a significant factor in the low level of digital development of regions is the low digital literacy of citizens and the existing cyber risks, threatening users with loss of money, personal and professional information, and deteriorating reputation [12, 17].

This research aims to identify the peculiarities of the digital development of Russian regions in terms of socio-economic uncertainty and determine the vector of regional policy to ensure accelerated digitalization of territories. The research objectives are as follows:

1. To analyze the indicators of digital development of Russian regions in the pre-pandemic period and at the stage of active pandemic development;
2. To systematize the factors contributing to and hindering the intensification of digital development of Russian regions under conditions of socio-economic uncertainty;
3. To propose a set of regional policy measures to overcome the current limitations of digital development.

The most common approach to assessing the level of digitalization of the territory is the calculation of the Network Readiness Index of the Portulance Institute [4], which is based on indicators of the availability of digital technologies, their use by the population, businesses, and public administration, as well as indicators reflecting the level of institutional regulation of digitalization processes and the impact of digital technologies on various aspects of society (including quality of life, sustainable development of territories, etc.) [4]. Another approach to assessing the digitalization of regions is a comparative analysis of territories based on the calculation of integral indicators of digitalization. However, the operational data of regional statistics are not sufficient for this method [2]. To reflect the dynamics of ongoing changes in the level of digitalization of Russian regions, the authors conduct a comparative economic analysis of the digitalization indicators of the federal districts, followed by an explanation of the reasons for these trends and the corresponding grouping of regions.

3   Results

The basic condition for the digital development of the region is the digital infrastructure. In this aspect, the availability of devices and Internet access to the population and businesses is important. Let us analyze what has changed in the state of the digital infrastructure of Russian regions due to the increased socio-economic uncertainty caused by the COVID-19 pandemic.

The ability of households to have access to a computer and the availability of alternative access through televisions or cell phones differed significantly in different

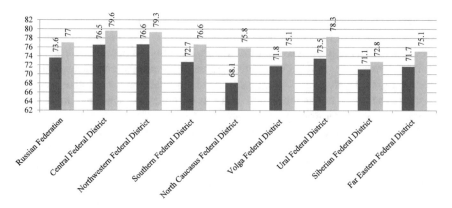

Fig. 1 Share of households with broadband Internet access, %, 2019–2020. *Source* Compiled by the authors based on [1]

regions of Russia before the COVID-19 pandemic. This is primarily caused by differences in income and education (the higher the level of education, the more likely people are to have access to ICTs) [16]. Other variables such as household size and type, age, gender, and location also play an important role (e.g., Internet access is greater in cities than in rural areas).

However, the pandemic contributed to the fact that to help citizens during the period of high alert deployed additional electronic services, regional authorities expanded existing services, negotiated with operators about the easing of communication services, and organized informational support, in some cases—direct financial and material.

The period of socio-economic uncertainty caused by the COVID-19 pandemic has benefited households' broadband connection dynamics (Fig. 1).

In Russia as a whole, the growth of this indicator amounted to 4.6%. The most significant increase in this indicator was achieved in the North Caucasus Federal district (+ 11.3% compared to 2019). Internet traffic increased in all regions of the district, as well as the time spent online. Simultaneously, the Internet coverage of socially important objects expanded in the region's cities and in the hard-to-reach mountainous regions. In the Chechen Republic, 60% of the region's residents were to have Internet access in 2020. Additionally, the quality of communication has also improved [20]. In 2021, the regions of the North Caucasus Federal district saw a significant increase in investment of telecommunications companies in infrastructure development, especially in tourist areas [13]. This expansion of the digital infrastructure has occurred because of the introduction of self-isolation regimes and the increasing need of the population to communicate, perform professional activities, and be included in learning processes.

In 2020, fixed Internet traffic growth was 34.2% over the same period in 2019. Despite a 51.9% year-over-year increase in mobile Internet traffic during the second quarter lockdown, overall mobile traffic growth for 2020 was 47%, below the trend of previous years. This confirms the assumption of the population's desire for reliable

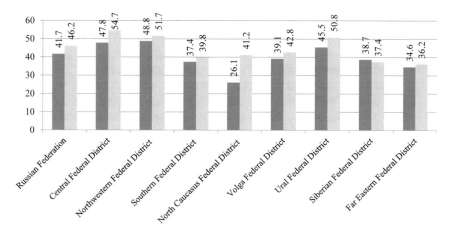

Fig. 2 Online purchase of goods and services by the population, % of the total population of the region, 2019–2020. *Source* Compiled by the authors based on [1, 6]

connectivity, which is provided by fast-developing fixed Wi-Fi, despite the changed working and living conditions during the pandemic [1].

Access to the Internet allows citizens to actively use digital platforms and services of digital services. This concerns the daily routine of citizens (paying utility bills, using online government services, purchasing goods), the realization of opportunities for online education, professional functions, and leisure activities. Naturally, the volume of goods and services purchased online increased during the pandemic due to the self-isolation of the population. In the pre-pandemic period, e-commerce services were most developed in the Northwestern Federal District (48.8% of households used them) and in the Central Federal District (47.8%). The most lagging region was the North Caucasian Federal District (26.1%). The situation changed in 2020 (Fig. 2).

In 2020, in all regions of Russia, except for the Siberian Federal District, there was an increase in the share of the population purchasing goods and services online. The new consumer habits of the population to use remote services to purchase goods and services have accelerated the development of online commerce. The greatest contribution to the development of the industry was made by online sales of convenience goods. The drop in purchasing power had an impact only on premium consumption. The maximum growth was achieved in the North Caucasian Federal District (57.9%); the decline was in the Siberian Federal District (− 3.4%) due to a decrease in consumer activity in the region due to mass layoffs.

In 2019, the level of broadband Internet access for organizations reached 93.1% in the Central Federal District, while the lowest value was in the North Caucasus Federal District (77.7%) [1, 6]. However, the pandemic had the opposite effect on the functioning of organizations. Due to the introduction of lockdowns across the country, there was a forced suspension of the production activities of most enterprises. Simultaneously, the management of enterprises transferred employees to remote work whenever possible. As a result, the indicators of connection of enterprises to the Internet decreased: by 15.6% in Russia as a whole, with a maximum reduction of 21.1% in the Southern Federal District and by 17.5% in the Volga Federal District [6].

In the pandemic year of 2020, there was also a decrease in the number of organizations that used electronic data exchange systems with external information systems (− 18.9%), used SCM systems (− 34.8%), placed orders for goods and services on the Internet (− 6%), having special software to manage the procurement of goods (works, services) (− 39.2%), having special software to manage the sales of goods (works, services) (− 38.5%), using ERP-systems, in the total number of surveyed organizations (− 22.3%), and using CRM-systems (− 22.3%). However, in 2021, the dynamics for the above indicators of organizations' digital development began to straighten out and return to the previous vector [2, 7].

The COVID-19 pandemic has also impacted organizations' use of data protection tools transmitted over global networks. For example, in 2019, the share of organizations that used data protection tools in the total number of organizations in Russia surveyed was 89.5%. During the lockdown period of 2020, it decreased to 75.3%; in 2021, it showed an increase of 1.6% and amounted to 76.5%.

Another aspect of the digital development of the region's economy is the level of digitalization of the public administration system. This refers to the readiness of state and local governments to use ICTs in their activities, as well as in the process of providing public services. Thus, the indicators of the use of fixed (wired and wireless) Internet by public authorities in 2020 and 2021 increased significantly compared to previous periods and amounted to 86.9% and 87.8%, respectively. The number of special software tools for providing access to organizations' databases via global information networks and the use of Intranet in the activities of public authorities has also increased. The Russian e-government platform includes the following three components:

- The Unified Portal of State Services and Functions;
- The Unified System of Identification and Authentication;
- The System of Interagency Electronic Interaction.

In particular, the introduction of social distancing regimes has created restrictions for the population to personally apply for public services in specialized organizations in the region. The result of this situation was the close attention of regional authorities to maintain the smooth operation of these online services of state organizations in the regions and achieve growth in the overall share of services received by the population. The leading region for the distribution of online government services in the pre-pandemic period was the Central Federal District (83.2% of the population). Only 63.6% of the population used such services in the North Caucasus Federal District [3]. In 2020, the share of the population using online public services in Russia as a whole grew by 4.5% compared to 2019 [6]. These dynamics were observed in all regions of Russia except for the Volga Federal District (− 2% compared with 2019). Additionally, in Russia, the share of organizations that used the Internet to obtain certain types of public and municipal services increased by 1.2% and reached 67.7% in 2020.

Digitalization requires regular investment in ICT development. These costs are shared among regional populations, businesses, and regional governments. The public's interest in a seamless Internet experience during the pandemic has only grown stronger. Thanks to the pandemic, there is a public demand for change in the existing way of life and established socio-economic processes. The basis of these changes is the rapid and mass transition of the population to online communication and autonomous remote work. The changes taking place form an increased demand on the part of the population and businesses for communication services, development of data processing and storage infrastructure, growth of traffic consumption, and increase of network bandwidth, ensuring the stability of network connection and provision of a wide range of demanded digital services.

As a rule, large regions with a strong scientific and technological base have the largest volumes of investment in ICT development—these are the regions of the Central and Northwestern Federal Districts. However, the Volga Federal District and the Ural Federal District are closer to the leading regions in terms of "capital expenditures on information and computer equipment."

Socio-economic uncertainty caused by the pandemic has led to unstable dynamics of the total amount of regional spending on ICTs:

- In 2019, it was 161.4 billion rubles;
- In 2020, it was 225.1 billion rubles (+ 39.5% against 2019);
- In 2021, it decreased to 205.1 billion rubles (− 8.9% against 2020) [18].

Simultaneously, against a background of growing socio-economic uncertainty due to the events in Ukraine and the beginning of special operations of the Russian army, the regions began to increase these costs systemically. In 2022, their value reached 243.4 billion rubles (an increase of 18.7% against 2021).

In 2021, the leaders in actual expenditures on ICT (development of online services on public service portals, improvement of the digital health platform, support and development of IT infrastructure in terms of connecting socially important objects to the Internet, and providing digital transformation of cities) were the regions of the Central and Northwestern Federal Districts: Moscow (in 2021, the ICT spending was 76.3 billion rubles), St. Petersburg (21.6 billion rubles), and the Moscow Region (6.9 billion rubles).

Thus, the overall picture for the regions of Russia in the context of assessing changes in the level of digitalization in the context of socio-economic uncertainty is as follows (Table 1).

Thus, we can systematize the regions characterized by the acceleration of digitalization in a period of socio-economic uncertainty (Central, Northwestern, Southern, North Caucasus, Ural, and Far Eastern Federal Districts), as well as the regions that face the limitations of digitalization (the Volga and Siberian Federal Districts). The first group of regions requires the implementation of state policy in the field of digital development with an emphasis on the following supporting measures:

Table 1 Changes in the level of digitalization of the regions of Russia in the context of socio-economic uncertainty, 2019–2020

Region/indicator	Percentage of households with the broadband Internet access	Organizations' access to the broadband Internet	Online purchase of goods and services by the population	Use of public online services by the population	Investing in the development of ICT	Groups of regions
Central Federal District	⇦	➡	⇦	⇦	⬅	1
North-Western Federal District	⇦	➡	⇦	⇦	⬅	1
Southern Federal District	⇦	➡	⇦	⇦	⇦	2
North Caucasus Federal District	⇦	➡	⇦	⇦	⇦	1
Volga Federal District	⇦	➡	⇦	➡	⇦	2
Ural Federal District	⇦	➡	⇦	⇦	⇦	1

(continued)

Assessment of the Level of Digitalization of Russian Regions Under …

Table 1 (continued)

Region/indicator	Percentage of households with the broadband Internet access	Organizations' access to the broadband Internet	Online purchase of goods and services by the population	Use of public online services by the population	Investing in the development of ICT	Groups of regions
Siberian Federal District	⇦	⇨	⇨	⇦	⇦	2
Far Eastern Federal District	⇦	⇨	⇦	⇦	⇦	1

Source Compiled by the authors

- Improving the quality characteristics of digital infrastructure (e.g., increasing Internet bandwidth, improving the digital educational environment;
- Introducing inclusive solutions for municipal and regional services and services for people with disabilities);
- Implementing a set of measures in the field of cybersecurity;
- Improving digital literacy.

For problem regions, it is necessary to implement specialized measures, including the following:

- Development of the necessary information technology and telecommunications infrastructure for the organization of secure interagency electronic interaction;
- Modernization of the portals of electronic services; increasing the amount of funding for the implementation of ICT;
- Development of incentive mechanisms for the accelerated digitalization of businesses in the region.

4 Conclusions

The analysis allows us to conclude about the multidirectional influence of the socio-economic uncertainty caused by the pandemic on the course of digital development processes in Russian regions. On the one hand, the transformation of consumer behavior and the new external conditions of socio-economic uncertainty have led to an accelerated digitalization of regional economic systems in terms of expanding and improving the quality of digital infrastructure. On the other hand, the socio-economic uncertainty of the 2020 period caused fundamental changes in the implemented business models, the formats of organizations and the nature of employment. This manifested itself in the reduction of most indicators of digital development of organizations in 2020, primarily in the use of ICT in the activities of organizations. However, already in 2021, the dynamics of organizations' digital development indicators began to return to the pre-digital period, indicating their adaptation to the challenges of uncertainty. Russian regions where digital development processes accelerated during the period of socio-economic uncertainty were identified, as well as regions with existing digitalization constraints. State policy measures for supporting and accelerating digital development are proposed for Russian regions.

References

1. Abdrakhmanova GI, Utyatina KE (2021) Internet-infrastructure of Russia during the pandemic. Digit Econ 46(208). Retrieved from https://issek.hse.ru/mirror/pubs/share/488807139.pdf. Accessed 14 Nov 2022
2. Bychkova N, Tavbulatova Z, Ruzhanskaya N, Tamov R, Karpunina E (2020) Digital readiness of Russian regions. In: Proceeding of the IBIMA 2020: 36th international business information management association conference. Granada, Spain, pp 2442–2461
3. Dolgikh EA, Parshintseva LS (2019) A statistical study of the use of the Internet by the population in the Russian Federation. Vestn Univ [Univ Herald] 1:108–112. https://doi.org/10.26425/1816-4277-2019-1-108-112
4. Dutta S, Lanvin B (eds) (2019) The network readiness index 2019: towards a future-ready society. Portulance Institute, Washington DC. Retrieved from https://networkreadinessindex.org/wp-content/uploads/2020/03/The-Network-Readiness-Index-2019-New-version-March-2020-2.pdf. Accessed 12 Nov 2022
5. European Commission (2020) The 2020 predict report. Key facts report. Publications Office of the European Union, Luxembourg. Retrieved from https://publications.jrc.ec.europa.eu/repository/bitstream/JRC121153/jrc121153_predict_key_facts_report_2020_final.pdf. Accessed 12 Nov 2022
6. Federal State Statistics Service and Higher School of Economics (2020) Information society in the Russian Federation, 2020: statistical collection. HSE University, Moscow, Russia. Retrieved from https://rosstat.gov.ru/storage/mediabank/lqv3T0Rk/info-ob2020.pdf. Accessed 14 Nov 2022
7. Federal State Statistics Service of the Russian Federation (2022) Monitoring the development of information society in the Russian Federation. Retrieved from https://rosstat.gov.ru/storage/mediabank/monitor.xlsx. Accessed 14 Nov 2022
8. Gukasyan ZO, Tavbulatova ZK, Aksenova ZA, Gasanova NM, Karpunina EK (2022) Strategies for adapting companies to the turbulence caused by the COVID-19 pandemic. In: Popkova EG (ed) Business 4.0 as a subject of the digital economy. Springer, Cham, Switzerland, pp 639–645. https://doi.org/10.1007/978-3-030-90324-4_102
9. Karpunin KD, Ioda JV, Ternavshchenko KO, Aksenova ZA, Maglinova TG (2022) The "Invisible hand" of digitalization: the challenges of the pandemic. In: Popkova EG (ed) Imitation market modeling in digital economy: game theoretic approaches. Springer, Cham, Switzerland, pp 162–173. https://doi.org/10.1007/978-3-030-93244-2_19
10. Karpunina EK, Tavbulatova ZK, Kuznetsov YV, Dzhabrailova ND, Anichkina OA (2021) The challenges of digitalization for economic relations of tourism industry subjects. In: Alpidovskaya ML, Karaseva LA, Mamagulashvili DI, Bogoviz AV, Krivtsov A (eds) Industry 4.0: implications for management, economics and law. De Gruyter, Berlin, pp 31–44. https://doi.org/10.1515/9783110654486-004
11. Karpunina E, Kharchenko E, Mikhailov A, Nedorezova E, Khorev A (2019) From digital development of economy to society 5.0: Why we should remember about security risks? In: Proceedings of the IBIMA 2019: 34th international business information management association conference. Madrid, Spain, pp 3678–3688
12. Karpunina E, Shurchkova J, Kochetkova E, Ponomarev S, Tretyak V (2020) Cybercrime in the system of economic security threats. In: Proceeding of the IBIMA 2020: 35th international business information management association conference. Seville, Spain, pp 2679–2690
13. Klyuchko S (2021) Traffic growth and constant online: trends and results of the telecom sector in the North Caucasus Federal District. RBC. Retrieved from https://kavkaz.rbc.ru/kavkaz/30/12/2021/61cc55299a79470dd79fbfce. Accessed 14 Nov 2022
14. Kukina EE, Fomenko NM, Alekhina OF, Smirnova EV, Pecherskaya OA (2022) Long-term effects of COVID-19: how the pandemic highlighted the global digital divide. In: Ostrovskaya VN, Boboviz AV (eds) Big data in the GovTech system, studies in big data. Springer, Cham, Switzerland, pp 137–148. https://doi.org/10.1007/978-3-031-04903-3_17

15. Molchan A, Karpunina E, Kochyan G, Petrov I, Velikanova L (2019) Effects of digitalization: new challenges for economic security systems. In: Proceedings of the IBIMA 2019: 34th international business information management association conference. Madrid, Spain, pp 6631–6639

16. OECD (2001) Understanding the digital divide. OECD Publications, Paris, France. Retrieved from https://www.oecd.org/sti/1888451.pdf. Accessed 12 Nov 2022

17. Pilipchuk N, Beilina A, Udovik E, Orlovtseva O, Karpunina E (2021) The development of digital competences of teachers in the higher education system. In: Proceeding of the IBIMA 2021: 37th international business information management association conference. Cordoba, Spain, pp 517–527

18. Rudycheva N (2022) Russian regions have planned a significant increase in ICT expenditures in 2022. Cnews. Retrieved from https://www.cnews.ru/articles/2022-03-09_ikt-rashody_regi onov_v_2022_godu_vyrastut. Accessed 14 Nov 2022

19. Sukhadolets T, Stupnikova E, Fomenko N, Kapustina N, Kuznetsov Y (2021) Foreign direct investment (FDI), investment in construction and poverty in economic crises (Denmark, Italy, Germany, Romania, China, India and Russia). Economies 9(4):152. https://doi.org/10.3390/economies9040152

20. TASS Russian News Agency (2020) North Caucasus expands Internet coverage and improves quality of communication. Retrieved from https://tass.ru/v-strane/8511077. Accessed 14 Nov 2022

Accounting and Analytical Management of Sustainable Business Development and ESG Management in Russia and East Asia

Lyudmila N. Usenko⊙, Victoriya A. Guzey⊙, and Natalia M. Usenko⊙

Abstract The current global economic environment demonstrates a high dynamic of global challenges, including the COVID-19 pandemic and a comprehensive recession, the consequences of which can be severe. This has raised interest in achieving the Sustainable Development Goals (SDGs) in terms of three aspects: whether it is possible to achieve the SDGs after the 2020 recession, the main ways out of the recession based on achieving the SDGs, and the prospects for adapting the SDGs and ESG management to the current global realities. The Russian Federation and the People's Republic of China have substantial backlogs and significant interest in achieving the SDGs. The present study examines the methodological aspects of including indicators in the SDGs in terms of traditional and new indicators.

Keywords Sustainable development goals · Inequalities · Health · Poverty · Pandemic · ESG management

JEL Classification D63 · I30 · O10 · O13 · O15

1 Introduction

Currently, the entire global community is seeking to explore the current problematic aspects, risks, and available opportunities to control them against the background of the COVID-19 pandemic and the global recession [1, 2]. In this context, discussions on the future of the post-pandemic development of the world are intensifying. The following main aspects are essential.

The main challenge now is maintaining partnerships that ensure the coordination of measures to overcome the COVID-19 pandemic. The full cooperation of international health organizations is essential because they are responsible for creating,

L. N. Usenko · V. A. Guzey (✉) · N. M. Usenko
Rostov State University of Economics, Rostov-On-Don, Russia
e-mail: UchetiStatistica@yandex.ru

© The Author(s), under exclusive license to Springer Nature Switzerland AG 2023   473
E. G. Popkova (ed.), *Smart Green Innovations in Industry 4.0 for Climate Change Risk Management*, Environmental Footprints and Eco-design of Products and Processes, https://doi.org/10.1007/978-3-031-28457-1_48

testing, registering, and producing vaccines and drugs to be used throughout the world to prevent infections from breaking out.

Additionally, an essential global challenge is to make the economic sector of the world more active, as well as to restore and intensify the international interactions that were impossible during the COVID-19 pandemic [9, 10].

Moreover, an essential aspect of overcoming the recession is to ensure the financial stability of individual countries and the world economy. In this connection, we should note the need to recover the industry's most prone to crisis changes. The global recession that began in 2020 represents the most significant downturn in the world economy since the early 1930s.

2 Materials and Methods

The global crisis of 2020 demonstrated the need to consolidate international efforts to overcome it. The threefold interaction of coordination, governance, and development is a simple and effective way of overcoming difficulties that were first announced at the Club of Rome. Currently, significant efforts to implement joint coordination, governance, and development activities have focused on issues related to the environmental agenda. In this regard, Sustainable Development Goals (SDGs) and ESG management are a way of managing humanity's global challenges.

The BRICS countries, among which the Russian Federation and the People's Republic of China are particularly important, have a significant impact on coordination, governance, and development [3]. These countries are the main driving forces in these processes. Their actions should be considered in achieving the SDGs [5].

The need to overcome the negative impact of the recession and the recovery of the world economy will require significant efforts; close international cooperation is required to achieve all SDGs [6–8]. The impact of the global crisis on the global economy can be formulated as a set of problematic aspects to be addressed by the world community [4].

The first and most urgent aspect is the need to ensure the achievement of the SDGs comprehensively. This is due to the need to consolidate the efforts of the world community in ensuring progressive development in this direction, considering the level of economic development achieved by the countries and the existing opportunities for sustainable development worldwide.

The next problematic point is the significant difference in the level of development of the economies, which determines the intensity of achieving the SDGs using different kinds of tools. During the recession, the level of socioeconomic development achieved by the countries has a significant impact on the economic activity, the adaptive capacity to overcome the COVID-19 infection, and the reserves available to solve emerging problems and further development.

The third, but not the most important, problem is the possibility of joint achievement of the SDGs in the context of the functioning of imperfect institutions and emerging budgetary constraints.

3 Results

The past five years have demonstrated an increased interest in achieving the SDGs on the part of the global community. The COVID-19 pandemic significantly affected the pace of approaching the SDGs and the fall in oil prices. By 2022, the global economy is still fragile. The achievement of the 2030 Sustainable Development Goals faces certain challenges.

Russia and China are actively involved in implementing the concept of sustainable development, viewing it as the primary development paradigm for all humankind. The most important United Nations summits on achieving sustainable development have been held with the direct participation of their representatives, including the UN Conference on Environment and Development held in Rio de Janeiro in 1992 and 2012 and Johannesburg in 2002. The main postulates of the final resolutions within the framework of these summits are as follows:

- Revealing the prospects for human development in the twenty-first century based on the concept of sustainable development (Rio de Janeiro, 2012);
- Formulating the 2030 Sustainable Development Goals (New York, 2015);
- Setting priorities for the international community as part of the commitments to stabilize climate and reduce the losses from climate change (Paris, 2015).

Nearly a decade after the Rio Conference, the international community is facing a COVID-19 pandemic and a global recession that has had a negative impact on the pace of achieving the SDGs. In this context, it is necessary to adjust the system of SDGs in terms of its main aspects and indicators. A combination of negative factors (i.e., the impact of sanctions, the crisis of global governance, and escalating trade conflicts) has had a negative impact on the development of partnerships worldwide. During this period, there was a significant decline in investment in fixed capital and in the level of capital accumulation. The specified decrease was even higher than in the period of the world recession of 2008–2009. Simultaneously, the developed world puts the main focus on climate preservation, the reduction of greenhouse gas emissions, and leveling the problem of poverty.

As confirmation, it is necessary to analyze poverty in the Russian Federation and the People's Republic of China and measures to reduce it. For several years, the poverty of a significant part of the population has continued to be one of the main social threats to the successful development of society. In recent years, analysis of changes in the living standards of the Russian population has shown that the persistence of low living standards of the majority of the population blocks its economic development and exacerbates its socio-political instability. Social policy in Russia remains passive and inadequate in the current tense situation. An increasing number of citizens and socio-political forces advocate a change in the course of socioeconomic transformations in the country.

Failure to meet the minimum needs of a person (family) is considered poverty. Failure to meet needs can lead either to a change in the normal life of a person or to the death of this person.

The method of measuring poverty officially adopted in Russia is based on the concept of absolute poverty when minimum needs (necessities) and the range of goods and services that satisfy these needs (the composition of the minimum consumer basket) are determined.

The most vulnerable groups are young people, women, people of retirement age, and low-skilled workers. Along with poverty and destitution (extreme poverty), there are also disadvantaged people, including children, the disabled, the unemployed, pensioners, and the chronically poor.

Nowadays, the threat of impoverishment is looming quite wealthy over the socio-professional strata of the population. The social bottom, which includes beggars and street children, is ready to absorb and is already absorbing low-skilled workers, engineers, technicians, teachers, and scientists. The main elements of this mechanism are economic reform, the criminal world, and the country incapable of protecting its citizens. Therefore, poverty is not only a minimum income but also a special way of life, norms of behavior, stereotypes of perception, and psychology passed on from generation to generation.

In China, the problem of poverty eradication is one of the priority tasks. Since the country's goal is to build a middle-income society, poverty eradication is the most important part of this plan. Over the past four decades, China's poor population has decreased by more than 850 million people. This is a very important image for China's political leadership.

In South Korea, the government has largely succeeded in solving the problem of poverty. In the past two decades, less than 2% of the population lived on less than $5.50 daily. However, according to the Republic of Korea's National Statistics Service, more than half of the country's poor are residents of retirement age, which increases social tensions.

Poverty in Japan reaches 16%, which means that every sixth Japanese belongs to the layer of relatively poor. An analysis of this situation cites declining incomes for families with children in the context of prolonged deflation in the economy, as well as the increasing number of single mothers working on a freelance basis for low wages, as reasons for the record high poverty rate.

The poverty model that has emerged in Russia is primarily the result of low income from employment and, consequently, through their taxation, a low level of social transfers. In this connection, the phenomenon of Russian poverty can be defined primarily in terms of categories of "market poverty"—poverty associated with the place of the (economically active) population in the labor market.

A society may eliminate absolute poverty but always retain relative poverty. After all, inequality is a constant companion of complex societies. Thus, relative poverty persists even when the living standards of all sectors of society have risen.

In today's conditions, there is a global problem of human development associated with the impossibility of its existence without an orientation on sustainable development. It is necessary to transform the SDGs and their set of indicators, which is essential for Russia and the People's Republic of China, in terms of compliance with the specifics of their development, adaptation, and implementation. To adjust and effectively implement the SDGs and ESG management in Russia and the People's

Republic of China, the following main directions within the "Agenda 2030" should be highlighted:

- Adjusting the SDGs in relation to the national and local conditions of the countries;
- Ensuring sub-regional cooperation to achieve the SDGs;
- Creating a database for in-depth research and monitoring its results.

In the traditional format, the SDGs are a system of interconnected economic, social, and environmental goals, targets, and indicators. The emergence of the SDGs is due to the need to ensure the existence of humanity in the long term. By the early 2000s, the global community had shown positive momentum toward the SDGs. Nevertheless, a number of factors have exacerbated national and global challenges. First, we should note the significant scale of poverty, the existence of a significant gap between the incomes of the poor and the rich, and the growing economic "chasm" in the financial situation of economically developed and developing countries. Additionally, there is an increase in environmental problems, including negative climate change, reduced access to clean water in the poorest countries, and a significant increase in poverty.

Within the framework of the Sustainable Development Agenda, the countries of the world made a voluntary commitment to developing their own strategies for sustainable development, identifying the main areas of implementation of the SDGs and a set of the most significant indicators. The BRICS countries have developed and submitted national voluntary reviews of the implementation of the 2030 Agenda for Sustainable Development. There are significant differences in these reviews due to the existing views on the elaboration of sustainable development issues, considering the specifics of the functioning of the countries and their national development goals. Some BRICS countries formed two reviews, as China did, including the preliminary one in 2016 and the main one in 2021. The Russian Federation formed an expanded national review in 2020. The President of the Russian Federation signed a decree "On the national development goals of Russia until 2030." The approval of this decree contributes to a comprehensive formulation of national goals of sustainable development, the increase of the intensity of their implementation, and the progressive movement toward achieving the formulated system of goals.

As articulated by the United Nations, the set of SDGs is intended to be implemented throughout the global community, subject to adjustments in accordance with national circumstances. As such, the set of SDGs is a significant milestone for shaping and implementing a sustainable future in a partnership world.

Some countries have not formed an exhaustive set of indicators for the national system of SDGs. This refers to the distribution of indicators according to basic parameters, including disaggregation by gender, income, age, ethnicity, race, geographic location, migration status, disability, and other indicators that determine the principles of the formation of official statistics.

The occurring significant technological transformations require the adaptation of national development strategies to meet current trends. If such adaptations of strategies are successful, countries will have additional opportunities to achieve significant progress in development. The Russian Federation and the People's Republic of

China are moving along the path of industrial development toward a post-industrial society. The growth and development of these countries are directly dependent on the efficiency of internal institutions and changes in the external environment. In the case of efficient functioning of internal institutions, supported by forward-looking policies, the countries can successfully strive to achieve sustainable development, including in the process of comparing progress in growth rates. This requires a significant consolidation of efforts in trade, education, investment, and advanced scientific developments. Simultaneously, each country has specific features of development.

Russia's economy is still affected by the problems of the transition to market relations. The main feature of economic development can be considered the presence of high technology, unique materials, and qualified human capital, the use of which is associated with a lack of efficiency in the functioning of various kinds of institutions. Thus, the Russian Federation needs to increase the efficiency of domestic institutions and predictability of governance and reduce income inequality in the population to achieve the SDGs.

As for the People's Republic of China, we should point out the existence of a national system of mass production and export, which allowed it to make a significant breakthrough from extreme poverty and form an efficient market economy, competitive production, and consumption, which can rightfully be considered an example of the formation of industrial society. The main characteristic feature of such a society is inequality and the transition to a post-industrial society, which, in this case, requires significant institutional changes.

4 Discussion

The current state of the global world and the stability of its existence are currently of particular concern to the global community. Governance at the global level is fragmented, and there is no partnership in its implementation. This is due to the problems that have arisen with the beginning of the COVID-19 pandemic and the issues of stabilization of the world economy, which has been affected by the recession. Since 2015, the global agenda of the entire world community has undergone significant changes, moving from planning for prosperity by 2030, increasing the pace of achieving the SDGs, and taking measures to prevent negative climate change to the need to overcome the negative impact of the recession on economies, save millions of human lives from COVID-19 contamination, and prevent the spread of the disease. In today's environment, there is a need to address the urgent challenges posed by the COVID-19 pandemic, the recession, and the pursuit of the SDGs and ESG management.

The current world situation raises the question of the need to transform the SDGs towards their relevance in today's realities. This transformation should be mainly focused on human development, which, in turn, is impossible without the full support of public institutions, decision-makers of the most important management decisions, and detailed scientific research.

The emergence of the threat from COVID-19 should change the idea of priorities in economic policy in terms of achieving the SDGs. The pandemic situation has forced the world community to prioritize saving human lives, shifting economic growth to the periphery. It has come to understand that the life of every human being is of the highest value. In this regard, it is necessary to assess development differently, putting human life in the first place. For example, the life expectancy of China's metropolitan population is about five years shorter than that of the residents of China's ecologically clean regions. This is explained by the impact of environmental problems of water, land, and air pollution. In this connection, for a more accurate calculation of possible scenarios for increasing life expectancy, one should consider the risks of mortality due to environmental pollution from coal combustion and emissions of pollutants into the air, water, and land. The risks of death from such pollutants are much higher than the losses from the COVID-19 pandemic.

In this regard, it seems advisable to introduce a set of key indicators that could be used in the monitoring of the SDGs in the study of the possibility of achieving the SDGs. In this context, it is possible to propose the use of the following indicators:

- The planned life expectancy of the population (years). This indicator can be considered the main one because, based on the calculated values, it clearly shows the level and conditions of life of the country's population;
- The percentage ratio of medical expenses and GDP, which will allow controlling the general level of readiness of countries to withstand the threats to the health of the population caused by different health and life-threatening factors;
- The total number of hospital beds calculated to eliminate the shortage of places for patients in medical institutions. This refers to inpatient and outpatient care;
- The aggregate index of human development, which includes a set of parameters conditioning the assessment of the social component of sustainable development. Scientists have developed a significant number of modifications to this human development index, which includes such factors as gender, inequality, various aspects of education, etc. It is difficult to overestimate its importance because it is based on the evaluation of social aspects as opposed to economic components. For example, countries with a high level of GDP, which have a low life expectancy, are evaluated significantly lower than those with a high life expectancy. When examining the human development index by country, the Russian Federation and the People's Republic of China are among the countries with medium and high levels of human development. However, this indicator cannot help address multi-dimensional sustainability issues because it does not contain the ability to assess the environmental component.

5   Conclusion

To summarize the above, in our view, the implementation of the transformation of the SDGs and ESG management, carried out to update them, seems to be the main vector of the interaction of the world community in the long term. In doing so, it may be easier to form an additional official document for the SDGs, within which the main opportunities and priorities for behavior during a global recession and pandemic should be disclosed in a protocol. The combined forces of the international community must now be focused on tackling the COVID-19 pandemic and the global recession. The recovery of the world economy due to the impact of the recession and the COVID-19 pandemic will only be possible under the conditions of global coordination of a set of measures for the efficient use of the world's limited resources, which will contribute to the achievement of the SDGs. The actualization of the SDGs will allow combining efforts to achieve them through global coordination and management of this process within the international community.

References

1. Grigoryev L, Morozkina A (2013) Different economies, similar problem. Russia Glob Aff 2:26–39. Retrieved from https://eng.globalaffairs.ru/articles/different-economies-similar-pro blems/?ysclid=l984hclhcp777685853. Accessed 9 Sept 2022
2. Grigoryev L (2020) Global social drama of pandemic and recession. Popul Econ 4(2):18–25. Retrieved from https://populationandeconomics.pensoft.net/article/53325. Accessed 9 Sept 2022
3. Jain-Chandra S, Khor N, Mano R, Schauer J, Wingender P, Zhuang J (2018) Inequality in China—trends, drivers and policy remedies. In: IMF working papers No. 18/127. IMF, Washington, D.C.. Retrieved from https://www.imf.org/en/Publications/WP/Issues/2018/06/05/Ine quality-in-China-Trends-Drivers-and-Policy-Remedies-45878. Accessed 9 Sept 2022
4. OECD (2020) Carbon dioxide emissions embodied in international trade. Retrieved from https://www.oecd.org/sti/ind/carbondioxideemissionsembodiedininternationaltrade.htm. Accessed 9 Sept 2022
5. Richard J (2020) Towards a new ecological and human type of national accounting for developing economies (The CARE/TDL model). BRICS J Econ 1(1):43–59. Retrieved from https://brics-econ.arphahub.com/article/24159/. Accessed 9 Sept 2022
6. UN (2020) Voluntary national reviews: database. Retrieved from https://sustainabledevelop ment.un.org/vnrs/. Accessed 9 Sept 2022
7. UN. Economic and Social Council (2020) Progress towards the sustainable development goals. Report of the secretary–general (E/2020/XXX). Retrieved from https://sustainabledeve lopment.un.org/content/documents/26158Final_SG_SDG_Progress_Report_14052020.pdf. Accessed 9 Sept 2022
8. UN General Assembly (2000) United nations millennium declaration. Retrieved from https://www.un.org/en/development/desa/population/migration/generalassembly/docs/ globalcompact/A_RES_55_2.pdf. Accessed 9 Sept 2022
9. Usenko LN, Usenko AM, Guzey VA, Bidzhieva AS (2022) Analysis of management paradigms in the contemporary finance theory. In: Trifonov PV, Charaeva MV (eds) Strategies and trends in organizational and project management. Springer, Cham, Switzerland, pp 645–650. https://doi.org/10.1007/978-3-030-94245-8_88
10. Usenko LN, Guzey VA, Bidzhieva AS (2022) Modern opportunities for optimizing business processes to achieve sustainable development. In: Trifonov PV, Charaeva MV (eds) Strategies and trends in organizational and project management. Springer, Cham, Switzerland, pp 138–150. https://doi.org/10.1007/978-3-030-94245-8_19

Targeting ESG Initiatives in the Formation of Financial Mechanisms for the Development of Bordering Territories of Kyrgyzstan

Chinara R. Kulueva⬤, **Kurmanbek K. Ismanaliev**⬤, **Jibek B. Seitova**⬤, **Gulsana P. Turganbayeva**⬤, and **Jazgul A. Tokosheva**⬤

Abstract This research focuses on the formation of financial mechanisms for the development of regions, in particular, border areas. Due to the current conditions associated with the strengthening of integration processes, the problems of socio-economic development of border territories are of particular interest to researchers-economists, where special attention is paid to the issues of regional labor resources as a significant factor in the sustainable economic development of any country. In the industrial economy, the volume of factors involved in the production was measured by the amount of production means, labor objects, and labor itself. In turn, nowadays, the measures are the mechanisms that regulate labor relations, considering the protection of workers' rights, and aim to ensure the financial and economic stability of regions. Due to the existing imbalance between the production and technological parameters of jobs and the existing structure of labor resources, it seems important to develop policy directions that would help achieve consistency between the area of employment in the context of sectors of the economy and in the context of regions. The socio-economic transformations occurring in Kyrgyzstan, particularly in its border areas, are closely related to the pace of migration processes, which show a relatively high activity compared to the more prosperous areas close to the country center. Therefore, targeted scientific research that examines issues related to reproduction, formation, and rational use of the able-bodied population, the labor market,

C. R. Kulueva (✉) · K. K. Ismanaliev · G. P. Turganbayeva · J. A. Tokosheva
Osh State University, Osh, Kyrgyzstan
e-mail: ch.kulueva@mail.ru

K. K. Ismanaliev
e-mail: kurmanbekismanaliev@gmail.com

G. P. Turganbayeva
e-mail: Tgp.1778@mail.ru

J. A. Tokosheva
e-mail: jtokosheva@mail.ru

J. B. Seitova
Kyrgyz National University Named After Jusup Balasagyn, Bishkek, Kyrgyzstan
e-mail: jseitova.88@mail.ru

© The Author(s), under exclusive license to Springer Nature Switzerland AG 2023
E. G. Popkova (ed.), *Smart Green Innovations in Industry 4.0 for Climate Change Risk Management*, Environmental Footprints and Eco-design of Products and Processes, https://doi.org/10.1007/978-3-031-28457-1_49

especially in depressed areas, and their interaction with appropriate financial policies and financial and credit mechanisms and instruments are undoubtedly relevant and timely. The authors attempted to recommend ESG initiatives (approximate model) for the formation of financial mechanisms for the development of border territories of Kyrgyzstan, designed for 10–15 years, the efficiency of which determines the sustainable development of the national economy and the country.

Keywords Border areas · Socio-economic development · Labor resources · Labor market · Financial mechanism · Financial policy · Living standards · Sustainable development · Economic welfare

JEL Classification R38 · O18 · J40 · J22 · G32 · P43 · I31 · O11 · Z18 · D60

1 Introduction

The comprehensive program of socio-economic development (SED) of a region is a large and archival large-scale document, which details the development strategy of the territory and methods of implementation of the set tactical tasks, defines the form of organization and management of economic life, and offers the optimal model of the organization of the life of the society in a particular territory. However, the question is in the correctness of the choice of a strategy and policy objectives proposed by the program aimed at the concentration of available resources using the economic and production potential, etc., through which it will be possible to achieve some growth in the region, in our case, the border areas of Kyrgyzstan. The development program of an individual territory must fit into the country's development program. Thus, its strategy must express its interests and its goals; this is an indisputable requirement that once again emphasizes the country's organic integrity.

Since Kyrgyzstan's independence (1991–2022), several constructive measures were taken aimed at regional economic development in conditions of adaptation of market mechanisms, which were to yield tangible results in the next 5–10 years after the declaration of independence. As practice shows, the model of socially-oriented development of market relations still does not yield the desired results due to some unreasonable and hasty decisions, where the border areas fall out of sight of the strategies of accelerated SED of the republic.

Nevertheless, the development of border areas is an evolutionary process of development of production relations and productive forces, where all conditions for developing the country, business, entrepreneurship, and individuals are concentrated at the qualitative level. It should be noted that it is impossible to achieve the desired sustainable development without appropriate education, science, innovation and labor policies, financial injections, and capital investments.

Kyrgyzstan has adopted several state development programs of the country, including the following:

- National development strategy of the Kyrgyz Republic for 2018–2040 (Approved by the Decree of the President of the Kyrgyz Republic of October 31, 2018 UP No. 221) [1];
- Concept of regional policy of the Kyrgyz Republic for 2018–2022 [2];
- Program for the development and support of small and medium entrepreneurship in the Kyrgyz Republic for 2019–2023 [3];
- Digital transformation concept "Digital Kyrgyzstan 2019–2023" [4].

These programs pay considerable attention to the scale and policy directions of sustainable development of the considered territories, given their rates and proportions of growth, giving much attention to depressed territories. The documents mentioned above pay special attention to structural and functional subsystems of the SED, aimed at the adoption of, first, the real tasks associated with increasing the efficiency of production management and, second, cardinal measures related to the improvement of labor relations and the quality of working life of the able-bodied population and the residents of border territories.

Financing the development of border areas is primarily an expense in the budget; it must be balanced with revenues. We consider it advisable to make a calculation of budget revenues after determining the additional production of industries, regions, and border areas. After this, it is necessary to consider the potential for the effective use of financial resources, including other additional forms of financing (e.g., investment). A significant principle of inclusion of an investment project in the program of development of border territories is to classify it by the totality of the relevant characteristics as a priority, thus forming the economic core of the country's development strategy.

Nowadays, the lack of financial resources is a considerable obstacle to implementing programs of the SED of border territories. As for the country's own sources, without external reciprocity or investment, the main revenue-generating taxes are income tax, value-added tax, and land tax; the main tax-forming parameter is the output volume. Accordingly, providing the necessary amount of output intersects with the necessary quantity and quality of the able-bodied population, which, in recent decades, has been stable at the level of migration activity due to the low level of quality of working life and financial security of households.

The real situation related to COVID-19 and its consequences exposed all existing problems in the border areas and caused an increase in migration processes at the end of 2021, continuing to the present.

2 Materials and Methods

Theoretical substantiation of issues of the SED of regions and border areas occupies a special place in the works of researchers, including Antonyuk and Kornienko [5], Bozhko [6], and Popkova et al. [7]. These authors try to work out a rational version of the scenario of the development of regions and border areas through the formation of mechanisms of sustainable growth. We also share the opinion of scholars engaged in studying the region, including Samofalova et al. [8], as well as Stepanov et al. [9]. These authors offer a methodology SAPSED (system of analysis and prediction of socio-economic development), contributing to the understanding of the real picture of the region and the adoption of optimal profitable decisions in the studied areas.

It is also necessary to highlight the theoretically substantiated views of Harrod and Domar [10], the supporters of neo-Keynesianism who preach the role of labor resources in the development of the territory through the effective use of investment resources. Representatives of neoclassicism Rostow [11], Nurkse and Leibenstein [12], as well as Hirschman and Singer [13], associated the importance of economic growth rates with the use of the latest generation technology, advanced technology, the development of priority sectoral economic structures, and increasing the share of productive accumulation in national income aimed at meeting the diverse needs of the population in the region. The adherents of institutionalism, Kondratiev [14], Murdahl [15], and Veblen [16], argue the need for a controlled economy, where the role of the working person is invaluable in achieving economic growth of the territory and ensuring its security.

In recent years, the issues of economic (financial) security in the regions have become the subject of notable attention and research by scientists, including Chernolutskaia [17], Illarionov [18], Kislaya [19], and Kuklin [20]. These authors focus on the problems of uncontrolled migration, deteriorating living conditions, the collapse of the educational system, the emergence of conflicts, radicalism, extremism, increased criminal activity, etc., which becomes a catalyst for the escalation of conflicts and economic crises. According to the most recent works, the demographic factor, including migration, is a pivotal factor in achieving sustainable SED in the border areas due to their high sensitivity to the impact mechanisms, thus determining the level of effectiveness of regional policy in the national economy of Kyrgyzstan. Consequently, a significant part of the research is aimed at identifying the level of determinants that contribute to the inclusion of the national economy of the Kyrgyz Republic in the framework of the international division of labor, thereby assessing the effectiveness of its participation and activity in the global economy.

The studies of Abdymalikov [21], Koichuev [22], and Kupuev [23] focus on the problems of economic development of regions, especially rural areas of different levels of Kyrgyzstan. These studies aim to find effective mechanisms of interaction between the subjects of the market economy at the level of the industry economy, macroeconomics, and the world economy to achieve some stability in socio-economic development, where a significant role belongs to contemporary problems of environmental protection, policy, and environmental issues.

An important place in the study of sustainable development of border areas belongs to the competent construction of appropriate strategic development programs. Failure to address and study the topical problems of border territories, particular regions, or the whole country can cause tangible damage and irreversible processes in the social and economic life of the studied territories [24].

The various theoretical and methodological doctrines considered in the work allow us to speak about them as the main determinants of national economic development, especially in the border areas, where the influence of many factors on the state of labor resources, as life practice shows, is complex, contradictory, and ambiguous. Based on the doctrines of the researchers mentioned above and our own field research, we note that Kyrgyzstan, especially for its border areas, requires an effective state financial and economic policy based on principles that ensure fairness in the revision of labor market functions to transition to contemporary, relatively non-standard forms and types of employment, functionally more flexible differentiated methods of wages and social protection of workers and citizens, a policy of adapting the price and quality of labor, its structure and quantity to cyclical fluctuations in supply and demand in the labor and financial markets, and the viability of the national [25] and cross-border economy.

3   Results and Discussion

Contemporary profound scientific research and theoretical substantiation of the problems of sustainable development of border territories require generating them towards the development of a new branch of economic theory—the theory of the economics of border territories, theory of regional economics, and theory of cross-border economy (economy of border territories of two or more bordering countries). Nevertheless, the study of the problem identified the need for innovation and investment-oriented development as one of the key mechanisms that drive all available resources of the regions.

According to the analysis, the current state of the development of territories is determined by such a sensitive indicator as the level of household finances of the able-bodied population, which is currently considered not only from the quantitative side but also focused on its qualitative component. The issues of professional [25] and competent attitude to labor require legal revision by the state, accompanied by tax benefits and vacations and the provision of relatively cheap funds for the construction of housing complexes, the development of livestock at home and as a family business, support for the able-bodied population in obtaining vocational education, etc.

The most significant economic levers to strengthen the attractiveness of border areas are the proposed ESG initiatives in the formation of financial mechanisms for the territorial development of Kyrgyzstan.

The model (Fig. 1) considers the timing, scope and direction, and program interests of the national and regional economies. This perspective must be controlled by the relevant authorities (in Fig. 1, they are marked with the symbol—**GO**). At the stage of implementation of the proposed ESG initiatives in the long term, it is necessary to implement purposeful work to achieve a relatively decent level and quality of working life for citizens living in the border areas and in the whole country (**R**). The proposed initiatives consist of the following specific tasks.

- 1R—population of territories and regions with a high level of housing improvement;

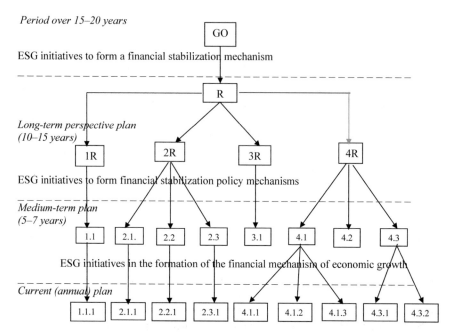

Fig. 1 ESG initiatives in the formation of financial mechanisms of territorial development in Kyrgyzstan. *Source* Developed by the authors based on [25])

- 2R—high degree of service to the population in socio-cultural areas;
- 3R—integrated development of subjects and objects of the economy of border territories;
- 4R—a relatively high degree of disposable and real income of residents of the territories.

The following initiatives have been considered at the five-year perspective stage:

1.1 A relatively higher rate of housing construction;
2.1 The creation of a relatively high level of medical care for residents of territories and regions, the activities of sanatoriums and rest homes to protect the health of workers;
2.2 A comparatively faster pace of construction of enterprises of strategic importance and development of public services;
2.3 All-level reform of education aimed at improving the training of competent professionals guided by advanced innovative technologies;
3.1 Construction of enterprises of various forms of ownership, considering the policy of support for gender equality;
4.1 Relatively high growth rates of real wages of able-bodied residents of the territories, 3–5 times higher than the national average;
4.2 Revision of the income policy in the direction of supporting the achievement of relatively high incomes for agricultural producers;

4.3 Supporting the activities of public consumption funds to provide guaranteed payments to the population of border territories.

At the level of the current (annual) plan, the authors proposed initiatives to ensure the implementation of plans of the first and second levels, affecting the development of the economy of border areas and the overall national economy in the following directions:

1.1.1 House construction;

2.1.1 Construction of social infrastructure, health care, education, etc.;

2.2.1 Construction of commercial and cultural facilities;

2.3.1 Construction of institutes, technical schools, vocational schools, and secondary schools;

4.1.1 Improvement of territorial regulation of wages and other forms of payments and remuneration;

4.1.2 A relative increase in the level of territorial coefficients to the wages of workers and employees, military personnel of border territories;

4.1.3 Review of the long-service salary increment for workers and employees of border territories;

4.3.1 Setting preferential rates for residents of the territories for social, cultural, and household services and transportation;

4.3.2 Introducing additional leave for workers and employees of border territories, depending on the length of service and continuous work at the same enterprise.

4 Conclusion

Thus, the effectiveness of the proposed ESG initiatives in the formation of financial mechanisms for the development of border areas of Kyrgyzstan, aimed at the regulation and management of human and financial resources in border areas to optimize them, suggests a systemic approach to solving this problem.

The degree of development of specialized financial and credit institutions provides a full basis for such institutional investors as insurance companies and investment funds to provide "long and cheap money" to the population living in the border areas and attract foreign direct investment resources, etc., which requires the development of state programs for socio-economic development of the considered territories.

The initiatives mentioned above are long-term. Therefore, they require a sufficient period of time and substantial financial investment. The implementation of cardinal programs requires step-by-step attention and solid investments. It is appropriate and more effective to solve problems for the next 3–5 years first and only then proceed to the implementation of tasks of a long-term nature.

Simultaneously, the ESG initiatives involve strengthening the positive and smoothing the negative effects in the process of economic development in the border areas, which can be caused by specific reasons and the intensification of international economic activity in the EAEU aimed at ensuring national security and the country's integrity.

References

1. President of the Kyrgyz Republic (2018) National development strategy of the Kyrgyz Republic for 2018–2040 (Approved by the Decree of the President of the Kyrgyz Republic of October 31, 2018 UP No. 221). Bishkek, Kyrgyzstan. Retrieved from https://mfa.gov.kg/uploads/content/1036/3ccf962c-a0fc-3e32-b2f0-5580bfc79401.pdf. Accessed 15 Aug 2022
2. Government of the Kyrgyz Republic (2017) Concept of regional policy of the Kyrgyz Republic for 2018–2020 (Approved by the Decree of March 31, 2017 No. 194). Bishkek, Kyrgyzstan. Retrieved from http://cbd.minjust.gov.kg/act/view/ru-ru/99907. Accessed 15 Aug 2022
3. Government of the Kyrgyz Republic (n.d.) Draft resolution "On the approval of Program for the development and support of small and medium entrepreneurship in the Kyrgyz Republic for 2019–2023." Bishkek, Kyrgyzstan. Retrieved from https://mineconom.gov.kg/ru/discussion/18. Accessed 16 Aug 2022
4. Cabinet of Ministers of the Kyrgyz Republic (2019) Roadmap for implementing the digital transformation concept "Digital Kyrgyzstan 2019–2023" (Adopted by Order of the Government of the Kyrgyz Republic of February 15, 2019, No. 20-r, as amended October 5, 2020 No. 347-r). Bishkek, Kyrgyzstan. Retrieved from https://www.tunduk.gov.kg/files/1/ДОРОЖНАЯ%20КАРТА%20по%20реализации%20Концепции%20цифровой%20трансформации%20Цифровой%20Кыргызстан%202019_2023.pdf. Accessed 15 Aug 2022
5. Antonyuk VS, Kornienko EL (2022) Economic development of Russia's old industrial border regions. J New Econ 23(2):45–63. https://doi.org/10.29141/2658-5081-2022-23-2-3
6. Bozhko LL (2011) Theoretical and methodological foundations for the study of the processes of economic development of border territories (synopsis of dissertation of doctor of economics). Ural State University of Economics, Yekaterinburg, Russia
7. Popkova EG, Oudah AMMY, Ermolina LV, Sergi BS (2021) Financing sustainable development amid the crisis of 2020. A research note. In: Popkova EG, Sergi BS (eds) Modern global economic system: evolutional development versus revolutionary leap. Springer, Cham, Switzerland, pp 773–780. https://doi.org/10.1007/978-3-030-69415-9_88
8. Samofalova EV, Kuzbozhev EN, Vertakova YV (2008) State regulation of the national economy: textbook, 4th edn. Knorus, Moscow, Russia
9. Stepanov AA, Zotova AI, Savina MV, Guznina GN, Kochetkov AA (2015) Efficiency estimation criteria of agro-industrial systems in post-industrial economy. Asian Soc Sci 11(14). Retrieved from https://pdfs.semanticscholar.org/4a46/6e482243ecb8c18f633431636bcc54600946.pdf . Accessed 19 Aug 2022
10. Zavelsky MG (1998) Economics and sociology of labor. Catallaxy, with the participation of Knorus, Moscow, Russia
11. Rostow WW (1961) The stages of economic growth (BP Martschenko Transl. from English). Frederick A. Praeger, New York, NY. Retrieved from https://vtoraya-literatura.com/pdf/rostou_stadii_ekonomicheskogo_rosta_1961__ocr.pdf. Accessed 12 Sept 2022
12. Brovkina VV, Sycheva AV (2017) The concept of a balanced set of investments by R. Nurkse. NovaInfo 74:263–266. Retrieved from https://novainfo.ru/article/14247. Accessed 16 Aug 2022
13. Kapitsa SP, Kurdyumov SP, Malinetsky GG (2001) Synergetics and predictions for the future, 2nd edn. Editorial URSS, Moscow, Russia
14. Grinina LE, Korotayeva AV, Bondarenko VM (eds) (2017) N. D. Kondratiev: crises and forecasts in the light of long wave theory. A view from modernity. Uchitel, Moscow, Russia
15. Murdahl G (1972) Contemporary problems of the third world. Progress Publishers, Moscow, Russia. Retrieved from https://www.booksite.ru/fulltext/eco/nom/iks/8.htm. Accessed 12 Sept 2022
16. Veblen T (1899) The theory of the leisure class: an economic study of institutions. Retrieved from https://gtmarket.ru/library/basis/5890. Accessed 12 Sept 2022
17. Chernolutskaia EN (2019) State development programs of the Far East and demographic dynamics in the Far Eastern border territories in the late 20—early 21 century, vol 22. In:

Proceedings of the institute of history, archaeology and ethnology FEB RAS pp 76–93.https://doi.org/10.24411/2658-5960-2019-10006

18. Illarionov AN (1998) Criteria for economic security. Vop Ekonomiki 10:35–58. Retrieved from http://iea.ru/article/publ/vopr/1998_10.pdf. Accessed 17 Aug 2022

19. Kislaya TN (2021) Theoretical and methodological approaches to the management of the economic security of the region: monograph. Publishing house "Sreda.", Cheboksary, Russia

20. Kuklin AA (2014) Economic security of regions: theoretical and methodological approaches and comparative analysis. Fundam Res 6–1:142–145. Retrieved from https://fundamental-research.ru/ru/article/view?id=34127. Accessed 12 Sept 2022

21. Abdymalikov KA (2010) The economy of Kyrgyzstan (in transition): textbook. Bijiktik, Bishkek, Kyrgyzstan

22. Koichuev TK (2007) Collection of selected essays. Volume II: post-soviet perestroika: theory, ideology, and realities. CEC under the PCR "Economists for Reform" Public Association, Bishkek, Kyrgyzstan

23. Kupuev PK (1995) Shaping the labor market: issues of theory and practice: a practical guide. Osh, Kyrgyzstan

24. Khandazhapova LM, Lubsanova NB (2015) The scientific basis for the study of sustainable economy of the border region. Reg Econ: Theor Pract, 13(15):40–47. Retrieved from http://213.226.126.9/re/2015/re15/re1515-40.pdf. Accessed 19 Aug 2022

25. Kulueva CR (2016) Economic and financial problems of labor resources in the southern region of the Kyrgyz Republic (synopsis of dissertation of doctor of economics). Osh State University, Bishkek, Kyrgyzstan

ESG Initiatives for the Managed Development of Light Industries in the Regions of Kyrgyzstan

Chinara R. Kulueva(iD)**, Zhannat K. Rayimberdieva**(iD)**,
Baktygul T. Maksytova**(iD)**, Cholpon A. Nuralieva**(iD)**,
and Clara T. Paiysbekova**(iD)

Abstract This research focuses on developing a controlled light industry in the regions of Kyrgyzstan. The strengthening of integration processes and the problems of socio-economic development of the country's regions are of particular interest to researchers-economists in the issues of the current state of Kyrgyz industrial policy, in particular, light industry, as the main factor in achieving sustainable economic development. The light industry plays a significant role in replenishing the country's budget, providing employment, meeting the population's needs in goods, using local raw materials, increasing exports and import substitution, and developing small and medium-sized businesses and in the regional policy of the state. In today's conditions, the domestic market for light industry products is one of the most popular in the CIS countries, especially for Russia and Kazakhstan, due to its popularity, high attractiveness, and relatively low cost for neighboring countries. The branches of the light industry ensure the country's strategic security, satisfying the needs of the subjects for clothing. In terms of consumption, light industry products are second only to food products, far ahead of the markets for other goods. Currently, the economic situation in the light industry is characterized by the challenges of globalization and the open market, goals and objectives, the achievement and solution of which requires new approaches in the short and long term. The authors attempted to recommend ESG initiatives for developing a controlled light industry in the regions of Kyrgyzstan,

C. R. Kulueva (✉) · Z. K. Rayimberdieva · B. T. Maksytova · C. T. Paiysbekova
Osh State University, Osh, Kyrgyzstan
e-mail: ch.kulueva@mail.ru; ch.kulueva@oshsu.kg

Z. K. Rayimberdieva
e-mail: barsbeknur@mail.ru

B. T. Maksytova
e-mail: makbak67@mail.ru

C. T. Paiysbekova
e-mail: kpajysbekova@gmail.com

C. A. Nuralieva
Kyrgyz-Russian Slavic University named after B. Yeltsin, Bishkek, Kyrgyzstan
e-mail: nur-cholpon@mail.ru

the effectiveness of which depends on the sustainable development of the domestic light industry, the national economy, and the country as a whole.

Keywords Light industry · Export and import · Industrial policy · Cluster economy · Regional development · Sectoral features of entrepreneurship · Organizational and economic mechanism · Competitiveness of enterprises · Production potential

JEL Classification B30 · L67 · O19 · L52 · L69 · R58 · M21 · F41 · O40 · D26

1 Introduction

The light industry has always played an important economic and social role in the country's economy. Currently, the Kyrgyz Republic has many opportunities to develop light industry, increasing exports to near and far abroad. However, not all light industry enterprises turned out to be prepared for market conditions. Independent economic activity in market conditions is associated with overcoming many difficulties. In this situation, it is necessary to find effective forms of organization and management of production.

The contemporary light industry operates in an open market, which requires an appropriate organization of business activities in industries and new approaches to solving problems in the long term, where the role of meeting the internal needs of the population with domestic products and then external consumers is invaluable [1]. For reference, Kyrgyzstan was the first post-Soviet country to join the WTO—it became its 132nd full member in November 1998.

Nowadays, most of the economic entities of the industry operate in the presence of two interrelated problems. On the one hand, there is a relatively low level of competitiveness of manufactured products due to the presence of obsolete technology in production. On the other hand, there is a lack of sufficient funds to upgrade the old technopark due to the provision of financial and credit institutions for these purposes only expensive money.

The orientation of production to a stable and efficient activity is aimed at improving the complexes of theoretical and methodological issues to strengthen the effective state of the external competitive environment, considering the internal features of production. Nevertheless, one should consider the long-term interests of the state in the development of domestic light industry, determined by the following factors:

- First, the light industry of Kyrgyzstan is traditionally focused on the domestic market; due to objective circumstances related to ensuring the conditions for human life, it always has a steady demand;
- Second, the light industry must solve the problem of self-sufficiency of the Kyrgyzstan Republic with the most important types of fabrics, clothing, and footwear for the population and clothing for law enforcement agencies, which will eliminate the threat to the country's economic security.

Currently, the economic environment can be characterized by the following features: an increase in the rate of change in environmental factors, increased competition in existing markets, the desire of business in IT technology, and penetration of business into the Internet and mobile technologies. A business based on unique ideas, knowledge, and information, focused on the creation of new products and services, is replacing the usual types of business that are based on the use of natural resources. In these conditions, there arises the question of adapting industrial enterprises to contemporary business conditions.

Along with the above, the light industry also has serious staffing problems, associated, above all, with the low level of prestige of working professions in the industry and the lack of infrastructure for training workers, insufficient competitiveness of manufactured products compared with foreign analogs, the observed rupture of economic ties between organizations and partners, underloading of production capacities, and, as a result, their wear and tear and the impossibility of producing competitive products [2].

Therefore, to solve the above problems, new approaches and tools are required for organizing the economic activity of the subject and doing business. Integration processes are becoming increasingly important in the economy. The creation of integrated structures in the light industry is one of the conditions for the further development of the economy, strengthening the competitive positions of business entities, and controlling changes in the external environment in the Kyrgyz and international markets.

The identification of stable and regular links between technical and technological innovations and fluctuations in the business activity of enterprises creates the grounds necessary for the development of effective mechanisms to overcome or reduce the impact of the negative consequences of the cyclical nature of market mechanisms and laws on the development of the economic life of a country, industry, region, and enterprise.

Therefore, the study, generalization, justification, and refinement of theoretical and methodological approaches to the development of light industry in an innovative economy are relevant and can be put forward as priorities in the structure of socio-economic research.

2　Materials and Method

The works of such foreign scientists as Feldman [3], Granberg [4], Laundhardt [5], Lösch [6], Weber [7], and others are devoted to the study and justification of the regional development of the industry.

Certain aspects of the development of industrial policy and innovative economy are considered in the works of Kyrgyz scientists, including Abdymalikov [8], Abdyrov and Toktogulov [9], Chernova [10], Koychuev [11], Kupuev [12], and Musakozhoev [13].

Simultaneously, the versatility and complexity of the problems of development of regional industry, including light industry, require further research. Since the beginning of the 1990s, namely from the declaration of the country's independence, the development of the national economy of the Kyrgyz Republic has been complex and ambiguous. The transition period of the economic system to new economic conditions—to market relations, the crises of 1998, 2008, and 2018, the COVID-19 pandemic that broke out in 2019–2021, and the deterioration of the foreign policy were a serious test for the sectors of the national economy and the scientific community.

Contemporary industrial policy, in particular, sectoral (light), is constantly subjected to critical reflection in view of the accumulated experience over the historical period. Therefore, we believe that it is necessary to consider the approaches of different economic schools to state regulation of the economy and industrial production, particularly the light industry. Although the country's light industry is developing relatively better, fully-fledged market mechanisms have not yet been formed in it because there are logistical, technological, organizational, and economic problems. The problems of business development in the Kyrgyz Republic also require deep research and can be solved in the long run. Therefore, a revision of government strategies to support entrepreneurship is required. In this case, it is necessary to consider the regional and sectoral characteristics of entrepreneurship.

3   Results and Discussions

The Kyrgyz light industry, in particular the clothing industry, is mainly export-oriented. The main part of the sales of products of Kyrgyz clothing enterprises falls on the CIS countries, namely Russia and Kazakhstan. According to official statistics, the export of garments in 2016 amounted to $97.2 million. By 2020, this figure has increased several times over the past four years. Approximately 80% of production is exported to Russia. Nowadays, Kyrgyzstan occupies about 6% of the Russian market for ready-made clothes. Approximately 40% of trade transactions are found to pass through the Dordoi market in Bishkek, the largest market in Central Asia, where transactions are carried out for cash. The main part of sales takes place in the clothing markets of various regions of Russia (cities where there are sales representatives of the Legprom association).

For the further development of the light industry, we consider it necessary to implement the following measures by the government:

- Creation of autonomous production bases of technopolises in the regions of the Kyrgyz Republic;
- Resumption of cotton sowing and silk production;
- Creation of industrial parks in the city of Osh and the Osh region using attracted investments;

- Adoption of a specialized program for the development of animal husbandry to create a domestic raw material base for wool and leather;
- Modernization and development of the textile industry and the republican base of laboratory expertise.

The global light industry is characterized by constant economic growth associated with an increase in the population of the Earth and an increase in its well-being and purchasing power. Accordingly, the world market for light industry products is developing dynamically. Over the past fifteen years, trade has more than doubled. The consumption of fabrics, clothing, and footwear increased by 90.5% in the EU countries, by 99.3% in the USA, and by more than two times in Japan.

Active innovation and investment activity play an important role in the global textile market and give them significant competitive advantages. The trading system determines global trends in the creation and development of the textile market. It should be noted that individual textile enterprises in neighboring Tajikistan have the opportunity to enter world markets with their products.

Depending on the nature of exports in the world market, countries engaged in production should be classified into three types:

1. Manufacturers of unbranded garments of lower quality in bulk, such as China, India, Bangladesh, and Vietnam;
2. Producing countries of the main brands with higher quality. These countries include Germany, France, Italy, and the USA;
3. A trading country with a small volume of production or added value of garments. These countries include Hong Kong, Belgium, and the Netherlands.

The compound annual growth rate of global garment exports was 6.7% between 2016 and 2020. In this indicator, China is by far the largest supplier in the world. It accounts for 30.3% of garment exports.

Based on the above, the following conclusions can be drawn:

- It is necessary to carry out an analysis and actions on the regional development of Kyrgyzstan, where the role of the government in supporting economic entities in light industry sectors is invaluable;
- We consider appropriate the country's desire for the uniform development of all sectors of the economy of Kyrgyzstan, in particular, the light industry. This requires the application of industrial policies aimed at specific goals.
- The success of the sustainable development of the regions of Kyrgyzstan largely depends on the coordinated actions of business structures, the regions' population, and the state and regional authorities.
- To achieve leading positions by light industry enterprises of the Kyrgyz Republic in the world textile market, it is necessary to master advanced foreign practices and intensify the introduction of innovations and advanced technologies;
- To win leading positions in the world market, it is necessary to increase the competitiveness of goods (quality, design, etc.).

At the beginning of the reforms, in the 1990s, in the light industry and its sectors, as well as in other sectors of the real sector, the artificially and illiterately carried out reform became the main reason for a significant decline in the volume of production of light industry products, which led to a halt in production and the closure of many enterprises, not only republican but also world significance: Osh cotton mill named after 50th anniversary of October, Osh silk factory, Kyrgyz shoe factory "Cholpon," Kyrgyz worsted and cloth factory, etc. The transition to market production mechanisms and the open market caused a massive influx of low-quality and cheap imported goods. As a result, the manufactured products turned out to be relatively uncompetitive compared to goods from Turkey, India, China, etc.

Simultaneously, there were problems with the supply of raw materials, especially cotton, from neighboring republics for the production of environmentally friendly products. Cotton in Kyrgyzstan is mainly grown in some areas of the Osh and Jalal-Abad regions. It is not grown in other areas of the southern region due to relatively harsh natural and climatic conditions, which led to a reduction in the share of light industry in the gross national product.

From 2011 to 2020, the industry has undergone ups and downs, which is mainly characterized by heterogeneity and unstable development. Only in 2020 did the production volume reach the level of 2011, amounting to a growth rate of 100.5%. Thus, the share of the light industry in the structure of the entire industry amounted to 2.4%, which shows the lowest result for the reviewed period [14, 15]. In the industrial production of the Kyrgyz Republic, the total volume of production in 2020 amounted to 319.4 billion soms; the production of precious metals traditionally dominates in terms of volume. If we exclude the "Production of basic metals" component from the total industrial production (50% in total industrial production in 2020), then the share of the light industry would be 7% of industrial production [14, 15].

For the reviewed period, the share of production by light industry branches has a slight decrease from year to year (Fig. 1).

First, this is explained by the prevalence of imported goods in regional and domestic markets. Second, it is explained by the weak intra-industry policy of

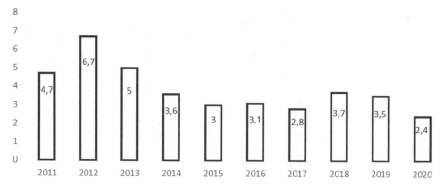

Fig. 1 The share of manufactured light industry products in the total volume of the industrial output of the Kyrgyz Republic for 2011–2020, in %. *Source* Compiled by the authors based on [14, 15]

import substitution and export orientation. Third, the consequences of the COVID-19 pandemic have caused irreparable damage to the entire national economy of the Kyrgyz Republic, including industry, especially light industry enterprises, which were forced to quickly restructure their production to the production of sanitary and hygienic products related to the prevention of problems with COVID-19.

Since the beginning of market transformations, Kyrgyzstan and its regions generally saw a tendency to reduce the number of light industry enterprises. In 2020, enterprises of domestic textile and clothing production in the Kyrgyz Republic as a whole decreased by 51.8% compared to 2004. In the regions of the Southern region, the decrease is 18.2%. The sector of production of footwear and leather products in the industry occupies a vulnerable place [15].

In terms of regions, the main share of light industry production falls on Bishkek (57.7%), Chui (29.7%), and Jalal-Abad (4.9%) Regions. In other regions, the share of the industry ranges from 0.01%; it reaches up to 4.3% in the Talas Region and 5.7% in the Osh Region. Until the beginning of 2020, the light industry of the Kyrgyz Republic had all signs of growth in production volumes, which were characterized by the following indicators:

- The index of the physical volume of garment products over the past three years has shown a positive trend of at least 124% annually;
- From 2017 to 2018, there was a positive trend in the export of garments, respectively, $119.8 million and $151.7 million [15].

Textile production occupies the second position after the clothing industry, which plays a significant role in the production life of each individual region. A significant share of textile production falls in the city of Bishkek (29.0%), followed by the Jalal-Abad Region (28.5%), and the Chui Region (23.0%). The middle position is occupied by the Osh Region (16.5%). The underdevelopment of textile production is observed in the Issyk-Kul Region (1.3%), the city of Osh (0.8%), and the Batken (0.2%), Naryn (0.5%), and Talas (0.1%) Regions [15]. Nevertheless, these regions have a sufficiently high resource potential for the production of the textile industry.

It is difficult to disagree with the opinion of economists who believe that the market mechanism is fundamentally unstable and that there is no automatic balancing that ensures full employment and a balance between production and consumption. A state close to full employment, price stabilization, equilibrium of the balance of payments, and public well-being is achieved only thanks to the corrective and orienting actions of the government in the economy.

In our opinion, the introduction of market mechanisms for improving and managing the organizational forms of competitive private enterprises in the light industry in the region depends on the coordinated actions of business structures, the population of the regions, state and regional authorities, as well as their attitude to innovation and investment activities. We believe that without the development of regional organizational and economic mechanisms for the management system of light industry enterprises (Fig. 2) and their adaptation, it will be impossible to revive the once (the 1990s) "artificially forcedly frozen" industry facilities and restore the former glory of the industry leader lost over the years.

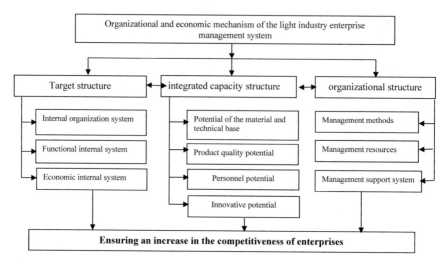

Fig. 2 Organizational and economic mechanism for ensuring the competitiveness of a light industry enterprise. *Source* Developed and compiled by the authors

Therefore, mechanisms indicated in Fig. 2 contribute to the development of a mechanism for attracting investments in the light industry, aimed at the formation of investment management systems, improving investment attractiveness and their effective use, mobilizing their own investment funds, and creating risk management systems at enterprises. In market conditions, these mechanisms will make it possible to ensure a high level of competitiveness of light industry enterprises [16].

4 Conclusion

Thus, the current state of the domestic light industry is experiencing a number of apparent problems that affect the development of potential cluster structures of industries. In our opinion, in the Kyrgyz Republic, there are territories that have untapped potential for the renewal of original technologies for production, marketing, processing, and adaptation to market conditions of individual resources requiring special attention, such as cotton, cocoon, leather, etc., which can provide and strengthen a significant part of the needs of enterprises in their own raw material base.

Consequently, to increase the production of products from local raw materials, it is necessary to re-equip and modernize existing enterprises using advanced processing technology.

We believe that the following ESG initiatives are proposed for the development of controlled light industry in the regions of Kyrgyzstan:

- Pursuing an administrative policy to legalize clothing enterprises and firms operating in the "shadow" (in the country, approximately over 30% of enterprises in the clothing industry continue to work in the "shadow economy");
- Provision of relatively "cheap leasing" to accelerate technical and technological re-equipment, the introduction of equipment systems that continuously perform all processing operations, capable of adapting to innovations;
- To develop a comprehensive program at the enterprises to improve the quality and competitiveness of products to ensure consistency between the price and quality of products;
- Direct the activities of enterprises to the strategy of "green economy" or "green industry," which will take their rightful place in the world market through the production and satisfaction of the needs of the population in "green clothes," "green shoes," and other products of the industry.

References

1. Government of the Kyrgyz Republic (2015) 20 years of the formation of the World Trade Organization and the membership of the Kyrgyz Republic in it. Retrieved from https://www.gov.kg/ru/post/s/20-let-obrazovaniya-vsemirnoy-torgovoy-organizatsii-i-chlenstvo-kyirgyizskoy-respubliki-v-ney. Accessed 2 Sept 2022
2. Government of the Kyrgyz Republic (2017) Concepts of the regional policy of the Kyrgyz Republic for the period 2018–2022 of the Kyrgyz Republic (Approved by Resolution No. 194 of March 31, 2017). Bishkek, Kyrgyzstan. Retrieved from http://cbd.minjust.gov.kg/act/view/ru-ru/99907. Accessed 26 Aug 2022
3. Feldman GA (n.d.) Model of economic growth. Retrieved from https://studme.org/1167032012238/politekonomiya/model_ekonomicheskogo_rosta_feldmana. Accessed 30 Aug 2022
4. Granberg AG (2004) Fundamentals of regional economics: textbook for universities (4th edn). HSE University, Moscow, Russia
5. Laundhardt W (n.d.) Launhardt triangle. Retrieved from https://spravochnick.ru/ekonomika/treugolnik_launhardta/#razmeschenie-predpriyatiy-vne-vershin-lokacionnogo-treugolnika. Accessed 28 Aug 2022
6. Lösch A (2007) Spatial organization of the economy. Nauka, Moscow, Russia
7. Weber A (n.d.) The theory of industrial location. Retrieved from https://zaochnik.com/spravochnik/ekonomika/ekonomika-predprijatija/razmeschenie-promyshlennosti/. Accessed 26 Aug 2022
8. Abdymalikov KA (2010) Economy of Kyrgyzstan (in transition): textbook. Biyiktik, Bishkek, Kyrgyzstan
9. Abdyrov TS, Toktogulov AK (2017) Problems of formation of innovative clusters in the Kyrgyz Republic. Eng Constr Bull Caspian Reg 2(20):58–63. Retrieved from https://cyberleninka.ru/article/n/problemy-formirovaniya-innovatsionnyh-klasterov-v-kyrgyzskoy-respublike. Accessed 26 Aug 2022
10. Chernova EP (1970) Labor resources of industry in Kyrgyzstan. Ilim, Frunze, USSR
11. Koychuev TK (2015) Kyrgyzstan—the choice of the path of development in the civilized world: options for development models. Ilim, Bishkek, Kyrgyzstan
12. Kupuev PK (1995) Formation of the labor market: issues of theory and practice. Practical Guide, Osh, Kyrgyzstan
13. Musakozhoev SM (2014) Modern problems of the economy: textbook for undergraduates of universities. Bishkek, Kyrgyzstan

14. National Statistical Committee of the Kyrgyz Republic (2016) Manufacturing industry of the Kyrgyz Republic 2011–2015. Bishkek, Kyrgyzstan. Retrieved from http://www.stat.kg/media/publicationarchive/a2f50872-b761-4e3e-afd5-cc9464c2c18a.pdf. Accessed 26 Aug 2022
15. National Statistical Committee of the Kyrgyz Republic (2021) Manufacturing industry of the Kyrgyz Republic 2016–2020. Bishkek, Kyrgyzstan. Retrieved from http://www.stat.kg/media/publicationarchive/9e01fb85-916f-40d7-b9c6-61b124c5c9de.rar. Accessed 26 Aug 2022
16. Litvinova AG (2014) Analysis of the situation in the light industry in Russia. RUDN J Econ 2:16–26. Retrieved from https://cyberleninka.ru/article/n/analiz-situatsii-v-rossiyskoy-legkoy-promyshlennosti. Accessed 2 Sept 2022

ESG Management of Climate Risks and Green Finance in Support of Combating Climate Change

Financial Aspects of Decarbonization of the Russian and Central Asian Economies in the Context of Climate Change: Comparative Analysis

Natalia G. Vovchenko⊙, Olga V. Andreeva⊙, Valeria D. Dmitrieva⊙, Natalya F. Zaruk⊙, and Svetlana A. Sulzhenko⊙

Abstract This research aims to analyze the current state of the problem of introducing financial instruments for the decarbonization of the economy in the context of changing climatic conditions. Now, in some regions, especially with intensive industrial production, excessive aridization, desertification, biodiversity reduction, etc. are observed. Additionally, there is an extremely negative impact on the nutritional status and health of the population. This research has an applied analytical nature. Its main task is to compare decarbonization strategies in Russia and its closest neighbors—the countries of Central Asia. It is apparent that the volumes of greenhouse emissions produced by Russian enterprises are several times higher than the total emissions of all countries in the Central Asian region. Nevertheless, considering the territorial proportions, it should be assumed that the problem of decarbonization of the economy is acute for all countries. To a certain extent, this research is important for understanding the problem of decarbonization in the region. The authors note that the targets for reducing greenhouse emissions by 2030 differ significantly among the countries of Central Asia. However, these targets are currently being revised toward a significant increase. Some countries can cope with this independently, others with international assistance. According to the experience of several countries, the introduction of a carbon tax seems to be the most promising mechanism for influencing producers to green production. However, of all the countries of the Central Asian region, such a tool is used only in Kazakhstan. The authors of this research believe that the task of achieving carbon neutrality requires a comprehensive solution. A powerful promotional campaign is required for the public and the business community to realize the importance of joint efforts. The full use of financial instruments to decarbonize the economy is required.

N. G. Vovchenko · O. V. Andreeva (✉) · V. D. Dmitrieva · S. A. Sulzhenko
Rostov State University of Economics, Rostov-On-Don, Russia
e-mail: olvandr@yandex.ru

N. F. Zaruk
Russian State Agrarian University—Moscow Timiryazev Agricultural Academy, Moscow, Russia
e-mail: zaruk84@bk.ru

503

Keywords ESG · Sustainable development · Climate changes · Decarbonization · Green economy · Green finance · Carbon taxes

JEL Classification Q28 · Q54 · Q57

1 Introduction

In 2018, giving his Nobel Lecture, William D. Nordhaus, characterizing the threat of global warming to the future of humanity, compared it with the Colossus in the Goya painting of the same name. Like the Colossus that sows destruction, global warming is just as massive and destructive, but, much worse, it was created over decades and centuries by the efforts of many people, and then the actions of an individual will never lead to changes [1].

Back in the 1970s, Nordhaus pointed to the limited capacity of the environment to absorb greenhouse gases. He sought to show that at the present stage of the development of society, worrying about the limited nature of natural resources as such does not make sense in view of their millennial reserves, and, even if cheap natural resources are once exhausted, this will not lead to a decrease in the average standard of living.

However, the problem of absorption of emissions from the use of energy resources remains. Economic growth, supported by an increase in production capacity, leads to an increase in carbon dioxide (CO_2) emissions, which, in turn, has a negative impact on the environment and triggers the mechanism of economic regulation. By introducing taxes, fines, and other regulatory instruments, the state finds a temporary solution to the problem. However, it is cheaper to pay a fine than to reduce emissions. Therefore, the whole process starts from the beginning. This logic underlies the dynamic integrated climate and economic model (DICE) developed by Nordhaus.

The only way known to the economy to eliminate the problem is the introduction of a corrective tax (Pigou tax). This tax raises private marginal costs to the level of common ones, thus forcing the producer to bear costs that explicitly include damage [2]. The current analog of such a corrective tax can be the so-called carbon tax, which is purely national in nature, as well as the Carbon Border Adjustment Mechanism (CBAM), which is an example of a tax on imports of "dirty" goods with a large carbon footprint. The latter is not formally a tax; rather, it largely corresponds to the emissions trading scheme.

Such regulation operates in a number of countries (Sweden, Norway, Finland, Switzerland, Liechtenstein, Iceland, Canada, Portugal, Great Britain, France, etc.) and has already shown quite good results. For example, according to the OECD, an increase in carbon tax by 1 euro per metric ton of CO_2 emissions leads to an average emission reduction of 0.73% [3]. Another example is given: in the UK, the amount of CO_2 emissions in the period 2012–2018 decreased by 73% due to the gradual increase in the carbon tax to 36 euros per metric ton of CO_2 [3].

In some countries, the carbon tax is set differentially depending on the industry affiliation of the manufacturer and the fuel used. The most significant carbon tax was set in Sweden. In 2021, this tax amounted to \$137 per metric ton of CO_2. Switzerland and Liechtenstein are in second place (\$101 per metric ton of CO_2) [4].

Russia is discussing its own version of the carbon tax. Simultaneously, major Russian manufacturers are pursuing green modernization, introducing the latest energy-efficient and resource-saving technologies (including the abandonment of fossil fuels and the transition to hydrogen) and installing advanced systems for capturing emissions into the atmosphere and water flows. However, not all companies are ready to invest in new capacities.

2 Materials and Method

This research aims to analyze the current state of the problem of introducing financial instruments for the decarbonization of the economy in the context of changing climatic conditions. Already now, excessive aridization, desertification, biodiversity reduction, etc. are observed in a number of regions, especially with intensive industrial production. Additionally, there is an extremely negative impact on the nutritional status and health of the population.

This problem is complex. Joint efforts of economists, ecologists, geographers, geologists, biologists, and climatologists are required to solve this problem. However, this requires the impact of administrative resources and the involvement of businesses and the general public in a joint decision. The efforts of individuals will not lead to particular results. Nevertheless, if the movement towards carbon neutrality becomes wide enough and the actions of enterprises that have a negative impact on the environment are condemned by society, this goal will probably become achievable.

This research is of an applied analytical nature. Its main task is to compare decarbonization strategies in Russia and its closest neighbors—the countries of Central Asia. Apparently, the volumes of greenhouse emissions produced by Russian enterprises are several times higher than the total emissions of all countries in the Central Asian region. However, considering the territorial proportions, it should be assumed that the problem of decarbonization of the economy is acute for all countries. To a certain extent, this research is important for understanding the problem of decarbonization in the region.

Statistical sources of this research are presented by official open reports of the UN, the World Bank, OECD, Climate Action Network, the Energy Center of the Skolkovo Moscow School of Management [5], open statistical resources Statista and Climate Watch, as well as passports of strategic documents of Kazakhstan, Kyrgyzstan, Tajikistan, Turkmenistan, Uzbekistan, and Russia, which reflect the main directions of development and measures to combat climate change.

3 Results and Discussion

Currently, the largest contribution to the generation of greenhouse gases is made by China, the USA, the countries of the European Union, India, Russia, and Japan, whose total CO_2 emissions amounted to 27.59 gigatons in 2019 [6]. The role of the countries of Central Asia (CA countries) in the production of emissions against their background is extremely small (Fig. 1).

EU Program "Fit for 55" is part of a strategy to green the economy and achieve neutral environmental impact by 2050. The intermediate target is to reduce emissions by 55% already by 2030.

These goals are set not only by the EU countries. Russia, China, Japan, Latin American countries, and many others have joined the movement to achieve carbon neutrality. Greenhouse emissions have a detrimental effect on the state of agriculture, causing damage to human life and health.

For the countries of Central Asia, the task of transitioning to carbon neutrality is of particular importance because climate change provoked by emissions leads to the melting of glaciers. The disappearance of glaciers over time will lead to a decrease in surface water runoff, which threatens an ecological catastrophe for the entire Central Asian region because the main freshwater reserves are located in Kyrgyzstan and Tajikistan, from where the water then descends to Kazakhstan, Turkmenistan, and Uzbekistan.

Experts note that the reduction in the region's water supply has been so much that by 2030, given the current rates of population growth in the region and the growing water consumption needs, the volume of water supply will reach a critical level of 1.7 thousand cubic meters per year while the minimum level of water consumption in Central Asia should be about 500–700 million cubic meters of water per year. Agriculture is also threatened because 50–90% of irrigated lands are already waterlogged or saline [7].

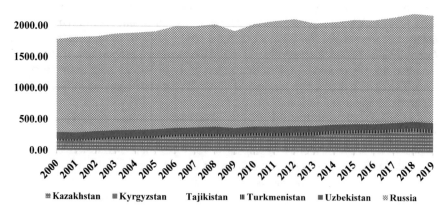

Fig. 1 Dynamics of CO_2 emissions by CA countries and Russia, 2000–2019. *Source* Compiled by the authors based on [6]

In the post-Soviet space in the Central Asian region, the Republic of Kazakhstan is the largest emitter of greenhouse gases [8].

In December 2020, the President of the Republic of Kazakhstan, K.-Zh. Tokayev, announced the country's intention to achieve carbon neutrality by 2060. As a result, a draft Doctrine (strategy) for achieving carbon neutrality of the Republic of Kazakhstan until 2060 was developed [9]. However, it is controversial in the business community. Nevertheless, the document reflects the idea that without measures to decarbonize the economy, it is impossible to achieve the forecast indicators of economic growth and the compliance of local producers' products with international requirements.

In addition to the adoption of a number of strategic documents aimed at greening the economy, the Republic of Kazakhstan has a system for trading greenhouse gas emissions quotas. The national plan for the distribution of quotas for greenhouse gas emissions provides for the distribution of quotas between six areas of activity: electric power, oil and gas industry, mining, metallurgy, chemical industry, and manufacturing industry (in terms of production of building materials). Thus, according to the National Plan for 2021, the largest quotas are provided to enterprises in the electric power industry (57.2%), metallurgy (18.5%), and the oil and gas industry (13.9%) [10].

Separately, it is worth mentioning the Concept for the transition of the Republic of Kazakhstan to a green economy, the implementation of which is planned in three stages [11]:

- 2013–2020—the main priority of this stage was the optimization of the use of resources and the creation of green infrastructure;
- 2020–2030—transition to production based on modern energy-efficient technologies, increasing the efficiency of environmental protection activities in general;
- 2030–2050—the functioning of the national economy on sustainability principles.

One of the projects demonstrating the activity of the Concept is the international environmental initiative of Kazakhstan, the Green Bridge Partnership Program, the idea of which was put forward by the First President of the country, N. Nazarbayev, back in 2010. The existence of the project yielded the following results [11]:

- Numerous partnership agreements on environmental issues were concluded with a number of foreign countries, including Russia, Kyrgyzstan, Georgia, Germany, Mongolia, Belarus, Latvia, Spain, Sweden, Finland, etc.,
- The "Green Standard KZ" for certification of residential real estate objects was adopted;
- The movement to create green offices in Kazakhstan, Kyrgyzstan, and Russia was expanded, etc.

The Republic of Uzbekistan ranks second among the countries of the Central Asian region in terms of greenhouse gas emissions. The main producers of emissions are the enterprises of the electric power industry, agriculture, and industry.

In 2018, the country ratified the Paris Agreement and set a new target to reduce CO_2 emissions by 35% by 2030 [12]. The achievement of the goal should be facilitated by measures to introduce energy-saving technologies, switch to alternative fuels, improve the efficiency of water resources and solid waste management systems, etc. The introduction of a carbon tax is seen as one of the most effective mechanisms to stimulate the reduction of emissions. Nevertheless, it is a more distant prospect.

The contribution of Kyrgyzstan, Tajikistan, and Turkmenistan to greenhouse gas emissions among the countries of Central Asia is the smallest, even if it does not exceed the contribution of Uzbekistan in total.

Like Uzbekistan, Kyrgyzstan has updated its CO_2 emission target. By 2030, the reduction should be 15.97% if the economy develops in the future according to the baseline scenario without significant measures to reduce emissions and 43.62% if international support is provided [13, 14].

Thus, the targets for reducing greenhouse emissions by 2030 differ significantly among the countries of Central Asia: from 10% in Tajikistan to 15% in Kazakhstan (for comparison, the target in Russia is a threshold of 20–30%). However, the nationally determined contributions to the Paris Agreement are subject to review and should, in the long term, amount to at least 20% for all countries of Central Asia [15, 16], which, in turn, should bring these countries closer to the desired goal—carbon neutrality by 2050.

The achievement of this goal is also seen in the introduction of a carbon tax. Thus, Russia currently considers three possible options:

1. Collection at a certain rate for exceeding threshold values for emissions;
2. Emissions trading;
3. A combination of two options.

In our opinion, the latter option is preferable, provided that various factors are considered to minimize the burden on market participants.

4 Conclusion

The pursuit of carbon neutrality is a powerful challenge for the economy and society of any country. The transition from fossil fuels to alternative clean types will lead to the compression of this sector, the reduction of jobs, and the death of some professions. Economic transformations should not lead to social upheavals. On the contrary, maximum opportunities should be extracted from this process to improve the quality of life of the population.

The polluted air of cities and countryside poisons people's lives and food, reduces the biodiversity of territories, and dries up the climate. Complete decarbonization of the economy is an apparent and necessary goal. However, for certain reasons, it is a very long-term goal. Whether countries will be able to achieve the stated goals of carbon neutrality by 2050 or 2060 is a big question. Nevertheless, if such transformations are not started now, in some countries, the unfolding of the situation will lead to an environmental and humanitarian catastrophe by the indicated period.

The authors of this research note that the task of achieving carbon neutrality requires a comprehensive solution. A powerful promotional campaign is needed for the public and the business community to realize the importance of joint efforts. It is necessary to introduce transparent mechanisms for monitoring [17] and evaluating the achievements of strategic guidelines in the field of adaptation to climate change and ensuring social and economic security. The full use of financial instruments to decarbonize the economy is required.

In many ways, the ideas presented can be seen as reflections on the unfolding processes of decarbonization of the economy, which opens up wide opportunities for further research. Nevertheless, some assumptions and analyses of successful foreign practices may contribute to rethinking and adjusting the current economic policy in the areas of decarbonization in Central Asia and Russia.

References

1. Nordhaus WD (2018) Climate change: the ultimate challenge for economics. Prize Lecture. Retrieved from https://www.nobelprize.org/uploads/2018/10/nordhaus-lecture.pdf. Accessed 1 May 2022
2. Pakhnin MA (2020) The economics of climate change: Nobel prize 2018 by William Nordhaus. Finan Bus 16(1):5–22. Retrieved from https://www.researchgate.net/publication/342955 086_Ekonomika_izmenenia_klimata_Nobelevskaa_premia_2018_goda_Uilama_Nordhausa. Accessed 11 May 2022
3. OECD (2021) Effective carbon rates 2021. Retrieved from https://www.oecd.org/tax/tax-pol icy/effective-carbon-rates-2021-brochure.pdf. Accessed 11 May 2022
4. Statista (2021) Carbon taxes worldwide as of April 2021, by country. Retrieved from https://www.statista.com/statistics/483590/prices-of-implemented-carbon-pricing-instru ments-worldwide-by-select-country/. Accessed 11 May 2022
5. Skolkovo Moscow School of Management (2020) Global climatic threat and Russian economy: searching for the way. Retrieved from https://energy.skolkovo.ru/downloads/docume nts/SEneC/Research/SKOLKOVO_EneC_Climate_Primer_EN.pdf. Accessed 11 May 2022
6. Climate Watch (2022) Historical GHG emissions: CO_2 emissions from 1998 to 2019. Retrieved from https://www.climatewatchdata.org/ghg-emissions?breakBy=countries&end_ year=2019&gases=co2&regions=TOP&source=PIK&start_year=1998. Accessed 11 May 2022
7. World Bank (2019) Water and energy program for Central Asia. Retrieved from https://the docs.worldbank.org/en/doc/864641567759267834-0080022019/original/CAWEPBrochure20 19ru.pdf. Accessed 11 May 2022
8. CSR Center for Strategic Research Foundation (2021) Russia's climate agenda: responding to international challenges. Retrieved from https://www.csr.ru/ru/news/klimaticheskaya-pov estka-rossii-reagiruya-na-mezhdunarodnye-vyzovy/. Accessed 11 May 2022
9. Ministry of Ecology, Geology and Natural Resources of the Republic of Kazakhstan (2021) Draft Presidential Decree of the Republic of Kazakhstan "On approval of Doctrine (strategy) of achieving carbon neutrality of the Republic of Kazakhstan until 2060" (developed September 14, 2021). Retrieved from https://legalacts.egov.kz/npa/view?id=11488215. Accessed 11 May 2022

10. Government of the Republic of Kazakhstan (2021) Decree "On approval of the national plan for allocation of quotas for greenhouse gas emissions for 2021" (January 13, 2021 No. 6). Nur-Sultan, Kazakhstan. Retrieved from https://adilet.zan.kz/rus/docs/P2100000006. Accessed 11 May 2022

11. Galushko M (2019) How the "Green Bridge" helps to solve environmental problems. Kapital News Agency. Retrieved from https://kapital.kz/economic/79694/kak-zelenyy-most-pomoga yet-reshit-ekologicheskiye-problemy.html. Accessed 11 May 2022

12. UNFCCC (2021) Republic of Uzbekistan: updated nationally determined contribu- tion. Retrieved from https://unfccc.int/sites/default/files/NDC/2022-06/Uzbekistan_Updated% 20NDC_2021_EN.pdf. Accessed 11 May 2022

13. UNECE (2017) Tajikistan: environmental performance reviews. UN, New York, NY; Geneva, Switzerland. Retrieved from https://www.un-ilibrary.org/content/books/9789210601696/read. Accessed 11 May 2022

14. UNFCCC (2021) Republic of Kyrgyzstan: updated nationally determined contribution. Retrieved from https://unfccc.int/sites/default/files/NDC/2022-06/ОНУВ%20ENG%20от% 2008102021.pdf. Accessed 11 May 2022

15. Climate Action Network (CAN) (2020a) Climate policy analysis of Eastern Europe, Caucasus and Central Asia. Retrieved from https://infoclimate.org/wp-content/uploads/2020/12/analiz-klimaticheskoj-politiki-vekcza.pdf. Accessed 11 May 2022

16. Climate Action Network (CAN) (2020b) Green recovery and climate action in Central Asia: position of CAN members from Central Asia. Retrieved from https://infoclimate.org/wp-con tent/uploads/2020/11/pozicziya-cza-na-russkom-1.pdf. Accessed 11 May 2022

17. Andreeva OV, Dmitrieva VD (2022) The necessity and directions for development of theoretical and methodological approaches to identifying the integration effect of the green and digital economy. CITISE 1(31):159–169. https://doi.org/10.15350/2409-7616.2022.1.13

Innovative Green Finance Tools

Olga B. Ivanova ⓘ**, Natalia G. Vovchenko** ⓘ**, Olga V. Andreeva** ⓘ**,**
Oksana S. Chernobay ⓘ**, and Elena D. Kostoglodova** ⓘ

Abstract The research is based on the theoretical approaches of the concept of green economy, suggesting the need to justify the use of new financial instruments. The research discusses the tools of green financing—emissions trading, environmental insurance, and carbon tax, which contribute to the development of effective provisions for increasing the sustainability of the environment. The authors used data from open sources published on official Russian and foreign websites. The authors analyzed the available scientific publications on the subject of green finance. The experience of various countries (China, Great Britain, New Zealand, Russia, etc.) on the implementation of the introduction of green financial instruments has been studied. The authors formulated a number of conditions that will contribute to the more active development of innovative green financial instruments.

Keywords Green economy · Green financial instruments · Pollution quotas · Carbon tax · Environmental insurance · Sustainable development

JEL Classification G21 · G28

1 Introduction

The study of directions for the development of green finance to ensure the sustainable development of the economy today is an urgent and priority task.

O. B. Ivanova · N. G. Vovchenko · O. V. Andreeva · O. S. Chernobay (✉) · E. D. Kostoglodova
Rostov State University of Economics, Rostov-On-Don, Russia
e-mail: Oksana.chernobay1982@yandex.ru

O. B. Ivanova
e-mail: sovet2-1@rsue.ru

O. V. Andreeva
e-mail: olvandr@ya.ru

© The Author(s), under exclusive license to Springer Nature Switzerland AG 2023
E. G. Popkova (ed.), *Smart Green Innovations in Industry 4.0 for Climate Change Risk Management*, Environmental Footprints and Eco-design of Products and Processes, https://doi.org/10.1007/978-3-031-28457-1_52

Green financial instruments are aimed primarily at the rational use of natural resources and maintaining an optimal state of the environment. They are also called upon to act as one of the necessary and important conditions for the sustainable development of society.

The development of the industrial complex, the increase in the number of vehicles, and the growth of energy consumption, waste, and air pollution have an extremely negative impact on the state of the environment and are global environmental problems. Some regions (especially those with intensive industrial production) face excessive aridization, desertification, biodiversity reduction, etc. Additionally, there is an extremely negative impact on the nutritional status and health of the population. As a result of the negative impact, there was an imbalance in the environment, in the "human-nature" system, which already requires appropriate costly measures to restore it [1].

The search for additional financial resources is still relevant to compensate for the damage caused by economic activity and ensure environmental safety.

Recently exposed to numerous challenges, economic and financial stability required rethinking, overcoming, and continuous monitoring of situations.

New opportunities and new innovative prospects have opened up due to the need to develop clean technologies. This new wave of innovation will inevitably lead to the merger and unification of the natural environment and business. Green finance can become a significant factor that will determine further development and ensure the progress of society.

Innovative green finance can transform information and financial flows into material innovations and rebuild the "human-nature" system to be optimal for preserving and improving the quality of people's lives.

Numerous programs and various environmental initiatives are developed at the level of governments in many countries. These measures are aimed at reducing the level of emissions of harmful substances into the atmosphere and hydrosphere, climate change, waste disposal, conservation of the number of plant and animal species, etc. These problems are also relevant to Russia.

This paper considers innovative green finance tools that contribute to developing effective environmental sustainability.

These questions are being developed by Russian and foreign scientists (T. A. Zambrovskaya, N. N. Semenova, O. L. Konyukova, B. S. Goryachkin, A. A. Tatuev, A. S. Fedoryashchenko, R. Mastini, B. Sutherland, and others).

Zambrovskaya et al. [2] identifies the problems of verification of green financial instruments. Semenova and Grebentsova [3] consider the use of green financial instruments in the interests of sustainable development. Konyukova and Korchebny [4] describe features of the development of environmental insurance in Russia. Goryachkin and Stryukov [5] prove that carbon quotas are an effective economic solution for resolving the carbon problem.

Mastini et al. [6] analyze the Green Deal and the possibility of its growth. Sutherland [7] studies the financing of the new Green Deal.

2   Materials and Methods

The research methods include general scientific (analysis, synthesis, generalization, and comparison) and special (expert assessments and statistical data analysis). The research is based on the theoretical approaches of the concepts of sustainable development and the green economy, suggesting the need to justify the use of new financial instruments, such as trading in greenhouse gas emissions, carbon taxes, environmental insurance, etc. The authors used data from open sources published on official Russian and foreign websites. Additionally, the authors analyzed the available scientific works on the subject of green finance. Moreover, the authors studied the experience of various countries (the UK, China, Russia, etc.) in developing green financial instruments.

3   Results

An innovative source of financing, which is actively used in many countries to solve environmental problems, is the trading of quotas for greenhouse gas emissions.

The Paris climate agreement was adopted on December 12, 2015. Its goal is to prevent the average annual temperature on a planetary scale from exceeding by more than 2 °C by 2100 from pre-industrial levels, as well as to take measures to keep warming within 1.5 °C.

The participants of this agreement assumed one of the obligations, which is to reduce CO_2 emissions into the atmosphere systemically. To realize this commitment by 2020, it was necessary to develop national strategies for the transition to a carbon-free economy [8].

Nowadays, there are 24 emissions trading systems in the world. At the beginning of 2016, there were only 17 such systems in the world, covering 35 countries [9].

A cost-effective policy tool that governments and companies can use as part of their broader climate strategy is carbon pricing. This creates a financial incentive to reduce emissions through price signals. By factoring the costs of climate change into economic decision-making, carbon pricing can drive changes in production and consumption patterns, thereby supporting low-carbon growth.

In developed countries, evidence suggests that carbon pricing promotes productivity and innovation and has no detrimental effect on economic development [10].

Planning measures to regulate emissions trading began in the early 2000s. Nevertheless, the development of many legal and technical issues took quite a long time. The first full-fledged system of trading in quotas appeared only in 2005 in the European Union.

Switzerland and New Zealand joined in 2008. In South Korea, cap-and-trade (or quotas) was launched in 2015. In China, the first pilot CO_2 trading systems started operating in 2013; the full system should be launched before 2025 [11].

Table 1 Advantages and disadvantages of the CO_2 emissions trading mechanism

Advantages	Disadvantages
• Maintaining a balance between production and environmental protection, equalizing the volume of harmful emissions from different companies or countries	• Many countries are not interested in reducing emissions, citing the fact that emissions are already low compared to neighboring countries
• Stimulating business, which makes it possible to make production less harmful	• The effect of "leakage of carbon"—the effect of emissions flowing to countries with less stringent environmental policies
• Voluntary responsibility of business for the environment	• Formation of a market for clean air and the size of available quotas
• Predictability. The company builds a plan, taking the number of quotas into account. This makes it possible to plan the expenses in advance	• The high cost of innovation. It is easier to buy quotas than to modernize production

Source Compiled by the authors

"Quotas" is the allowable volume of emissions of pollutants into the atmosphere.

The operation scheme of the emissions trading mechanism is as follows. Special bodies or the government sets the standard for greenhouse emissions per year from all types of pollution sources. Polluting companies are assigned a quota based on the established norm. In this case, the quota is the volume of emissions that pollutants can release into the atmosphere. Next, the quota is sold to specific pollutants—the subjects of pollution (fuel and energy companies, the chemical industry, mechanical engineering, and others).

Pollutants are required to have an emission permit in the amount that will be determined based on the actual level. The number of quotas should not exceed the level of CO_2 emissions that corresponds to the enterprise's production program [12].

This mechanism has advantages and disadvantages (Table 1).

Currently, CO_2 emissions trading operates in 24 national and subnational markets. There are 20 more such marketplaces under development. Since 2021, three new markets with considerable money turnover have already been launched—in China, the UK, and New Zealand.

China has launched its national CO_2 trading system, which will regulate more than 2000 power plants with combined emissions of 4.3 billion tons.

This does not rule out an increase in the growth of prices for quotas. This is likely to happen as soon as the government of China sets an upper limit on emissions and systemically lowers it annually.

Between 2021 and 2025, the New Zealand government has set a cap on total emissions from sectors covered by the New Zealand ETS of 40 Mtpa.

New exchange and marketplace for carbon credits were planned to be launched in Singapore by the end of 2021, supported by banks DBS (DBSM.SI), Standard Chartered (STAN.L), Singapore Exchange (SGXL.SI), and state investor Temasek Holdings. There are more and more such sites in Asia [13].

From May 2021, the UK cap-and-trade system began issuing free emissions permits for the UK's traditionally largest emitters in the steel and cement industries.

The UK Emissions Market has shifted the CO_2 emission charges levied by the EU on UK businesses to England.

Trading in quotas for CO_2 emissions is also an important factor that affects the export and import of oil and gas to different countries of the world. In some countries, exchanges for trading in such goods have been operating for more than ten years; in some, they are being launched only now.

Trading occurs not only in the so-called carbon units but also in credits for them when one company resells unused quotas (because it has not exceeded the allowable emission standard) to another. Moreover, in some countries, futures for CO_2 emission allowances are already being traded [13].

Another financial innovation is greenhouse gas taxes (carbon taxes).

In early 2021, the European Parliament approved the carbon tax law, which is a de facto duty on imported goods into the EU. It also allows issuing CO_2 emission allowances free of charge to European companies. The European Commission should implement a bill by 2023.

Similar to the EU Emissions Trading Scheme (EU ETS), carbon taxes use the price of carbon but with a different pricing mechanism.

On carbon exchanges, prices are set by the market based on physical limits set by the authorized state body on greenhouse gas emissions. These limits are gradually reduced, stimulating an accelerated transition to more energy-efficient and environmentally friendly technologies and industries.

A carbon tax aims to force individuals and firms to pay the full social cost of carbon pollution. Theoretically, the tax would reduce pollution and encourage cleaner alternatives. However, some critics argue that a carbon tax will increase business costs, reduce investment, and slow economic growth.

According to the experience of several countries, introducing a carbon tax seems to be the most promising mechanism for influencing producers to green production.

There are several arguments in favor of a new innovative financial instrument.

First, the carbon tax will significantly affect the environment. With higher taxes, businesses, firms, and organizations will reduce pollution and look for alternatives with minimal environmental impact. Alternative sources of electricity can serve as an example. This may make it more feasible to produce electricity from environmentally friendly sources (e.g., solar energy). If we develop environmentally friendly sources, it will make us less dependent on oil. Solar energy can become even more competitive than traditional fossil fuels.

Second, it will help ease the transition to a post-oil economy.

The high price of carbon emissions will encourage businesses and organizations to develop and implement cleaner technologies, more efficient engines, carbon consumption alternatives, etc.

It is possible that this may be a call for society to become more responsible consumers. More people may be interested in switching from cars to bicycles and walking more. This would have a health benefit, such as a reduced risk of a heart attack.

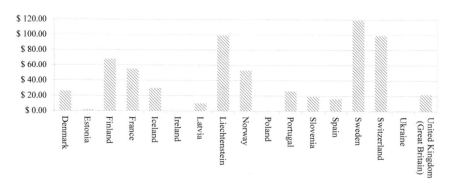

Fig. 1 Carbon tax rates (US dollar) in Europe. *Source* Developed by the authors based on [17]

Third, there are opportunities to increase carbon tax revenues, which can then be used to subsidize alternatives (e.g., clean electricity) or to repair damage caused by pollution.

This will certainly lead to a socially significant result, making people pay the social costs and overcome the excess consumption that we more often see nowadays [14].

Compensation for the costs of environmental protection measures, as well as compensation for damages from losses, can be made at the expense of funds accumulated through such a financial instrument [15].

Many European countries, as well as Mexico and Japan, use carbon taxes. As of April 2015, the annual global carbon tax collection was estimated at $14 billion [16].

In a number of countries, the carbon tax is set differentially depending on the industry affiliation of the manufacturer and the fuel used.

The amount of carbon tax established in European countries (as of April 1, 2020) is shown in Fig. 1.

In 2021, the largest carbon tax was set in Sweden and amounted to $137 per ton of CO_2, followed by Switzerland and Liechtenstein ($101 USD/t of CO_2) [18].

All countries of the European Union (as well as Iceland, Liechtenstein, and Norway) are part of the EU Emissions Trading Scheme (EU ETS), a market set up to trade a limited amount of greenhouse gas credits. Except for Switzerland, Ukraine, and the UK, all European countries that impose a carbon tax are also part of the EU ETS. Switzerland has its own emissions trading system, which has been linked to the EU ETS since January 2020. After Brexit, the UK introduced its own British ETS from January 2021 [17].

Russia is discussing its own version of the carbon tax. Simultaneously, major Russian manufacturers are pursuing green modernization, introducing the latest energy-efficient and resource-saving technologies (including the abandonment of fossil fuels and the transition to hydrogen) and installing advanced systems for capturing emissions into the atmosphere and water flows. However, not all companies are ready to invest in new capacities.

Another promising financial instrument that can fully or partially compensate for the damage caused to the environment in case of accidental pollution is environmental insurance [19].

Environmental insurance is one of the methods of economic regulation in the field of environmental protection [20].

A prerequisite for the emergence of such insurance relations is environmental risk. Environmental risks have their own classification and are characterized by the following types: individual, technical, ecological, social, and economic.

Discussing the expected environmental damage and economic losses is a serious job of assessing environmental risks.

According to the Federal law "On environmental protection," environmental risks are understood as "the probability of the occurrence of an event that has adverse consequences for the natural environment and is caused by the negative impact of economic and other activities, natural and human-made emergencies" [21].

Environmental insurance is a tool for ensuring the financial sustainability of various types of environmental activities. It is one of the methods for managing environmental risks in nature management.

Environmental insurance has two forms: voluntary and mandatory.

Environmental insurance, which is carried out at the expense of budgetary funds and is carried out by virtue of Federal law No. 225-FZ, is compulsory. Special norma-tive acts determine the types, conditions, and procedures for its implementation. An example is the current Federal law "On compulsory insurance of civil liability of the owner of a hazardous facility for causing harm as a result of an accident at a hazardous facility" (July 27, 2010 No. 225-FZ) [22].

Potentially dangerous objects are subject to compulsory insurance. If the owner of such an object does not fulfill his or her insurance obligations, the operation of the enterprise is not allowed.

Voluntary environmental insurance is comparable to voluntary certification of products. It increases the enterprise's prestige and is an excellent competitive advantage for the company.

The reasons for the insured event, the list of pollutants, as well as compensation for damage under voluntary environmental insurance between the insurer and the insured are negotiated upon the conclusion of the insurance contract in each case individually. The insured pays insurance premiums at tariff rates depending on the enterprise's annual turnover.

Environmental insurance is important for Russia as a mechanism for managing environmental risks. Russia has many facilities that carry a potentially serious environmental hazard.

Insurance funds are the financial basis for environmental insurance [23].

The sources of the formation of the total insurance fund in the Russian Federation are still primarily the federal and regional budgets, then the reserves of commer-cial insurance companies and personal savings of citizens. In Russia, environmental insurance began to form back in the 1990s; this process continues to this day.

4   Conclusion

The authors of this research identified several conditions that will contribute to the more active development of innovative green financial instruments.

Thus, the main instruments for stimulating the reduction of carbon emissions are emissions trading systems, carbon tax, and environmental insurance.

The analysis of the creation and stages of development of emissions trading in the EU showed that this mechanism still faces numerous difficulties. Accordingly, this requires finding ways to overcome them and opens up a wide range of possibilities for future research.

Many conditions are required for the successful implementation of emission quotas.

First, it is necessary to continue the development and gradually introduce pilot projects and innovative regional programs to control and reduce CO_2 emissions. Programs must be flexible enough to respond to the many changes occurring in the country's economy. One example is the National Project "Ecology," in particular the Federal project "Clean Air," aimed at reducing the level of air pollution. Within the framework of the project, work related to air was planned, such as creating an effective system for monitoring and controlling the quality of atmospheric air. Similar projects are needed in all regions of Russia, especially those with intensive industrial production.

In our opinion, under the current conditions of uncertainty, the emissions regulation system in Russia (CO_2 emissions quotas) requires the following:

- Well-established methodology,
- Detailed relevant regulatory legal acts,
- Creating carbon disclosure standards,
- Formation at the national level of independent bodies to control carbon emissions,
- Improvement of the monitoring and verification system for emissions.

The emission regulation system should be as simple and transparent as possible for the economic and technological greening of the country. We also believe that polluters should decide for themselves how they will reduce their CO_2 emissions.

High carbon taxes can cause tax evasion by polluting firms, reducing investment, and slowing down economic growth.

It is possible that firms can move their production to countries where there is no carbon tax. In the current situation, Russia is of great interest to the experience of those countries where the mechanism is already operating and producing positive results.

The carbon tax should become integral to the country's large-scale low-carbon development strategy. First, for a carbon tax to be introduced, it must be built in as a means of incentivizing cleaner energy production, while the system of taxation of the energy complex will need to be transformed. Second, it is necessary to introduce mechanisms for measuring the "carbon footprint" and work out opportunities to reduce carbon emissions through the modernization of production.

Second, there is no full-fledged normative document on environmental insurance in Russian legislation that would regulate public relations in this area.

The development of environmental insurance in the Russian Federation will reduce the burden on the federal and regional budgets. By making environmental insurance affordable, providing a choice of insurance programs and policies, and ensuring lower insurance premiums, businesses can be encouraged to take on voluntary insurance.

The accumulated funds received as a result of compulsory and voluntary environmental insurance will subsequently be directed to the modernization of production and preventive environmental measures.

We consider the reviewed green financial instruments and mechanisms for influencing climate problems to be extremely promising for Russia. Nevertheless, they require additional detailed study.

References

1. Cheshev AS, Alieva NV (2017) Ecological problems of the natural environment in modern conditions. Econ Ecol Territorial Entities 1:6–10
2. Zambrovskaya TA, Grishchenko AV, Grishchenko YI (2022) Verification issues "green" financial instruments. Financ Manage 3:100–109
3. Semenova NN, Grebentsova RA (2022) Using green financial instruments for sustainable development. Financ Life 2:96–102
4. Konyukova OL, Korchebny IA (2021) Development of environmental insurance in Russia. J Econ Bus 2–1(72):108–112. https://doi.org/10.24412/2411-0450-2021-2-1-108-112
5. Goryachkin BS, Stryukov NA (2022) Carbon quotas as an effective economic solution to resolve the carbon problem. Naukosfera 3(1):237–243
6. Mastini R, Kallis G, Hickel J (2021) A green new deal without growth? Ecol Econ 179:106832. https://doi.org/10.1016/j.ecolecon.2020.106832
7. Sutherland B (2020) Financing a green new deal. Cell Press 4(6):1153–1155. https://doi.org/10.1016/j.joule.2020.06.002
8. UN (2015) Paris agreement (adopted December 12, 2015). Paris, France. Retrieved from https://unfccc.int/files/essential_background/convention/application/pdf/english_paris_agreement.pdf. Accessed 16 July 2022
9. International Carbon Action Partnership (ICAP) (2016) Emissions trading worldwide: ICAP status report 2016. ICAP, Berlin, Germany. Retrieved from https://icapcarbonaction.com/system/files/document/icap_status_report_2016_online.pdf. Accessed 8 July 2022
10. World Bank (2021) State and trends of carbon pricing: report. World Bank, Washington, DC. https://doi.org/10.1596/978-1-4648-1728-1
11. Ministry of Economic Development of the Russian Federation (2021) Greenhouse gas emissions trading systems in the Asia-Pacific region: report. Retrieved from https://www.economy.gov.ru/material/file/d8d7071b90d7af3818ec3a836355244f/ETS_ATP.pdf. Accessed 16 July 2022
12. Doroshenko SV, Mingaleva AD (2020) Carbon exchanges: European experience in emissions trading mechanism development. Financ J 12(4):52–68
13. Oil Capital (2021) CO_2 emissions trading—pay or decarbonize? Retrieved from https://oilcapital.ru/article/general/31-05-2021/torgovlya-kvotami-na-vybrosy-so2-zaplatit-ili-dekarbonizirovatsya. Accessed 8 July 2022
14. Pettinger T (2020) Carbon tax—pros and cons. Retrieved from https://www.economicshelp.org/blog/2207/economics/carbon-tax-pros-and-cons/. Accessed 8 July 2022

15. Problems of development of environmental insurance in Russia and ways to overcome them. Retrieved from https://cyberpedia.su/5x1d07.html. Accessed 1 Apr 2022
16. The Economist (2015) Dooms day scenario. Retrieved from https://www.economist.com/democracy-in-america/2015/03/05/doomsday-scenarios. Accessed 1 Apr 2022
17. Asen E (2021) Carbon taxes in Europe. Tax Foundation. Retrieved from https://taxfoundation.org/carbon-taxes-in-europe-2021/. Accessed 11 July 2022
18. Statista (2021) Carbon taxes worldwide as of April 2021, by select countries. Retrieved from https://www.statista.com/statistics/483590/prices-of-implemented-carbon-pricing-instruments-worldwide-by-select-country/. Accessed 8 July 2022
19. Malyshenko AV (2004) Legal regulation of environmental insurance in the Russian Federation (Synopsis of Dissertation of Candidate of Legal Sciences). Orenburg State University, Orenburg, Russia
20. Bogolyubov SA, Khludeneva NI (2018) Commentaries on Russian Federal Law No. 7-FZ "On environmental protection" on January 10, 2002 (item-by-article). Yustitsinform, Moscow, Russia
21. Interparliamentary Assembly of Member Nations of the Commonwealth of Independent States (IPA CIS) (2003) Model law "On environmental insurance" (new edition) (Adopted by regulation on November 15, 2003 No. 22–19). St. Petersburg, Russia. Retrieved from https://docs.cntd.ru/document/901898832. Accessed 27 Mar 2022
22. Russian Federation (2010) Federal law "On compulsory insurance of civil liability of the owner of a hazardous facility for causing harm as a result of an accident at a hazardous facility" (July 27, 2010 No. 225-FZ). Moscow, Russia. Retrieved from http://www.consultant.ru/document/cons_doc_LAW_103102/. Accessed 4 May 2022
23. Muratov VR (2010) System interaction model of ecological insurance entities. Bull Don State Tech Univ 10(8):1265–1273

The Global Financial Security and Financing for Sustainable Development Interaction: The Role of the ESG Factors

Natalia N. Reshetnikova⬤, Zhanna V. Gornostaeva⬤,
Yulia S. Chernysheva⬤, Inna V. Kushnareva⬤,
and Ekaterina S. Alekhina⬤

Abstract The paper aims to solve the scientific problem of the formation of global financial security in the context of the transition to sustainable development, addressing social, economic, managerial, and environmental challenges. The research task is to develop a research methodology for a new scientific direction – the relationship between financing sustainable development and global financial security. The methodological basis of this research includes the tools of institutional and other economic theories: the methods adopted in contemporary financial theory, adapted to the sphere of economic security and applied adequately to their functional capabilities (systemic method, abstraction, analysis of information arrays, generalization and logical study of socio-economic phenomena, and comparative analysis). A conceptual model of the relationship between mechanisms and instruments for financing the transition to sustainable development and ensuring global financial security will be a new scientific result due to the absence of Russian and international studies that combine sustainable development and financial security. The scientific novelty consists in the study of a new idea of the relationship between financing sustainable development and global financial security based on the development of a new methodology adequate to the research objectives, which will provide a more complete and systemic understanding of the relationship of these processes and the consequences of financing sustainable development for ensuring global financial security.

Keywords Global financial security · National financial security · Financial globalization · Global financial market · Threats and challenges · Sustainable development · Risks of transition to sustainable development · ESG factors

N. N. Reshetnikova (✉)
Rostov State University of Economics, Rostov-on-Don, Russia
e-mail: nata.dstu@yandex.ru

Don State Technical University, Rostov-on-Don, Russia

Z. V. Gornostaeva · Y. S. Chernysheva · I. V. Kushnareva · E. S. Alekhina
Institute of Service and Entrepreneurship (Branch) of the Don State Technical University, Shakhty, Russia

JEL Classification B26 · F5 · F52 · G20 · G32 · Q01 · Q56

1 Introduction

The scientific significance of solving the problem of countering threats to global financial security in the context of the transition to sustainable development consists of the following:

- The formation of a new direction in studying financial security, namely the methodology for studying global financial security in the context of the sustainable development paradigm;
- The definition of the directions for further development of global financial security and tools for developing countries with emerging markets (including Russia) and developing countries into the global financial security network.

The scientific novelty of this research consists in the study of a new idea of the relationship between financing sustainable development and global financial security based on the development of a new methodology adequate to the research objectives, which will provide a more complete and systemic understanding of the relationship of these processes and the consequences of financing sustainable development for ensuring global financial security.

First, the currently accumulated environmental problems have a planetary scale, which suggests the need for global solutions.

Simultaneously, the issues of assessing the financial consequences of environmental problems, as well as determining the sources of financing for the relevant global solutions, are closely linked to ensuring global financial security.

Second, the existing mechanism for implementing global measures related to climate change, which provides for various institutions and mechanisms for financing countries' expenditures on the transition to sustainable development, practically does not work. The reason for its dysfunction is the lack of financial capabilities for the transition to sustainable development in several countries and their groups and significant shortcomings in coordination, interaction, and mutual understanding between countries, including contradictions between developed and developing countries. Such shortcomings are well described in terms of institutional theory (externalities, information asymmetry, the free-rider problem, etc.), which creates opportunities for developing ways to overcome these shortcomings on a given theoretical basis.

Simultaneously, these shortcomings are similar to the problems accompanying the formation of a global financial security network, which is the institutional core of global financial security: constructive unevenness, lack of access for developing countries to network tools, and, as a result, ineffective solutions to the problem of preventing and resolving crises. Accordingly, it is necessary to search for common solutions and trajectories for forming global financial security, considering the development of the mechanism for implementing global measures in connection with climate change.

Third, the transition to sustainable development, accompanied by continuing weather anomalies and natural disasters, is a source of new risks, such as transitive (i.e., risks of adaptation to climate change, risks of impact on climate, and others) and natural (physical) risks. Although these new risks need identification, assessment, and management, their relationship with financial risks is apparent. Therefore, there is a theoretical and practical need for forming financial security indicators that can capture these risks.

2 Materials and Method

Methodologically, the research is based on the tools of institutional and other economic theories.

The study of dysfunctions, barriers to implementing mechanisms, and instruments for sustainable financing development are planned through the prism of well-known institutional effects (the free-rider problem, institutional arbitration, path dependence, and others) and game theory. These studies are presented by Hardin and Cullity [1] and Puffer and McCarthy [2].

The ESG factors of global financial security are assessed based on the complementarity of the institutional and network approach since the existing global financial security network is the institutional core of global financial security. Albekov et al. [3], Evlakhova [4], Nivorozhkina et al. [5], Reshetnikova et al. [6, 7], and Vovchenko et al. [8] are among the scientists who have accumulated significant scientific experience in the field of studying the financing of the green economy, global financial security, and its relationship with the sustainable development concept.

Contemporary research in the field of sustainable development has the following main directions.

The first direction is the analysis of the main aspects of the concept of sustainable development. After the global financial and economic crisis of 2008–2009, the documents of the UN and OECD included the concepts of green economy and green growth. The socio-ecological and economic aspects of the sustainable development concept came to the fore. Clearly defined goals, objectives, and indicators have appeared [9, 10]. In the past four years, the concept of sustainable finance has been actively developed. Defining sustainable development as an integrated concept with three aspects (economic, social, and environmental), the developers of the sustainable finance concept see its essence in the patterns of interaction of financing processes with economic, social, and environmental problems, as presented in work of Schoenmaker and Schramade [11].

The second direction is the development of sustainable development indicators. There is a significant volume of international research by Böhringer and Jochem [12] and Moran et al. [13]. The purpose of sustainable development indicators is to review and compare indices developed to measure sustainable development.

The Ecological Footprint (EF) concept was developed by Wackernagel. This is an attempt at quantifying sustainability.

Environmental scarcity indicates how much a population lives beyond nature's means. The per capita ecological footprint of economic activity shows a large divergence in the demands on nature from people in different societies, from Qatar at the top level (15.5 hectares per person) to Haiti at the bottom level (0.7), with the USA (8, 4), Germany (5.1), China (3.7), and other intermediates (data for 2014). These figures are the basis for the assertion that if humanity consumed like the average American, it would take about five Earths. EFs also vary greatly within a country based on wealth levels. In contrast to the increase in biological load from human activity, the Earth's biocapacity has not increased at the same rate as the ecological footprint. The increase in biocapacity averaged only 0.5% per year. Due to agricultural intensification, biocapacity was 9.6 billion hectares in 1961 and increased to 12.2 billion hectares in 2016.

The performed literature review showed that there is currently no comprehensive indicator of sustainable economic development that would allow identifying macroeconomic variables for further analysis and construction of a mechanism for transferring impulses from the development of the green economy to economic growth and global financial security.

The third direction is the analysis of mechanisms for financing the transition to sustainable development.

The issues of financing the costs of combating environmental pollution and achieving the corresponding goals of sustainable development began to be addressed already at the first stages of the adoption and implementation of international programs in the field of climate change [14]. The problem of moving toward sustainable development of society depends on the availability of a sufficient number of financial resources to implement the necessary measures, as well as on the effectiveness of the allocation of funds between directions, actors, and objects of funding for programs to combat negative climate change [15].

Green financing and various approaches to this concept have become widespread in the last decade.

According to UN experts, a comprehensive transition to sustainable growth requires $22 trillion annually. However, there are currently not enough green bonds and loans issued [16]. In 2011, the German government proposed a strategic approach involving the financial sector in the transition to a low-carbon and resource-efficient economy in the context of climate change [17]. In 2015, the People's Bank of China put forward policy and institutional mechanisms to attract private investment in green areas such as environmental protection, energy conservation, and clean energy production through financial services, including lending, private equity funds, bonds, stocks, and insurance [18].

The applicability of the financial mechanisms proposed under the Kyoto Protocol and the United Nations Framework Convention on Climate Change has been largely constrained by the lack of adequate national and international institutions to regulate the accumulation and allocation of financial resources [19]. Bouwer and Aerts [20] described the low activity of developed and developing countries in implementing Sustainable Development Goals and financing activities to combat climate change as constraints to actively implementing financing programs for the transition to sustainable development [21].

In 2008, the Adaptation Fund, for the purpose of financing sustainable development programs and solving the problems of climate regulation, began its operation. For some time, it solved the issue of redistributing funds received from the implementation of highly developed countries of measures to combat environmental pollution in favor of financial support for the least developed countries exposed to the greatest negative impacts of climate change [22]. However, later, the funds of the Adaptation Fund began to be formed mainly through contributions provided voluntarily by developed countries [23, 24]. The institutions underlying international financing mechanisms are characterized by internal and external contradictions, have a low ability to implement, and show low attractiveness for investors [25].

International and Russian authors conducted no studies to analyze the relationship between sustainable development and financial security.

3 Results

The expected results of the research will be characterized by the following elements of scientific novelty:

1. A conceptual model of the relationship between mechanisms and instruments for financing the transition to sustainable development and ensuring global financial security;
2. Combinations of new solutions, conditions, and factors in the search for a balance of market forces and the state in financing sustainable development based on institutional effects;
3. New methodological tools for studying the relationship of the analyzed processes, the boundaries of its use, and heuristic possibilities as an element of the original methodological approach;
4. The original concept of minimizing the financial risk of reducing the income of a group of countries with carbon-intensive economies and a high level of budget revenues from the production and export of hydrocarbons during the transition of the world economy to clean energy sources.

The expected research results are as follows:

1. Conceptual model of the relationship of mechanisms and instruments for financing the transition to sustainable development and ensuring global financial security;
2. A theoretical approach based on the application of the provisions of neo-institutional theory to study the balance of the market and the state in financing the transition to sustainable development;
3. Methodological approach to the formation of an indicative assessment of ESG factors of global financial security:

(a) Identification of the system of indicators of global financial security related to ESG factors;

(b) Substantiation and development of an integral ESG indicator of global financial security.

4. The concept of minimizing the financial risk of reducing the income of a group of countries with carbon-intensive economies and a high level of budget revenues from the production and export of hydrocarbons during the transition of the world economy to environmentally friendly energy sources.

Scientific and social significance (compliance of the expected results with the world level of research and the possibility of practical use of the expected results of the chapter in the economy and social sector, including for the creation of new or improvement of manufactured products (goods, works, services) and applied technologies).

The scientific novelty of research forms the heuristic significance of the expected results of the chapter. The new theoretical and methodological foundations, developed as a result of researching the idea of the relationship between financing sustainable development and global financial security, can be used to analyze other research objects of economic science, including economic growth, socio-economic inequality of the population, global financial stability, the problem of equalizing countries belonging to different groups in terms of economic and social parameters, the transformation of the world financial architecture, and others.

The scientific problem of the formation of global financial security in the context of the transition to sustainable development has a global scale since its study is related to the implementation of the Sustainable Development Goals developed by the United Nations in 2015. Therefore, the formulation of a scientific problem and the expected results of its research corresponds to the world level.

The expected results have the potential for theoretical and practical implementation.

The Russian market for sustainable development tools performed well in 2021 [16]. The country provided new regulations for business. Companies and banks successfully tested the proposed formats, as described by Gornostaeva et al. [26], Gornostaeva et al. [27], and Mezinova et al. [28].

The largest share of the sustainable finance market is still held by green bonds (80% of the market in value terms). Simultaneously, the sector of social bonds demonstrated significant growth, which amounted to almost 60 billion rubles at the end of the year.

Nowadays, countries are responsible for coordinating their work and enhancing cooperation to reduce environmental risks. Achieving the goals specified in the concept of sustainable development will allow the state to overcome the global challenges in the world economy. However, the question remains regarding the methodology for studying the financing of the green economy, as well as its relationship with the concept of sustainable development. The study is aimed at the development and implementation of environmental programs and proposals for improving the national strategy to attract civil and business communities to a joint dialogue and cooperation in the field of the green economy.

4 Conclusions

In today's conditions, it is necessary to improve socio-economic mechanisms that contribute to the reduction and prevention of negative environmental changes and the provision of a more rational use of natural resources.

The green economy can reduce the negative environmental impact and preserve natural resources for future generations. However, the existing provisions of the green economy require the actualization of scientific knowledge in the context of the sustainable development of countries.

Therefore, studies aimed at comprehensively analyzing world experience in implementing the green economy, its financing, and active implementation are becoming relevant. A comprehensive analysis of all the factors that will ensure the dynamic growth of the economy, considering the rational use of natural resources and environmental changes, can be carried out using Russia as an example.

The scientific novelty of this research lies in the improvement of theoretical provisions and the development of practical recommendations for the formation and financing of the green economy in the country within the framework of the sustainable development concept.

The scientific problem to be solved by the research is the formation of strategies and conditions for the transition to sustainable development and the solution of social, economic, managerial, and environmental problems. Under the new conditions, the Russian economy has undergone changes associated with an increase in environmental priorities in the context of the formation of the green economy, which has not been properly represented in the regulatory framework. In this regard, it is necessary to continue the further scientific analysis of the considered problems, which are reflected in the goals and objectives of the study.

Further research will be based on the above principles and approaches.

References

1. Hardin R, Cullity G (2003) The free rider problem. In: Zalta EN (ed) The Stanford encyclopedia of philosophy (revision Oct 13, 2020 Winter 2020 Edition). Stanford University, Stanford. Retrieved from https://plato.stanford.edu/archives/win2020/entries/free-rider/. Accessed 8 Aug 2022
2. Puffer SM, McCarthy DJ (2017) Two decades of Russian business and management research: an institutional theory perspective. Acad Manag Perspect 25(2):21–36. https://doi.org/10.5465/amp.25.2.21

3. Albekov AU, Parkhomenko TV, Polubotko AA (2018) Green economy: a phenomenon of progress and a concept of environmental security. In Grima S, Thalassinos E (eds) Contemporary issues in business and financial management in Eastern Europe. Emerald Publishing Limited, Bingley, pp 1–8. https://doi.org/10.1108/S1569-375920180000100002
4. Evlakhova YS (2022) Analysis of systematic risk of global ETFs with maximum ESG rating. Siberian Financial School 2:18–23. https://doi.org/10.34020/1993-4386-2022-2-18-23
5. Nivorozhkina LI, Alifanova EN, Evlakhova YS, Tregubova AA (2018) Indicators of financial security on the micro-level: approach to empirical estimation. Eur Res Stud J 21(1):324–332. Retrieved from https://www.um.edu.mt/library/oar/bitstream/123456789/33986/1/Indicators_of_Financial_Security_on_the_Micro_Level_2018.pdf. Accessed 8 Aug 2022
6. Reshetnikova NN, Magomedov MM, Zmiyak SS, Gagarinskii AV, Buklanov DA (2021) Directions of digital financial technologies development: Challenges and threats to global financial security. In: Shakirova OG, Bashkov OV, Khusainov AA (eds) Current problems and ways of industry development: equipment and technologies. Springer, Berlin, pp 355–363. https://doi.org/10.1007/978-3-030-69421-0_38
7. Reshetnikova N, Magomedov M, Zmiyak S, Chernysheva Y (2022) Digital technologies adoption in the agro-industrial complex as a priority of regional development in the conditions of global macroeconomic changes. In: Beskopylny A, Shamtsyan M (eds) XIV international scientific conference "INTERAGROMASH 2021". Springer, Berlin, pp 3–12. https://doi.org/10.1007/978-3-030-81619-3_1
8. Vovchenko NG, Andreeva OV, Dmitrieva VD, Galazova SS, Krstev TV (2021) Modelling of economic clusters in the Russian regions in the context of formation of the Industry 4.0. In: Trifonov PV, Charaeva MV (eds) Strategies and trends in organizational and project management. Springer, Berlin, pp 667–673. https://doi.org/10.1007/978-3-030-94245-8_91
9. OECD (2009) Declaration on green growth (adopted at the meeting of the council at ministerial level on 25 June 2009 [C/MIN(2009)5/ADD1/FINAL]). Paris, France. Retrieved from https://www.oecd.org/env/44077822.pdf. Accessed 8 Aug 2022
10. Leal Filho W, Manolas E, Pace P (2015) The future we want: key issues on sustainable development in higher education after Rio and the UN decade of education for sustainable development. Int J Sustain High Educ 16(1):112–129. https://doi.org/10.1108/IJSHE-03-2014-0036
11. Schoenmaker D, Schramade W (2018) Principles of sustainable finance. Oxford University Press, Oxford. Retrieved from https://ssrn.com/abstract=3282699. Accessed 8 Aug 2022
12. Böhringer C, Jochem PE (2007) Measuring the immeasurable: a survey of sustainability indices. MPRA Paper No. 91562. Retrieved from https://mpra.ub.uni-muenchen.de/91562/1/MPRA_paper_91562.pdf. Accessed 8 Aug 2022
13. Moran DD, Wackernagel M, Kitzes JA, Goldfinger SH, Boutaud A (2008) Measuring sustainable development—nation by nation. Ecol Econ 64(3):470–474. https://doi.org/10.1016/j.ecolecon.2007.08.017
14. Smit B, Burton I, Klein RJ, Street R (1999) The science of adaptation: A framework for assessment. Mitig Adapt Strat Glob Change 4(3–4):199–213. https://doi.org/10.1023/A:1009652531101
15. Weiler F, Klöck C, Dornan M (2018) Vulnerability, good governance, or donor interests? The allocation of aid for climate change adaptation. World Dev 104:65–77. https://doi.org/10.1016/j.worlddev.2017.11.001
16. UN (2015) Report of the intergovernmental committee of experts on sustainable development financing. New York. Retrieved from https://www.un.org/esa/ffd/wp-content/uploads/2015/03/ICESDF_Ru.pdf. Accessed 8 Aug 2022
17. Schaefer J (2011) Green finance: an innovative approach to fostering sustainable economic development and adaption to climate change. GIZ, Bonn and Eschborn. Retrieved from https://www.greengrowthknowledge.org/sites/default/files/downloads/resource/Green_finance_GIZ.pdf. Accessed 8 Aug 2022
18. Research Bureau of the People's Bank of China & UNEP (2015) Theoretical framework of green finance. In Establishing China's green financial system: report of the green finance task force. Beijing, China, p 37. Retrieved from https://unepinquiry.org/wp-content/uploads/2015/12/Establishing_Chinas_Green_Financial_System_Final_Report.pdf. Accessed 8 Aug 2022

19. Flåm KH, Skjærseth JB (2009) Does adequate financing exist for adaptation in developing countries? Clim Policy 9(1):109–114. https://doi.org/10.3763/cpol.2008.0568

20. Bouwer LM, Aerts JC (2006) Financing climate change adaptation. Disasters 30(1):49–63. https://doi.org/10.1111/j.1467-9523.2006.00306.x

21. Katasonova Y, Mitrofanov P (2022) The future of the sustainable finance market: maintain and strengthen national expertise. Expert RA. Retrieved from https://raexpert.ru/docbank/5fc/ba4/29e/5da3d697880e5ea15091c6e.pdf. Accessed 8 Aug 2022

22. Ayers JM, Huq S (2009) Supporting adaptation to climate change: what role for official development assistance? Dev Policy Rev 27(6):675–692. https://doi.org/10.1111/j.1467-7679.2009.00465.x

23. Berrang-Ford L, Biesbroek R, Ford JD, Lesnikowski A, Tanabe A, Wang FM, Chen C, Hsu A, Heymannet SJ, Pringle P, Grecequet M (2019) Tracking global climate change adaptation among governments. Nature Clim Change 9(6):440–449. https://doi.org/10.1038/s41558-019-0490-0

24. Berrang-Ford L, Ford JD, Lesnikowski A, Poutiainen C, Barrera M, Heymann SJ (2014) What drives national adaptation? A global assessment. Climatic Change 124(1–2):441–450. https://doi.org/10.1007/s10584-014-1078-3

25. Mingaleva ZA (2020) Institutional features of international financing for climate change adaptation programs. Fin J 12(4):10–25. https://doi.org/10.31107/2075-1990-2020-4-10-25

26. Gornostaeva ZV, Alekhina ES, Kushnareva IV, Malinina OY, Vasenev SL (2018) Barriers on the path of information economy's formation in modern Russia. In: Sukhodolov AP, Popkova EG, Litvinova TN (eds) Models of modern information economy. Emerald Publishing Limited, Bingley, pp 89–98. https://doi.org/10.1108/978-1-78756-287-520181011

27. Gornostaeva ZV, Alekhina ES, Tregulova NG, Chernysheva YS (2020) Scientific and methodological approach to systemic analysis of the circular economy from the positions of interested parties. In: Popkova EG, Bogoviz AV (eds) Circular economy in developed and developing countries: perspective, methods and examples. Emerald Publishing Limited, Bingley, pp 47–55. https://doi.org/10.1108/978-1-78973-981-720201008

28. Mezinova I, Balanova M, Bodiagin O, Israilova E, Nazarova E (2022) Do creators of new markets meet SDGs? Anal of platform companies. Sustainability 14(2):674. https://doi.org/10.3390/su14020674

Financing the Green Economy in the Context of Global Sustainable Development

Natalia N. Reshetnikova⬛, Zhanna V. Gornostaeva⬛,
Anastasia V. Khodochenko⬛, Marina V. Bugaeva⬛,
and Natalia G. Tregulova⬛

Abstract

Purpose The research conducted in this article is devoted to the consideration and analysis of the features of financing the green economy in the context of global sustainable development in the world. *Design/methodology/approach* The scientific problem raised in the article—the formation of strategies and conditions for the transition to sustainable development—is associated with the solution of social, economic, managerial and environmental problems. A number of reasons that determine the relevance of the designated problem are given by the authors in the article. Namely, environmental problems have a planetary scale and imply the need for global solutions. *Findings* The existing mechanism for implementing global measures to deal with climate change, which provides for various institutions and mechanisms for financing expenses of countries for the transition to sustainable development, practically does not work. The reason for its dysfunction is not only the lack of financial opportunities for a number of countries and their groups to implement the transition to sustainable development, but also significant shortcomings in coordination, interaction and understanding between countries, including contradictions between developed and developing countries. Such shortcomings are well described in terms of institutional theory. *Originality/value* Based on the leading international practices, the main instruments of financing the green economy were studied, allowing for the effective implementation of green projects in both developed and developing countries, the authors concluded that financing the green economy contributes to the economic growth of the country, taking into account the reduction of the negative impact on the environment and the conservation of natural resources for future

N. N. Reshetnikova (✉) · A. V. Khodochenko
Rostov State University of Economics, Rostov-on-Don, Russia
e-mail: nata.dstu@yandex.ru

N. N. Reshetnikova
Don State Technical University, Rostov-on-Don, Russia

Z. V. Gornostaeva · M. V. Bugaeva · N. G. Tregulova
Institute of Service and Entrepreneurship (Branch) of the Don State Technical University,
Rostov-on-Don, Russia

generations. The existing provisions of the "green" economy and the understanding of its place at a new stage in the development of world economic relations require the actualization of scientific knowledge in the context of sustainable development of countries.

Keywords Green economy · Sustainable development · Green finance · Ecology · Green projects · Financial security

JEL Classification F36 · F53 · F63 · F64 · P18 · P28 · P45 · Q01 · Q56 · Q57 · O13 · O41 · O43 · O44 · O47

1 Introduction

With the development and improvement of each technological mode, the relationship of a person with the environment increases, which is expressed through a direct influence on environmental processes. Global problems of humanity are manifested in many areas of the economy, therefore, the green economy is able to act as an effective mechanism for ensuring the well-being of citizens within the framework of environmental and socio-economic strategies of the state. Taking into account the goals of the concept of sustainable development, the country is able to reduce and prevent negative changes in the environment. There is already a practice of financing green initiatives at the national and international level.

The need to analyze the financing instruments of the green economy is related to the differences in existing national strategies. Identification of priority tasks and tools for the development of national economies in modern conditions together will allow us to offer a unified approach to the sustainable development of the country's economy, taking into account the specifics of financial policy. The relevance of the study is caused by the need to analyze international practices on financing the green economy.

In the context of the deteriorating environmental situation on the planet, the economies of many countries have undergone changes associated with increased attention to environmental conservation through effective financing of the green economy. The purpose of the article is a practical study of the international practice of using financial instruments in the field of green finance. Within the framework of the designated goal, the following tasks were set: (1) to analyze the specifics of financing the green economy in the leading countries; (2) to identify trends in the development of the green economy at the present stage of the world economy; (3) to propose common principles and instruments for financing the green economy in the new realities.

The hypothesis of this study is based on the analysis of financing instruments for the green economy, which will contribute to the rapid achievement of Sustainable Development Goals. At the same time, the research hypothesis is based on the authors' assumption that active financing of green projects will involve the elaboration of financial approaches that expand the knowledge of the practical significance and prospects for the development of the green economy in the world.

2 Methodology

In the process of preparing the article, the works of domestic and foreign scientists in the field of sustainable development, green economy and its financing were analyzed. The research methodology was based on a system-functional approach, including generalization and systematization of scientific knowledge. The following methods of analysis were used: dialectical logic method of cognition of phenomena and processes, methods of induction and deduction; method of system analysis and generalization; methods of analysis, synthesis and comparative analysis, etc.

The works of foreign authors Turner [37], Ha and Byrne [21] are devoted to the definition of the concepts of green economy, green growth in the context of sustainable energy and the fuel and energy complex.

In the works of such Russian authors as Alexandrin [3], Bobylev et al. [6], Nikulina and Khit [25], methods of improving tools for stimulating the introduction of environmentally friendly technologies into production and increasing the financial stability of industrial enterprises are proposed.

Modern research in the field of sustainable development has several main directions. First of all, we are interested in analyzing the main aspects of the concept of sustainable development.

Initially, the concept of sustainable development was considered as an environmental one, as evidenced by the works of such authors as Meadows et al. [24], Alexandrin [3], as well as the Declaration of the United Nations Conference on the Human Environment adopted in 1972 [9] and the Declaration on Green Growth adopted at the Meeting of the Council at Ministerial Level on 25 June 2009 [10]. After the global financial and economic crisis of 2008–2009, the concepts of green economy and green growth appeared in the UN and OECD documents, the socio-ecological and economic aspect of the SD concept came to the fore, clearly defined goals, tasks and indicators appeared [20, 32]. In the last four years, the concept of sustainable finance has been actively developed. The concept of sustainable development is an integrated model of three key aspects—economy, society and ecology. The authors of this concept, Smit et al. [31], including the concept of sustainable finance, emphasize the importance of the interaction of the processes themselves in the field of finance, as well as interaction with the economy, social and environmental problems.

Danilov [8] reveals the current state of the concept of sustainable financing in Russia, highlights the main obstacles and the feasibility of its implementation. In the work of the authors Hamdouch and Depret [22], it is proclaimed that "the principles of sustainable finance imply the achievement of sustainable development goals by using the opportunities of financial markets and financial organizations to transform savings into responsible investments, to manage (transform and redistribute) environmental and social risks, and to distribute public resources based on the criterion of maximizing aggregate (financial, environmental and social) cost."

Over the past 10 years, green financing and various approaches to this concept have become widespread. According to UN experts, $22 trillion a year is needed for a comprehensive transition to sustainable growth, but today the issued bonds and loans belonging to the green category are far from sufficient, as noted by authors such as Schoenmaker and Schramade [30].

In his work, Bobylev [5] considers such problems of sustainability as: the relationship between welfare and sustainable development; quantitative interpretation, SD indicators; transformation of the energy sector as the key to sustainable development.

A number of works by scientists of the Rostov scientific school, who have accumulated significant scientific experience in studying the financing of the green economy; global financial security and its relationship with the concept of sustainable development, deserve mention: Albekov et al. [2], Nivorozhkina et al. [26], Gornostaeva et al. [15, 16], Balanova et al. [4], Fursovet al. [13], Vovchenko et al. [38], Reshetnikova et al. [27–29].

3   Results

This study is based on the results obtained by the American consulting company Dual Citizen LC. For the first time, this company compiled The Global Green Economy Index (GGEI), the results of which revealed the best practices for the implementation of most of the financing instruments for the development of the green economy in the country. The methodology of evaluation and ranking of 160 countries consists in the analysis of 18 quantitative and qualitative indicators [36]. Thus, according to the results of the last rating (2022), Switzerland, France, Austria, Great Britain and Luxembourg have become the leaders (Table 1).

Thus, we see that European countries occupy leading positions in this rating, thereby demonstrating their commitment to the implementation of green principles in their national strategies.

First, it is necessary to consider the experience of developed countries in shaping the green agenda for sustainable development of the world economy and financing green projects.

For a long time, the European Union (EU) has been an active supporter of promoting the principles of a green economy in Europe. The European environmental policy can be traced in all national strategies of the participating countries within the framework of the concept of sustainable development. The course of action and goals aimed at the transition to a new eco-friendly model of the economy are presented in many EU regulatory documents. The Roadmap to a Resource Efficient Europe reflects the ways in which this transition is planned, including financing of clean energy [33]. €17.5 Billion Just Transition Fund acts as a tool to support green projects. The Fund is included in the "fair transition" mechanism, where the

Table 1 The ranking of the top 10 countries in the green economy, 2022

№	Country	Result
1	Switzerland	0.781
2	France	0.744
3	Austria	0.711
4	Great Britain	0.704
5	Luxembourg	0.696
6	Belgium	0.693
7	Monaco	0.645
8	Romania	0.623
9	Greece	0.617
10	Bulgaria	0.604

Source Developed by the authors on the basis of the Global Green Economy Index [33]

main task is to overcome the socio-economic consequences of the transition to a green economy in European countries that are gradually abandoning the production and use of coal, oil shale, etc. The mechanism also includes a special initiative of InvestEU and a credit line for the public sector [19].

Thanks to effective market and political incentives, conditions will be created to encourage eco-friendly innovations and new green directions in all spheres of society. For example, the introduction of renewable energy sources into production processes will contribute to an early transition to a low-carbon economy, taking into account environmental and economic benefits, i.e. green jobs, reduction of harmful emissions into the atmosphere, and more. Thus, the EU pays special attention to the green economy in the process of discussing reforms aimed at protecting the environment.

Within the framework of the Energy Union, European countries have formed a policy aimed at providing sustainable, affordable and safe energy for their citizens [1]. Among the key mechanisms for the implementation of the above goal, it is worth highlighting the program "Clean Energy for all Europeans", which is aimed at supporting and developing the renewable energy sources (RES) for the transition to a low-carbon economy [7]. By 2030, it is planned to achieve a 55% reduction in CO_2 emissions. The European Commission is committed to the complete decarbonization of energy by 2050 in all countries of the Union [35].

In general, European countries, such as Germany, are actively engaged in the development of environmental policy, both at the national and international level [14]. The national policy of Germany reflects the key objectives for achieving sustainable development of the country [34]. As for the green economy, the Strategy emphasizes that in order to improve the standard of living of future generations, significant changes are required in the traditional economy, and first of all, in the habits of people [14]. In Germany, waste incineration can be singled out among the priority green industries. Thus, 68 waste incinerators operate in the state [12]. Electricity is generated during heat treatment due to waste incineration. The country demonstrates a unique experience of interaction between government and business in order to create an effective waste management model [18].

In European countries, small and medium-sized enterprises (SMEs) play a key role in the formation of a sustainable economy. Today, the competitiveness of European countries in green areas (green technologies, eco-labeling, etc.) provides them with leadership in many areas in the green economy. New business models are becoming relevant, especially startups that promote green innovative ideas. About six thousand green German companies are able to make a significant contribution to environmental protection through innovative ideas and products [19].

At the initial stages of the implementation of the Sustainable Development Goals, the United States actively promoted a policy on the introduction of renewable energy. The Renewable Electricity Production Tax Credit (PTC) and The Business Energy Investment Tax Credit (ITC) were used as a tool [11].

Modern economic models can no longer cope with the current global challenges that are changing the living conditions and way of life of people on the planet.

Within the framework of economic development, countries are developing strategies to reduce the negative impact on the environment. In 2015, the countries adopted the Paris Agreement to replace the Kyoto Protocol. The agreement is aimed at reducing carbon dioxide emissions into the atmosphere while maintaining the average temperature on the planet by more than 2 degrees Celsius.

A striking example of the active use of tools for financing the green economy is South Korea, where funding is provided annually to prevent the effects of climate change [23]. Green financing in the country is coordinated, for example, by the Green Climate Fund (GCF). The Foundation actively invests its resources in green projects in the field of climate change [17].

Among developing countries, China is a leader in the field of sustainable development and promotion of the idea of "Ecological Civilization". Within the framework of the 13th Five-year plan, China proclaims the principles of a green economy and supports the introduction of high-tech and environmentally friendly innovations in all industrial sectors [36]. Thanks to the increased investment attractiveness and public policy, the country was able to partially abandon traditional energy sources to mitigate the effects of climate change.

The current global crisis has affected the well-being of the population. Against the background of a decline in production and the closure of businesses, incomes are falling, unemployment is rising, etc. This leads to a reduction in the consumption of expensive environmental goods and services. The solution of socio-economic problems becomes a priority for the state, and environmental projects, unfortunately, fade into the background.

The researched topic will remain relevant, including for Russia. For example, in his speech at the St. Petersburg International Economic Forum 2022, the President of the Russian Federation paid special attention to the attitude to the environment. The country will continue to implement green projects aimed at preserving the climate. Existing initiatives to finance the green economy will remain relevant and will continue to develop.

The global environmental trend is aimed at maintaining the scale of production activities while providing for rational use of natural resources and reducing the strain on the environment. International practice provides for various types of environmental protection activities, as well as investments in environmental protection activities.

4 Conclusion

The process of degradation of the ecological situation on the planet is more active than ever. The reduction of natural resources and biological diversity is reaching unprecedented proportions. Prevention of global environmental problems, development of quantitative and qualitative ecological and economic indicators and assessment of the dynamics of ecologization in all spheres of the economy are strategic directions of any modern state. The solution lies in the formation of a new model of human behavior, namely, an economy that is able to unite all spheres of human activity to combat environmental consequences in the interest of future generations. The green economy is the way to achieve the Sustainable Development Goals.

The task of improving existing economic models in order to prevent global problems affecting everyone is facing every nation. Among the options for solving the problem is the formation of a green economy aimed at improving the welfare of the population and ensuring the preservation of the environment. The current economic model does not fully meet the modern needs of society, and even harms future generations, which is reflected in environmental degradation and the emergence of social inequality.

Every year, countries develop and strive to introduce new carbon-free mechanisms to identify and negative consequences for the environment and human health. This study provides examples of those countries that are actively implementing tools, including financial ones, for the development of the green economy. Every year the number of ongoing green projects in all regions of the world is growing. One of the leaders is China, which is actively fighting the deterioration of the environmental situation in the country. The development and implementation of effective taxation in the USA allowed the population to rationally dispose of electricity based on renewable energy sources. Thus, international experience and best practices will accelerate the process of transition to the green economy in other countries, which will contribute to improving the environment on the planet.

Nevertheless, the barriers to financing the green economy still remain the profitability of the investment in resource efficiency, the uncertainty of the policy to support the transition to the green economy as well as the lack of confidence in it. This, in turn, increases financial risk, and financial markets focused on short-term results often do not look kindly upon long-term investments.

References

1. A Framework Strategy for a Resilient Energy Union with a Forward-Looking Climate Change Policy (2015) The European Commission. https://eur-lex.europa.eu/resource.html?uri=cellar:1bd46c90-bdd4-11e4-bbe1-01aa75ed71a1.0001.03/DOC_1&format=PDF. Data accessed 8 Aug 2022
2. Albekov AU, Parkhomenko TV, Polubotko AA (2018) Green economy: a phenomenon of progress and a concept of environmental security. In: Contemporary issues in business and financial management in Eastern Europe. Emerald Publishing Limited. https://doi.org/10.1108/S1569-375920180000100002
3. Alexandrin YN (2020) Improvement of tools for stimulating the introduction of environmentally friendly technologies into production. Econ Theor Pract 1(57):54–63
4. Balanova M, Bodiagin O, Mezinova I, Zhelev P (2020) I-business firms in the Industry 4.0 global value chains. Proc Int Conf Econ Manag Technol 2020:212–216. https://doi.org/10.2991/aebmr.k.200509.039
5. Bobylev SN (2017) Sustainable development: paradigm for the future. World Econ Int Relat 61(3):107–113
6. Bobylev SN, Minakov VS, Soloviova SV, Tretyakov VV (2012) Ecological and economic index of the regions of the Russian Federation. Methodology and calculation indicators. WWF Russia, RIA Novosti
7. Clean energy for all Europeans (2019) Publication office for the European Union. https://op.europa.eu/en/publication-detail/-/publication/b4e46873-7528-11e9-9f05-01aa75ed71a1/language-en. Data accessed 8 Aug 2022
8. Danilov YA (2021) The concept of sustainable finance and prospects for its implementation in Russia. Economic Issues (5):5–25. https://doi.org/10.32609/0042-8736-2021-5-5-25
9. Declaration of the United Nations Conference on the Human Environment (1972) Official website of the UN. https://www.un.org/ru/documents/decl_conv/declarations/declarathenv.shtml. Data accessed 8 Aug 2022
10. Declaration on Green Growth (2009) Adopted at the meeting of the council at ministerial level on 25 June 2009. OECD. https://www.oecd.org/env/44077822.pdf. Data accessed 8 Aug 2022
11. Doing Business with DOE. Office of Small and Disadvantaged Business Utilization. https://www.energy.gov/osdbu/doing-business-doe. Data accessed 8 Aug 2022
12. Environmental policy for a sustainable society. Sustainable Development Report by the Federal Environment Ministry on the implementation of the United Nations 2030 Agenda. Federal Ministry for the Environment, Nature Conservation and Nuclear Safety (BMU). https://www.bmuv.de/fileadmin/Daten_BMU/Pools/Broschueren/umwelt_nachhaltige_gesellschaft_en_bf.pdf. Data accessed 8 Aug 2022
13. Fursov VA, Lazareva NV, Takhumova OV, Semenova LV, Kushch EN (2021) Qualitative assessment of an industrial business entity: economic potential and maximum performance. In: Bogoviz AV (ed) The challenge of sustainability in agricultural systems. Lecture notes in networks and systems, vol 206. Springer, Cham, pp. 139–149. https://doi.org/10.1007/978-3-030-72110-7_14
14. German Sustainable Development Strategy (2017) The federal government. https://www.bundesregierung.de/resource/blob/998220/455740/7d1716e5d5576bec62c9d16ca908e80e/2017-06-20-langfassung-n-en-data.pdf?download=1. Data accessed 8 Aug 2022
15. Gornostaeva ZV, Alekhina ES, Kushnareva IV, Malinina OY, Vasenev SL (2018) Barriers on the path of information economy's formation in modern Russia. In: Sukhodolov AP, Popkova EG, Litvinova TN (eds) Models of modern information economy, pp 89–98. Emerald Publishing Limited, Bingley. https://doi.org/10.1108/978-1-78756-287-520181011
16. Gornostaeva ZV, Alekhina ES, Tregulova NG, Chernysheva YS (2020) Scientific and methodological approach to systemic analysis of the circular economy from the positions of interested parties. In: Circular economy in developed and developing countries: perspective, methods and examples, pp 47–55. https://doi.org/10.1108/978-1-78973-981-720201008

17. Green Climate Fund (2020) https://www.greenclimate.fund/home. Data accessed 8 Aug 8 2022
18. Green growth in action: Germany (2002) OECD. https://www.oecd.org/germany/greengrow thinactiongermany.htm. Data accessed 8 Aug 2022
19. Green Startup Monitor (2018) Borderstep institute for innovation and sustainability. https://www.borderstep.de/wp-content/uploads/2019/06/GreenStartupMonitor2018_ EN.pdf. Data accessed 8 Aug 2022
20. Greening the European Semester (2022) The European Commission. https://ec.europa.eu/env ironment/integration/green_semester/index_en.htm. Data accessed 8 Aud 2022
21. Ha YH, Byrne J (2019) The rise and fall of green growth: Korea's energy sector experiment and its lessons for sustainable energy policy. Wiley Interdisc Rev Energy Environ 8(4):e335. https://doi.org/10.1002/wene.335
22. Hamdouch A, Depret MH (2010) Policy integration strategy and the development of the 'green economy': foundations and implementation patterns. J Environ Planning Manag 53(4):473–490. https://doi.org/10.1080/09640561003703889
23. Just Transition Fund (2020) The European Commission. https://ec.europa.eu/info/funding-ten ders/find-funding/eu-funding-programmes/just-transition-fund_en. Data accessed 8 Aug 2022
24. Meadows DH, Meadows DL, Randers J, Behrens WW III (1972) The limits to growth. Universe Books, New York
25. Nikulina OV, Khit YA (2020) Increase in financial stability of the industrial enterprises (on the example of the oil and gas industry). Econ Sustain Dev 2(42):242–247
26. Nivorozhkina LI, Alifanova EN, Evlakhova YS, Tregubova AA (2018) Indicators of financial security on the micro-level: approach to empirical estimation. Eur Res Study J XXI(Special issue):324–332. https://doi.org/10.35808/ersj/1183
27. Reshetnikova N, Magomedov M, Buklanov D (2021) Digital finance technologies: threats and challenges to the global and national financial security. In: IOP conference series: earth and environmental science, vol 666, p 062139. https://doi.org/10.1088/1755-1315/666/6/062139
28. Reshetnikova N, Magomedov M, Zmiyak S, Chernysheva Y (2022) Digital technologies adoption in the agro-industrial complex as a priority of regional development in the conditions of global macroeconomic changes. In: Beskopylny A, Shamtsyan M (eds) XIV international scientific conference "INTERAGROMASH 2021". Lecture notes in networks and systems, vol 246. Springer, Cham, pp 3–12. https://doi.org/10.1007/978-3-030-81619-3_1
29. Reshetnikova NN, Magomedov MM, Zmiyak SS, Gagarinskii AV, Buklanov DA (2021) Directions of digital financial technologies development: challenges and threats to global financial security. In: Shakirova OG, Bashkov OV, Khusainov AA (eds) Current problems and ways of industry development: equipment and technologies. Lecture notes in networks and systems, vol 200. Springer, Cham, pp 355–363. https://doi.org/10.1007/978-3-030-69421-0_38
30. Schoenmaker D, Schramade W (2018) Principles of sustainable finance. Oxford University Press, Oxford
31. Smit B, Burton I, Klein RJT, Street R (1999) The science of adaptation: a framework for assessment. Mitig Adapt Strat Glob Change 4:199–213
32. The future we want (2012, June 20–22) The final document of the UN conference on sustainable development. Rio de Janeiro, Brazil
33. The Global Green Economy Index (2022) Dual citizen. https://dualcitizeninc.com/global-green-economy-index. Data accessed 8 Aug 2022
34. The Strategy of Sustainability (2022) The Federal Government. https://www.bundesregierung. de/breg-en/issues/sustainability/the-strategy-214722. Data accessed 8 Aug 2022
35. The Treaty on the Functioning of the European Union (2022) EUR-lex. Access to European Union Law. https://eur-lex.europa.eu/legal-content/EN/TXT/?uri=CELEX%3A1 2012E%2FTXT. Data accessed 8 Aug 2022
36. The 13th five-year plan for economic and social development of the people's Republic of China (2016) Central compilation and translation Press, Beijing. Beijing, China. https://www. un-page.org/files/public/china_five_year_plan.pdf. Data accessed 8 Aug 2022

37. Turner GM (2008) A comparison of the limits to growth with 30 years of reality. Glob Environ Chang 18(3):397–411. https://doi.org/10.1016/j.gloenvcha.2008.05.001
38. Vovchenko NG, Andreeva OV, Dmitrieva VD, Galazova SS, Krstev TV (2022) Modelling of economic clusters in the Russian regions in the context of formation of the Industry 4.0. In: Trifonov PV, Charaeva MV (eds) Strategies and trends in organizational and project management. DITEM 2021. Lecture notes in networks and systems, vol 380. Springer, Cham, pp 667–673. https://doi.org/10.1007/978-3-030-94245-8_91

Russian Sustainable Finance Market: Dynamics and Outlook of Development in the Context of Digitalization

Inna A. Chekunkova⊙, Yulia S. Evlakhova⊙, Elena I. Brichka⊙,
Elena N. Alifanova⊙, and Ekaterina N. Gruzdneva⊙

Abstract The paper aims to identify the features of the development of the Russian sustainable finance market in the context of the determination of factors of its dynamics. The research hypothesis is that the dynamics of the sustainable development sector of the Russian securities market are not closely linked with the macroeconomic conjuncture. In the research, the verification of this assertion proceeds by correlation analysis. To test the research hypothesis, the authors choose bonds of the sustainable development sector, bonds with volatility between December 2021 and June 2022, and some fundamental indicators of the real and financial sectors of the economy. The research hypothesis was partial confirmed, which is indicated by the obtained results in two key findings: (1) the development of the real sector of the economy weakly correlates with the dynamics of the sustainable financing market (i.e., the basic hypothesis is confirmed); (2) despite this, the Russian sustainable bond market is not currently developing independently from the market of traditional bonds. As it turned out, this close relationship is due to the lack of regular and well-regulated reporting on the intended use of funds received from the issuance of sustainable bonds. Based on the research results, the authors suggest a financial market regulator, considering the need for the development of established reporting forms and the accelerated development of digital technologies in the monitoring of sustainable finance market indicators.

Keywords Green bonds · Social bonds · Bond market factors · Correlation · Digital agenda

JEL Classification G10

I. A. Chekunkova (✉) · Y. S. Evlakhova · E. I. Brichka · E. N. Gruzdneva
Rostov State University of Economics, Rostov-on-Don, Russia
e-mail: finmonitor.rsue@gmail.com; iakolesnik@mail.ru

E. N. Alifanova
Financial University Under the Government of the Russian Federation, Moscow, Russia

1 Introduction

The transformation of the economy to sustainable development involves a steady increase in the role of projects aimed at rationalizing the economic mechanism and improving the standard of living of the population as the most important link in reproductive relations. To make such processes focused, the concept of sustainable development should be mainstreamed into the strategic goals of enterprises. Monitoring the effectiveness of the implementation of ESG projects necessarily involves their provision not only at their own expense, which in itself is a resource-saving motive but also with the involvement of borrowed funds. The expansion of the opportunities for investment activities of existing enterprises in environmental and socially significant projects was the development of a special segment of the financial market—the market of sustainable financing instruments.

A set of documents developed by the International Capital Market Association (ICMA) and Climate Bonds Initiative (CBI) has become a driver for developing the global sustainable financing market. In particular, most Russian sustainable finance bond issues are verified against the Green Bond Principles [1] and the Social Bond Principles [2]. In 2021, the National Taxonomy of Greens and Adaptation Projects [3] was approved to simplify the verification system for sustainable development projects financed, among other things, through the issuance of bonds. According to these standards, financial instruments for sustainable development include social financial instruments, green financial instruments aimed at financing unique green projects in terms of international taxonomies, and adaptation financial instruments aimed at financing projects that are not included in the first group but correspond to Russia's priorities in the field of ecology and climate change. Three Russian rating agencies (ACRA, Expert RA, and NRA) have developed relevant methodologies and offer their services to verify the compliance of sustainable financing instruments with these standards. This tendency led to the fact that six bond issues were implemented in the sustainable development sector of the Moscow Exchange in 2020 and ten issues in 2021. Thus, a comparison with the 2016–2019 period, when bond emissions were recorded 1–2 times a year, is informative. This rapid growth of the market volume of bonds intended to finance sustainable projects determines the relevance of studying the dynamics and development prospects of this market. Simultaneously, in the first half of 2022, there were only two issues of sustainable bonds. This fact is perhaps explained by the geopolitical tensions that developed during this period.

In fact, the first quarter of 2022 marked a turning point for the Russian economy due to sanctions pressure from countries with which Russia has close economic relations, the crisis of energy markets, and clear signs of a recession in the entire global economy. Under such conditions, the creation and development of import-substituting production require significant investment from the state and the corporate sector. Does this cancel the significance of transforming the Russian economy in accordance with the sustainable development goals? Central banks, as regulators of financial markets, do not plan to change course to create conditions for financing projects for the transition to a low-carbon economy. On the contrary, the intensification of the energy crisis becomes an additional motivation to continue the development of sustainable development projects. Along with solving the problems of supporting price and financial stability in a difficult geo-economic situation, the Russian mega-regulator and central banks of Russia's potential partner countries are pursuing a policy of promoting economic restructuring. Therefore, it is relevant to study current trends in the Russian market of sustainable financing and the driving forces of its development in the context of geo-economic turbulence.

Green bonds are regarded as effective tools for environmental change together with green loans [4–6] because they permit to involve institutional and retail investors in the transformation processes [7]. The COVID-19 pandemic also led to the development of the social bond market because funds were needed to rebuild and support the economy and ensure public health after the lockdown [8]. Green bonds are inherently a financial instrument similar to traditional bonds. Nevertheless, green bonds have a special purpose—fundraising to finance climate or environmental projects [9]. Social bonds are also fixed-income securities, the funds of which are directed to social initiatives. A comparison of the trading dynamics of green bonds with the trading dynamics of non-green bonds of the same issuers shows that green bonds are not more attractive than traditional ones in the aspect of investment [10, 11]. Along with this, the green bond market has a significant positive impact on the corporate and government bond markets of developed countries [12]. In the literature on sustainable financing mechanisms, a significant question now is to identify determinants of green bond market dynamics. The results of the analysis [13] show a strong influence of macroeconomic factors stimulating the issuance of green bonds. Most economic factors other than GDP are important for developing sustainable finance; environmental factors are slightly less important [14]. The weak influence of investors' environmental preferences on bond prices was found [15]. The impact of macroeconomic factors, in particular the stock index, exchange rate, and oil prices, on the parameters of the green bond market were studied by Ortolano and Angelini [16], using the securities included in the Borsa Italiana index as an example. The dependence of the green bond price dynamics of market indicators was studied by Reboredo and Ugolini [17]. These authors indicate that the green bond market is closely related to currency and is weakly dependent on the stock market, energy resources, and high-yield corporate bonds. The importance of appropriate information and attention for directing financial flows toward sustainable investment was identified by Pham and Huynh [18].

The literature review showed a gap in the study of the Russian sustainable bond market; namely, there is no analysis of factors affecting its dynamics. This question is relevant because sustainable financing bonds are inherently another form of debt collateral. Therefore, issuing sustainable financing bonds is likely driven by mostly the same factors as issuing traditional bonds. On the other hand, since the loans from sustainable bonds are directed to specific projects, the impact of the macroeconomic environment may be insignificant.

Sustainable development projects are characterized by low profitability, regardless of the level of economic development; expected results are more qualitative in nature.

Therefore, we assume that the dynamics and prospects for the development of the sustainable development sector in the Russian securities market, designed to finance projects in the ecology, environmental protection, and socially significant projects, are not closely related to the macroeconomic situation. The research aims to identify the features of the development of the Russian market of sustainable financing based on identifying the relationship between its dynamics and the macroeconomic conjuncture.

This is an original study for Russian science because the data is analyzed not for securities whose issue complies with ESG standards but for securities whose issuer aims to raise borrowing funds for special purposes—the financing of environmental projects, projects in the field of environmental protection, and socially significant projects.

2 Methods

The research goal predetermined the choice of the research object—Russian bonds of the sustainable development sector of the Moscow Exchange. The research subject is the relationship between the dynamics of the bond market of the green, social, and national and adaptation projects segments with macroeconomic factors. This study proposes to analyze specific securities rather than the entire index for several reasons. Securities included in the RUEUESG (Eurobond ESG Index) and the MRSVT (Stock Index Sustainability Vector) are targeted at financing sustainable development projects. The basis for calculating these indices includes traditional securities of companies that follow ESG principles and, therefore, can be influenced by the same factors as the general corporate securities indices. In another work by the authors [19], when studying the MRSVT stock index, it was found that the conditions of its dynamics and development trends are similar to the general market stock index MOEX. It was also found that the bases for calculating the MRSVT and MOEX indices are the same in most companies. In this connection, the authors found it inappropriate to study the trends in the sustainable financing market using the above indices. In the current research, the authors propose analyzing the prices of securities intended specifically for financing sustainable development projects. According to the data of the Moscow Exchange, there are 15 issuers of sustainable financing of the Moscow Exchange in the Russian market [20].

For research purposes, it is preferable to exclude issuers from the third level of listing (i.e., the non-quoted list) because these are issuers recognized as unreliable by the exchange due to non-compliance with several listing requirements or due to the recent inclusion of their securities in exchange trading, which makes it impossible to determine the reliability of this issuer fully. It is also proposed to exclude issuers with no volatility (i.e., the entire period of circulation of the bonds has the same price). Thus, 9 out of 15 issuers with a total of 14 (out of 24) issues fall into our research. Nevertheless, the issues selected for the analysis account for 83% of the entire sustainable development bond market on the Moscow Exchange by nominal volume of the issue in rubles. Therefore, the analysis of the dynamics of 9 issues can be extrapolated to the entire market of such bonds.

The trend toward sustainability has become widespread recently, as evidenced by the issue dates of the first sustainability instruments. The earliest issue took place in 2016. Simultaneously, factors that can be recognized as specific to sustainable bonds are reflected in indicators for which data are published annually. Accordingly, it is too early for the Russian market to draw conclusions about the impact of such factors on the dynamics of bond prices because the amount of data will not allow us to obtain statistically significant results. At the same time, the bonds under discussion have a maturity of up to 30 years; the data accumulated over such a period will help conduct a qualitative study on the relationship between specific ESG factors and sustainable financing securities. It is now necessary to determine the extent to which the overall macroeconomic environment concerns the targeted financing of environmental, conservation, and socially significant projects. To this end, it is required to study the relationship of the sustainable bond market with macroeconomic factors that have long influenced the yield dynamics of traditional bonds: industrial production index (IP), Brent crude oil prices (OIL), RUB/USD, interest rate (R), Moscow Exchange Index (MOEX), and Moscow Exchange Corporate Bond Index (CBITR).

The relationship between the price dynamics of bonds in the green, social, and national and adaptation projects segments and monthly data on macroeconomic indicators is analyzed using a statistical analysis tool—paired linear correlation. The significance of the obtained correlation coefficients is examined using Student's t-test. The results of correlation analysis are interpreted using the Chaddock scale.

3 Results

The results of calculating the correlation coefficients between the dynamics of bonds of the sustainable development segment and macroeconomic indicators are represented in Table 1.

Based on the correlation analysis, we obtained statistically significant results about the relationship between the dynamics of bonds of the sustainable development sector and the main Russian corporate bond index. Most of the coefficients are positive and indicate a high degree of relationship. The correlation coefficient between the green bonds of PJSC "KAMAZ" and the CBITR index has the lowest level of significance; it characterizes the strength of this relationship as moderate. In general, the result on this factor does not agree with the results of the research by Reboredo and Ugolini [17], who found a weak dependence of foreign markets for green and traditional bonds. Simultaneously, a moderate relationship has been established between some bonds of the sustainable development sector and the main stock index of the Moscow Exchange; in other cases, the relationship is not statistically significant. This result is consistent with the dependencies in foreign markets identified in the works of Ortolano and Angelini [16] and Reboredo and Ugolini [17]. Additionally, it was

Table 1 The results of the correlation analysis of the dynamics of bonds of the sustainable development sector and macroeconomic indicators

	CBITR	MOEX	IP	OIL	R	USD
AEK	0.87***	0.54*	0.15	− 0.49**	− 0.74**	− 0.45
INFR1	0.87***	0.27	− 0.26	− 0.023	− 0.20	− 0.77**
INFR2	0.85***	0.51*	0.17	− 0.423	− 0.78**	− 0.41
INK	0.78**	0.48*	0.23	− 0.51*	− 0.90***	− 0.28
KMAZ	0.66*	0.38	0.21	− 0.41	− 0.97***	− 0.20
MGOV	0.97***	0.38	− 0.12	− 0.20	− 0.60*	− 0.72**
MTS	0.82***	0.63*	0.18	− 0.48*	− 0.66**	− 0.46
SBER	0.94***	0.44	− 0.02	− 0.25	− 0.70**	− 0.62*
SINTR	0.93***	0.40	0.01	− 0.31	− 0.78**	− 0.55*

Note 1 *** The significance of the correlation coefficient at 1% level
** The significance of the correlation coefficient at 5% level
* The significance of the correlation coefficient at 10% level
Note 2 AEK JSC Atomenergoprom; *INFR1, INFR2* Specialized Project Financing Company "Infrastructure Bonds"; *INK* JSC "INK-Capital"; *KMAZ* PJSC "KAMAZ"; *MGOV* The Government of Moscow; *MTS* PJSC "MTS"; *SBER* PJSC "Sberbank of Russia"; *SINTR* JSC "Sinara Transport Machines"
Source Calculated by the authors based on [20–26]

revealed that the price dynamics of bonds of the sustainable development sector are completely unrelated to the development of the real sector of the economy. Therefore, it can be considered a feature of such bonds because it is known from earlier studies that the index of industrial production, being significant for the formation of the yield of traditional bonds, affects its dynamics negatively [27].

Another feature of sustainable development bonds is that the dynamics of most of these securities are not related to the dynamics of oil prices, in contrast to the volatility of traditional bonds [28]. Additionally, three significant correlation coefficients indicate only a moderate correlation of stable bonds of "Atomenergoprom," "INK-Capital," and "MTS" with oil prices. Our result is similar to the research by Reboredo and Ugolini [17], in which a weak relationship between the foreign green bond market and energy prices is found.

The results of Table 1 demonstrate the existence of a relationship between the dynamics of the bonds of the sustainable development sector and the key rate, except for the first issue of the SPFC "Infrastructure bonds." Therefore, this macroeconomic indicator can be recognized as a general factor in the dynamics of the sustainable and traditional bond markets, as Kolesnik and Gruzdneva [29] earlier found a high correlation between the dynamics of the corporate bond index and the key rate of the Bank of Russia.

An ambiguous result was obtained on the correlation coefficients of the currency pair USD/RUB and bonds of the sustainable development sector: some securities have a significant correlation, but more than half of the securities studied have no statistically significant correlation with the currency pair.

4 Discussion

We investigated the relationship of the sustainable bond market with other macroeconomic variables not previously used in the analysis of foreign green bond markets: economic growth (IPI) and key rate. The obtained results partially confirm the research hypothesis and reveal the ambivalent nature of the bonds of the sustainable development sector. On the one hand, the attractiveness of a debt financial instrument with an interest income depends on the course of the central bank's interest rate policy. On the other hand, these financial instruments are aimed at a qualitative transformation of the economic structure and are mainly intended for investors who follow the principles of responsible investment. Therefore, monthly (short-term) fluctuations in industrial production volumes do not play a key role in the market pricing of such instruments.

However, the results of the correlation analysis showed a significant positive relationship between the markets for sustainable development bonds and traditional bonds. The key role of the dynamics of the market for traditional corporate debt securities in the pricing of special purpose bonds is explained by the fact that this factor is so far the only logical guideline in decision-making for investors when investing in sustainable bonds because the lack of regular unified reporting on the performance of the projects to which it is issued blurs the boundaries between the factors of the dynamics of traditional and sustainable bonds. The report on the intended use of funds was published only by PJSC "MTS" in the year of the issue [20]. Moreover, such issuers as JSC "Atomenergoprom," JSC "Sinara-Transport Machines," and SPFC "Infrastructure Bonds" did not publish reports on the targeted spending of raised funds even in the year following the issue.

The above trends indicate the lack of identity of unique instruments of sustainable financing in the securities market and the lack of reporting information on such securities. The co-directional movement of the traditional and sustainable bond markets tells us that the crisis phenomena affecting the overall market are now moving into the sustainable segment. The shock of the market should not be fully reflected in ESG instruments. Responsible investment is focused not on short-term economic gain but rather on the long-term development of projects, environmental protection, and high social standards. Thus, the refusal of investors from excess profits should be compensated by understanding the volumes and structure of the development of funds by the issuer. This requires regulatory action and support for the initiative of the providers of financial data and financial services to create platform solutions for monitoring and evaluating ESG data on alternative assets (sustainable finance instruments). We assume that the creation of such digital solutions will make this market segment more accessible and transparent for investors. That is why we consider it appropriate to create an analytic web service aimed at aggregating, monitoring, and comparing indicators of the use of invested funds, as well as evaluating the ESG efficiency of such investments. To scale up such a service, the participation of the regulator and the state is necessary in terms of standardizing the information provided for the service, as well as ensuring its accessibility to a wide range of people.

5 Conclusion

The results of this research allow us to deduce two key outcomes. First, the development of the real sector of the economy weakly correlates with the dynamics of the sustainable financing market. It follows that the basic hypothesis is validated. Second, despite this, the Russian sustainable bond market is not currently developing independently from the traditional bond market. Consequently, the development of the sustainable financing market is influenced by shocks in the bond market, which depend on the macroeconomic environment. It has been found that such a close correlation is caused by the lack of regular and regulated reporting on the use of funds raised through sustainable bonds. Relying on the results obtained, the authors recommend considering the accelerated development of digital technologies in the financial market. The creation of specialized web services will help reduce transaction costs for ESG investors and make financing of environmental projects, projects in the field of ecology, and socially important projects more transparent during the public offering of ESG securities.

References

1. International Capital Market Association (ICMA) (2021) The green bond principles (GBP) 2021 (with June 2022 Appendix 1). Retrieved from https://www.icmagroup.org/assets/doc uments/Sustainable-finance/2022-updates/Green-Bond-Principles_June-2022-280622.pdf. Accessed 5 Sept 2022
2. International Capital Market Association (ICMA) The social bond principles (SBP) 2021 (with June 2022 Appendix 1). Retrieved from https://www.ifc.org/wps/wcm/connect/46f 2b384-a7dc-4067-8170-ee5c3aa14712/Social+Bonds+Principles+June+2022.pdf?MOD= AJPERES&CVID=o6LJ0Mu. Accessed 5 Sept 2022
3. Government of the Russian Federations. (2021) Decree "On approval of criteria for projects of sustainable (including green) development in the Russian Federation and requirements for the system of verification of sustainable projects (including green) development in the Russian Federation" (September 21, 2021 No. 1587). Moscow, Russia. Retrieved from http://static.gov ernment.ru/media/files/3hAvrl8rMjp19BApLG2cchmt35YBPH8z.pdf. Accessed 5 Sept 2022
4. Gianfrate G, Peri M (2019) The green advantage: exploring the convenience of issuing green bonds. J Clean Prod 219:127–135. https://doi.org/10.1016/j.jclepro.2019.02.022
5. Kazlauskiene V, Draksaite A (2020). Green investment financing instruments. In Kuna-Marszałek A, Kłysik-Uryszek A (eds) CSR and socially responsible investing strategies in transitioning and emerging economies. IGI Global, Hershey, pp 189–213. https://doi.org/10. 4018/978-1-7998-2193-9.ch010
6. Sartzetakis ES (2021) Green bonds as an instrument to finance low carbon transition. Econ Chang Restruct 54(4):755–779. https://doi.org/10.1007/s10644-020-09266-9
7. Daszyńska-Żygadło K, Marszałek J (2018) Green bonds—sustainable finance instruments. Proceedings of the 15th International scientific conference: European financial systems 2018. Masaryk University, Brno, pp 77–85
8. Torricelli C, Pellati E (2022) Social bonds and the "Social Premium". CEFIN working papers No. 85. University of Modena and Reggio Emilia, Modena. https://doi.org/10.13140/RG.2.2. 11206.34883
9. Liaw K (2020) Survey of green bond pricing and investment performance. J Risk Finan Manag 13(9):193. https://doi.org/10.3390/jrfm13090193

10. Hachenberg B, Schiereck D (2018) Are green bonds priced differently from conventional bonds? J Asset Manag 19(6):371–383. https://doi.org/10.1057/s41260-018-0088-5

11. Wu Y (2022) Are green bonds priced lower than their conventional peers? Emerg Mark Rev 52:100909. https://doi.org/10.1016/j.ememar.2022.100909

12. Alifanova E, Evlakhova Y, Brichka E (2022) The role of the global stock market in the transition to the green economy in the context of the climate risks. In Popkova EG, Vovchenko NG, Andreeva OV (eds) Climate-smart innovation: social entrepreneurship and sustainable development in the environmental economy. World Scientific Publishing, Singapore

13. Tolliver C, Keeley AR, Managi S (2020) Drivers of green bond market growth: the importance of nationally determined contributions to the Paris agreement and implications for sustainability. J Clean Prod 244:118643. https://doi.org/10.1016/j.jclepro.2019.118643

14. Azarova E, Jun H (2021) Investigating determinants of international clean energy investments in emerging markets. Sustainability 13(21):11843. https://doi.org/10.3390/su132111843

15. Zerbib OD (2018) The effect of pro-environmental preferences on bond prices: Evidence from green bonds. J Bank Finance 98:39–60. https://doi.org/10.1016/j.jbankfin.2018.10.012

16. Ortolano A, Angelini E (2021) Green bonds capital returns: the impact of market and macroeconomic variables. In: La Torre M, Chiappini H (eds) Contemporary issues in sustainable finance. Palgrave Macmillan, Cham, pp 91–116. https://doi.org/10.1007/978-3-030-65133-6_4

17. Reboredo JC, Ugolini A (2019) Price connectedness between green bond and financial markets. Econ Model 88:25–38. https://doi.org/10.1016/j.econmod.2019.09.004

18. Pham L, Huynh TLD (2020) How does investor attention influence the green bond market? Financ Res Lett 35:101533. https://doi.org/10.1016/j.frl.2020.101533

19. Evlakhova YuS, Chekunkova IA (2022) The correlation of Russian companies' orientation towards sustainable development in the environmental, social and economic realm with trends in stock market indices. Finan Credit 28(7):1470–1492. https://doi.org/10.24891/fc.28.7.1470

20. Moscow Exchange (n.d.) Sustainability sector. Retrieved from https://www.moex.com/s3019. Accessed 5 Sept 2022

21. Central Bank of the Russian Federation (n.d.) Key rate of the Bank of Russia. Retrieved from https://www.cbr.ru/hd_base/KeyRate/. Accessed 5 Sept 2022

22. Investing.com (n.d.) Futures for brent oil. Retrieved from https://www.investing.com/commodities/brent-oil. Accessed 5 Sept 2022

23. Investing.com. (n.d.). USD-RUB exchange rate. Retrieved from https://ru.investing.com/currencies/usd-rub. Accessed 5 Sept 2022

24. Joint Economic and Social Data Archive. (n.d.). Dynamic series of macroeconomic statistics of the Russian Federation. Retrieved from http://sophist.hse.ru/hse/nindex.shtml. Accessed 5 Sept 2022

25. Moscow Exchange (n.d.) MOEX index. Retrieved from https://www.moex.com/ru/index/IMOEX. Accessed 5 Sept 2022

26. Moscow Exchange (n.d.) Moscow exchange corporate bond index. Retrieved from https://www.moex.com/ru/index/RUCBITR. Accessed 5 Sept 2022

27. Erofeeva TM (2019) Research of factors and building a model for forecasting the yield spread of corporate bonds in the Russian market. Finan Bus 15(4):54–76

28. Nazlioglu S, Gupta R, Bouri E (2020) Movements in international bond markets: the role of oil prices. Int Rev Econ Financ 68:47–58. https://doi.org/10.1016/j.iref.2020.03.004

29. Kolesnik IA, Gruzdneva EN (2019) Identification of the relationship between indicators of the development of the money market and the corporate securities market. In Berdnikov SV, Kuznetsov NG (eds) Global challenges, new risks and priorities of economic systems. RINH, Rostov on Don, pp 290–297

Prospects for the Use of AI and IT Technologies for the Purpose of Sustainable Financing and Development of Healthcare

Lyudmila V. Bogoslavtseva⬤, Oksana Yu. Bogdanova⬤, Oksana I. Karepina⬤, Svetlana N. Meliksetyan⬤, and Vera V. Terentyeva⬤

Abstract *Purpose*: The study reveals the genesis of the implementation of digital healthcare products and services and the prospects for the development of AI and IT technologies in the context of medical services. *Structure/Methodology/Approach:* The structure of the study is determined by an integrated approach to the disclosure of the topic, which predetermined the need to identify problems of digital transformation of the industry, as well as formulate recommendations for improving the accessibility of modern digital healthcare technologies. The methodological apparatus of the research is traditional scientific methods that ensure the logic, consistency and reliability of the conclusions obtained as a result. *Results:* The classification of the levels of AI and IT technologies in healthcare has been refined. The deterministic influence of financing on the development of AI and IT technologies in healthcare is substantiated. The analysis of the TOP 5 popular online services of medical information systems in Russia is presented. The features of the main projects for healthcare development of the Rostov region within the framework of the strategy of digital transformation of the region until 2024 are characterized. *Practical implications:* Proposals for further development of digitalization of healthcare in the region can be taken into account in the practice of sustainable financing of strategic directions in the field of digital transformation of healthcare. *Originality/Value:* The main contribution of this research lies in the modern argumentation of the objective need for the development of digitalization of medicine in the country and its regions.

Keywords Digital healthcare · Healthcare technologies · Strategies for healthcare digitalization · Financing · Efficiency of medical services

JEL Classification O33 · I15

L. V. Bogoslavtseva (✉) · O. Yu. Bogdanova · O. I. Karepina · S. N. Meliksetyan ·
V. V. Terentyeva
Rostov State University of Economics, Rostov-On-Don, Russia
e-mail: b_ludmila@bk.ru

S. N. Meliksetyan
e-mail: m.s88@bk.ru

1 Introduction

In 2005, the World Health Assembly adopted a resolution on e-health, which called on the member States of the World Health Organization (WHO) "to design and develop information and communication technology infrastructure in the interest of public health… promote equitable, cheap and universal access to the benefits they provide." Since that time, WHO has been continuously implementing policies aimed at developing a nationwide e-health strategy.

In 2018, the experts of the World Health Assembly concluded that IT was necessary to accelerate the pace of development and implementation of IT technologies in the healthcare sector, as a result of which the Resolution WHA71.7 on digital healthcare was signed.

In 2021, the WHO Director-General made the reports on the issues of up-to-date information on IT management and sustainable health financing, in which he noted the problem of mobilizing funds to cover budget costs. In addition, after the declaration of the COVID-19 pandemic, the WHO secretariat noted an increase in the number of targeted cyber attacks, which required additional funding.

In the Russian Federation, as a WHO member country, global trends in improving the quality of medical services are also of practical interest. In 2018, the national program "Digital Economy of the Russian Federation" was developed. However, "the complex digital transformation of healthcare in Russia was hindered by both factors at the national level (insufficient state support, rejection of the project "Digital Healthcare") and regional factors (differences in the level of digital technology implementation in different regions, lack of digital personnel and different levels of investment by region, etc.)" [12, p.117].

By 2021, the subjects of the Russian Federation have developed and approved regional digital transformation strategies, including for healthcare. For example, the Rostov region has been implemented a strategy in the field of digital transformation of economic, social and public administration sectors since August 2021.

However, due to the incompleteness of regulatory documents and different financial opportunities, the regions developed and implemented diverse digitalization projects. Consequently, there is an objective need for the integrated implementation of digitalization of healthcare throughout the country. At the end of December 2021, the decree of the Government of the Russian Federation approved the strategic direction of digital transformation of the healthcare system in Russia in order to standardize the organization and procedure for providing medical services to all citizens of the country.

The argument in favor of the all-Russian strategy of digitalization of the healthcare industry was the research of foreign and domestic scientists on new global trends in the development of the digital healthcare market and the financing of medical services [9].

In the context of the development of digital technologies in the domestic healthcare system, according to the authors of the presented study, it would be useful to identify the features of the use of AI (Artificial Intelligence) and IT (Information Technology) technologies in order to improve the quality of medical services and reduce the cost of these services within the framework of state guarantees.

In accordance with this goal, the following tasks were identified: to study the problems and risks of using AI and IT technologies for medical services and their role in financing domestic healthcare.

The theoretical and practical significance of the article is to substantiate the impact of digital technologies in healthcare on improving the efficiency of medical services, which is deterministically interrelated with the financing of these services.

2 Methodology

During the research, the following methods were used: analysis of regulatory and methodological documents of the relationship between digitalization and healthcare financing, generalization, systematization of experience on healthcare financing; monitoring of indicators, comparison; data analysis and table-driven method.

Since the purpose of the article is to identify the features of the use of AI and IT technologies in healthcare, the authors studied the following program and conceptual materials:

- The reports of the 74th session of the World Health Assembly "Updated information on information management and technology" on January 11, 2021 and "Sustainable financing" on May 14, 2021;
- The National Project of the Russian Federation "Healthcare";
- The report on the results of the expert-analytical event "Analysis of the current state of healthcare informatization in the context of the concept of creating a single digital circuit in healthcare", approved by the Collegium of the Accounts Chamber of the Russian Federation on 31.05.2022;
- State programs of the Rostov region "Information Society" for 2021–2022 and "Development of healthcare" for 2019–2030;
- The strategy of digital transformation in the Rostov region, etc.

During the preparation of the article, the authors used foreign and domestic sources, textbooks, statistical reference books and chronological collection of normative legal acts regulating the development of digital technologies in the Russian Federation [14, p. 47].

3 Results

Healthcare technologies are medical devices, robotics, registry data storage technologies and wireless communication technologies, virtual and augmented reality technologies, GLONASS/GPS-based technologies, IT systems, algorithms, artificial intelligence (AI technologies), cloud technologies and blockchain, big data, real-time analytics, contactless electronic payment systems designed for the development of healthcare organizations.

Modern digital trends in the world of medicine, which in the future can reduce the total cost of medical services while maintaining the quality of these services, include:

(1) development of telemedicine. Online medical consultations are becoming one of the main trends in the field of medicine;
(2) an individual approach in the interaction of a medical institution and patients using IT technologies, for example, to determine and control blood sugar levels, blood pressure or individual recommendations to patients;
(3) wearable digital medical devices and medical gadgets that help patients monitor their health;
(4) the use of AI technologies is one of the digital innovations in medicine. Artificial intelligence successfully replaces time-consuming tasks in the field of healthcare. Medicine is now available remotely, providing real-time solutions to patients. The role of chatbots has gone far beyond prepared responses to messages and has begun to offer intelligent answers;
(5) protection of patients' personal data, which will be strengthened in 2022. One of the leading trends in the field of healthcare, which medical institutions and insurance agencies cannot ignore, is the protection of customer privacy. As it develops, medical institutions should implement a robust data protection program to detect potential threats and avoid them. Thus, platforms based on artificial intelligence cannot only improve the efficiency of medical services, but also ensure the information security of patients;
(6) the website of a medical institution, including: the content and design of the website; navigational simplicity; compatibility with various devices; website's SEO; accessibility; round-the-clock interaction tools such as chatbots, automatic support by email and phone.

Table 1 shows the TOP 5 popular online services of medical information systems necessary for automation of work, accounting, cost reduction, billing for medical services and improving the quality of work with patients.

The variety of online services of medical information systems confirms the need for state support, development and implementation of modern AI and IT technologies that provide high-quality medical care and data protection in order to disseminate best practices in healthcare institutions.

Healthcare institutions, as well as the industry as a whole, are changing and improving extremely rapidly—modern and competitive medical centers should provide high-level care to patients, regardless of the market situation and subject to strict requirements from regulatory authorities.

To help such clinics provide the best medical care and treatment, modern analytical and cognitive technologies play an important role in the implementation of the daily use of data, with the help of which predictive models can be developed to help doctors choose the best treatment methods for individual patients, introducing analytical algorithms into the practice of medical decision-making.

Table 1 TOP 5 popular online services of medical information systems

№	Name	Functions
1	Medesk	This medical platform used for the modern functioning of the digital environment of private clinics. It covers the registry, doctors' offices, laboratories, telemedicine, online appointment system, reporting, financial issues, VMI, integration with booking portals—ProDoctorov, Docdoc, Napopravku. mobile application
2	MedElement	This is a medical platform that combines cloud services (MIS) and reference systems for doctors, clinics and patients. It provides a convenient program for medical practice management: patient database, appointment schedule, keeping of medical records, communication with patients via the Internet
3	INFOCLINICA	This service forms an individual configuration of the information system, taking into account the characteristics of the medical institution. It is integrated with the portal "Gosuslugi", as well as with other participants in the organization and provision of medical services. It includes a mobile application
4	IDENT	This is a cloud service that was created for the digitalization of dental clinics. The main functions include online patient appointment system, paperwork for patients, maintaining a cash register and generating reports, analytics. It includes a mobile application
5	MedWork	There are two areas of activity: the sale of MIS and automation of medical institutions, which includes, in whole or in part, automation of offices and laboratories, interaction with patients, online cash register management and reporting, as well as a mobile application

Source Compiled by the authors based on [8]

Based on the above, the following levels of AI and IT technologies can be classified and developed in healthcare:

- Medical services at the stages of prevention, diagnosis and treatment;
- Organization of both the healthcare system in general and the medical institution in particular;
- Program and project budgeting in the conditions of digitalization of the budget process at the stages of planning, financing, stimulating and analyzing the activities of healthcare institutions [3, p. 950];
- Directions of information technologies in healthcare on digitalization of data collection, processing and analysis.

Further, it is necessary to characterize the features of the introduction of AI and IT technologies in the context of the implementation of the Strategy of digital transformation in healthcare of the Rostov region from August 19, 2021. The funding of the measures provided for by the Digital Transformation Strategy is carried out within the framework of the Rostov Region state program "Information Society", taking into account conceptual approaches to ensuring effective and responsible management of state programs [11, p.261]. The main sources of financing are the federal budget and the regional budget of the Rostov region.

The key results of the Strategy by 2024 should be the creation of a single digital circuit and services in healthcare through the implementation of digitalization projects of healthcare in the Rostov region, presented in Table 2, which was compiled by the authors on the basis of the Strategy. The data contained in the table reflect the names and goals of projects as priorities for digitalization of the region.

As part of the implementation of the Strategy, during 2021, healthcare institutions of the Rostov Region conducted 5439 telemedicine consultations with municipal medical organizations and 966 telemedicine consultations with national medical

Table 2 Priority projects of the strategy of digital transformation in healthcare of the Rostov region until 2024

№	Name of the project	Project objective
1	Creation of a single digital circuit based on the unified state information system in the healthcare sector (USISHS)	Improving the functioning of the healthcare system by creating mechanisms for the interaction of medical organizations based on on the unified state information system in the healthcare sector and the introduction of digital technologies and platform solutions that form the USISHS
2	Reliable infrastructure	Automation of medical workers' workplace; Introduction and use of medical information systems in polyclinics and hospitals
3	"My health"—on the portal "Gosuslugi"	Creation and development of digital services for patients
4	Development of interdepartmental cooperation	Expansion of interaction between medical organizations and authorities involved in the organization of medical care
5	Creation and maintenance of a unified register system	Creation and development of interaction of medical institutions with subsystems of the USISHS
6	Organization of patient flows and provision of medicines	Creation and development of centralized subsystems with the state information system of subjects
7	Creation of federal-level healthcare platform solutions	Creation and implementation of integrated medical information systems on the profiles of treatment or medical care
8	Personal specialized medical assistants	Extensive implementation of preventive measures and remote monitoring of patients
9	Artificial Intelligence (AI) in healthcare	Multiple growth of AI solutions and medical products using artificial intelligence technologies for healthcare

Source Compiled by the authors based on strategy in the field of digital transformation of sectors of the economy, social sphere and public administration of the Rostov region [13]

research centers. Telemedicine consultations are conducted for patients of inpatient departments, especially those in serious condition, including as a result of emergencies and accidents.

Telemedicine technologies are used both in the provision of primary health care and in the provision of emergency medical care.

In order to improve the efficiency of medical services, since 2022, the Ministry of Health of the Russian Federation has been taking measures to unify the digitalization of regional medicine by changing the functions and principles of the medical information and analytical center (MIAC), which ensures the implementation of a single digital circuit within the framework of the national project "Healthcare".

In general, digital technologies of healthcare organization and financing should contribute to the achievement of the goal of the national project "Healthcare", namely: to reduce the cost of medical services; to improve the quality of medical services and, at the same time, meet one of the WHO goals: universal health coverage.

4 Discussion

The problem of financial provision of medical services hinders the development of digitalization of healthcare. Therefore, it is necessary to improve the efficiency of management of both medical institutions and costs for treatment. In this regard, it is advisable to take into account the position of Doctor of Economics, Professor Chernov, who believes "the latest developments in the field of digital systems, for all their intellectual significance, are more concerned with the technical side of business management…" [5, p. 286].

It is impossible not to mention the textbook "Digital Medicine: A Primer on Measurement", in which the authors note that many technologies move seamlessly between research and treatment. The authors' position on the differentiation of the meaning of the terms "digital health" and "digital medicine" is interesting. In their opinion, it is the term "digital medicine" that should be used as a term to describe digital products that measure human health indicators and are designed to promote health and prevent diseases [6].

In 2020, the WHO secretariat presented a reviewed version of the strategy for the period 2019–2023, which aims not only to empower people to receive medical services, but also to automate and optimize the digital healthcare platform. According to WHO recommendations, taking into account the coronavirus pandemic, health care expenditures to fully meet the needs of the population for quality medical care should be at least 12% of GDP, and to ensure the basic needs of the population at an acceptable level of quality—at least 6–8% of GDP [1, p.31].

Based on the above, it is necessary to analyze the expenditures of the budgets of the budgetary system of the Russian Federation on healthcare according to Fig. 1 compiled by the authors according to the Main directions of budget, tax and customs tariff policy for the current year and the planning period. The data in Fig. 1 reflect that the volume of expenditures of the budgets of the budgetary system of the Russian

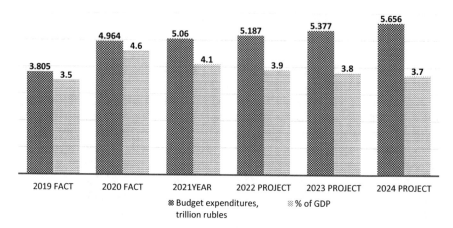

Fig. 1 Expenditures of the budgets of the budgetary system of the Russian Federation on healthcare in 2019–2024. *Source* Compiled by the authors based on "The main directions of the budget, tax and customs tariff policy for 2022 and for the planning period of 2023 and 2024" [15]

Federation on healthcare in absolute terms is growing to 5.656 trillion rubles by 2024, and as a% of GDP it is falling to 3.7% by 2024. However, the funding has not yet reached the values recommended by WHO, which justifies the conclusion that it is possible to ensure the rights of citizens to health protection by improving the efficiency of medical services and automated process management, that is, due to a competent digital transformation of the industry.

In September 2021, it became known about the increase in funding for the national project "Healthcare". In the period 2022–2024, it is planned to spend over 687 billion rubles on its implementation, it is noted in the explanatory note to the draft federal budget. The change in volumes is due to the revision of some financing parameters. For example, while implementing the International Classification of Diseases of the eleventh revision, the budget of the federal project "Creation of a single digital circuit in healthcare based on the unified state information system in the healthcare sector (USISHS)" will be changed.

When researching the financing of digital products in healthcare, scientists and practitioners note, in addition to the risks of financing, some risks of digitalization of healthcare and fragmentation of financing and medical care. "The COVID-19 pandemic has increased the need for well-functioning primary health care (PHC)" [7, p. 715]. For example, the risks to the reliability and security of information systems and infrastructure. The problem also lies in the fact that the design and development of competitive domestic software products requires significant funding [4, p.16].

In addition, in 2021, the Accounts Chamber of the Russian Federation, based on the results of the analysis of the current state of digitalization of healthcare, revealed an increase in the additional burden on medical personnel due to multiple entry of information. As a result, there is a low involvement of health system participants in information interaction. The report of the Accounts Chamber indicates the need to develop uniform requirements for the transition to electronic medical document management, as well as the need for a deep transformation of automation of medical processes through their formalization and optimization. During the audit, the auditors found that "the indicators of the federal program "Creation of a single digital circuit" do not sufficiently characterize the expected results of digitalization of healthcare and do not allow to assess the final effect" [10].

In order to solve the problems of using artificial intelligence in the healthcare sector, a survey "2022 AI in Healthcare Survey" was conducted, in which more than 300 respondents from around the world took part. Based on the survey results, we have formulated some of the main trends in the use of AI and IT technologies [2].

1. The trend of ease of use and democratization of artificial intelligence; the use of low-code/no-code tools.
2. The fundamental technologies planned for implementation by the end of 2022 were for almost half of the respondents—data integration, as well as:

   - for technical leaders—data integration, NLP (Natural Language Processing), BI (Business intelligence) and data annotation;
   - for respondents from mature organizations—data integration, data annotation, data science platform and NLP.

These tools make it possible to solve important tasks such as supporting medical decision-making, selection of medicines and evaluating a treatment plan.

3. Increased security threats. Data loss or intervention in medical device's operation directly affects the lives of patients, which justifies the financing of digitalization in healthcare, taking into account security and ethical issues.

5 Conclusion

Taking into account the world experience and the results of domestic practice, it seems necessary to coordinate the activities of federal and regional ministries in terms of budgetary expenditure allocated for the digitalization of healthcare and the use of AI and IT technologies to improve the quality of life, including monitoring internal and external factors influencing the digitalization of healthcare, as well as the study and dissemination of best practices of digital technologies in industries.

The main directions of the development of domestic healthcare in the regions at the present stage include:

1. The transition of all medical organizations to a single software.
2. Creation of technological infrastructure in the field of healthcare.
3. Introduction of digital services in the field of telemedicine.
4. Implementation of interdepartmental electronic document management standards.
5. Exclusion of citizen's participation in information exchange between social institutions.

6. Implementation of speech recognition systems based on self-learning neural networks for administrators and call center operators when consulting on the procedure for providing medical care.
7. Implementation of speech recognition systems based on self-learning neural networks to automate the process of recording laboratory instrumental research and medical manipulations.
8. The use of artificial intelligence mechanisms to automate typical operations and elementary processes.
9. Introduction of electronic assistants based on artificial intelligence.
10. Creation of a unified regional medical data bank for decision-making based on artificial intelligence.

Thus, the digital transformation of healthcare in the country and its regions should be implemented within the framework of the concept of creating a single digital circuit in the industry in order to improve the efficiency of medical services.

References

1. Afyan AI, Polozova DV, Gordeeva AA (2021) Russian health system digitalization: opportunities and contradictions. Digital Law J 2(4):20–39. https://doi.org/10.38044/2686-9136-2021-2-4-20-39
2. AI in healthcare: three trends worth paying attention to (2022) Open Systems Publishing House. https://www.osp.ru/medit/2022/07/13056234.html. Data Accessed 11 July 2022
3. Bogoslavtseva LV, Karepina OI, Bogdanova OY, Takmazyan FS, Terentieva VV (2020). Development of the program and project budgeting in the conditions of digitization of the budget process. In: Popkova E, Sergi B (eds) Digital economy: complexity and variety versus rationality. ISC 2019. Lecture notes in networks and systems, vol 87. Springer, Cham, pp 950–959. https://doi.org/10.1007/978-3-030-29586-8_108
4. Bogoslavtseva LV (2022) Modern paradigm of financial support for digitalization of state and municipal services to the population. Finan Res 1(74):16–24. https://doi.org/10.54220/finis.1991-0525.2022.74.1.002
5. Chernov VA (2020) Implementation of digital technologies in financial management of economic activity. Econ The Reg 16(1):283–297. https://doi.org/10.17059/2020-1-21
6. Coravos A, Goldsack JC, Karlin DR, Nebeker C, Perakslis E, Zimmermann N, Erb MK (2019) Digital medicine: a primer on measurement. Digit Biomark 3(2):31–71. https://doi.org/10.1159/000500413
7. Hanson K, Brikci N, Erlangga D, Alebachew A, De Allegri M, Balabanova D, Blecher M, Cashin ., Esperato A, Hipgrave D, Kalisa I, Kurowski C, Meng Q, Morgan D, Mtei G, Nolte E, Onoka C, Powell-Jackson T, Roland M, Sadanandan R, Stenberg K, Vega Morales J, Wang H, Wurie H (2022) The Lancet global health commission on financing primary health care: putting people at the centre. Lancet Glob Health 10(5):e715–e772. https://doi.org/10.1016/S2214-109X(22)00005-5, https://www.thelancet.com/action/showPdf?pii=S2214-109X%2822%290 0005-5. Data Accessed 08 July 2022
8. Kuzovkov S (2020) Overview of medical information systems (MIS) in 2020. Blog of a medical marketer. Symmetry. http://symmetria-med.ru/blog/obzor-meditsinskih-informatsionnyh-sistem-mis-v-2020-godu.html#25. Data Accessed 15 July 2022
9. Pugachev PS, Gusev AV, Kobyakova OS, Kadyrov FN, Gavrilov DV, Novitsky RE, Vladzimirsky AV (2021) Global trends of digital transformation of the healthcare industry. National Health Care 2(2):5–12. https://doi.org/10.47093/2713-069X.2021.2.2.5-12

10. Report on the results of the expert-analytical event "Analysis of the current state of healthcare informatization in the context of the concept of creating a single digital circuit in healthcare" (2022) Approved by the Board of the Accounts Chamber of the Russian Federation on May 31, 2022. https://ach.gov.ru/upload/iblock/b2e/1wl5z0qtvef2puoaywx1a7xm8pgu63qx.pdf. Data Accessed 15 July 2022

11. Romanova TF, Bogoslavtseva LV, Karepina OI, Bogdanova OY (2018) Conceptual approaches in providing the effective and responsible management of state programs management. Europ Res Stud J 21(4):261–272. https://doi.org/10.35808/ersj/1177

12. Romanova NV (2020) Key trends in digital transformation of the healthcare sector: similarity of global and domestic trends. In: Proceedings of the II international scientific conference "Actual problems of management, economics and economic security" (Kostanay, September 28, 2020). Cheboksary, ID "Wednesday", pp 117–120

13. Strategy in the field of digital transformation of sectors of the economy, social sphere and public administration of the Rostov region (2022) The official portal of the Government of the Rostov region. https://www.donland.ru/activity/2760/. Data Accessed 11 July 2022

14. Takmazyan AS, Samoilova KN (2022) Higher education in the conditions of digital transformation: organizational and financial aspects. Finan Res 1(74):46–53. https://doi.org/10.54220/finis.1991-0525.2022.74.1.005

15. The main directions of the budget, tax and customs tariff policy for 2022 and for the planning period of 2023 and 2024 (2022) Legal reference system "ConsultantPlus". https://www.consultant.ru/cons/cgi/online.cgi?req=doc&base=LAW&n=396691#TtSPuJTWWenLDUis. Data Accessed 11 July 2022

ESG Management and Its Accounting and Analytical Reflection in the Practice of Sustainable Development in Russia at Different Territorial Levels

Karina F. Mekhantseva⬤, **Nikolay G. Kuznetsov**⬤, **Lidia N. Roschina**⬤, and **Natalya D. Rodionova**⬤

Abstract The ideas of Sustainable Development Goals gain more traction and distribution and create a new platform for management decisions at different levels of the economy. The research aims to analyze the practice of assessing the achievement of sustainable development goals at different levels of the economy. Information and statistics were brought from the official sites of the United Nations, the Governments of the Russian Federation, Moscow, and Rostov Regions, and the Federal State Statistics Service. The authors carried out a statistical and comparative analysis of the practice of assessing the achievement of sustainable development goals at different levels of the economy. It is proposed to highlight cities as a separate level. It is necessary to continue research to determine sets of vertically comparable indicators of sustainable development at the global, regional, city, and organization levels with a different number of personnel, while the number of indicators in the set may differ upwards and downwards. The main trends in the development of statistical support for the analysis of sustainable development are determined—systematization, harmonization, and standardization. A mechanism has been laid out—an open information channel with reporting on the achievement of sustainable development goals at all levels of the economy.

Keywords Sustainable development goals · Indicators · Sustainable development assessment · National level · Regional level · Millionaire cities · Moscow

JEL Classification Q010 · O110 · C010 · F630 · R110

K. F. Mekhantseva (✉) · N. G. Kuznetsov · L. N. Roschina · N. D. Rodionova
Rostov State University of Economics, Rostov-on-Don, Russia

N. G. Kuznetsov
e-mail: kuznecov@rsue.ru

© The Author(s), under exclusive license to Springer Nature Switzerland AG 2023
E. G. Popkova (ed.), *Smart Green Innovations in Industry 4.0 for Climate Change Risk Management*, Environmental Footprints and Eco-design of Products and Processes, https://doi.org/10.1007/978-3-031-28457-1_57

1 Introduction

The contemporary information economy is increasingly linking economic objects into a unified hierarchy due to the exponentially increasing technological transformation of its main resource–information [1]. In response, economic objects change their form and complement their content with the information, which becomes increasingly available, grows in volume, and gets new application technologies. A human, an individual, as an origin of the main resource–information—becomes a more important economic object in such a type of economy. Human influence is manifested differently at different economic levels and divides economic levels into parts depending on the information support of management decisions, inserting novel terrestrial (areal, large cities) levels of transnational organizations and households into the three basic levels—international, national, and organization level. The formats of economic levels alter too. Thus, an international level acquires the formats of territorial unions between the bordering states, agreements on national interests, or transnational corporations. The new post-2015 Sustainable Development Goals (SDGs) are clearly human-centered.

Millennium Development Goals were formulated by the United Nations General Assembly in 2000 for the period up to 2015. The new purposes and indicators were determined in 2015 at the UN Summit [2]. The number of goals significantly altered—8 goals and 48 indicators were transformed into 17 goals and 231 indicators (initially—169 targets). However, a closer look at the content of goals and indicators shows that there is continuity between the old and new goals, that they are formulated more precisely and better reflect the current economic, social, and environmental situation in the world. The orientation of the developing countries is expanded to the whole world, though the three-direction structure of goals remained constant. In clarifying goals and indicators, three economic levels were defined— the national level, the regional and sub-regional level, and the global level. At the national level, sustainable development goals and their indicators are supposed to be used; at the regional level, the usefulness of experience-sharing is identified; at the global level, the efforts in achieving the intended goals are brought together.

At the national level, 193 countries, including the Russian Federation, adopted the Declaration with a new agenda up to 2030. Besides the indicators and achievement of goals in sustainable development monitoring, the UN releases annual reports about the achievement of SDGs. In 2019, the UN Secretary-General called on all sectors of society to mobilize for a decade and highlight three levels:

- Traditional global level;
- Local level with its policies, budgets, institutions, and regulatory frameworks of governments, cities, and local authorities;
- The level of people—youth, civil society, the media, the private sector, unions, academia, and other stakeholders.

Such an expansion of the realization of SDGs is understandable, justifiable, and consistent with the current COVID-19 situation. Nevertheless, it requires the harmonization of the existing legal basis for expanding the process at least. Due to the COVID-19 pandemic, the interest in different economic objects to the goals and indicators of sustainable development has grown. The economic situation must be reviewed, considering the global risks at all economic levels and putting humans at its center. Thus, this paper aims to analyze the practice of assessing the achievement of sustainable development goals at different levels of the economy. The result of such analysis can create a basis for wider dissemination of sustainable development ideas, practices of assessment and analyzing the SDGs, and defining perspectives of effort synergy at different economic levels in combating the COVID-19 pandemic.

2 Materials and Method

During the research, the authors used official open sources containing legal and regulatory information and statistical and research data. Legal regulatory information was brought from the United Nations Department of Economic and Social Affairs [3], the Analytical Center for the Government of the Russian Federation [4], the Moscow Investment Portal [5], and the Rosstat Regional Office of Rostov Region [6]. Calculated results have been obtained on the statistical data from the website SDG tracker [7], Federal State Statistic Service of the Russian Federation (Rosstat) [8], Rosstat Regional Office of Rostov Region [6], and other sources mentioned in the research.

3 Results

3.1 Assessment of the Achievement of SDGs at the National Level

The goals of sustainable development were discussed in the Government of the Russian Federation and in public circles, which naturally led to their adoption at the state level and then the reflection in the policy of the Russian Federation and its strategic documents. In 1996, the first President of Russia, B. N. Yeltsin, announced the fundamental possibility of supporting the sustainable development concept of Russia and enshrined it in his Decree. Then, a presidential decree was adopted in 2018, which supported all the goals until 2024 and reflected the decision to participate in the monitoring of indicators that coincide with those observed at the national level. In 2020, the Decree on the convergence of national goals with sustainable development goals up to 2030 was signed. The adopted documents have become a logical and fundamental continuation of the previous document, having outlined more specifically the goals and indicators of sustainable development that will be supported in Russia. These decisions are reflected in the strategic documents of the Russian Federation, which have been developed and are being updated, including the Environmental Security Strategy of the Russian Federation until 2025, the Energy Strategy of the Russian Federation until 2035, the Industry Development Strategy for the Processing, Disposal, and Neutralization of Production and Consumption Waste until 2023, the Climate Doctrine of the Russian Federation, as well as a whole set of strategic documents in the field of environment and ecology and other doctrines, concepts, strategies, and government programs. Twelve national projects are aimed at achieving the SDGs in the following areas: demography, healthcare, education, housing, and the urban environment, ecology, safe and high-quality roads, labor productivity and employment support, science, digital economy, culture, small and medium-sized businesses and support of the individual entrepreneurial initiative, and international cooperation and export. Thus, all 17 SDGs and 107 out of 169 tasks are covered. However, only in 2017, "Indicators of achieving the Sustainable Development Goals for the Russian Federation" was developed, discussed, and accepted by the Federal State Statistics Service in subsection 2.8. Wherein it now includes only 90 indicators that nevertheless meet all the goals. In 2019, all global goals were adopted, and a national set of 160 indicators was developed for them. They fully overlap with national projects, goals of the Russian Federation development, and 44 regional projects. The national project "Demography" reflects 5 out of 17 SDGs, the national project "Ecology" reflects 6 SDGs, and the national project "Science" reflects 10 SDGs. Thus, a statistically comparable vertical has been created: from the UN Goals to the national system of indicators, which is combined with national projects and goals, and further to the goals and projects of the region.

In the Russian Federation, monitoring of the Millennium Development Goals achievement has been carried out since 1995 and was reflected in the Human Development Reports, each of which is devoted to a specific topic [9]. The reports are quite systemic and detailed within the framework of the UN sustainable development concept. These reports give an idea of the state and trends of sustainable development in three classical directions. A detailed statistical analysis of the indicators selected for this report is also being carried out. There are limited possibilities for comparing statistical information in reports. Among all reports, the 2005 report stands out not so much for its results as for the Appendix "Russia in 2015: Development Goals and Priorities," in which the Millennium Development Goals were adopted, considering the situation in Russia.

After the transition to 17 SDGs in the Russian Federation, two statistical yearbooks "SDGs in the Russian Federation" have already been published. In June 2020, the UN website published the first Voluntary National Review of the Russian Federation's achievement of SDGs and the implementation of the 2030 Agenda for Sustainable Development, prepared by the Analytical Center for the Government of the Russian Federation in collaboration with the Ministry of Economic Development Russia, the Russian Ministry of Foreign Affairs, Rosstat, and other departments, organizations, and companies. The review was prepared to determine the current position of Russia on the path to achieving the SDGs. On March 17, 2020, the Government of the

Russian Federation approved a plan of priority measures (actions) to ensure sustainable development of the economy in the context of a worsening situation due to the spread of COVID-19.

Generally, the situation for monitoring the achievement of goals at the national level has significantly improved (Table 1).

There is a discrepancy in the number of SDG indicators declared by the UN [10], accepted for measurement in the Russian Federation, and measured in the Russian Federation annually. First, the Russian Federation provides statistics on 179 national indicators for all SDGs for analysis at the global level, which is reflected in an open country database. Of these, 126 indicators are presented in dynamics, and the rest—in single values after 2015. This statistical plateau provides a base for general conclusions and comparisons. Second, since 2019, the number of indicators declared for sustainable development monitoring has increased and coincides with or is close enough to 12 goals; for 9 goals, the number of RF SDG indicators exceeds the number of UN SDG indicators. Third, the statistical authorities continue to conduct special

Table 1 Comparative analysis of the number of indicators used to assess sustainable development at the global and national levels

SDG No	Number of indicators	Russian Federation			
		Number of SDG indicators		Number of indicator values of the list	
		(Measured)	(Declared)	(One)	(More than one)
1	13	3	13	6	7
2	14	4	13	3	10
3	27	36	27	6	21
4	15	7	15	12	3
5	14	5	10	7	3
6	10	6	7	2	5
7	5	5	4	1	3
8	16	18	12	1	11
9	12	28	16	2	14
10	14	6	10	1	9
11	14	14	1	1	0
12	13	8	12	2	10
13	8	2	8	2	6
14	10	3	4	2	2
15	14	5	13	3	10
16	24	5	5	2	3
17	24	5	9	0	9
Total	247	160	179	53	126

Source Created by the authors

observations on sustainable development indicators that have not been declared by the Russian Federation for constant monitoring, which also gives rise to conclusions on the achievement of individual SDGs.

Thus, monitoring the SDG indicators in the Russian Federation has become more systemic in nature, and its spectrum has expanded, providing a real basis for regional practice.

3.2 Assessment of the Achievement of Sustainable Development Goals at the Regional Level

The relevance of ensuring sustainable development at the regional level is due to the fact that, in federal state conditions, the achievement of socio-economic and environmental balance in the development of regions makes it possible to achieve a synergistic effect on the scale of the national economy. In this regard, regional authorities need effective methods to assess the level of sustainable development of a constituent entity of the country and tools that allow making such an assessment.

With regard to the socio-economic system of the region, sustainability can be considered as the ability to maintain the specified parameters (in this case, the rates of economic growth and the indicators determined by them, characterizing the level of socio-economic development) for a certain time in the face of challenges of endogenous and exogenous nature. The balanced socio-economic development of the region involves the achievement of a dynamic balance between the needs and resources required to satisfy them by maintaining the stability of economic growth indicators. Being predetermined by the interconnection of the structural components of the regional economic system controlled by the regional authorities, balance can be considered a manifestation form for the proportionality of the region's economy. The latter can be represented as an optimal ratio of all types of economic activity characteristic of the region, its socio-economic results, and interregional economic ties. For example, in the Russian Federation, sustainable development is assessed by some indicators in many regions, but the practice of applying goals and corresponding indicators in the regions is rare [11, 12]. Statisticians from the Rostov Region were the first in Russia to conduct a comprehensive analysis of indicators of 17 SDGs in dynamics at the level of the subject of the Russian Federation. In 2019, they released the first collection of articles "Rostov Region—Movement towards the Sustainable Development Goals," which aroused great interest from completely different groups of users.

In 2021, the second statistical collection "Rostov Region—Moving towards Sustainable Development Goals: 2016–2020" was published, which, for the first time, provided an analysis of the achievement of some SDG indicators at the level of cities and districts. A comparative analysis of the use of sustainable development indicators has shown that indicators are analyzed for all goals; for some (SDGs 3,

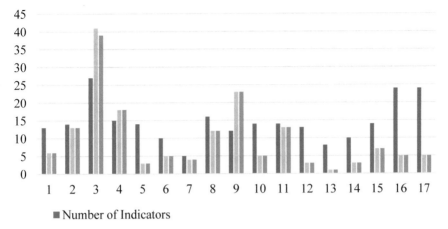

Fig. 1 Comparative analysis of the number of indicators used to assess sustainable development at the global and regional levels. *Source* Created by the investigators

4 and 9), a wider set is used that meets national goals and the Strategy until 2035 (Fig. 1).

Turning to the analysis of SDG indicators at the national level, it should be said that it was decided to statistically monitor only 71 indicators out of 160 indicators at the national level of the Russian Federation and 89 indicators at both the regional and, respectively, the national level. Thus, the Rostov Region demonstrates the best practice of SDG monitoring because it presents 167 indicators in its monitoring, which exceeds even the national level in terms of the indicators number.

3.3 Assessment of the Achievement of Sustainable Development Goals for Cities with a Population of More Than a Million People

Currently, the Organization for Economic Cooperation and Development (OECD) is implementing a pilot project "Territorial Approach to the Sustainable Development Goals." Since 2019, Moscow has been one of the cities in the Russian Federation that has joined the pilot project "Territorial Approach to Achieve the SDGs." The project aims to focus on a comprehensive system for assessing the city's long-term development based on the principles of UN sustainable development. In total, cities from nine countries of the world participate this project: Russia, Denmark, Germany, Norway, Iceland, Belgium, Argentina, Brazil, and Japan.

Moscow has a 100% achievement on all indicators of the six SDGs: quality education, gender equality, affordable clean energy, decent work and economic growth, industrialization, innovation and infrastructure, and the fight against climate change. Comparative analysis of metrics to reflect SDG implementation shows that targets 7, 8, 11, 12 and 16 are best met by metrics. In the course of the analysis carried out in 2021, OECD experts highlighted the special results of the city on such goals as SDG 4 "Quality education," SDG 8 "Decent work and economic growth," and SDG 9 "Industry, innovation, and infrastructure" [13].

Moscow implements several projects at once to implement SDGs (Fig. 2). One of the key projects is the integrated development of industrial territories "Industrial Quarters." It is directly related to achieving several sustainable development goals at once. Under the program, the city will receive at least 500,000 high-tech jobs, which will contribute to the achievement of SDG 8 "Decent work and economic growth." Additionally, the city will have new growth points, technological solutions, more environmentally friendly and efficient industrial enterprises, and improved infrastructure. To reduce emissions of pollutants and greenhouse gas (CO_2) from vehicles, in May 2021, Moscow issued green bonds. The International Capital Markets Association has included these bonds in the register of financial instruments for sustainable development.

Moscow, as the capital of the Russian Federation, has become an example of best practices for other cities with a population of over one million. VEB.RF, based on the experience of Moscow and the OECD, creates a system for assessing the quality of life in Russian cities [14]. At the first stage, it will provide an opportunity to assess the quality of life in 120 Russian cities and compare the city's development by certain indicators with the cities of the OECD countries. This is the first evaluation system

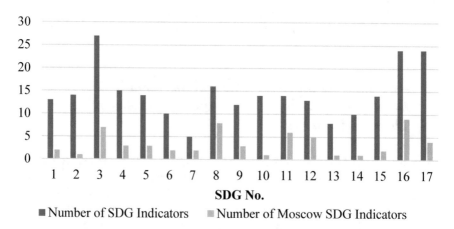

Fig. 2 Comparative analysis of the number of indicators used to assess sustainable development at the global level and the level of the millionaire city Moscow. *Source* Composed by the authors

of its kind in Russia, which includes 80 international OECD indicators (SDGs and the Better Life Index).

3.4 Assessment of the Achievement of Sustainable Development Goals at the Organization Level

Russian enterprises and organizations have been participating in events and initiatives to achieve the SDGs for a long time and on a large scale. Individual enterprises and organizations implement SDG business projects, while associations and unions of the business community take part in large-scale initiatives of an economic, social, and environmental nature that support the SDGs. When developing their strategies, large Russian organizations and enterprises rely on the national SDGs and monitor their sustainable development, reflecting their achievements in open reporting.

Sustainable development reporting among enterprises and organizations is becoming popular primarily among large companies that have been developing and updating their strategies for more than five years, since 2005. Russia is no exception.

The Russian Union of Industrialists and Entrepreneurs (RUIE) has formed a set of tools for the practical application of the principles of responsible business conduct, increasing its transparency and introducing advanced standards of business culture (including in the field of employment, creating decent working conditions, increasing labor productivity, etc.). The RUIE implements the following SDG achievement assessment initiatives—the annual compilation of Sustainable Development Indices based on public reports of the 100 largest companies that identify leaders in sustainable development (two indices—"Responsibility and Openness" and "Vector of Sustainable Development"—are included in the international database of indices and ratings the Reporting Exchange in this area); implementation of a joint project with the Moscow Exchange to calculate stock indices for sustainable development. The RUIE line of indexes and ratings is further developed—the Corporate Sustainability Index in cooperation with the Analytical Credit Rating Agency and the Sustainable Development Goals Integration Rating in cooperation with PWC [15].

According to RUIE polls in 2020, more than half of enterprises and organizations that generate sustainability reporting based on GRI standards consider this presentation of enterprise data for stakeholders to be the most complete and the most demanded in the context of the spread of the COVID-19 pandemic.

The practice of assessing and analyzing the achievement of SDG will continue to spread. Simultaneously, the main methodological issue will remain the comparability of achievements at the level of the enterprise and the region or country.

4   Conclusions

The practical activity of assessing and analyzing sustainable development at multiple levels shows the following main trends.

First, the content of sustainable development continues to diversify, with the UN's global enterprise to work for future generations expanding and gaining support at all economic levels–global, national, regional, and at the organization and enterprise levels–resulting in a growing body of information used in the process. Second, such exponential data growth at the level of purposes and indicators will require starting a constant cycle of its systematization, harmonization, and standardization on all economic levels. Third, sustainable development, more than any other process, requires transparency because it affects the interests of every human on the planet–future and present, which explains the total digitalization of sustainable development plans and results, providing its availability and simplicity for everyone to use. The trends described here make clear the necessity of creating and keeping the unified information resource working on the chosen economic levels–global (UN), national (the government of the state), and areal and local (local government), as well as on the levels of organizations and enterprises.

References

1. Inozemtsev VL, Gorbachev MS (Eds) (1998) The constitution of the post-economic state: Post-industrial theories and post-economic trends in the contemporary world, 1st ed. London, UK, Routledge. https://doi.org/10.4324/9780429440687
2. UN General Assembly (2015) Transforming our world: the 2030 agenda for sustainable development (adopted by Resolution on September 25, 2015 No. A/70/L.1). New York, NY, UN. Retrieved from https://www.un.org/en/development/desa/population/migration/generalassembly/docs/globalcompact/A_RES_70_1_E.pdf. Accessed 12 Nov 2021
3. UN DESA (n.d.) Sustainable Development Goals. Retrieved from https://www.un.org/sustainabledevelopment/sustainable-development-goals/. Accessed 12 Nov 2021
4. Analytical Center for the Government of the Russian Federation (2020) Voluntary national review of the implementation of the 2030 Agenda for Sustainable Development. Retrieved from https://ac.gov.ru/uploads/2-Publications/analitika/DNO.pdf. Accessed 15 Nov 2021
5. Moscow Investment Portal (2019) Pilot project "A territorial approach to the Sustainable Development Goals." Retrieved from https://en.investmoscow.ru/about-moscow/cur/main. Accessed 15 Nov 2021
6. Rosstat Regional Office of Rostov Region (n.d.) Statistical materials on the Rostov Region. Retrieved from https://rostov.gks.ru/folder/183080. Accessed 15 Oct 2021
7. SDG Tracker (n.d.) Measuring progress towards the sustainable development goals. Retrieved from https://sdg-tracker.org/. Accessed 20 Nov 2021
8. Federal State Statistic Service of the Russian Federation (Rosstat) (n.d.) On the sustainable development goals. Retrieved from https://rosstat.gov.ru/sdg. Accessed 15 Oct 2021
9. Analytical Center for the Government of the Russian Federation (2018) Man and innovation. Human Development Report in Russian Federation. Retrieved from https://ac.gov.ru/archive/files/publication/a/19663.pdf. Accessed 15 Oct 2021
10. UNSD (2022) SDG Indicators. Retrieved from https://unstats.un.org/sdgs/indicators/indicators-list/. Accessed 20 Nov 2021

11. Agency for Strategic Initiatives (n.d.) National rating of the investment climate in the subjects of the Russian Federation. Retrieved from https://old.asi.ru/investclimate/rating/. Accessed 20 Nov 2021
12. Institute for Statistical Studies and Economics of Knowledge of the HSE University (2019) Rating of innovative development of the Russian Federation constituent entities (6 issue). Retrieved from https://issek.hse.ru/rirr2019. Accessed 20 Nov 2021
13. OECD (2021) A territorial approach to the sustainable development goals in Moscow, Russian Federation. OECD Regional Development Papers, No. 23. Paris, France, OECD Publishing. https://doi.org/10.1787/733c4178-en
14. City Life Index Project (n.d.) Quality of life index. Retrieved from https://citylifeindex.ru/database. Accessed 15 Oct 2021
15. Russian Union of Industrialists and Entrepreneurs (RUIE) (n.d.) ESG indices and RUIE sustainability rankings. Retrieved from https://rspp.ru/activity/social/indexes/. Accessed 20 Nov 2021

Structural Elements of Modern Financial Management of State and Municipal Forestry Institutions

Tatyana F. Romanova⬤, Marina O. Otrishko⬤, Galina V. Popova, and Lyudmila S. Medvedeva⬤

Abstract The research highlights functional elements of financial management of state and municipal institutions and presents the profile of financing state and municipal institutions of the forest industry in the South of Russia. Moreover, the research considers the problems of financial support for the activities of state and municipal institutions in the context of forestry organizations. The authors present recommendations to improve the quality of financial management of state and municipal institutions and highlight the differences in patterns of financing depending on the type and the conditions and rules for the management of financial resources.

Keywords Financial management · State and municipal institutions · Budget planning · Financing of institutions · Forestry

JEL Classification H50

1 Introduction

The effective solution of the state tasks is directly related to its financial activity, which is the financial matter of the state, economy, and society.

The stability of the country's financial position guarantees the financial stability of business entities and, as a result, the satisfaction of society (as the main consumer of everything created within the country's economy) with the economic strategy implemented.

T. F. Romanova · M. O. Otrishko (✉) · G. V. Popova
Rostov State University of Economics, Rostov-on-Don, Russia
e-mail: starka13@mail.ru

T. F. Romanova
e-mail: kafedra_finance@mail.ru

L. S. Medvedeva
Don State Technical University, Rostov-on-Don, Russia
e-mail: milla1988@mail.ru

E. G. Popkova (ed.), *Smart Green Innovations in Industry 4.0 for Climate Change Risk Management*, Environmental Footprints and Eco-design of Products and Processes, https://doi.org/10.1007/978-3-031-28457-1_58

To optimally choose the regulation of the financial and economic activities of state and municipal institutions (the Institutions), the country utilizes various tools for the distribution and allocation of financial resources between all levels of the country's budget system aimed at ensuring a constant and uninterrupted reproduction of financial, credit, and monetary resources in the economy.

The choice of financial and economic strategy and the implementation of the impact on financial relations by public authorities, local governments, and business entities is carried out through the management of financial flows at the macro level and within each state organization.

In the financial sector, the objects of management are a variety of financial relations, in connection with which financial management is a process of purposeful influence on financial relations and the corresponding types of financial resources. Such an impact is aimed at effectively solving problems and implementing the functions of state authorities and business entities using a combination of certain techniques and methods.

Thus, financial relations are the object of management, i.e., an economic category that requires objective, specific, and clear knowledge and study. As developed historically, the subjects of financial management in financial science are considered as organizational structures that are directly involved in managing public finances.

The paper aims to develop recommendations to overcome the issues in the current financial and economic activities of state and municipal institutions of the forest industry.

2 The Status of the Problem

In connection with the features of the activities in the forestry sector in Russia and abroad, the authors consider various approaches to financial management.

Nipers, Pilvere, Sisenis, and Feldmanis [1] emphasize the importance of financial support for forestry. As a result of the assessment of funding from the Latvian Rural Development Program (RDP) for 2014–2020, the implemented measures, and the return from the forest industry in Latvia, the analyzed support measures were found to be effective.

Blagoev talks about the role of innovation in the development and growth of companies, which are especially important in an unstable economic environment and in times of financial and economic crises [2].

Santiago, Forero-Montaña, Melendez-Ackerman, Gould, and Zimmerman consider the feasibility of involving the private, non-profit, and public sectors to find a model of social acceptability of forest management practices that will work for the majority of stakeholders sides [3].

Gordeev, Pyzhev, and Yagolnitser studied regional differences in the forestry industry and management using multivariate analysis, which led to the conclusion that there was an increase in forest loss and a lack of control and finance due to the vast territory [4].

Shvets, Shevts, Markov, Fitisov, and Didenko analyzed the financial and economic activities of forestry enterprises in conditions of economic and environmental instability and identified the main causes of changes in forestry activities [5].

Thus, in the contemporary scientific literature, we have not come across works devoted to the study of financial management processes, specifically of state and municipal forestry institutions.

This research is one of the first studies of the problems of financial management of institutions in the forest industry, which requires further scientific work, including in the areas of corporate social responsibility and the introduction of ESG.

3 Methodology

Methodologically, we proceed from the fact that at the state level (macro level), the system includes participants in the budget process that manage state (municipal) finances—state authorities, local governments, and other state structures. The totality of such bodies forms the financial management bodies.

Since the objects of financial management are financial relations and the corresponding financial resources, the activities of financial management bodies are multifaceted, aimed at the sectors and links that make up financial relations—the objects of state financial management.

The management process includes functional elements that provide financial forecasting and planning, operational management and execution, and, finally, financial control.

The first group of these functional elements (forecasting and planning) is intended to scientifically substantiate the prospects for further development of the country's economy, including through a clear analysis of the current and past situation. Additionally, at the forecasting stage, financial authorities at all levels ensure the preparation of forecasts and justifications for the indicators of the relevant plans in the context of the actually possible volumes of financial resources, sources of their formation, and key areas of use.

The second group of functional elements of state and municipal finance management (execution and financial control) is the implementation of a set of measures.

Finally, another classification of financial management is the nature and period of management: strategic financial management and operational management [6, 7].

Since state and municipal institutions are the "hands" of the state in implementing state policy and the economy, it is the aggregate result of the financial activities of the Institutions that reflects the real situation and problems in the public administration system [8].

The concept of the financial activities of state and municipal institutions is understood as the implementation by the Institutions of the functions of formation, distribution, and use of financial resources (decentralized monetary funds) of institutions to implement the state and municipal task in providing a wide range of socially and economically significant services in accordance with the legal status and powers of the Institution [9].

4 Results

State and municipal institutions in Russia could be divided into state-owned, autonomous, and budgetary ones. The activities of the relevant types of institutions are regulated by federal laws and other documents developed within the framework of the concept of building an effective, economically and socially oriented country [10, 11].

The different order of financing of Institutions and the conditions and rules for managing financial resources are caused by their different statuses [12, 13]. To implement the powers of state authorities or local self-government bodies, the financing of such institutions should be strictly controlled, transparent, and available for any type of audit. Additionally, in view of the estimated funding, from a legal point of view, state employees are participants in the budget process.

Since budgetary and autonomous institutions are called upon to implement socially significant activities—to effectively provide high-quality services to the population and perform socially useful work, determining the required volume and managing the financial resources of the Institutions is the most important task of public authorities at each level. Thus, the funds allocated in the form of subsidies for the tasks indicated by the founders of the Institutions are the financial resources of this type of Institution. Additionally, borrowed funds and funds from income-generating activities are considered as financial sources for the activities of a budgetary and autonomous institution.

For each type of subsidy, an agreement is concluded between the public authority and the Institution on the provision of such a subsidy. Further financing and responsibility for the financial result are carried out following the provisions of such agreements. Thus, for example, financing of the implementation of the powers transferred in the field of forest relations is allocated by the state to the regional Ministries, Forestry Committees, or Forest Administrations.

In accordance with the approved structures, staffing tables, and internal procedures, the state authorities authorized in the field of forest relations distribute budgetary funds according to expenditures: directly administrative and managerial apparatuses, subordinate state institutions, and subordinate autonomous and budgetary institutions that carry out economic activities.

According to the data shown in Fig. 1, the year 2021 saw a decrease in planned funding for the activities of state institutions due to the need to optimize federal spending because of the impact on the country's economy of the spread of COVID-19 in connection with which:

- Federal funding was clarified (reduced) by 10%, excluding expenses related to wages;

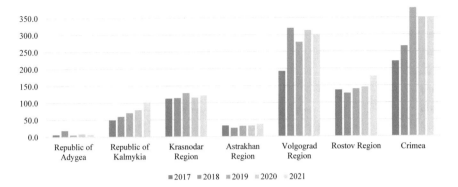

Fig. 1 Dynamics of budget financing of state-owned forestry institutions of the constituent entities of the Russian Federation of the Southern Federal District in 2017–2021, million rubles. *Source* Compiled by the authors

- The annual indexation of budget allocations for the annual increase in wages to the forecast level of inflation was canceled from October 1, 2021.

Figure 1 presents the dynamics of budget financing of state-owned forestry institutions of the constituent entities of the Russian Federation of the Southern Federal District in 2017–2021 (million rubles).

As shown in Fig. 2, the costs of forest management by budgetary and autonomous institutions of the constituent entities of the Southern Federal District remained practically unchanged compared to 2020, despite the goals set by the National Project "Ecology" in terms of the implementation of the Federal Project "Preservation of Forests" to increase the area of reforestation and afforestation. This trend is also explained by the budget constraint on reducing federal spending by 10% and the refusal to index the wage fund for forest industry workers.

Underfunding the implementation of measures to protect forests from fires carries risks of an environmental nature—a failure to carry out or incomplete protection measures can cause an increase in the number of forest fires and an increase in the area covered by fire, which poses a threat to the infrastructure of settlements bordering on forest due to the possible spread of fire.

Measures for ground patrolling of forests on the territory of the Southern Federal District are carried out by subordinate budgetary and autonomous institutions subordinate to the executive authorities authorized in the field of forest relations.

Additionally, the decline and the lack of positive dynamics in financing activities to increase the area of reforestation and afforestation contradict the idea of the Federal Project "Forest Preservation"—the annual increase in the planned planting area and conversion of land to forested areas. It is necessary to increase funding for forest care work in accordance with natural processes and the requirements of the production process.

The main element of managing the financial activities of autonomous and budgetary institutions is the state and municipal task (the Task).

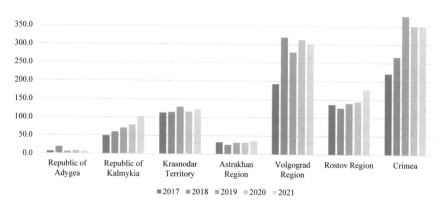

Fig. 2 Dynamics of budget financing of budgetary and autonomous forestry institutions of the constituent entities of the Russian Federation of the Southern Federal District in 2017–2021, million rubles. *Source* Compiled by the authors

Under the state and municipal task, it is customary to understand a document fixing the requirements for the quality, composition, procedure, and conditions for the provision of state and municipal services. It is important that, in accordance with the current legislation, neither budgetary nor autonomous institutions have the right to refuse to complete the Task.

Financial support for fulfilling the Task is carried out by bringing subsidies to the Institutions by the founder. The standards of costs for the provision of services and work are the basis for calculating the need for financial support for the state or municipal task.

Another peculiarity of funding the implementation of state assignments is that if the Task is completed in full, the Institutions can direct the savings of the current financial year to similar work in the next year without returning these savings to the budget [14].

Thus, a state or municipal task is the main motivating and managerial tool administered by a public authority that exercises the powers of the founder of a budgetary or autonomous institution, since the quality and completeness of its implementation determine the effectiveness of the managerial activity of the founder– the public authority.

A problematic issue arises: underfunding of the forest industry according to the approved standards.

In view of the budget restrictions established for 2021 (a reduction in budget allocations to ensure a balance of the federal budget by 10% and the cancelation of the annual indexation of budget allocations for the annual increase in wages to the forecast level of inflation from October 1, 2021), the acute problem of lagging behind the actually allocated volumes of financing for forestry activities became even more negative.

We have carried out an analysis. According to the results, we can track the lag of federal funding for implementing the transferred powers in the field of forest relations (except the acquisition of specialized forest fire and forestry machinery and equipment) in accordance with the federal law on the federal budget for 2021 from the estimated one, made in accordance with the order of Federal Agency for Forestry (Rosleskhoz), which is:

- 756.8 million rubles (64%) for financing the implementation of state management;
- 771.1 million rubles (121%) for financing the implementation of forestry activities.

Simultaneously, in 2021, according to the approved cost standards authorized in the field of forest relations by the executive authorities of the constituent entities of the Russian Federation of the Southern and North Caucasian Federal Districts, 1099.6 million rubles were required to carry out forestry activities in these subject of Russia, excluding costs for the maintenance of state institutions and the formation of reserves for the prevention and suppression of forest fires, the purchase of firefighting equipment and inventory, the maintenance of forest fire units, fire equipment, and communication and warning systems, and the creation of a reserve of fire equipment, firefighting equipment and inventory, and fuel-lubricants. In the regions of the North Caucasus Federal District, according to the standards approved in the constituent entities, the total need is 496.7 million rubles, which is lower than the calculated one by the Order of the Federal Forestry Agency (527.4 million rubles) but three times higher than planned in budget designs—166.8 million rubles [15].

Due to the chronic underfunding of the forest industry, the authorized executive authorities of the constituent entities were forced to apply correctional reduction factors. In accordance with the budget projections, the constituent entities allocated only 469.2 million rubles to implement activities.

Thus, the position of the forest industry on the need for funding in the subjects and at the federal level is very different from the possibilities of the economy.

The problematic issue of underfunding the forest industry started to be actively discussed in 2020; the actual funding still needs to be increased.

5   Discussion

One of the areas of financing the activities of state and municipal institutions are subsidies for other purposes, which are provided to the Institutions based on the General Requirements for regulatory legal acts at the state level and municipal legal acts that establish the procedure for the volume and conditions for granting subsidies to the Institutions for other purposes.

This type of subsidy has a strictly targeted nature and clearly measurable results of its use. Thus, within the framework of the Federal project "Forest Conservation," institutions purchase relevant machinery and equipment to implement the state-set task to equip forestry institutions of the constituent entities of the Russian Federation and institutions that perform the functions of forest fire units with a specialized forest fire and forestry equipment by 2024 by 100% (for forest fire) and 70% (for forestry). Purchases are financed by attracting subsidies for other purposes.

Simultaneously, subsidies for other purposes could be provided for the following purposes:

- To compensate for damage in the event of an emergency;
- To organize one-time events within the framework of state programs, the costs of which are not included in the Task;
- To provide major repairs and the acquisition of fixed assets, the calculations for which were not included in the cost standards when justifying the funding needs of a state or municipal task.

The founder makes calculations for such subsidies only after checking the documents confirming the occurrence of monetary obligations.

A new type of financial resource for implementing activities by state and municipal institutions are subsidies for capital investments in capital construction projects of state or municipal property, as well as the acquisition of real estate in state ownership. This type of subsidy implies the investment of the state by financing the activities of the Institutions in increasing the state (municipal) property at the budget expense. It is important that any property acquired by the Institution at the expense of state investments is state (municipal) property.

Finally, subsidies in the form of grants are provided to Institutions on a gratuitous and non-refundable basis for the implementation of specific scientific research, scientific and technical programs, and innovative projects on strictly defined conditions by the grantor. For example, Russia currently has state support programs for small and medium-sized businesses aimed at industries identified by the state as priorities.

Several other aspects of the financial activities of budgetary and autonomous institutions should be noted.

In accordance with the current legislation and in view of the activities carried out, budgetary and autonomous institutions are payers of taxes and fees, like other legal entities, i.e., the participants in tax legal relations (taxes on income, property, and land tax). That is, budgetary and autonomous institutions may be granted tax incentives for regional and local taxes, which will eliminate the artificial overestimation of revenues and expenditures of the regional or municipal budgets.

After the reform of the system of the Institutions, in accordance with their new legal status, the founder no longer acts as the key manager of budgetary funds. Nevertheless, it is important that the founder approves the state and municipal tasks for the Institution and determines the goals for obtaining subsidies for other purposes in accordance with the federal, subject, or municipal order for federal, regional, and municipal institutions, respectively.

6 Conclusions and Recommendations

To improve the quality of financial management of state and municipal institutions, the following is necessary:

1. To use various methods of financial forecasting and planning, the most appropriate for the current situation, and ensure sufficient justification of forecasts and the fulfillment of planned appointments;
2. To conduct regular monitoring of the financial condition of state and municipal institutions for the timely adoption of management decisions in the field of financial and economic activities;
3. To prevent shortfalls in budget financing, which must be fully secured, being expenditure obligations;
4. To summarize the experience of various institutions in the same industry to identify the best financial management practices to recommend changing their type to a more optimal option;
5. To clarify existing and develop new cost standards based on deep scientific study and existing experience in their use.

The implementation of these recommendations could be useful not only in the forest industry but also in other sectors of the economy where state and municipal institutions operate.

References

1. Nipers A, Pilvere I, Sisenis L, Feldmanis R (2020) Support measures of the rural development program for the forest industry in Latvia. In: Proceedings of the SGEM '20: 20th international multidisciplinary scientific GeoConference "Surveying Geology and Mining Ecology Management", Sofia, Bulgaria, pp 731–739. https://doi.org/10.5593/sgem2020/3.1/s14.094
2. Blagoev D (2021) Innovativeness and innovation potential function at forest companies. In: Proceedings of the WoodEMA 2021: 14th international scientific conference "The Response of the Forest-Based Sector to Changes in the Global Economy", Koper, Slovenia, pp 245–250
3. Santiago LE, Forero-Montaña J, Melendez-Ackerman EJ, Gould W, Zimmerman JK (2022) Social acceptability of a sustainable forestry industry in Puerto Rico: views of private, public, and non-profit sectors. Forests 13:576. https://doi.org/10.3390/f13040576
4. Gordeev RV, Pyzhev AI, Yagolnitser MA (2021) Drivers of spatial heterogeneity in the Russian forest sector: a multiple factor analysis. Forests 12(12):1635. https://doi.org/10.3390/f12121635
5. Shvets M, Shvets M, Markov F, Fitisov A, Didenko P (2020) Analysis of financial and economic activity of state forest enterprises in the conditions of economic and ecological crisis. Scient Horizons 6(91):92–100
6. Borovitskaya MV (2021) On the organization of internal control in state (municipal) institutions. Azimuth of Scientif Res: Econ Adminis 10(1):390–394
7. Ignatova MA, Lukyanchenko AV, Pavlov DS (2020) Finance management theory and practice. Colloquium-J 8—6(60):145—149
8. Botasheva ZR, Kyabishev ZU, Chazhaev MI (2019) Organization of an effective system of financial resource management. Bullet Acad Knowled 35(6):80–85

9. Pisareva EG (2011) State institutions in the system of subjects of financial law (Synopsis of Dissertation of Doctor of Law). Saratov, Russia, Saratov State Law Academy
10. Mikhalenok NO, Litvinova DD (2020) On the issue of analyzing the mechanism of spending funds in state (municipal) institutions. Scient Vector of the Balkans 4(3):82–84. https://doi.org/10.34671/SCH.SVB.2020.0403.0020
11. Romanova TF, Andreeva OV, Meliksetyan SN, Otrishko MO (2017) Increasing of cost efficiency as a trend of public law entitie's activity intensification in a public administration sector. Europ Res Stud J 20(1):155–161. https://doi.org/10.35808/ersj/605
12. Fedoseeva KP (2020) On the issue of imposing restrictions on budgetary (autonomous) institutions that carry out purchases in order to fulfill state (municipal) tasks. Finan Managem 4:14–23. https://doi.org/10.25136/2409-7802.2020.4.34612
13. Kuramshina AV, Nikitina NN (2020) Economic analysis of the financial and economic activities of state bodies and institutions: opportunities and legal support. Innov Investm 9:183–187
14. Kuryaeva GY, Tagirova AA (2022) Modern problems of financial and economic activity of the institution. Interact Sci 2(67):48–50. https://doi.org/10.21661/r-555996
15. Federal Forestry Agency of the Russian Federation (2020) Order "On approval of cost standards for the provision of state works (services) for the protection, protection, reproduction of forests, afforestation and Forest Management and on Invalidation of the Order of the Federal Forestry Agency dated June 19, 2019. No. 762" (June 29, 2020 No. 607). Moscow, Russia. Retrieved from https://base.garant.ru/74947552/. Accessed 12 Sept 2022

Towards Sustainable Finance Through Environmental and Social Responsibility of Business

Sergey M. Nikonorov⬤, Sergey G. Tyaglov⬤, Raisa M. Bogdanova, Alexander A. Khokhlov, and Pavel A. Degtyarev

Abstract Sustainable finance based on a green agenda and focused on projects in the field of ecology and social aspects of development cannot yet find application due to the difficult geopolitical situation in Russia. Corporate environmental responsibility of business appeared along with corporate social responsibility (CSR) and was the basis for the emergence of such an element as environmental, social, and corporate governance (ESG). ESG is more than ecology. ESG is a new business philosophy in Russia. ESG ratings also enable investors to identify an object for investing their money. ESG is distinct from but correlates with sustainability. In this research, we focused on several painful points that the real sector of the Russian economy needs to overcome in today's conditions, which will last from three to ten years.

Keywords Sustainable finance · Green trend · ESG transformation · Sustainability rating · Green bonds · Green investments · Corporate environmental responsibility

JEL Classification M14 · Q01 · Q56

S. M. Nikonorov (✉)
Lomonosov Moscow State University, Moscow, Russia
e-mail: nico.73@mail.ru

S. G. Tyaglov
Institute for Sustainable Development and Environmental Protection, Rostov State University of Economics, Rostov-on-Don, Russia

R. M. Bogdanova · A. A. Khokhlov
Rostov State University of Economics, Rostov-on-Don, Russia

P. A. Degtyarev
Sochi Institute (Branch) of the RUDN University, Sochi, Russia
e-mail: degtiarev.pa@rudn-sochi.ru

© The Author(s), under exclusive license to Springer Nature Switzerland AG 2023
E. G. Popkova (ed.), *Smart Green Innovations in Industry 4.0 for Climate Change Risk Management*, Environmental Footprints and Eco-design of Products and Processes, https://doi.org/10.1007/978-3-031-28457-1_59

1 Introduction

Sustainable finance based on a green agenda and focused on projects in the field of ecology and social aspects of development cannot yet find application due to the difficult geopolitical situation in Russia.

The banking sector, which by all means clung to the global green trend, is now 80% under severe sanctions and is looking for all sorts of options to stay competitive in the market.

In this research, we will focus on several painful points that the Russian banking sector needs to overcome in today's conditions, which, in our opinion, will last from three to ten years.

The first problem is the problem with money transfers, which is starting to gain momentum. It is becoming increasingly expensive and difficult to transfer currency from Russia abroad; sometimes, it is completely impossible to do it in a legal and civilized way. The largest banks of Russia that fell under the sanctions (e.g., SBER, VTB, and Alfa-Bank) have temporarily suspended foreign currency transfers abroad and within the country. It should be noted that the problem was clearly identified in state-owned banks (with state participation) and in 100% of private banks.

Clients of private banks that have not been sanctioned also have problems. For example, in June 2022, Tinkoff Bank increased the fees for incoming cross-border transfers to 3% of the transfer amount, but not less than $200 in the account currency and not more than the transfer amount.

Tinkoff also changed the terms of SWIFT transfers. Now, the minimum transfer amount is $20,000. Raiffeisenbank announced plans to increase the commission for foreign exchange transactions through the SWIFT system; it will be the same—3% of the transfer amount in the bank branch and 2% of the transfer amount in the Raiffeisenbank Online system.

The second problem is the problem with Russian exchanges. The turnover of the Moscow and St. Petersburg stock exchanges fell catastrophically. If an increase in interest in domestic securities is still expected in the medium term, then it is not clear how potential investors should treat foreign securities. For example, in July 2022, the turnover of trading in foreign shares on the St. Petersburg Stock Exchange (St. Petersburg Exchange) was 77% lower compared to July of the previous year. The number of active investor accounts also decreased significantly. At the end of 2014, the St. Petersburg Exchange launched trading in foreign securities. In 2016, it laid down a certain basis for them—the so-called routing mechanism, which allowed Russian investors to buy securities around the world on Western sites through Russian brokers. The year 2018 turned out to be the most successful year for Russian exchanges (Moscow Exchange and St. Petersburg Exchange), which saw a serious increase in turnover and an influx of new private investors. This happened mainly due to lower deposit rates. Investors began to look for an alternative to passive investing. In 2021, the St. Petersburg Exchange overtook the Moscow Exchange in terms of the trading volume. However, they also began to compete with each other. Moscow Exchange began to trade in foreign securities, and St. Petersburg Exchange began to trade in Russian securities. Nowadays, there is no competition. The Central Bank has blocked trading in some foreign securities in Russia for half a year. Most of them previously passed the primary listing in the USA and were stored in the National Settlement Depository. St. Petersburg Exchange is currently at a dead end in its development because its main activity is foreign securities. The sharply increased risks of blocking foreign shares have significantly reduced interest in them from retail investors. Simultaneously, the Russian Government already discusses the issue of a ban on the purchase of foreign securities by unqualified investors. It is possible to continue to act through Cyprus, but this country is an EU member. It is also possible to try to turn the Astana site into a kind of analog of the St. Petersburg Exchange, but it is still questionable in situations when any assets held abroad can be blocked. Distributing securities across multiple depositories can help reduce risks for investors. In June 2022, SPbExchange launched trading in securities from the Hong Kong Stock Exchange. By September 2022, 50 securities of companies from Hong Kong and China should be available. It is planned to expand this number to 200 by the end of 2022 and to 1000 by the end of 2023. However, the Bank of Russia is going to raise the threshold to 60.0 million rubles, upon reaching which it will be possible to buy MBS, which will greatly limit operations with them. Potential investors are waiting for the situation to recover. Nevertheless, this situation is long-term, and the factors that have already led to a decrease in the shareholder value of stock exchanges will continue. Since one of the goals of Western sanctions is to cut off all possible financial connections of Russians and remove all big and small bridges, it is probably possible to repeat this success with other foreign markets, for example, neutral ones. All this is also under the threat of secondary sanctions.

An interesting and promising direction could be the development of B2B in the form of an electronic platform for public procurement (similar to the RTS-Tender platform). It is possible to repeat the path of the St. Petersburg Exchange, which operates on the domestic market or go the way of commodity exchanges. A logical direction for developing the St. Petersburg Exchange would be a turn to the East. The exchange could work with Asian stock markets. Another option is more active work on the domestic market with new shares of Russian issuers. For example, medium and small companies could come to the site. A good and interesting direction could be more effective work with exchange-traded funds or creating a marketplace for stock instruments.

Foreign stocks remain an interesting asset for trading, investment, and portfolio diversification, including for private investors. However, the issue of accounting for rights is now becoming particularly acute. "The one who can level the risks will be able to become an interesting platform for Russian investors. Including for those who are ready to invest in securities of companies from friendly and neutral countries. The mechanism may be different, for example, the creation of a counter mutual interdepository accounting of rights or the maintenance of registers of owners" [10].

The third problem is not so much in the tariffs but in the unpredictability of payment results. SWIFT does not charge commissions for transfers; it is only a communication channel between banks. By limiting payments by any means, credit institutions try to cope with the increase in the number of checks by Western partners and the reduction in the number of banks that use SWIFT payments operating in Russia. The costs of the banking sector are growing due to the increase in terms of transfers, the freezing of assets, and the loss of part of the business. Therefore, banks try to pass on the part of the losses to the consumer. Nowadays, there is a disconnection of Russian banks from SWIFT and a reduction in the number of external banks that could or would like to work. This is important for any transactions with currency because such transactions always require a correspondent bank in the country where this currency is issued. Not all external banks are ready to completely break off relations, but making payments difficult in every possible way is the norm for doing business. Some foreign exchange payments pass quickly, while others are rejected or, at best, slowed down. This translates into an increase in the tariff, but no tariff guarantees results. Each sanctioned bank got its own list of restrictions, sometimes even in local currencies. For most counterparties, it is easier to end the relationship rather than to sort through the various formulations of the EU, US, or UK sanctions in force at that moment for each transaction. In fact, many sanctioned banks make payments in Yuan. Nevertheless, there is still a big question of how long the banks can rely on it [12].

2 Materials and Method

To determine the state of the Russian economy and its predisposition to the further adoption and strengthening of ESG standards in the country, Kostin [3], Nikonorov [8, 11] used the method of synthesis, deduction, and comparison, as well as media data from Russia and the CIS [1, 10, 12].

The analysis of the Russian companies from the point of view of sustainable development [7] was based on 24 indicators divided into three blocks (eight indicators in each block). Next, the authors formed the SWRP matrix (similar to the SWOT analysis) (Table 1).

3 Results

We propose the following actions. In Russia, an analog of SWIFT was developed in 2014 when, at a meeting of the European Commission, it was first proposed to disconnect the Russian Federation from this system. Currently, there is a Financial Message Transfer System (SPFS) developed by the Central Bank of Russia. This system began full-fledged work back in 2017. In 2020, the share of messages transmitted via SPFS exceeded 20% of Russian SWIFT traffic. However, this is not a full-fledged replacement because this system is ineffective for external transfers.

Sberbank has also begun to create an international payment system, an alternative to SWIFT; it is planned to complete the setup of this system within a year. However, since the BEAC itself is under sanctions, it is unlikely that it will be able to bring

Table 1 SWOT analysis of PJSC MMC Norilsk Nickel

Strengths	Weaknesses
(1) The total volume of emissions of pollutants into the atmospheric air to the total costs (2) The share of electric energy produced using renewable energy sources in the total volume of electric energy production (3) The coefficient of environmental friendliness of production (4) The proportion of neutralized waste subject to neutralization and disposal transferred to the federal operator (5) The real progress index (6) The average monthly nominal accrued salary of employees of the organization (7) The real disposable income of the average employee of the company (8) The proportion of polluted land in the area of the country (9) The coefficient of environmental hazard of products (product revenue/environmental footprint) (10) The share of contributions to state socio-ecological funds in relation to total costs 11) The availability of emission reduction programs 12) The share of financing of socio-ecological measures and the development of the company's infrastructure to finance socio-ecological measures by the country	(1) The coefficient of environmental efficiency (total costs/volume of waste generation) (2) The coefficient of utilization of production waste (3) LTIFR (Lost time injury frequency rate) (coefficient of frequency of injuries with temporary disability) (4) The assessment of working conditions in accordance with the requirements of labor legislation (5) The number of injuries with temporary disability per one million human hours worked by Norilsk Nickel (6) The share of accrued fines for environmental pollution to the total costs (7) Environmentally hazardous objects (8) The share of social and reputational capital to the annual profit

Source Compiled by the authors [8, 11]

its system to the international level. On the example of Iran, when the country was disconnected from SWIFT, it led to the use of a larger number of financial intermediaries, which, in turn, led to an increase in commissions because each bank is a commercial institution that charges its own commission. Therefore, cross-border transfers are becoming more expensive. Any system similar to SWIFT requires reliable protection and responsible personnel. Another possible solution could be to increase the number of transfers in digital currencies or based on the blockchain, as this direction is being actively developed by the Government of the Russian Federation. Operations with a "Digital Ruble" will make SWIFT unnecessary. A complete block of the SWIFT system may not happen, as this is fraught with difficult-to-predict consequences for the energy market and energy supplies, which creates huge risks for the countries that have imposed sanctions against Russia.

The second possible solution is Yuan bonds. The Yuan will rapidly replace the Euro and the US Dollar on the Moscow Exchange. The trading volume of the Chinese currency in August 2022 exceeded that of the US Dollar and Euro. Russia has now risen to third place in the world in terms of using the Yuan outside of China (in June

2022, it was in the seventh position). Russia accounted for 3.9% of all Yuan transactions in the world in July (only the UK and Hong Kong have more). Demand for Yuan has also increased significantly among ordinary citizens who want to diversify their savings. According to the Central Bank, in June 2022, the share of individuals was 13.9% in the volume of purchases and 5.5% in sales. Nowadays, Russian businesses are looking for new partners from Asian countries. Together with them, China is also humanizing trade turnover. As a result, Russia can serve trade flows with friendly and neutral countries through the Yuan. However, there remains the question of transitioning Russian rubles in foreign trade operations. While the bet is on the Chinese currency, it is necessary to remember an essential rule: "Concentration on simply one method of payment carries danger and risks" [2].

Despite the difficulties that have arisen, some Russian companies are gradually integrating ESG standards into their activities. Norilsk Nickel holds a leading position in this direction.

In terms of total emissions of pollutants into the atmosphere for total costs, Norilsk Nickel ranks third with an indicator of 0.218133968. In terms of the share of electricity produced using renewable energy sources in the total volume of electricity production, Norilsk Nickel ranks third with an indicator of 0.643274854. In terms of environmental expenditures on biodiversity conservation and protection of natural areas as a percentage of revenue, Norilsk Nickel ranks sixth with an indicator of 0.10375939. Nornickel has the second place after Gazprom (49.1 billion rubles) in terms of costs for environmental protection companies (34.6 billion rubles), but Nornickel's environmental friendliness ratio for production is an order of magnitude better than Gazprom (9.89% versus 0.87%). This indicator was determined as the ratio of the costs of environmental protection measures to the total costs of the enterprise and multiplied by 100%. For Norilsk Nickel, the total costs amounted to 350.0 billion rubles, and for Gazprom, for the same period—5665.8 billion rubles.

In terms of environmental friendliness of production, Norilsk Nickel ranks second with an indicator of 0.301013209. In terms of total production, Nornickel ranks sixth after Rosatom, T Plus, Bashneft-Polyus, Rosseti, and Zarubezhneft (2019 million tons). In terms of output, Nornickel ranks fifth (743 million tons), while Rosatom is far behind in this indicator (215.7 million tons). According to the degree of environmental friendliness of products with an indicator of 0.510997784, Nornickel takes sixth place. In terms of the environmental efficiency ratio (total costs/volume of waste generation), Norilsk Nickel ranks 12th with an indicator of 0.479410614. In terms of the share of neutralized waste subject to neutralization and disposal, transferred to the federal operator, Norilsk Nickel ranks third with an indicator of 0.892678034. Norilsk Nickel ranks 13th in terms of the utilization factor with an indicator of 0.419045455.

According to the group of environmental and economic indicators, consisting of eight elements, PJSC MMC NORILSK NICKEL took first place with an integral indicator of 0.0648. PJSC T Plus took second place with an indicator of 0.0620. PJSC Gazprom Neft took third place with an indicator of 0.0613. The fourth place with an indicator of 0.0515 was taken by the company Branch "RUSAL Kandalaksha." The fifth place with an indicator of 0.0459 was taken by the State Corporation Rosatom.

According to LTIFR (Lost time injury frequency rate), Norilsk Nickel is in 11th place with a score of 0.429. In terms of export of goods in value terms, the growth rate to revenue of Norilsk Nickel ranks 9th with an indicator of 0.739375911. According to the real progress index, Norilsk Nickel ranks 3rd with an indicator of 0.91011236. According to the assessment of working conditions in accordance with the requirements of labor legislation, Norilsk Nickel ranks 22nd with an indicator of 0.4. In terms of the share of funds that represent overdue wage arrears to employees of the organization from the total annual income (volume of debt/EBITDA), Nornickel ranks 8th with an indicator of 0.069358886. In terms of the number of lost time injuries per one million human hours worked, Norilsk Nickel ranks 10th with an indicator of 0.293333333. According to the average monthly nominal accrued wages of employees of the organization, Norilsk Nickel ranks fourth with an indicator of 0.505166475. In terms of the real disposable income of an employee of the company, Norilsk Nickel ranks first with an indicator of 1.002231598. According to the group of socio-economic indicators, PJSC MMC Norilsk Nickel takes sixth place with an integral indicator of 0.2069. The first place is occupied by Gazprom Neft PJSC (0.2354), second place—PJSC "T plus" (0.2297), third place—PJSC AK Alrosa (0.2231), fourth place—PJSC Rosneft (0.2172), and fifth place—PJSC Novatek (0.2163). In terms of the proportion of contaminated land in the area of the country, Norilsk Nickel ranks fourth with an indicator of 0.003142174. In terms of the share of accrued fines for environmental pollution to total costs, Norilsk Nickel ranks 19th with an indicator of 0.30632381. In terms of environmentally hazardous facilities, Norilsk Nickel ranks 17th with an indicator of 0.229885057. In terms of product environmental hazard ratio (revenue/environmental footprint), Norilsk Nickel ranks fourth with an indicator of 0.186378945. In terms of the share of contributions to state social and environmental funds in relation to total costs, Norilsk Nickel ranks fifth with an indicator of 0.185871766. In terms of emission reduction programs, Norilsk Nickel ranks first with an indicator of 0.8. In terms of the share of social and reputational capital in annual profit, Norilsk Nickel ranks 11th with an indicator of 0.091089349. In terms of the share of financing of social and environmental measures, the development of the company's infrastructure for financing social and environmental measures by the country, Norilsk Nickel ranks first with an indicator of 1. In terms of a group of social and environmental indicators, PJSC MMC NORILSK NICKEL ranks first with an integral indicator of 0.1057.

4   Conclusion

The following conclusion can be drawn. Cash flow in the Russian financial system is slowing down. Due to the increase in global and domestic tensions, lending to the population fell significantly in May 2022, which did not allow for maintaining the financial balance. However, banks still own large amounts of money and even have an increase. The amount of money of bank customers is increasing; the structural liquidity surplus of the financial (banking) sector has risen to around 2.5 trillion rubles. This suggests that the banking system is faced with an excess of money. The money "hung up" because of the conservative policy of commercial banks. However, already in the third quarter, we can see an increase in loans for foreign trade transactions. In general, most of the expert community assesses the situation as critical but notes that the current changes in lending to the population and firms are characteristic of the current "crisis of relations." The population is cautious; there is nowhere to invest money. In March 2022, there was an increase in citizens' investments in securities, but it quickly came to naught. Nowadays, there is a decrease in final demand; due to the aggravated situation, imports are at a minimum level. Accordingly, the compression of lending is a natural result of the ongoing processes. Most likely, this situation will last for another year and a half. Additionally, we are in a state of structural pause in the change of suppliers and consumers; although there is a parallel import, it is not fast. Due to the changing geopolitical situation, markets are also changing: the Great Silk Road and the Asian region are replacing Europe. If the probable markets become real, then it is worth assuming changes in the elements of banks' assets in the sector of increasing loans for foreign trade transactions by the end of 2022. The steady movement of financial resources correlates with the situation in the Russian economy. There are two points of view on this issue. The first point of view is pessimistic. According to this point of view, the Russian economy is slowing down. Production, export–import operations, and savings are unbalanced. The problem is not finances. Finance is only a reflection of monetary relations and a real reflection of the process of economic slowdown [5]. For the last 20 years, Russia has been fighting for a favorable investment climate, which is hopeless within the next ten years [4]. There will be no external investors, with rare exceptions. As for domestic investors, it is unclear who is willing to risk their money in the current difficult situation. If there is no investment, then there is no development. To engage in import substitution and transition to the best available technologies (BAT), we need financial resources and new equipment. Companies ask the following questions:

- Where to find money?
- If there is money, then where to buy?

The second point of view is optimistic. In the conditions of currency restrictions and declining confidence in the Russian stock market, the base of alternative forms of savings has decreased; that is, citizens have even fewer options. This may mean that there will be an influx of funds to banking institutions in the short and medium term, even with a further reduction in interest rates. The forecast for the Central Bank's key rate is 8.0–8.5% until the end of 2022, and the weighted average interest on deposits is 7.0–7.5%. By the end of 2022, we can expect a revival of the economic situation with an increase in loans to legal entities by up to 5% and to individuals—by up to 1–2%. The Bank of Russia predicts a full recovery of credit activity of subjects of the national economy in 2023 [1, 9].

We can draw the following conclusion. Cash flow in the Russian financial system is slowing down. Due to the growth of global and domestic tensions in May 2022, lending to the people decreased sharply, making it impossible to maintain the financial balance.

Under the conditions of sanctions pressure in March 2022, only two companies ventured to bring new bonds to the market: VTB and Aston (a financial management company). In April, there were already eight placements; mainly specialized financial companies entered the market (Aurum, MIP-1, Capital Plus, and Titan). They were placed by closed subscription among qualified investors; then, the placements were not of a market nature. The situation began to improve in May 2022, when major players returned to the primary market, including Avtodor, Ingrad, MTS, Rostelecom, etc. Slavneft, RESO Leasing, Russian Post, and others placed their bonds in June. The Russian Post has placed bonds in the amount of 30.0 billion rubles, and Rostelecom—for 15.0 billion rubles. Sibur also placed its bonds. The company borrowed 15.0 billion rubles for 3.5 years at 9.9%. In terms of the volume of bonds issued, May–June 2022 has already overtaken the indicators of January–February 2022. In January 2022, Russian companies placed bonds worth 16.08 billion rubles, in February—66.3 billion rubles, while most of them were attracted by BEAC and VTB. Most likely, the upward trend in the bond market will continue until the end of 2022. We see this as a development prospect, especially if Russian companies place bonds among unqualified investors (i.e., among Russian citizens). In this, we see the stability of the development of financial relations in the conditions of partial financial autarky.

According to the results of the study of Russian companies and, in particular, Norilsk Nickel, the rating of Norilsk Nickel served as one of the factors of the company's transition to the concept of a national enterprise, where 25% of ordinary voting shares will belong to the company's employees and freely circulate on the market. The authors consider these measures to be effective management of sustainable finances as a manifestation of corporate environmental and social responsibility of business in Russia.

There is no unambiguous interpretation of the concept of "social responsibility of a Russian company" (CSR). All interested parties in this area (i.e., government authorities, representatives of business and public organizations) try to interpret this concept in their own interests. One of the classic definitions of corporate social responsibility, "CSR is a system of ethical norms and values of the company, as well as consistent economic, environmental, and social measures implemented based on constant interaction with interested stakeholders, which are aimed at reducing non-financial risks, long-term improvement of the company's image, business reputation, growth of capitalization and competitiveness, profitability, and sustainable development" [3].

Corporate environmental responsibility of business appeared along with CSR and was the basis for the emergence of such an element as ESG [6]. ESG is a new philosophy of doing business in Russia. The presence of this philosophy in the company has a strong influence on the investor's decision-making because, despite all difficulties associated with the geopolitical and economic situation, investors are aware of the importance of corporate environmental and social responsibility.

References

1. Dolzhenkov A (2022) Rubles took a break. Expert. Retrieved from https://expert.ru/expert/2022/28/rubli-vzyali-pauzu/ (Accessed 29 Aug 2022)
2. Ivanitskaya EV, Buinovsky SN, Nikonorov SM, Sitkina KS (2019) Industrial safety as the main element of the sustainable development of the Russian Arctic zone. Occup Saf Ind 3:34–44. https://doi.org/10.24000/0409-2961-2019-3-34-44
3. Kostin AE (2013) Corporate responsibility and sustainable development. Moscow, Russia: Institute for Sustainable Development of the Civic Chamber of the Russian Federation and Center for Environmental Policy of Russia
4. Nikonorov SM (2021) Development of ESG—tools for financing the circular economy. In: Sintsova EA, Davydova OA, Manina MV, Savelyeva MM, Ivanova EV (eds) Development of financial relations in the circular economy: material of the national scientific and practical conference. Asterion, St. Petersburg, Russia, pp. 122–126
5. Nikonorov SM, Tyaglov SG (2022) Implementation of ESG standards and the planning of carbon neutrality until 2050 for Russian companies. In: Makarenko EN, Berdnikova SV (eds) Implementation of ESG principles in the strategy of sustainable development of the Russian economy. Publishing and printing complex of Rostov State University of Economics, Rostov-on-Don, Russia, pp 3–12
6. Nikonorov SM, Papenov KV, Talavrinov VA (2022) Business transition to ESG-strategies: innovative approaches in Russian and international experience. Strategizing: Theory Pract 2(1):49–56. https://doi.org/10.21603/2782-2435-2022-2-1-49-56
7. Nikonorov SM, Papenov KV, Krivichev AI, Sitkina KS (2019) Issues of the sustainable development measurement of the Arctic region. Moscow Univ Econ Bull 4:107–121
8. Nikonorov S, Utkina E (2021) Industrial symbiosis as an element of sustainable development of arctic companies. In: SHS web of conferences, vol 112, p 00027. https://doi.org/10.1051/shsconf/202111200027
9. Nikonorov S, Krivichev A, Sidorenko V (2021) NSR and NSTC projects: environmental and economic efficiency assessment of the goods' overseas transportation. In: SHS web of conferences, vol 112, p 00025. https://doi.org/10.1051/shsconf/202111200025
10. Obukhova E, Stolyarov A (2022) Clouds gathered over St. Petersburg. Expert. Retrieved from https://expert.ru/expert/2022/35/nad-piterom-sgustilis-tuchi/ (Accessed 29 Aug 2022)
11. Russia Matters (2022) Russia analytical report. Retrieved from https://www.russiamatters.org/news/russia-analytical-report/russia-analytical-report-aug-8-15-2022 (Accessed 16 Dec 2022)
12. Turuntsev A (2022) Banks are looking for a new SWIFT. Expert. Retrieved from https://expert.ru/expert/2022/26/banki-ischut-noviy-swift/ (Accessed 29 June 2022)

Protectionist Measures to Support RES Expansion (Spanish Experience)

Tatiana V. Kolesnikova⬡, Elina V. Maskalenko⬡, and Anatoly A. Ovodenko⬡

Abstract The paper aims to determine the significance of Spain's protectionist policy in the energy sector to ensure its technological competitiveness. The research methods are the method of comparative analysis to determine the place of Spain in the European and global energy complex; the method of dynamic analysis to determine the trend of the Spanish energy production structure; the method of critical analysis in researching the results of existing studies on the subject; the method of SWOT analysis to outline the directions of the Spanish RES industry and provide suggestions regarding necessary measures; the method of generalization and cause-effect analysis to draw conclusions. The research object is the protectionism in Spain's alternative energy as a factor ensuring its technological competitiveness. This research has proved that the Spanish renewable energy industry has significant potential, which is important in ensuring energy security and technological competitiveness. The industry is likely to develop in the face of increasing international competition if protectionist tools are applied to it. Particularly, it is necessary to ensure sufficient protection of the industry from foreign competition to strengthen it, as well as to provide state support to producers and consumers of green energy. Developing and expanding Spain's renewable energy industry by applying protectionist instruments create conditions to diversify the economy's structure and increase its economic sustainability. This is essential to make Spain more self-sufficient in terms of energy and economy.

Keywords Protectionism · Technology · Governance · Public administration · Spain · European Union · Green protectionism · Environmental protectionism · Technological transition · Energy security · Energy transition · Technological competition

JEL Classification O14 · Q43 · Q48

T. V. Kolesnikova (✉) · E. V. Maskalenko · A. A. Ovodenko
Saint-Petersburg State University of Aerospace Instrumentation, St. Petersburg, Russia
e-mail: kolesnikova-tv@mail.ru

© The Author(s), under exclusive license to Springer Nature Switzerland AG 2023
E. G. Popkova (ed.), *Smart Green Innovations in Industry 4.0 for Climate Change Risk Management*, Environmental Footprints and Eco-design of Products and Processes, https://doi.org/10.1007/978-3-031-28457-1_60

1 Introduction

In a situation when the price of traditional energy resources correlates with the unstable foreign policy situation, the development of renewable energy sources (RES) is of particular relevance for countries dependent on imported energy. It is a problem faced by the European Union (EU) countries that, despite the positive effect of using alternative energy, are not ready to completely abandon foreign energy supplies. In this regard, there is a need to look for additional protectionist tools capable of accelerating the EU's move to the replacement of traditional energy resources with renewable energy. This would ensure regional economic security in the face of exogenous factors. Spain ranks second in the EU's production of the main types of RES—solar and wind energy. This determines a special interest in the study of protectionism, which is seen as an instrument for developing the Spanish alternative energy industry.

The paper aims to determine the significance of Spain's protectionist energy policy in ensuring its technological competitiveness.

The research objectives are as follows:

- To analyze trends in the energy production structure and determine the technological potential of Spain's renewable energy industry for the decarbonization of the economy;
- To identify key strategic areas for the development of Spain's renewable energy industry by assessing the strengths and weaknesses of the industry, as well as the impact of exogenous factors that shape the opportunities and threats for the development of the industry;
- To identify the importance of protectionist measures in Spain's public administration that are aimed at protecting and developing the alternative energy industry in the domestic market and increasing its technological competitiveness in the international market;
- To evaluate the effectiveness of the proposed measures by illustrating the ongoing projects in the field of alternative energy development in Spain.

2 Methodology

Over the past five years, the interest in RES development research has significantly grown in different fields of science. In ecology, it is connected with global warming and the impact of an anthropogenic factor on the environment. In the energy sector, it is in line with the need to diversify energy resources and power. In economy, it concerns the feasibility study of the transition to alternative resources and the economic efficiency of producing new types of energy [1]. In public administration, first, it is about the reconciliation of interests among the state, business, and society in implementing the green transition. Second, it is about the state strategies in the environmental, energetic, and economic spheres. Third, it is related to the mechanisms and methods of support and promotion of the renewable energy industry.

This research is based on publications dedicated to Spain's technological possibilities in the field of renewable energy and the development and application of different types of energy to ensure decarbonization within the EU.

The analysis of the RES technological potential has shown that the Spanish economy has all resources necessary to implement decarbonization and the transition of households and industries to alternative energy sources [2].

Households are among the main consumers of energy in Spain. The decarbonization of residential heating systems is an important part of Spain's energy transition. The combination of political instruments, together with raising public awareness of environmental issues, leads to reductions in energy bills and costs [3].

Being widespread in the south of Spain, hybrid power systems based on wind and solar energy have the lowest cost of electricity generation because they do not require large storage volumes, compensating for each other in various weather conditions [4].

Biomass and solar energy can be used to generate electricity in the coal mining areas of Spain and become an alternative to coal. The implementation of combined energy systems based on biomass and solar energy will accelerate the energy transition of the EU countries producing coal [5].

Spain, along with China, has significant export capacity in wind power, which has allowed the countries to build and realize the export potential in related industries, such as adopting artificial intelligence in photovoltaics [6]. This has enabled these countries to become leaders in the green technological transition [7].

The development of hydrogen energy in Spain has high potential. The comparative research in vehicles powered on electric and fuel energy has shown that the use of hydrogen-based hybrid systems can provide a flexible costing system and be an alternative in a decarbonized environment [8].

Recent research pays much attention to the issue of providing Spain's domestic market with alternative energy. It has been scientifically proven that, given a properly developed energy policy, tax regime, and citizens' awareness, the country is capable of ensuring a quick transition to a green economy [9]. Nevertheless, the issue of the application of protectionist tools in the development of Spain's technological potential has not been sufficiently studied. In this regard, this research focuses on the evaluation of Spain's green transition under the ongoing trend of a changing energy production structure. A SWOT analysis has been applied to identify perspective solutions in developing alternative energy in Spain. The importance of applying protectionist measures in developing the renewable energy industry to ensure its technological competitiveness has been highlighted.

3   Results

Spain is actively moving forward to transforming its energy policy and replacing traditional energy with alternative energy. According to the European Association for the cooperation of transmission system operators for electricity (ENTSO-E), in

2021, Spain became the second country in Europe in terms of electricity production (with Germany in the first place), producing almost 47% of all energy produced in the EU. Figure 1 shows the changes in the energy structure in Spain's total electricity production.

According to Fig. 1, the share of renewables in Spain's energy balance has been increasing significantly since 2019 despite the influence of exogenous and endogenous factors. At the current growth rate, Spain is projected to reach zero total pollution by 2031. Figure 2 shows Spain's projected electricity production structure from 2019 to 2031.

Spain will be significantly ahead of other EU countries, all of which must make a complete transition to alternative energy by 2050 as stipulated in the "European Green Deal" adopted on December 11, 2019 [11]. With Spain's existing potential, this will bring the country to a new level of technological competition and solve some macroeconomic problems such as structural unemployment.

A SWOT analysis of the Spanish renewable energy industry has been conducted for the purpose of this research. The analysis allowed us to identify the key areas of

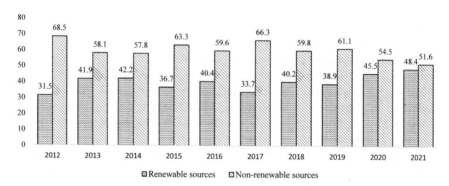

Fig. 1 Electricity production structure in Spain from 2012 to 2021, in %. *Source* Compiled by the authors based on Statista [10]

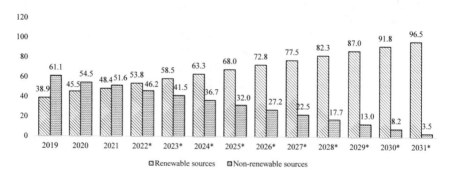

Fig. 2 Forecast of electricity production structure in Spain from 2021 to 2031, in %. *Source* Calculated and built by the authors

development strategy and highlight the importance of protectionism in implementing the energy strategy (Table 1).

The SWOT analysis of the Spanish renewable energy industry suggests that the opportunities and threats to the development of alternative energy are related to two exogenous factors. The first factor is Spain's participation in the EU. The second

Table 1 SWOT analysis of the Spanish renewable energy industry

	Opportunities (O)	Threats (T)
	Participation in the development and coordination of EU energy policy	Uncoordinated policies of the national and supranational bodies within the EU
	The use of EU integration opportunities to access the markets of other member states	High dependence of the Spanish market on exogenous factors
	Low technological competition in the renewable energy industry	Tougher technological competition in the renewable energy industry
Strengths (S)	*Measures (S–O)*	*Measures (S-T)*
Interstate contacts in the EU dialogue	Joint RES projects with other EU countries	Expansion of bilateral cooperation with EU countries
Geographical location allowing for high alternative energy generation from solar and wind power		
High level of protectionism in the Spanish renewable energy industry	Expanding state protectionist tools for the renewable energy industry to strengthen its competitiveness in the global market	Using trade barriers to reduce foreign technological competition in the RES industry
Spain is one of the leaders in the development of hydrogen energy	Gaining a foothold and increasing the share of the global hydrogen energy market	Setting trade barriers in Hydrogen Energy Development
High share of alternative sources in other industries and projects	Increasing EU alternative energy market share	
Participation in international energy projects (HyDealAmbition)		
Weaknesses (W)	*Measures (W–O)*	*Measures (W-T)*
High level of unemployment in the country (15.3% in 2021)	Creating new workplaces and reducing inflation through the intensification of RES production	Applying protectionist tools to support domestic production and fend off foreign competition to create jobs and reduce inflation
High inflation rate in the country (10.8% in 2022)		

Source Developed by the authors

factor is a rapidly growing, though not yet high, level of technological competition in this sector. Spain's strengths stem from its geographical location, its high level of inclusion in the international economy, and the significant achievements that it has already gained in the field of alternative energy. The industry's weaknesses are generally related to the macroeconomic problems in Spain. The combination of strengths and weaknesses predetermines the priority of RES in the economic and energy-related development strategy of Spain. Alternative energy can potentially increase competitiveness within the EU and the global market in a technologically competitive environment. The realization of the potential will create new jobs, which will help to reduce unemployment and inflation.

To implement this strategy, it is necessary to apply protectionist tools that can be seen in two forms [12, 13]: first, as the state support of the Spanish RES industry and, second, as protection from technological competition coming from foreign companies.

Protectionism, which supports the development of the renewable energy industry in Spain, is seen in raising quality requirements for greenhouse gas emitted from industrial production sites, transport, and households into the environment. Emissions trading associated with it leads to emissions reduction and provides a better position for companies that develop alternative energy rather than conventional energy. Concerning the issue of reducing foreign competition, the position of national production can be regarded as preferential due to state support [14]. In particular, it includes certain funding projects and providing tax incentives and preferential conditions for production implementation.

The Spanish government pays great attention to developing renewable energy sources and the transition to green transport, allocating funds from the state budget and attracting a new flow of investment. For instance, "Strategic Project for the Recovery and Economic Transformation of Renewable Energy Sources, Renewable Hydrogen and Its Storage" involves more than 3.5 billion euros of public investment, 5.3 billion euros of private investment, and support for R&D related to the development and implementation of electric vehicles [15].

Strong investment support for R&D is accelerating the transformation of the global energy system towards renewable sustainability and hence decarbonization [16]. Europe accounted for 28% of total investment in 2021, with more than half in Germany, France, the UK, and Spain [17]. In recent research, it has been confirmed that the level of clean energy investments will prevail in those OECD countries that are more vigorously implementing their green projects to combat climate change [18].

In this regard, sufficient support for developing clean hydrogen energy is particularly important. In 2020, the EU adopted the "Hydrogen Strategy," the priority of which is the development of pure hydrogen to reduce the use of conventional energy sources. Energy production with the use of pure hydrogen, obtained by electrolysis, is one of the promising directions in alternative energy.

The report "Ten-Year Network Development Plans" (TYNDPs) submitted to the European Commission confirms that one of the scenarios of transition to carbon neutrality, which would contribute to the energy security of the EU, could be the development of the use of electricity production technology by electrolysis. More-over, if operated on a stand-alone basis, the output could reach 30% of capacity by 2050. However, such production is directly dependent on the availability of wind and solar power [19]. In this regard, Spain has enough competitive advantages to implement the scenario with the appropriate use of RES.

Spain is included in the HyDealAmbition project, which brings together 30 companies covering the entire production chain from extraction and processing to financing clean hydrogen. It is based on the idea of a competitive fuel, which will be presented on the market as a mass product along with the products of the oil and gas industry. The first phase of the HyDealSpain project will involve a large industrial complex in Asturias, Spain, built to produce hydrogen by electrolysis at a competitive price compared to other alternative sources. The facility's construction is scheduled for 2025, with the project to be launched in 2030. This will result in an expected generation capacity of 9.5 GW and a consumption capacity of 7.4 GW.

The Spanish government is interested in the development and promotion of projects related to the extraction, storage, and use of pure hydrogen for industrial and economic purposes. In this regard, on December 22, 2021, the Incentive Program for Innovative and Unique Renewable Hydrogen Projects was approved, with 150 million euros allocated in its first phase [20].

One of the priorities in developing the competitiveness of Spain's energy sector will be a project connecting Spain, Portugal, and France with the rest of the EU energy network by pipeline. The Green Corridor plan was proposed by the European Commission and supported by the heads of government at the European Council meeting on October 20, 2022 [21]. It is part of a larger Hydrogen Highway project that will link 31 European Transmission System Operators (TSO) to develop a network of hydrogen corridors between North Africa and Southern Europe, Southwest Europe and North Africa, Scandinavian countries and the Baltic region, Eastern and South-Eastern Europe, and the North Sea [22].

Spain also witnesses an infrastructural transition of economic entities to the use of hydrogen fuel. With the Spanish government funding this direction, green hydrogen plants are beginning to appear. For example, the Catalina project aims to produce green hydrogen by using combined wind and solar energy installations. ViamedSan-José Hospital in Alcantarilla (Murcia, Spain) will be the first healthcare facility that will be able to produce and consume oxygen and hydrogen through electrolysis with the help of RES. This will make it possible to provide the hospital with electricity and use hydrogen for medical purposes. A hydrogen-powered train equipped with a hybridized fuel cell and battery system is also being tested. This emission-free vehicle concept involves an autonomous operation on sections without an electricity grid. The project, worth 14 million euros, is financed by the Clean Hydrogen Partnership and the European Commission's agency for developing and promoting hydrogen and fuel cells.

4 Conclusion

The research identifies the high technological potential of the alternative energy industry in Spain. Realization of this potential at the existing pace will allow Spain to achieve decarbonization of the economy faster than in other EU countries. A set of competitive advantages of the industry is associated with the state support of the industry, manifested in funding R&D projects, introducing them into production, and creating preferential conditions for producers and consumers of green energy products. Increasing international competition in solar, wind, and hydrogen energy predetermines the need to strengthen the use of protectionist tools in Spain. Diversification of mechanisms and methods of state protection and support of the industry will expand markets for electricity and RES industry equipment in Spain. Ultimately, the strengthening of the industry whose products will be in demand in foreign markets will lead to the diversification of Spain's GDP structure, creating additional jobs, reducing unemployment and inflation, and increasing tax revenues for the country's budget. Thus, the stability of the Spanish economy and its competitiveness in the world can be improved.

References

1. Stepnov IM, Kovalchuk JA, Gorchakova EA (2019) On assessing the efficiency of intracluster interaction for industrial enterprises. Stud Russ Econ Dev 30:346–354. https://doi.org/10.1134/S107570071903016X
2. Lisbona P, Frate GF, Bailera M, Desideri U (2018) Power-to-Gas: analysis of potential decarbonization of Spanish electrical system in long-term prospective. Energy 159:656–658. https://doi.org/10.1016/j.energy.2018.06.115
3. López-Bernabé E, Linares P, Galarraga I (2022) Energy-efficiency policies for decarbonising residential heating in Spain: a fuzzy cognitive mapping approach. Energy Policy 171:113211. https://doi.org/10.1016/j.enpol.2022.113211
4. Hwang H, Kim S, García ÁG, Kim J (2020) Global sensitivity analysis for assessing the economic feasibility of renewable energy systems for an off-grid electrified city. Energy 216:119218. https://doi.org/10.1016/j.energy.2020.119218
5. Paredes-Sánchez BM, Paredes-Sánchez JP, García-Nieto PJ (2022) Evaluation of implementation of biomass and solar resources by energy systems in the coal-mining areas of Spain. Energies 15(1):232. https://doi.org/10.3390/en15010232
6. Scheifele F, Bräuning M, Probst B (2022) The impact of local content requirements on the development of export competitiveness in solar and wind technologies. Renew Sustain Energy Rev 168:112831. https://doi.org/10.1016/j.rser.2022.112831
7. Peshkova GY, Bondar EG (2022) Energy cooperation between Russia and China in a low-carbon economy. Actual Dir Sci Res XXI Century: Theory Pract 10(1):180–190. https://doi.org/10.34220/2308-8877-2022-10-1-180-190
8. Villar J, Olavarría B, Doménech S, Campos FA (2020) Costs impact of a transition to hydrogen-fueled vehicles on the Spanish power sector. Utilities Policy 66:101100. https://doi.org/10.1016/j.jup.2020.101100
9. Telegina EA, Khalova GO (2022) Geoeconomic and geopolitical challenges of energy transition. Implications for world economy. World Econ Int Relat 66(6):26–34. https://doi.org/10.20542/0131-2227-2022-66-6-26-34

10. Statista (2022) Percentage distribution of electricity generation in Spain in 2021, by type. Retrieved from https://es.statista.com/estadisticas/993747/porcentaje-de-la-produccion-de-energia-electrica-por-fuentes-energeticas-en-espana/ (Accessed 30 Oct 2022)
11. European Commission (2019) The European green deal. Communication from the commission to the European Parliament, the European Council, the Council, the European Economic and Social Committee and the Committee of the Regions (December 11, 2019COM(2019) 640 final). Brussels, Belgium. Retrieved from https://ec.europa.eu/info/sites/default/files/european-green-deal-communication_en.pdf (Accessed 30 Oct 2022)
12. Afontsev SA (2020) Politics and economics of trade wars. Zhournal Novoi Ekonomicheskoi Associacii [J New Econ Assoc] 45(1):193–198. https://doi.org/10.31737/2221-2264-2020-45-1-9
13. Fajgelbaum PD, Goldberg PK, Kennedy PJ, Khandelwal AK (2020) The return to protectionism. Q J Econ 1(135):1–55. https://doi.org/10.1093/qje/qjz036
14. Kovalchuk JA, Stepnov IM, Melnik MV, Petrovic T (2022) Public goals and government expenditures: are the solutions of the "modern monetary theory" realistic? Finan: Theory Pract 26(3):6–18. https://doi.org/10.26794/2587-5671-2022-26-3-6-18
15. Government of Spain (2021) Plan de recuperación, transformación y resiliencia [PERTE] de energías renovables, hidrógeno renovable y almacenamiento [Recovery, Transformation and Resilience Plan [PERTE] of renewable energies, renewable hydrogen and storage]. Retrieved from https://www.lamoncloa.gob.es/presidente/actividades/Documents/2021/151221-Informe-pertes-renovables-2021.pdf (Accessed 30 Oct 2022)
16. Churchill SA, Inekwe J, Ivanovski K (2021) R&D expenditure and energy consumption in OECD nations. Energy Econ 100:105376. https://doi.org/10.1016/j.eneco.2021.105376
17. International Renewable Energy Agency (IRENA) (2022) World energy transitions outlook 2022: 1.5°C Pathway. Abu Dhabi, UAE: IRENA. Retrieved from https://irena.org/publications/2022/mar/world-energy-transitions-outlook-2022 (Accessed 30 Oct 2022)
18. Chen X, Fu Q, Chang C-P (2021) What are the shocks of climate change on clean energy investment: a diversified exploration. Energy Econ 95:105136. https://doi.org/10.1016/j.eneco.2021.105136
19. ENTSOG and ENTSO-E (2022) TYNDP 2022: scenario report. Retrieved from https://2022.entsos-tyndp-scenarios.eu/wp-content/uploads/2022/04/TYNDP2022_Joint_Scenario_Full-Report-April-2022.pdf (Accessed 30 Oct 2022)
20. Instituto para la Diversificación y Ahorro de la Energía (IDAE) (2021) Programa H2 Pioneros. Ayudas Para Proyectos Pioneros y Singulares de Hidrógeno Renovable [H2 pioneers program. Grants for pioneering and singular renewable hydrogen projects]. Retrieved from https://www.idae.es/ayudas-y-financiacion/programa-h2-pioneros-ayudas-para-proyectos-pioneros-y-singulares-de-hidrogeno (Accessed 30 Oct 2022)
21. La Moncloa (2022) Declaraciones del presidente del gobierno a su llegada al Consejo Europeo [Statements by the president of the government upon arrival at the European Council]. Retrieved from https://www.lamoncloa.gob.es/presidente/intervenciones/Paginas/2022/prsp20102022.aspx (Accessed 30 Oct 2022)
22. ENAGAS (2021) Plan Estratégico 2022–2030. Reliable energy for a decarbonized future: a 2030 strategy for a new stage in Europe. Retrieved from https://www.enagas.es/content/dam/enagas/es/ficheros/accionistas-e-inversores/informacion-economico-financiera/informes-resultados-presentaciones/informacion-publica-periodica/Plan%20Estrategico%202022-2030_vDef_ES.pdf (Accessed 30 Oct 2022)

Global Green Bond Market Amid Global Turbulence

Olga V. Khmyz⊙, Daria R. Pastukhova, and Anna A. Prudnikova

Abstract Amid the recent turbulence in the global economy, it seems very important to continue following the global environmental course. The COVID-19 pandemic, disruptions and overhaul of global supply chains, geopolitical tensions, and energy inflation of 2022 are holding back the progressive development of the world economy, which needs financial resources to continue the energy transition and other sustainable transformations. Green bonds attracted significant funds even before the pandemic. Thus, it seems timely to assess the impact of the main factors of global turbulence on the global green bond market and, based on the available data, predict the probable direction of the development of the global green bond market. The methods of analysis include general scientific, statistical, and econometric methods based on data from the Climate Bond Initiative, the World Bank, and S&P Global— power trend, logarithm, and the least squares method. The analysis showed that the challenges of adapting green bonds to the new global environment are out of the question. During COVID-19, there was a large-scale increase in issues and interest in green bonds caused by government incentives and the global financial market trends during the pandemic. In 2022, additional force majeure and more fundamental factors, such as global inflation and changes in the direction of monetary regulation frameworks in developed countries, have been added to the need to fight global warming. They had the opposite effect. However, econometric modeling shows an upward trend for the global green bond market, at least in the short term.

Keywords Green bonds · Sustainable development goals · Sustainable global market · Global green bond market · Global economic turbulence

JEL Classification G12 · C12

O. V. Khmyz (✉) · D. R. Pastukhova
MGIMO University, Moscow, Russia
e-mail: khmyz@mail.ru

A. A. Prudnikova
Financial University Under the Government of the Russian Federation, Moscow, Russia
e-mail: AAPrudnikova@fa.ru

© The Author(s), under exclusive license to Springer Nature Switzerland AG 2023
E. G. Popkova (ed.), *Smart Green Innovations in Industry 4.0 for Climate Change Risk Management*, Environmental Footprints and Eco-design of Products and Processes, https://doi.org/10.1007/978-3-031-28457-1_61

1 Introduction

Even in the pre-pandemic years, green bonds proved their advantages and promise as a tool for financing the fight against climate deterioration [1–4]. Therefore, governments, corporations, and the financial community are now increasingly considering green bonds to support a sustainable and environmentally sound recovery from the pandemic [5–7].

In the middle of the second and beginning of the third decade of the new century, an increasing number of investors become responsible, making green bonds an attractive investment vehicle. One of the most important investors in this market is the International Finance Corporation, which seeks to support new issuers of green bonds and prepare them for subsequent issues. Mutual funds and other types of investment funds that buy green bonds are also important players; their number is steadily growing. Pension funds seeking to minimize risks and support environmental initiatives are also active investors in the green bond market.

However, there are several problems in the green bond market, which may negatively affect its further development. Currently, the existing rules and standards for issuing green bonds are predominantly voluntary, which may potentially discourage investors seeking to minimize risks and direct money to invest in truly green projects. This also leads to the problem of greenwashing, when companies unreasonably declare their commitment to green principles, initiatives, and standards. Additionally, the issuance of green bonds is quite expensive, partly reducing the attractiveness of these securities for issuers.

Nevertheless, the world community, concerned about the existence of the above problems, is gradually taking steps towards their solution because the global Green Deal is based on sustainable financing, one of the leading elements of which are green bonds. Green bonds correlate with the global green trend. Since green bonds are used to improve the environment, they also affect economic growth. Therefore, it is important to analyze the impact of global factors of instability on the global green bond market and predict the trend of its development.

During the COVID-19 period, government incentives (fiscal and financial, which began even before the COVID-19 pandemic) boosted the green bond market; securities of pharmaceutical and other companies with a green orientation were progressively growing in price, which was due to the need to increase the production of medical and complementary goods, while shares and other securities of companies from different sectors of the economy were rapidly depreciating.

2 Methodology

The green bond segment is rapidly developing in the global financial market. These securities occupy an increasing share of the total global debt market, attracting ESG-oriented investors and conventional investors who have not greened their portfolios before [8, 9]. Issuers, investors, and investment projects are becoming more diversified. The physical risks of climate change are becoming more visible in the form of extreme and changing weather patterns. Simultaneously, the effects of the COVID-19 pandemic continue to cause economic disruption [10]. Under these circumstances, it is vital that an increasing percentage of government and corporate bonds be labeled green.

Several factors could drive the growth of the green bond market in the future.

First, increased implementation of regulation related to sustainable finance is a factor that will positively influence growth. Governments are gradually making more commitments to improve the environmental situation. Currently, 70 countries, which account for two-thirds of global carbon emissions, have zero emission targets to be reached by 2050 [11]. New targets for biodiversity, reduced coal use, and reduced methane emissions were introduced at COP26. The amount of clean energy spending in the governments' economic recovery plans is increasing. These initiatives should eventually be reflected in changes to the regulatory framework.

Second, the European Union's green bond program as part of the Green Deal [12] will significantly affect the market. The program was created as a tool to support Europe's recovery from the COVID-19 pandemic and provide further funding for green projects and the transition to a sustainable economy. By the end of March 2022, 14.5 billion euros had been raised through the issuance of green bonds [13]. By the end of 2022, the issuance is expected to exceed 50–75 billion euros [14]. Europe is expected to remain a key region driving the development and growth of the green bond market worldwide.

Third, more participants from developing countries will enter the global green bond market. One of the main problems of the global environment is that developing countries account for about two-thirds of energy-related carbon dioxide emissions. However, this group of countries does not have the necessary finance or innovation base to invest or invent their own way to a cleaner energy system. However, a few emerging market countries (including Indonesia, Egypt, and Chile) have stepped up efforts to issue green, social, and sustainable bonds, which indicates a high probability of further market development in emerging countries in the near future [15]. For example, in Chile, Colombia, the Dominican Republic, Costa Rica, and other countries, stock exchanges have published guidelines for issuing green bonds. Regulators have contributed to increasing the transparency of this process. In the Asia–Pacific region, demand for green bonds outstrips supply, which could lead to greater issuance in the region in the second quarter of 2022 after some issuance projects were postponed at the end of the first quarter due to adverse market conditions [16].

Fourth, pressure on politicians will accelerate green growth. Thus, the Bank of England announced a change in its mandate to include environmental sustainability issues and the transition to zero energy consumption from fossil sources. By the end of 2021, green bonds were included in its bond-buying program. The European Central Bank takes further steps to integrate climate risks by seeking to include climate specialists in its team.

Fifth, the global energy crisis of 2022, caused by rising energy prices, which has hurt households, businesses, and economies worldwide, could be another growth driver for the green bond market. Governments, companies, and investors should take much more action and provide the world with more affordable and clean energy as quickly as possible. According to the roadmap to achieve zero emissions by 2050 [17] published by the International Energy Agency, reducing global demand for fossil fuels is possible only if there is a massive increase in investment in renewable energy sources, energy efficiency, and other green energy technologies.

Sixth, the need to implement the Paris Agreement [18], aimed at fighting climate change and its negative consequences, currently joined by 192 countries and the European Union, will also significantly affect the further growth of the green bond market.

Additionally, the appetite of institutional investors for green bonds can be expected to grow, considering increased attention to climate risks and opportunities in investment portfolios. Moreover, the number of private investors and responsible qualified investors is steadily growing. For responsible qualified investors, it is important that the provided funds are directed to environmentally friendly projects and used without harm to the environment. Companies that use green bonds for financing also benefit from an improved market image.

However, in 2022, the growth rate of the green bond market will be lower than observed a year earlier; issuance volumes will remain at about the same level. This is due to the negative impact of the current geopolitical situation, soaring inflation, monetary tightening, and the expectation of a further increase in rates after a period of unprecedented easing by many central banks to support the economy, which discourage issuers from issuing green bonds due to rising financing costs, as well as creates uncertainty for investors.

According to the forecasts of the international rating agency Moody's, the issue of green bonds will be at the level of $550 billion in 2022 [19], which is significantly lower than the previous estimate of $750 billion.

3 Results

Since the green bond market has been growing rapidly in recent years, this paper will use a power trend to model the time series trend. To find the trend equation for the green bond market and use the least squares method, the equation should be taken in natural base logarithms and linearized. Additionally, it is necessary to calculate the deviations of the calculated values from the actual observations.

The analysis of the initial data from the Climate Bond Initiative and the World Bank on the value of green bond issuance for ten years (2012–2021) and the calculation of the power trend equation provided the following intermediate results (Table 1).

Substituting the obtained data into the derived equations, we find the following:

Table 1 Estimation of the parameters of the power trend equation for green bond issuance

t	y_i	$\ln(t_i)$	$\ln(y_i)$	$\ln(t_i)^2$	$\ln(t_i)\ln(y_i)$	y_p	$E(t)$
1	2.6	0.0	1.0	0.0	0.0	2.6	0.0
2	11.5	0.7	2.4	0.5	1.7	12.0	−0.5
3	37.0	1.1	3.6	1.2	4.0	29.6	7.4
4	46.2	1.4	3.8	1.9	5.3	56.2	−10.0
5	84.5	1.6	4.4	2.6	7.1	92.3	−7.8
6	159.5	1.8	5.1	3.2	9.1	138.4	21.1
7	172.4	1.9	5.1	3.8	10.0	195.0	−22.6
8	269.4	2.1	5.6	4.3	11.6	262.5	6.9
9	298.1	2.2	5.7	4.8	12.5	341.1	−43.0
10	517.4	2.3	6.2	5.3	14.4	431.1	86.3
55	1598.6	15.1	43.0	27.7	75.8	1560.8	37.8

Source Calculated by the authors

$$b = \frac{10 \cdot 75.8 - 15.1 \cdot 43.0}{10 \cdot 27.7 - 15.1^2} = 2.224,$$

$$\ln a = \frac{1}{10} \cdot 43.0 - \frac{1}{10} \cdot 2.224 \cdot 15.1 = 0.946.$$

By potentiating, it turns out that $a = 2.574$. Thus, the power trend equation has the following form:

$$y = 2.574t^{2.224}.$$

To assess the reflection by this model of the regularity of change in the studied indicator, it is necessary to evaluate its quality using the determination coefficient:

$$R^2 = 1 - \frac{\sum (y_i - y_p)^2}{\sum (y_i - \overline{y_i})^2}.$$

The results of the calculation are presented in Table 2.
Using the values obtained in Table 2, we can find the coefficient of determination

$$R^2 = 1 - \frac{10516.1}{239535.2} = 0.956.$$

Since this coefficient is close to one, this model has a high level of model accuracy.
To assess the accuracy of this model, we calculate the average relative approximation error, which reflects the average relative deviation of the calculated values from the actual ones:

Table 2 Calculation results

t	y_i	y_p	$(y_i - y_p)^2$	$(y_i - \bar{y}_i)^2$
1	2.6	2.6	0.0	24,730.7
2	11.5	12.0	0.3	22,010.7
3	37.0	29.6	54.8	15,094.6
4	46.2	56.2	100.0	12,918.6
5	84.5	92.3	60.8	5679.1
6	159.5	138.4	445.2	0.1
7	172.4	195.0	510.8	157.3
8	269.4	262.5	47.6	11,999.0
9	298.1	341.1	1849.0	19,110.3
10	517.4	431.1	7447.7	127,834.9
55	1598.6	1560.8	10,516.1	239,535.2

Source Calculated by the authors

$$E_{rel} = \frac{1}{n} \sum_{i=1}^{n} \frac{|E_t|}{y_i} \cdot 100\%,$$

$$E_{rel} = \frac{1}{10} \cdot \frac{205.6}{1598.6} \cdot 100\% = 1.28\%.$$

The allowable approximation error should not exceed 10%. In this case, the allowable approximation is equal to 1.28%. Thus, the regression equation can be considered satisfactory and can be used as a trend.

A point forecast for k steps ahead is obtained by substituting the parameter into the model $t = N + 1, \ldots, N + k$. When predicting two steps:

$$Y_p(11) = 2.574 \cdot 11^{2.224} = 532.9 (k = 1, t = 11),$$
$$Y_p(12) = 2.574 \cdot 12^{2.224} = 646.7 (k = 2, t = 12).$$

The predicted confidence interval has the following limits:

- Upper prediction limit: $Y_p(N + k) + U(k)$;
- Lower prediction limit: $Y_p(N + k) - U(k)$.

The U(k) has the following form:

$$U(k) = S_e \cdot t_a \sqrt{1 + \frac{1}{n} + \frac{(n + k - t_{av})^2}{\sum (t - t_{av})^2}}, \tag{1}$$

where

k   lead time;

t_a tabular value of the two-tailed Student's test for significance level a = 0.1 and
for the number of degrees of freedom equal to n−2;

S_e standard deviation from the trend line: $S_e = \sqrt{\frac{10508.1}{8}} = 36.2$.

$$\text{If } k = 1, U(k) = 36.2 \cdot 1.86\sqrt{1 + \frac{1}{10} + \frac{(10 + 1 - 5.5)^2}{404.3}} = 71.0.$$

$$\text{If } k = 2, U(k) = 36.2 \cdot 1.86\sqrt{1 + \frac{1}{10} + \frac{(10 + 2 - 5.5)^2}{404.3}} = 71.1.$$

As a result, calculations of predictive estimates based on the confidence interval for time t = 11 with step k = 1 means $Y_p(t) = 532.9$, the lower limit is 461.9 and the upper limit is 603.9. For time t = 12 $Y_p(t) = 646.7$, the lower limit is 575.6, and the upper limit is 717.8.

Thus, based on the constructed econometric model of the time series trend, it was revealed that the dynamics of green bond issuance are characterized by the equation $y = 2574t^{2.224}$. Considering the adequacy of this model, it can be argued that while maintaining the established patterns of development, the predicted value will fall into the interval formed by the lower and upper limits with a probability of 90%. Thus, as a result of a point forecast, it was revealed that the issue of green bonds could reach a value of $532.9 billion in 2022. As a result of building a confidence interval of the forecast, it can be argued with a high probability that the issue will be in the range of $461.9 to $603.9 billion. In 2023, the point forecast is $646.7 billion, and the forecast interval varies from $575.6 to $717.8 billion.

However, in 2022, the market is facing unprecedented pressure and a fairly large number of constraints, which is why its growth rate is declining. Nevertheless, due to the existing environmental challenges and threats and the need to address them, there is huge potential for developing the green bond market, which plays an important role in meeting internationally agreed climate standards.

4 Conclusion

The research showed that despite the COVID-19 pandemic and new geopolitical factors, the outlook for the global green bond market seems to be quite favorable. Nevertheless, there are some problems, including the following:

• Problems of national regulation of green bonds issuance and circulation and stimulation of green financing;
• Uneven implementation of the energy transition in the EU countries in recent years due to energy inflation and technical problems in the use of alternative energy capacities;

- Possible increase in the cost of financing due to global monetary changes.

However, several circumstances signal a further expansion of the green bond market. First, this is the active implementation of regulation and the commitment of governments to improve the environment. Second, the impact of the new EU program. Third, the active involvement of developing countries in the issuance of green bonds. Fourth, further pressure on the authorities in the field of ecology. Fifth, rising energy prices. Additionally, the need to implement the Paris Agreement and further interest from investors and issuers will also have an impact. Therefore, green bonds have provided and will continue to provide an opportunity to address the challenges directly related to the spread and impact of the COVID-19 pandemic, as well as to achieve planetary sustainable development goals.

Thus, while the global green bond market is still relatively small and investors' understanding of green investment is still limited, with the increasing importance of the fossil energy issue, the potential for further growth in the green bond market is huge. A number of measures can be taken to stimulate the growth of the green bond market. It is essential to enhance the market mechanism to improve the convenience of issuing and investing in green bonds. Directive decisions by the relevant authorities can help boost the supply of green bonds and formulate support policies that promote the development of the renewable energy sector. The providers of public capital can mitigate the risks of green assets and provide seed money, demo issues, and capacity building to support green bonds. Institutional investors can help by aligning their own capabilities and investment goals with long-term sustainability mandates.

However, growth in the green bond market slowed down in 2022 due to the current unfavorable geopolitical environment and monetary policy tightening. Nevertheless, the market expansion trend continues. There is significant potential for further development, which has been proven econometrically.

When modeling the time series, a power trend was used since the green bond market has been characterized by a sharp increase in recent years. While maintaining current development trends, it is highly probable that the volume of green bond issuance will amount to \$532.9 billion in 2022 and \$646.7 billion in 2023. Additionally, based on the confidence interval of the forecast, it can be asserted with a 90% probability that the issue will be in the range from \$461.9 to \$603.9 billion in 2022 and from \$575.6 to \$717.8 billion in 2023.

Therefore, the outlook for the global green bond market appears to be quite favorable despite the 2020–2021 pandemic and the energy crisis of early 2022.

References

1. Foulis P (2021) Energy investment needs to increase—so bills and taxes must rise. The Economist. Retrieved from https://www.economist.com/the-world-ahead/2021/11/08/energy-investment-needs-to-increase-so-bills-and-taxes-must-rise (Accessed 11 Nov 2022)
2. International Energy Agency (IEA) (2021) Net Zero by 2050: a roadmap for the global energy sector. Retrieved from https://www.iea.org/reports/net-zero-by-2050 (Accessed 15 Nov 2022)

3. Liu M (2022) The driving forces of green bond market volatility and the response of the market to the COVID-19 pandemic. Econ Anal Policy 75:288–309. https://doi.org/10.1016/j.eap.2022.05.012

4. Pego A (2023) Climate change, world consequences, and the sustainable development goals for 2030. IGI Global, Hershey, PA. https://doi.org/10.4018/978-1-6684-4829-8

5. Adekoya OB, Oliyide JA (2021) How COVID-19 drives connectedness among commodity and financial markets: evidence from TVP-VAR and causality-in-quantiles techniques. Resour Policy 70:101898. https://doi.org/10.1016/j.resourpol.2020.101898

6. Kreuzer C, Priberny C (2022) To green or not to green: the influence of board characteristics on carbon emissions. Financ Res Lett 49:103077. https://doi.org/10.1016/j.frl.2022.103077

7. Piñeiro-Chousa J, López-Cabarcos MÁ, Caby J, Šević A (2021) The influence of investor sentiment on the green bond market. Technol Forecast Soc Chang 162:120351. https://doi.org/10.1016/j.techfore.2020.120351

8. Cui T, Suleman MT, Zhang H (2022) Do the green bonds overreact to the COVID-19 pandemic? Financ Res Lett 49:103095. https://doi.org/10.1016/j.frl.2022.103095

9. Khmyz OV, Oross TG, Prudnikova AA (2023) Factors of attractiveness of green bonds as a financing tool for countering adverse climate change. In: Popkova E, Sergi B (eds) Current problems of the global environmental economy under the Conditions of climate change and the perspectives of sustainable development. Adv Glob Change Res, vol 73. https://doi.org/10.1007/978-3-031-19979-0_4. Springer, Cham, Switzerland

10. Altig D, Baker S, Barrero JM, Bloom N, Bunn P, Chen S, Davis SJ, Leather J, Meyer B, Mihaylov E, Mizen P (2020) Economic uncertainty before and during the COVID-19 pandemic. J Public Econ 191:104274. https://doi.org/10.1016/j.jpubeco.2020.104274

11. European Commission (2022) NextGenerationEU: commission launches a tool to provide details on the use of green bond proceeds. Retrieved from https://ec.europa.eu/info/news/nextgenerationeu-commission-launches-tool-provide-details-use-green-bond-proceeds-2022-mar-31_en (Accessed 11 Nov 2022)

12. European Commission (2019) A European green deal: striving to be the first climate-neutral continent. Retrieved from https://ec.europa.eu/info/strategy/priorities-2019-2024/european-green-deal_en (Accessed 11 Nov 2022)

13. Naeem MA, Farid S, Ferrer R, Shahzad SJH (2021) Comparative efficiency of green and conventional bonds pre- and during COVID-19: an asymmetric multifractal detrended fluctuation analysis. Energy Policy 153:112285. https://doi.org/10.1016/j.enpol.2021.112285

14. Bos B (2021). Buoyant green bond outlook for 2022 driven by EU issuance and initiatives. NN Investment Partners. Retrieved from https://www.nnip.com/en-INT/professional/insights/articles/buoyant-green-bond-outlook-for-2022-driven-by-eu-issuance-and-initiatives (Accessed 11 Nov 2022)

15. Argandoña B, Carolina L, Rambaud SC, López PJ (2022) The impact of sustainable bond issuances in the economic growth of the Latin American and Caribbean countries. Sustainability 14(8):4693. https://doi.org/10.3390/su14084693

16. Isjwara R, Tanchico E, Ahmad R (2022) Global green bond issuance slows amid rising interest rates, inflation. S&P Global Market Intelligence. Retrieved from https://www.spglobal.com/marketintelligence/en/news-insights/latest-news-headlines/global-green-bond-issuance-slows-amid-rising-interest-rates-inflation-69914070 (Accessed 11 Nov 2022)

17. Guo D, Zhou P (2021) Green bonds as hedging assets before and after COVID: a comparative study between the US and China. Energy Econ 104:105696. https://doi.org/10.1016/j.eneco.2021.105696

18. United Nations (UN) (2015) The Paris agreement (Adopted on 12 Dec 2015 at United Nations Climate Change Conference). Paris, France. Retrieved from https://www.un.org/en/climatechange/paris-agreement (Accessed 13 Nov 2022)

19. Segal M (2022) Moody's slashes sustainable finance market forecast as intensifying headwinds pressure volumes. ESG Today. Retrieved from https://www.esgtoday.com/moodys-cuts-sustainable-bond-forecast-as-intensifying-market-headwinds-pressure-volumes/ (Accessed 13 Nov 2022)

Facilitating SDG-Related Investment to Overcome Crisis Phenomena: The Key Tools of "Green" and "Pooled" Finance

Evgeniya A. Starikova

Abstract The majority of developed countries are currently formulating and adopting legislative norms that oblige companies to prevent or reduce negative externalities arising from doing business. However, government regulation of the activities of transnational business in this area is currently insufficient worldwide. In recent decades, some leading multinational corporations (MNCs) have acted on their own initiative, taking specific steps to minimize the negative impact of their business operations on society and the environment. In particular, the practice of participation in innovative mechanisms for financing the SDGs, which will be discussed in more detail in this research, has become increasingly demanded in the corporate environment. The research aims to identify and analyze the best practices of contributing to the achievement of SDGs performed by the private sector. Particular attention is paid to the experience of developing innovative institutions, including "green" finance instruments and "pooled" finance mechanisms, which are currently one of the most promising SDGs-related investment instruments that may contribute to the recovery of national economies.

Keywords Sustainable development · Economic development · Recovery plans · Green finance · Sustainable Development Goals · Agenda 2030 · Climate change · Innovative finance tools · Private finance · Blended finance

JEL Classification F23 · F640 · G24 · G38 · H12 · O16 · Q54

1 Introduction

In recent years, the most authoritative international organizations, on the sidelines of the summits held and within the framework of the documents they adopt, point to the ever-growing role of private investment in solving global problems and challenges

E. A. Starikova (✉)
MGIMO University, Moscow, Russia
e-mail: evan3132@gmail.com

© The Author(s), under exclusive license to Springer Nature Switzerland AG 2023
E. G. Popkova (ed.), *Smart Green Innovations in Industry 4.0 for Climate Change Risk Management*, Environmental Footprints and Eco-design of Products and Processes, https://doi.org/10.1007/978-3-031-28457-1_62

formulated in the current agenda, as well as the need to apply innovative approaches to financing the Sustainable Development Goals (SDGs), including SDG 13 that deals with the issue of climate change and the methods of adaptation to its post-effects. In particular, this is stated in such documents as the "Addis Ababa Action Agenda" [1], "The 2030 Agenda for Sustainable Development" [2], a document entitled "From billions to trillions" [3] issued by the world's largest financial institutions, etc.

International organizations and high-profile scientists consider "green" finance instruments and mechanisms of "pooled" financing as one of the most promising ones in terms of facilitating the recovery of national economies because they ensure the flow of additional private capital for development, make it possible to accumulate funds from various sources of financing (public, private, and philanthropic), and direct these funds to solve specific problems or the SDGs. In this research, the author analyzes the possibility of using these types of instruments to overcome economic crisis phenomena in the interest of implementing the SDGs under the current circumstances.

2 Methodology

In this research, the author used general scientific methods of induction, deduction, generalization, comparison, and classification, as well as methods of logical and statistical analysis.

The information and analytical basis of the research are the statistical materials and data of the United Nations, UNCTAD, World Bank, OECD, IEA, Sustainable Development Solutions Network (SDSN), Convergence, Business and Sustainable Development Commission, Global Commission on the Economy and Climate at the Brookings Institution, Accenture, and the Ministry of Economic Development of Russian Federation. The research also analyzes the data provided in financial and non-financial reports of the MNCs.

3 Results

From the very inception of the sustainable development concept in the 1960s–1970s, it posed a set of challenges for the activities of MNCs. Under pressure exerted on the corporate sector by political elites, human rights, and environmental organizations, the CEOs of large corporations began to participate in the decision-making process on the sidelines of international forums and conferences.

While the capabilities of governments and international institutions are limited when it comes to tackling global problems, large corporations possess massive financial, technological, expert, and organizational resources. That is the key reason why MNCs acquired the status of an important actor of the "green" pivot in world politics and economics [4].

It is important to note that environmentally friendly activities are non-philanthropic for corporations due to several reasons [5]. First, there is a whole range of benefits that an ecologically responsible company acquires that manifest in its increased credit ratings, investor confidence, customer loyalty, cost savings, the possibility of entering new "environmentally sensitive" sales markets, the ability to attract qualified specialists, etc. Thus, minimizing ecological footprint is currently a factor triggering the company's strategic development and strengthening business competitiveness.

For example, the German multinational company Siemens has been implementing its "environmental portfolio" over the past few years, producing the world's most energy-efficient gas turbines, cutting-edge technologies for water purification and desalination, and wind turbines that can reduce carbon dioxide emissions by 4 million tons annually [6]. Currently, the products and services produced by the company within its "environmental portfolio" represent the fastest-growing manufacturing segment of the business. From 2011 to 2019, more than 43% of the company's revenues came from sustainable products and services. In the fiscal year 2019, this segment accounted for 44.2% of the company's total revenues (Fig. 1) [6, 7].

Before the COVID-19 pandemic, UNCTAD estimated that the cumulative annual investment required to implement the SDGs in developing countries was $3.9 trillion, while the current volumes of allocated ODA and FDI covered about $1.4 trillion of these investment needs. Thus, the annual SDG financing gap in developing countries was estimated at $2.5 trillion [8], but it grew to $4.2 trillion in 2020 due to the COVID-19 crisis [9].

The calculations of the Global Commission on the Economy and Climate, made before the coronavirus pandemic, indicate that only 50% of this investment gap could be closed by the government sources of funding [10]. Thus, the annual volume of private investment required for achieving the SDGs is estimated at $1–1.5 trillion [11].

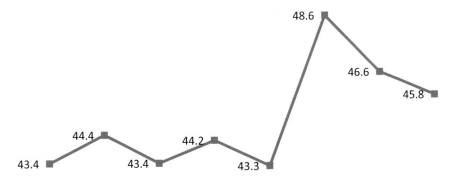

Fig. 1 Share of income received from the "environmental portfolio" in the total income of Siemens, %. *Source* Compiled by the author based on [6, 7]

Currently, the following two ways of business activities related to implementing the SDGs considering climate change—namely, "green" bonds and "blended" finance deals—seem to be the most promising in terms of private capital mobilization.

Since 2007, the largest investment banks and multilateral development banks have, for the first time, initiated the issuance of "green" bonds—the bonds in which proceeds are invested in projects aimed at environmental protection [12]. Since 2014, there has been a significant increase in private issuers of "green" bonds represented by transnational banks and MNCs. By the end of 2017, revenues from the sale of "green" bonds exceeded $380 billion [13]; today, "green" bonds account for about 2% of the debt securities market [14].

Many scientists assume that the financial sector, in general, and "green" banks, in particular, may become the "locomotives" of a sustainable recovery in the near future. Financing "green" power plants, promoting social investment in the basic infrastructure, or providing preferential prices for companies that promote the SDGs are what sustainable financing is, and it is a market that has grown significantly lately [15]. According to the IEA's data, "green" projects overtook hydrocarbon fundraising in 2021 for the first time ever [16].

It should be noted that institutional and legislative improvement in the formation of "green bonds" is needed in many countries to increase the feasibility and quantity of "green" projects and, as a result, the number of knowledge-intensive "green" jobs with an increased share of digital capital and high productive human capital. It could stimulate investments in the knowledge economy; the structural disbalances of the global labor market could be diminished [17, 18].

The Russian Federation also follows the global trend: in September 2021, Russian Prime Minister Mikhail Mishustin approved the criteria for selecting green projects, as well as projects in the field of sustainable development, for concessional financing through special bonds or loans. The Government Decree No. 1587 of September 21, 2021, prescribes specific quantitative and qualitative parameters that these projects must comply with. The document notes that a "green" project can be launched in the field of waste management, energy, construction, industry, transport, water supply, agriculture, biodiversity, and the environment [19].

In 2020–2021, VEB.rf, as a methodological center authorized by the Government of the Russian Federation for sustainable financial instruments [20], together with a number of ministries and departments, the Bank of Russia, the expert and business community, developed a national system of "green" financing. The basis of this system is the taxonomy of "green" projects and the "green" financing standard. In accordance with this taxonomy, projects of nuclear energy, ecotourism, hydrogen and gas motor transport, forest-climatic and agricultural projects, transport infrastructure projects, and projects that reduce carbon dioxide emissions will receive the "green" status.

Federal law "On conducting an experiment to limit greenhouse gas emissions in certain subjects of the Russian Federation" (No 34-FZ) [21] came into force on September 1, 2022. This event marked the beginning of an experiment to limit green-house gas emissions in the Sakhalin Region from September 1, 2022, to December 31, 2028. The Sakhalin Region is the first Russian region that has launched a pilot project to achieve carbon neutrality, which involves the creation of a hydrogen cluster, the construction of geothermal power plants and a wind park, and the production of "green" coal and "green" natural gas. It is planned to achieve carbon neutrality by introducing such mechanisms as emission quotas, mandatory carbon reporting by enterprises, and inventory of emissions and absorptions of greenhouse gases. Simultaneously, the relevant data on transactions with the fulfillment of quotas will be entered into a specially created national database of carbon units.

As for the "blended" finance deals, they seem to be increasingly important in the international development assistance agenda. According to the definition of the Global Commission on the Economy and Climate, "blended" finance involves the "strategic use of capital allocated from government or philanthropic sources in order to attract additional private funding needed to implement the SDGs." Under the Commission's calculations, this instrument can mobilize private capital up to $1–1.5 trillion annually, that is, the amount of funding required to implement the key SDGs in developing economies [10]. The key "suppliers" of commercial capital are institutional investors (transnational banks, insurance companies, asset managers, pension funds, sovereign wealth funds, etc.).

Notably, a number of "blended" finance institutions have already been established in the international climate finance architecture since climate change considerably affects the economies of developing countries. In the early 1990s, under the auspices of the World Bank, the Global Environment Facility (GEF) was established. It is a global mechanism that brings together resources from 183 countries and 18 specialized institutions, including the UN agencies and multilateral development banks.

Additionally, a global "blended" finance facility such as the Green Climate Fund (GCF) was established under the auspices of the United Nations, which funds low-carbon and climate-resilient projects to combat climate change in developing economies. With an emphasis on private funding sources, the GCF has established a private sector facility to mobilize private investment. However, it should be noted that the above-mentioned mechanisms have a very limited amount of accumulated funds.

Considering that the total capital volume based on GCF and GEF is about $5.5 billion [22, 23], it can be concluded that to create a full-fledged "pooled" finance mechanism to combat climate change, it is necessary to significantly increase the volume of capital mobilized from private sources of funding.

Simultaneously, in general, according to the estimates of scientists [24], if the total accumulated capital of these institutions increases to $6.6 billion, it seems possible to launch a pilot version of the global "pooled" finance mechanism aimed at combatting global climate change.

It is also worth mentioning that the climate solutions for developing countries initiated by the World Bank have been repeatedly criticized in recent years. The need to reform the activities of the World Bank and, particularly, its financial assistance to developing countries was discussed at COP27 held in November 2022 in Sharm el-Sheikh.

For nearly three decades, developing countries have sought financial assistance to compensate for "loss and damage" due to natural disasters and climate change-induced cataclysms. It should be noted that earlier, in the rhetoric of the international community, special attention was paid to the issue of providing developing countries with financial and technical assistance to reduce greenhouse gas emissions. However, at the past COP26 and COP27 conferences, the emphasis was placed precisely on the issue of compensation for losses and damage caused by natural disasters as a result of climate change and on the allocation of funds for special adaptation programs.

High-income developed countries have pledged to pay $100 billion annually to climate-affected developing countries starting in 2020. Nonetheless, until now, these obligations have not been fulfilled, and, in total, only $20 billion has been allocated for these purposes. In the final agreement of the summit in Sharm el-Sheikh, this financial obligation was surprisingly reaffirmed.

4   Conclusion

Currently, "green" investment and the creation of funds and mechanisms for "blended" or "pooled" financing is one of the most promising formats for inter-action between governments and MNCs in terms of implementation of the SDGs and overcoming the current economic crisis developments.

Apparently, "green" finance will gain momentum in the near future. The ambitious goal of transition to carbon-free energy by 2050–2060 committed by the countries that participated in the COP26 will require measures aimed at a radical transformation of the country's production, energy, and socio-economic models. "Green" projects are becoming increasingly demanded in Russia since the country confirmed its zero-carbon long-term commitment and is now planning to scale up such a phenomenon as "carbon-free clusters."

If the high rates of private capital mobilization are ensured, "blended" finance can be seen as one of the key instruments to overcome the financing gap for the 2030 Agenda in many countries. It is not for nothing that the largest transnational banks and corporations included in the most representative global sustainable development and CSR ratings are already among the most active institutional investors involved in "blended" finance deals.

In addition to stimulating extra funding from public sources, the issue of combatting climate change in the developed and developing world can be addressed by the creation of large "pooled" finance mechanisms. These mechanisms aimed at attracting private capital seem to be one of the most demanded formats of "blended" financing. In our opinion, these cooperation models between the state and the private sector deserve special attention. The research found that for the more efficient functioning of these institutions in the future, it is necessary to increase the volume of commercial capital mobilized compared to funds allocated from government sources.

Additionally, it is necessary to "consolidate" existing "blended" financing funds, which will specialize in creating public goods in the basic sectors of production and services such as agriculture, healthcare, education, water supply and sanitation, and renewable energy. It is assumed that if 25–30 large funds are formed, with a total investment of $1–10 billion, it will be possible to ensure the annual attraction of private capital at the rate of $1 trillion [25].

In this regard, the special role of "pooled" funding mechanisms is noted. When these kinds of environmental and climate partnerships are built, they can become an important financing tool for the SDGs complementing the efforts of the governments and business players.

The agreement on payments for losses and damage caused by global warming, adopted during COP27, was a breakthrough in resolving one of the most contentious issues in the UN climate negotiations. The parties are currently discussing the establishment of a committee with representatives of 24 countries who will work on creating a special fund to collect funds, as well as decide which countries and financial institutions should contribute there and determine for what purposes exactly these funds should be directed. Hopefully, this fund will become the embodiment of the long-awaited highly efficient "pooled" finance mechanism that will accumulate joint public and private efforts aimed at combatting the global issue of climate change.

References

1. UNCTAD (2015) Addis Ababa action agenda of the third international conference on financing for development (Addis Ababa Action Agenda) (adopted by Resolution on the General Assembly on 27 July 2015. 69/313.). New York, NY. Retrieved from https://unctad.org/mee tings/en/SessionalDocuments/ares69d313_ru.pdf (Accessed 7 Oct 2022)
2. United Nations (2015) Transforming our World: the 2030 agenda for sustainable development. United Nations, New York, NY. Retrieved from https://sdgs.un.org/2030agenda (Accessed 7 Oct 2022)
3. World Bank Group and IMF (2015) From billions to trillions: transforming development finance. Post-2015 financing for development: Multilateral development finance. Retrieved from https://olc.worldbank.org/system/files/From_Billions_to_Trillions-Transform ing_Development_Finance_Pg_1_to_5.pdf (Accessed 7 Oct 2022)
4. Starikova EA (2022) Sustainable development in a changing world. The role of government and business. KnoRus, Moscow, Russia
5. Afontsev SA (2005) Multinational companies and problems global management. In: Solovyov EG (ed) TNCs in the world politics and the world economy: problems, trends, prospects: collection of articles. IMEMO RAN, Moscow, Russia, pp 4–13. Retrieved from http://www.imemo.ru/files/File/ru/publ/2005/05003.pdf (Accessed 25 Nov 2022)
6. Siemens (2020) Siemens annual report 2020. Retrieved from https://assets.new.siemens.com/siemens/assets/api/uuid:45446098-6c39-45ba-a5fc-e5f27ebfa875/siemens-ar2020.pdf (Accessed 15 Oct 2022)
7. Statista (2020) Siemens AG's environmental portfolio revenue from FY 2011 to FY 2019. Retrieved from https://www.statista.com/statistics/292397/environmental-portfolio-revenue-of-siemens/ (Accessed 5 Oct 2022)
8. UNCTAD (2014) World investment report 2014. Investing in the SDGs: an action plan. UNCTAD, New York, NY; Geneva, Switzerland. Retrieved from https://unctad.org/system/files/official-document/wir2014_en.pdf (Accessed 20 Oct 2022)

9. OECD (2020) Global Outlook on financing for sustainable development 2021: a new way to invest for people and planet. OECD Publishing, Paris, France. https://doi.org/10.1787/e3c30a 9a-en

10. Bhattacharya A, Oppenheim J, Stern N (2015) Driving sustainable development through better infrastructure: key elements of a transformation program. Global Economy & Development Working Paper 91. Global Commission on the Economy and Climate, Washington, DC. Retrieved from https://newclimateeconomy.report/workingpapers/wp-content/uploads/sites/5/2016/04/Driving-sustainable-development-through-better-infrastructure.pdf (Accessed 12 Oct 2022)

11. Business & Sustainable Development Commission (Blended Finance Taskforce) (2018) Better finance better world: consultation paper of the blended finance taskforce. London, UK. Retrieved from http://s3.amazonaws.com/aws-bsdc/BFT_BetterFinance_final_01192018.pdf (Accessed 9 Oct 2022)

12. Zavyalova E, Studenikin N (2019) Green investment in Russia as a new economic stimulus. In: Sergi BS (ed) Modeling economic growth in contemporary Russia. Emerald Publishing Limited, Bingley, UK, pp 273–397

13. SEB Group (2017) The green bond 4Q 2017. Stockholm, Sweden. Retrieved from https://www.greenfinancelac.org/wp-content/uploads/2018/01/SEB_The_Green_Bond_December_2017-1.pdf (Accessed 18 Oct 2022)

14. Climate Bonds Initiative (2018) Green bonds market summary Q3 2018. Retrieved from https://www.climatebonds.net/files/reports/q3_2018_highlights_final.pdf (Accessed 8 Oct 2022)

15. Ivanov OV, Shamanina EA (2021) PPP as a tool to achieve sustainable development goals and implement the concept of "Quality infrastructure investments." In: Zavyalova EB, Popkova EG (eds) Industry 4.0: exploring the consequences of climate change. Palgrave Macmillan, Cham, pp 309–322. https://doi.org/10.1007/978-3-030-75405-1_28

16. IEA (2021) World energy outlook 2021. Retrieved from https://iea.blob.core.windows.net/assets/888004cf-1a38-4716-9e0c-3b0e3fdbf609/WorldEnergyOutlook2021.pdf (Accessed 7 Nov 2022)

17. Assylbayev AB, Safronchuk MV, Niiazalieva KN, Brovko NA (2023) Green transformation and the concept of energy efficiency in the housing sector. In: Lazareva EI, Murzin AD, Rivza BA, Ostrovskaya VN (eds) Innovative trends in international business and sustainable management. Springer, Singapore, pp 567–577. https://doi.org/10.1007/978-981-19-4005-7_61

18. Safronchuk MV, Ivanitskaya NV, Baibulov AK (2022) Global labor market and challenges of digitalization. In: Popkova EG (eds) Imitation market modeling in digital economy: game theoretic approaches. Springer, Cham, Switzerland, pp 142–150. https://doi.org/10.1007/978-3-030-93244-2_17

19. Government of the Russian Federation (2021) Decree "On approval of the criteria for sustainable (including green) development projects in the Russian Federation and the requirements for the verification system for sustainable (including green) development projects in the Russian Federation" (21 Sept 2021 No. 1587). Moscow, Russia. Retrieved from http://publication.pravo.gov.ru/Document/View/0001202109240043 (Accessed 28 Oct 2022)

20. Government of the Russian Federation (2020) Decree "On the coordinating role of the Ministry of Economic Development of Russia on the development of investment activities and attraction of extra-budgetary funds in sustainable (including green) development projects in the Russian Federation" (18 Nov 2020 No. 3024-r). Moscow, Russia. Retrieved from http://publication.pravo.gov.ru/Document/View/0001202011200033?index=1&rangeSize=1 (Accessed 27 Nov 2022)

21. Russian Federation (2022) Federal Law "On conducting an experiment to limit greenhouse gas emissions in certain subjects of the Russian Federation" (6 Mar 2022 No. 34-FZ). Moscow, Russia. Retrieved from https://www.consultant.ru/document/cons_doc_LAW_411051/ (Accessed 25 Nov 2022)

22. Global Environment Facility (GEF) (n.d.) Funding. Retrieved from https://www.thegef.org/about/funding (Accessed 21 Oct 2021)

23. United Nations (2019) United Nations, inter-agency task force on financing for development: Financing for sustainable development report 2019. United Nations, New York, NY. Retrieved from https://inff.org/assets/resource/financing-for-sustainable-development-2019.pdf (Accessed 6 Oct 2022)

24. Sachs JD, Schmidt-Traub G (2015) Financing sustainable development: implementing the SDGs through effective investment strategies and partnerships. Working Paper. SDSN, New York, NY. Retrieved from https://irp-cdn.multiscreensite.com/be6d1d56/files/uploaded/150619-SDSN-Financing-Sustainable-Development-Paper-FINAL-02.pdf (Accessed 15 Oct 2022)

25. Convergence (2018) Blended finance & SDG alignment. Data Brief. Retrieved from https://assets.ctfassets.net/4cgqlwde6qy0/5tk19IqmMEiiwsqUaKgqYk/a0e34cc2a5538069248f6eb41b6c5988/Convergence__SDG_Alignment__2018.pdf (Accessed 8 Oct 2022)

Global "Green Bond" Market Analysis and Perspective of Its Development

Alexey V. Logutov⊙

Abstract The paper aims to determine the relationship between the green bonds market development and the future of environmentally friendly development in the world. The author uses the methods of statistics and comparative analysis to determine the character of change in transparency of green bond issuers and its effect on the rate of market growth and the possibility of meeting global environmental goals. The research objects are markets of developed and newly industrialized countries. It is determined that as of 2022, the global green bond market is not developed to the necessary extent for achieving the goals of the Paris Agreement and UN Sustainable Goals. Most issuers do not disclose enough information on the green use of investor funds. In turn, there is a lack of trust from the investors toward the issuers of these bonds, thus not providing enough funds to fuel further market growth. Based on the analysis of the dynamics of change of the statistics, it is substantiated that under the proper influence of all bond issuers and supporters of sustainable growth in developed and developing countries, there is a good prospect of reaching environmentally friendly goals in the near future. All efforts should be put into the effective development of the legal structure of the market and the implementation of strict characteristics that green bond issues should correspond to. This will reduce institutional risks and make the market more transparent, thus attracting new investment and funding to drive the future development of ecological projects.

Keywords Green bonds · Energy transition · Green transformation · Sustainable growth · Climate change · Debt market

JEL Classification G23 · G24 · G28 · L38 · L51 · Q01 · Q42

A. V. Logutov (✉)
MGIMO University, Moscow, Russia
e-mail: a.v.logutov@gmail.com

1   Introduction

A "green bond" is a financial asset, the proceeds from the placement of which are directed exclusively to the financing of new or existing green projects. These bonds must comply with the four key elements of the Green Bond Principles.

The green bond market was established in 2007 when the World Bank issued the first climate bond. From there on, the market and its infrastructure gradually developed. However, the main driver for its development came in 2015–2016. Over the past decade, the green bond market has evolved from a minor to a generally accepted method of financing used in international capital markets. This tool is crucial for achieving the goals stated in the Paris Agreement on Climate Change and the UN Sustainable Development Goals.

In just ten years, the green bond market has become one of the fastest-growing markets. Green bonds are currently used worldwide as one of the main sources of financing for a wide range of issuers, including companies from the renewable energy sector, sovereign organizations, and issuers wishing to comply with the ESG principles.

Despite the relevance and prospects for the development of the market, many concerns could weaken confidence in green bonds.

2   Materials and Method

Due to the problem's versatility, the degree of its elaboration is considered in several directions. Contemporary trends in the development of the financial market were considered by such scientists as Berzon [6], Ershov [11], Mirkin [18], Shokhin et al. [22], and others.

Most of the relevant coverage and detailed analytics of specific topics (e.g., sustainable energy and green energy) is covered by research conducted by Western investment banks—in particular, research of Bank of America [4], Barclays [5], BNP Paribas [7], Credit Suisse [10], Goldman Sachs [13], HSBC [14], JPMorgan Chase & Co. [15], Morgan Stanley [19], and others.

Additionally, works of foreign economists are devoted to the theory and practice of the formation and development of the financial market: Bailey [3], Keynes [17], and others.

Theoretical studies of the functioning and regulation of financial markets are contained in the works of Akerlof [1], Volker and Harper [23], and others. The publications of Eugene et al. [12], Kemme [16], and Shiller [21] are devoted to the analysis of the macroeconomic role of financial markets.

3 Discussion

In recent years, the green bond market has been growing rapidly. In 2020, the total volume of global green bond issuance was around $300 billion (compared to $167.3 billion in 2018) (Fig. 1). The Paris Agreement and the UN Sustainable Development Goals were the main drivers of market growth.

The largest volume of green bond issues falls on Europe, about $450 billion (Fig. 1). The second and third largest regions are North America ($230 billion) and the Asia-Pacific region ($220 billion).

However, at the country level, the leader is the USA, which has issued green bonds worth more than $200 billion. Simultaneously, most of the funds raised from green issues are directed to the energy, construction, and transportation sectors (about 80% of the total volume). Issuers of green bonds are mainly corporations, developers, and governments (Fig. 2).

Fig. 1 The volume of the global green bond market by region. *Source* Compiled by the author based on Climate Bonds Initiative 2021 [9]

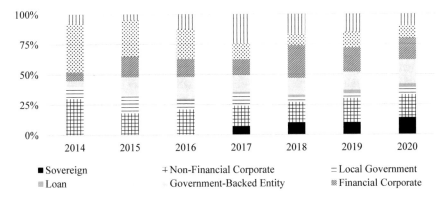

Fig. 2 The volume of the global green bond market by issuers. *Source* Compiled by the author based on Climate Bonds Initiative 2021 [9]

Today, the green bonds market has significant problems and risks of further growth. Despite the degree of development of the market, there is no generally accepted legal definition of a green bond. The International Capital Market Association (ICMA) has created principles that bond issues must comply with: disclosure of information about the future "green" use of proceeds, a report of the actual use of proceeds, and an official secondary opinion by an independent third party confirming the environmental friendliness of the bond.

The market is still young, and its institutional formation lags behind the rapid growth dynamics. The market developed so that half of the issuers satisfied none of the above criteria, which is important for a green bond. Disclosure of necessary and critically important information for investors is still not included as a mandatory parameter in the issue documents. Therefore, the non-use of bond proceeds for the declared green projects is not an event of a violation. On the contrary, the issue documents often prescribe that failure to report on the use of borrowed funds is not an event of default.

The institutional gap creates several problems for investors. For example, the owner of a green bond, who purchased it as a part of a specific investment strategy, may not suspect that he or she could actually buy an asset that violates his or her investment declaration, thereby creating the possibility of regulatory risk. To avoid a violation, the investor might be forced to immediately sell the asset and potentially incur losses caused by current market conditions.

In this regard, the main risk of the green bond market is investor trust and the issuer's reputation. As long as there are no contractual conditions guaranteeing that bonds sold as green remain truly green throughout their life, there is much room for mis-selling by issuers.

4 Results

As part of the risk analysis of the industry, the author identified two main factors that form the trust and faith of investors in the green bonds market: transparency of issuers and false positioning.

Reporting on green bonds is based on two simple principles: disclosure of information about the intended use of the funds raised and the actual use of proceeds.

The green bonds principles do not determine which use of capital will be considered green, thus creating legal and institutional problems. Companies have their own criteria for determining an acceptable green project with different detalization levels, expanding the range of possible projects. In turn, the investor finds out what his or her money was actually used for only post factum. This is not transparent and is a constraint for an investor concerned about not corresponding to his or her investment principles.

Despite the fact that reporting the actual use of funds is one of the principles of ICMA, market practice in relation to the frequency and level of disclosed information suggests the opposite. A recent study by the Climate Bonds Initiative showed that only

Table 1 Reporting statistics of raised funds

Reporting scope		UoP reporting	Impact reporting	Both
Number of issuers	Reporting	251	194	172
	Non-reporting	116	173	195
	% reporting	68%	53%	47%
Number of bonds	Reporting	715	1514	501
	Non-reporting	1190	391	1404
	% reporting	38%	79%	25%
Amount issued ($ billion)	Reporting	223	219	186
	Non-reporting	58	62	95
	% reporting	79%	78%	66%

Source Compiled by the author based on Climate Bonds Initiative 2019 [8]

68% of green issuers provide regular reporting (Table 1). Solving these problems at the legislative level would reduce institutional risks. This would increase the quality and feasibility of green projects and, as a result, the number of green jobs with an increased share of digital capital and highly productive human capital [20].

However, many areas of SDG implementation are subject to the incompleteness of the legislative elaboration, such as the inclusion of new indicators, standards, and criteria for energy efficiency and energy profitability in the legislative framework of energy use [2].

5 Conclusion

To achieve the goals of the Paris Agreement and EU Goals of Sustainable Development, it is necessary to maintain a high level of financing for environmentally friendly projects, which can only be achieved through the success and popularity of green bonds as an investment instrument. Currently, the green bond market is not developed enough to provide issuers with sufficient financing for green projects.

Concerns about confidence in green bonds risk jeopardizing the sustainable financing project as a whole, which could have serious consequences for everyone.

First, the industry should be regulated to a greater extent: the issuer must confirm compliance with the principles of green ICMA bonds, be included in the index of green bonds, or have confirmation of the bond's environmental friendliness.

It is also necessary to attract public sector enterprises with longer investment horizons and clear mandates to the market. As soon as the market has several large issuers that consistently adhere to all the principles of ICMA, smaller players will look less environmentally friendly compared to them and try to follow the best. In other words, the mechanism of positive selection should be activated, thus increasing institutional attractiveness (i.e., institutional risks could be reduced).

As soon as issuers become more transparent, investors will begin to develop trust. More investors will invest in green bonds as part of their strategies and principles, thereby providing funds for developing new projects.

In turn, the state will be able to introduce tax incentives for issuers and investors. Nowadays, this has not been implemented in any major jurisdiction. One of the arguments put forward against such a tax incentive is precisely the fact that bonds marked as green have no guarantees that the proceeds will be used in accordance with the disclosed green way. This argument can be easily overcome by providing tax benefits only for those bonds that have clear legal conditions focused on environmental friendliness. This would have a double effect: it would strengthen the green bond market and help maintain clear public policy goals.

References

1. Akerlof GA (2009) An economic theorist's book of tales. Cambridge University Press, Cambridge, UK. https://doi.org/10.1017/CBO9780511609381 (original work published 1984)
2. Assylbayev AB, Safronchuk MV, Niiazalieva KN, Brovko NA (2023) Green transformation and the concept of energy efficiency in the housing sector. In: Lazareva EI, Murzin AD, Rivza BA, Ostrovskaya VN (eds) Innovative trends in international business and sustainable management. Springer, Singapore, pp 567–577. https://doi.org/10.1007/978-981-19-4005-7_61
3. Bailey SJ (2003) Strategic public finance: textbook. Palgrave Macmillan, London, UK. https://doi.org/10.1007/978-1-4039-4394-1
4. Bank of America (2022) ESG-themed issuance. Retrieved from https://investor.bankofamerica.com/fixed-income/esg-themed-issuances. Accessed 2 Dec 2022
5. Barclays (2021) Barclays green issuance framework. Retrieved from https://home.barclays/content/dam/home-barclays/documents/investor-relations/debtinvestors/creditratings/20211021-Barclays-Green-Issuance-Framework-July-2021.pdf. Accessed 1 Dec 2022
6. Berzon NI (ed) (2019) Stock market: textbook, 5th edn. Urait, Moscow, Russia. Retrieved from https://mx3.urait.ru/uploads/pdf_review/5ED8DECE-964E-440A-AF76-E8210628C0BF.pdf. Accessed 1 Dec 2022
7. BNP Paribas (2021) BNP Paribas green bond. Retrieved from https://invest.bnpparibas/document/green-bond-reporting-methodology-notes-2021. Accessed 2 Dec 2022
8. Climate Bonds Initiative (2019) Climate bonds standard version 3.0. Retrieved from https://www.climatebonds.net/files/files/climate-bonds-standard-v3-20191210.pdf. Accessed 4 Dec 2022
9. Climate Bonds Initiative (2021) Sustainable debt global state of the market. Retrieved from https://www.climatebonds.net/files/reports/cbi_global_sotm_2021_02h_0.pdf. Accessed 1 Dec 2022
10. Credit Suisse (2021) Green bond report. Retrieved from https://www.credit-suisse.com/media/assets/about-us/docs/investor-relations/debt-investors/green-bonds/cs-green-bond-report-august_2021.pdf. Accessed 2 Dec 2022
11. Ershov MV (2019) Ten years after the global crisis: risks and prospects. Vopr Ekon 1:37–53. https://doi.org/10.32609/0042-8736-2019-1-37-53
12. Eugene F, Fama F, Miller M (1971) The theory of finance. Dryden Press, Hinsdale, IL. Retrieved from http://www.library.fa.ru/files/Fama_theory.pdf. Accessed 3 Dec 2022
13. Goldman Sachs (2021) Sustainability issuance report. Retrieved from https://www.goldmansachs.com/investor-relations/creditor-information/sustainability-issuance-report-2021.pdf. Accessed 1 Dec 2022

14. HSBC (2021) HSBC green bond report. Retrieved from https://www.hsbc.com/-/files/hsbc/investors/fixed-income-investors/green-and-sustainability-bonds/pdfs/211214-hsbc-green-bonds-report-2021.pdf?download=1. Accessed 2 Dec 2022

15. JPMorgan Chase & Co. (2021) Green bonds annual reports. Retrieved from https://www.jpmorganchase.com/content/dam/jpmc/jpmorgan-chase-and-co/documents/green-bond-annual-report-2021.pdf. Accessed 4 Dec 2022

16. Kemme DM (2005) Financial structure and economic growth: a cross-country comparison of banks, markets and development. Comp Econ Stud 47:710–712. https://doi.org/10.1057/palgrave.ces.8100125

17. Keynes JM (1936) The general theory of employment, interest and money. Biznesom Publishing, Moscow, Russia. Retrieved from https://publications.hse.ru/mirror/pubs/share/folder/jifpw3hgvj/direct/118199413?ysclid=lbxmwjnslc742115999. Accessed 3 Dec 2022

18. Mirkin YM (2002) Stock market: textbook. Finance Academy under the Government of the Russian Federation, Moscow, Russia. Retrieved from http://www.mirkin.ru/_docs/metodika3.pdf. Accessed 1 Dec 2022

19. Morgan Stanley (2021) Social bond impact report. Retrieved from https://www.morganstanley.com/assets/pdfs/Morgan_Stanley_2021_Social_Bond_Impact_Report.pdf. Accessed 1 Dec 2022

20. Safronchuk MV, Ivanitskaya NV, Baibulov AK (2022) Global labor market and challenges of digitalization. In: Popkova EG (ed) Imitation market modeling in digital economy: game theoretic approaches. Springer, Cham, Switzerland, pp 142–150. https://doi.org/10.1007/978-3-030-93244-2_17

21. Shiller RJ (1998) Macro markets: creating institutions for managing society's largest economic risks. Oxford University Press, Oxford, UK

22. Shokhin AN, Akindinova NV, Astrov VYu, Gurvich ET, Zamulin OA, Klepach AN et al (2021) Macroeconomic effects of the pandemic and prospects for economic recovery. Voprosy Ekonomiki 7:5–30 (Proceedings of the roundtable discussion at the XXII April international academic conference on economic and social development). https://doi.org/10.32609/0042-8736-2021-7

23. Volker PA, Harper C (2018) Keeping at it: the quest for sound money and good government. PublicAffairs, New York, NY

Monitoring of Strategic Criteria in the Form of an ESG Investment Accounting System

Elena V. Zeninaⓘ**, Marina V. Safronchuk**ⓘ**, Anna I. Kramarenko**ⓘ**, and Natalia A. Brovko**ⓘ

Abstract In this research, the authors monitored strategic criteria in the form of an ESG investment accounting system with the aim of applying it to companies in the construction industry in Kyrgyzstan. First, the authors presented an analysis of the external environment to identify predictable opportunities and threats, with the disclosure of Ansoff's method for constructing the product-market matrix. Corresponding matrices for construction enterprises were constructed. The identified threats and opportunities of the external environment were distributed according to the degree of their influence on the life of enterprises in the construction industry. The probability of using the identified opportunities is determined. Further, using M. Poter's mechanism of the model of competitive forces, the authors studied relevant tools for analyzing the internal environment. The application of governance standard for ESG investment is considered through a tool for effective monitoring of the quality of services provided, considering the problems and risks associated with the human factor. The authors confirmed that monitoring strategic criteria allows for the necessary adjustments in the activities and functioning of enterprises in today's environment and increases their adaptability and competitiveness. The authors concluded on the high degree of competition in the construction industry of Kyrgyzstan. The authors also confirmed the economic necessity and technical feasibility of applying governance standard for ESG investment for construction companies in the Kyrgyz Republic.

E. V. Zenina · A. I. Kramarenko
International University of Kyrgyzstan, Bishkek, Kyrgyzstan
e-mail: elena.zenina@iuk.kg

A. I. Kramarenko
e-mail: anna.kramarenko@iuk.kg

M. V. Safronchuk (✉)
MGIMO University, Moscow, Russia
e-mail: msafronchouk@gmail.com

N. A. Brovko
Kyrgyz-Russian Slavic University named after B. Yeltsin, Bishkek, Kyrgyzstan
e-mail: nbrovko@list.ru

© The Author(s), under exclusive license to Springer Nature Switzerland AG 2023
E. G. Popkova (ed.), *Smart Green Innovations in Industry 4.0 for Climate Change Risk Management*, Environmental Footprints and Eco-design of Products and Processes, https://doi.org/10.1007/978-3-031-28457-1_64

Keywords Strategic criteria · ISO standards · ESG investment · Accounting system · Construction market · Construction industry · Construction company · SWOT analysis · Porter's model · Product-market matrix · Competitive advantages · Threats · Opportunities

JEL Classification F63 · F64 · L10 · L15 · L16 · L53 · L70 · L74 · M2 · M15 · M12 · M14 · M20 · Q56

1 Introduction

New strategic goals and objectives related to the implementation of the concept of sustainable development and the need to respond to external shocks (e.g., the global financial and energy crises, the need for an energy transition, the COVID-19 pandemic, and exacerbation of geopolitical instability) require adjustments in the activities of enterprises. For successful entrepreneurship, it is especially important to expand the social and environmental responsibility of an enterprise in today's environment, flexibility, and dynamism [11]. Accordingly, it requires adjustments in the development strategy, parameters and aspects of annual reports, and monitoring tools.

"A development strategy is a program of actions that describes a specific and ... sustainable sequence of strategic actions of an enterprise in order to develop it" [5].

There are many successful examples of high adaptability to the external and internal environment among small and new companies. For example, the closure of catering companies during the quarantine period prompted the owners to work for delivery. The number of food delivery services began to grow.

The world news is already talking about a fast food delivery service in 15 min in California. To create it, the owner of the service, Russian Vitaly Aleksandrov, attracted $2 million in investments. He explains the choice of a new startup by the crisis in the restaurant market that arose at the beginning of the pandemic. After losing customers, he focused on the grocery delivery segment, which is growing despite the crises [8]. The company's strategy is to profit from the organization of super-fast delivery: users order food delivery in a mobile application, and the courier brings the order to their home in 10–15 min.

After the strategy of a single enterprise is developed and fixed in the strategic plan, the implementation of specific activities begins to achieve the enterprise's strategic goals.

During the global COVID-19 pandemic, businesses faced major challenges, especially related to sanitation, safe spaces, and the transformation of physical products into digital ones. The main tasks of the companies were to maintain the client base and indicators of income and profit. However, the restrictions of 2020 also brought new opportunities.

Some enterprises immediately adapted to the new requirements of the environment, developed new skills among employees, and changed the priorities and needs of customers. If the company does not analyze how to use these opportunities faster, it will surely lose out to competitors.

Considering the specifics of the construction market in Kyrgyzstan, the authors propose their own development of the stages of monitoring strategic criteria in the form of ESG investment in the accounting system of an enterprise in this industry.

2 Methodology

The research methodology is based on a theoretical and statistical analysis of the development of the construction industry in Kyrgyzstan. The authors studied the competitive environment and the position of construction companies. The method of comparative analysis was applied.

A financial analysis of the company was made beforehand. SWOT analysis was used to identify strengths, weaknesses, threats, and opportunities for the construction company. The investigation involved methods for assessing the internal and external competitiveness of the company. In particular, for the analysis of the microenvironment, the authors applied the model of competitive forces by M. Porter; the method of constructing I. Ansoff's product-market matrix was also used. To develop the stages of implementation of strategic criteria in the form of ESG investment in the accounting system, the authors studied ISO standards and the PDCA algorithms embedded in them.

To study the features of green building in the world, its main advantages and trends, foreign sources were analyzed, in particular research articles, reports of consulting companies, the World Green Building Council, and the Report of the World Bank Office in the Kyrgyz Republic. To conduct an institutional analysis, the authors studied the legislative framework of the Kyrgyz Republic, which regulates the work of companies in the construction industry.

3 Discussion

Considering the large volume of the construction market and high internal competition, as well as the risks of the human factor, the authors offer the following stages of monitoring the strategic criteria of the accounting system for ESG investment for a construction market enterprise:

1. Stage 1: to analyze the external environment to identify predictable opportunities and threats;
2. Stage 2: with Ansoff's method of constructing the product-market matrix to identify the possibilities of the external environment and rank them according to the degree of their influence on the life of enterprises and the probability that enterprises would be able to use this opportunity;

3. Stage 3: to analyze the internal environment of a construction company using the mechanism of the model of competitive forces by M. Porter;
4. Stage 4: application of the Governance standard in ESG investment through a tool for effective monitoring of the quality of services provided, considering the problems and risks associated with the human factor.

The stages of implementing strategic criteria in the form of ESG investment in the accounting system include the following algorithms (PDCA—Plan, Do, Check, Inform), incorporated in the ISO standards: plan (strategy development), execute (strategy implementation), control (introduction of a scorecard), and react (informing) [15, p. 30].

The charter of the enterprises of the construction industry reflects the goals of the creation of the enterprise and its subject. It is determined that the purpose of creation in the first place is to make a profit in the Kyrgyz Republic through the implementation of commercial activities not prohibited by the legislation of the Kyrgyz Republic. It is noted that the main activity is the construction of civil, industrial, and socially significant facilities.

When enterprises enter the construction market, we suggest using the classic SWOT analysis model, which presents four basic positions [6, p. 156]:

1. Strengths of the company:

   • Mobility, universality, and flexibility;
   • Knowledge and understanding of the country's construction market;
   • Individual approach to each client.

2. Weak sides of the company:

   • Financially "illiterate" state;
   • Dependence on the client's budget;
   • Freelance system of work of employees—waiting for their free time.

3. Possibilities for the company:

   • Loyal customer base;
   • New and small and medium-sized enterprises to meet small needs for construction services;
   • Construction enterprises need high-quality materials and highly qualified personnel for the planned budget.

4. Threats for the company:

   • Highly competitive environment;
   • Dependence of construction budgets on the political and economic situation in the country.

Most enterprises in Kyrgyzstan may not have a developed strategy. Their activities are usually concentrated on the current client base, composed and promoted by the experience and competence of external managers. However, the problem is that external managers usually demand too high compensation, which is much higher than the market estimates.

When independently developing a new strategy for entering the construction market, it is necessary to consider aspects of the external environment, including political, legal, economic, socio-cultural, and technological factors. The legal framework of the developed strategy should include an assessment of the loyalty of the legal environment and an analysis of the relevant laws.

Thus, the legislative framework of the Kyrgyz Republic contributes to the development of small and medium-sized businesses—the Law "On urban planning and architecture of the Kyrgyz Republic" [1], which regulates the activities of the enterprise, and the Law "On competition of the Kyrgyz Republic" [2], in which the state supports competition as a free competitiveness of business entities on the market, stimulating the production of goods required by the consumer. The foreign economic policy of the state in the National Development Strategy of the Kyrgyz Republic for 2018–2040 [3] aims to promote the export of domestic products to foreign countries and attract foreign investment, advanced technologies, and innovations to the national economy of the country, which also corresponds to favorable conditions for the emergence and development in the country of international goods and services, as well as investment and social projects.

Unfortunately, the Kyrgyz Republic is highly dependent on the global political and economic situation and crises. One of the most important factors for business development is tracking the country's economic factors. In general, inflation in the Kyrgyz Republic amounted to 9.7% in 2021, which is the highest rate among the countries of the Eurasian Economic Union, where the inflation target for 2021 was 8.7% [14]. The main impact on the acceleration of inflation in the Kyrgyz Republic, as noted on the website of the Eurasian Economic Commission, is the increase in food prices—almost 5.7% [14].

In today's world, the qualities of a successful enterprise are flexibility and reaction to the external environment and the ability to focus on the enterprise's strengths and be in constant development. For the reaction to the external environment not to be delayed, it is necessary to build trusting and partnership relations with the team. Each employee must be a manager—a manager of work and department, taking responsibility for the quality and delivery time of the services provided [9, p. 122]. In such a team, it is easier to keep abreast; information is not delayed or hidden, and it is openly discussed within the team.

The political crisis in the Kyrgyz Republic in October 2020 and news about the COVID-19 pandemic increased the influence of the media as a source of the first signals of possible threats.

The practice of identifying new opportunities or threats helps managers make adequate strategic decisions. As a recognition of external threats by weak signals, the authors propose an action plan for assessing the first signals of possible threats presented in Table 1.

Table 1 shows that online publications and Telegram news channels, where news spreads faster, become the source of information about new opportunities or threats. Signal strength is determined by analyzing publications in the media. If information

Table 1 Evaluation of the first signals of possible threats

Kind of signal	The nature of measures as instability and the real danger is increasing					
	Permanent monitoring	Determination of signal strength	Reducing strategic vulnerability	To increase response flexibility	Development of preparatory plans and programs	Plans of practical measures and its implementation
Possible political aggravation in the country. News is first published on Telegram channels and sent via WhatsApp groups						
Possible political aggravation in the country. News is first published on Telegram channels and sent via WhatsApp groups						
The scale of the danger or the new opportunity takes concrete shape—the problem is discussed in traditional media and evening news on TV						
Ways to solve the problem have been established; the company has plan B. Employees go to remote work. Signals come from all media						

Source Compiled by the authors

about the same problem is published several times in different sources of information, the problem becomes serious and apparent. Further, the enterprise's management begins to take measures to reduce strategic vulnerability. If the news is broadcast in all media, including in evening news, the company moves to an increased level of flexibility and begins to develop preparatory protective programs. With clear signals about possible threats, the enterprises have already developed plans for practical actions and measures. There is a redistribution of resources, saving of current assets, and expectation of an improvement in the external environment.

4 Results

Nowadays, in view of the activities of competitors, customers, and suppliers in the given socio-economic condition of the Kyrgyz Republic, the main goal of construction enterprises is to increase business activity, considering the preservation of operating assets. On the other hand, in creating content, it is important to decentralize management and reduce restrictions when the production process is underway. In times of crisis, a company with a limited budget can break through with its product, stand out from the competition, and make itself known.

"A successful strategy for a construction enterprise should be based on a certain course of development formed by the mission of the construction enterprise. The mission contributes to the formation or consolidation of a certain image of the enterprise in the construction industry in the representation of the subjects of the external environment. As a possible version of the mission, the following wording is proposed:

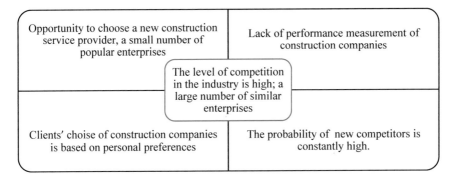

Fig. 1 Analysis of the competitive environment of enterprises in the construction industry. *Source* Developed and compiled by the authors based on [10]

"Construction Holding" contributes to the development and promotion of clients' business, creating original products, using advanced technologies, and developing effective innovative strategies. The key values of the construction industry enterprise "Construction Holding" are as follows: attention to the client, professionalism of the employees of the construction enterprise, creativity, and focus on results" [7].

The inclusion of strategic ESG investment criteria into the company's reports and monitoring of their implementation involves the company in the active promotion of Sustainable Development Goals, making this company more transparent and attractive. Multilateral trust in such a company is increased on the part of authorities, partners, and clients. Therefore, the introduction of criteria for ESG investment and ESG reporting is becoming an important factor in the company's competitiveness, as evidenced by the experience of corporations in other industries [13].

A relevant tool for analyzing the microenvironment is the model of competitive forces developed by Porter [10] (Fig. 1).

The analysis presented in Fig. 1 revealed a high risk of new competitors entering the industry. It was found that there are practically no barriers to the entry of new competitors into the industry. Due to the highly competitive environment, the struggle for numerous customers is greatly increased. Enterprises have low profitability from existing transactions. The customer struggle between competitors can go in different directions: costs, product/service quality, implementation time, application of innovative technologies, staff qualifications, etc. There is already a threat of competitive substitutes appearing on the market. This threat lurks in new products and services developed by competitors, which can be innovative or revolutionary when the attention and budgets of the client can switch from placement on traditional sites to the online environment. Therefore, there is a need for continuous training of employees in the skills of selling Internet resources.

The power of buyers lies in manipulating their budget. Knowing that there are many enterprises in the Kyrgyz Republic and much fewer real clients interested in promotion, the client begins to look for the best offer based on the price. Kyrgyz enterprises have to conclude contracts at cost or even lower, which means dumping.

We must consider that some companies may be intermediaries, the profitability of which is a margin or percentage of the transaction between the client and the supplier. Suppliers may unilaterally reduce the interest rate or discount for the agency.

An analysis of the existing client base shows that companies can work with local and international companies, commercial companies, and social projects. The services of construction enterprises are used by many consumers, especially legal entities interested in a long-term cooperation.

The implementation of an effective development strategy will entail the following changes:

- An increase in market share;
- An increase in market share in its segment;
- A change in management methods and principles;
- An increase in the profitability of a construction enterprise;
- An improvement in the liquidity of a construction industry enterprise.

To successfully achieve one's goal, it is also advisable to apply the method of constructing the "product-market" matrix by Ansoff [4, p. 108], according to the chosen strategy. The matrix of the market development strategy was built by the method of expert assessment by the management of the construction industry enterprise. The possibilities of the external environment were distributed according to the degree of their influence on the life of the construction enterprise (strong, moderate, and small) and the probability that the company will be able to take advantage of the opportunity (high, medium, and low).

It was determined that the new market development strategy is suitable for construction enterprises. Entry opportunities are great. The company must continue to develop a new market for its growth.

Governance standard for ESG investment was used to monitor the quality of services provided effectively. On this basis, the strategy for developing new services was worked out.

The success of achieving the strategic goal of construction enterprises depends on many factors. For example, sales effectiveness and, as a result, the number and budget of existing contracts depend on the human factor, particularly the management and qualifications of employees. There may be such problems as unpreparedness and incompetence of managers of various management levels to new standards and the tasks set to increase the main indicators of construction industry enterprises. The dependence of construction enterprises on employees is generally very high. For a more accurate definition, the authors conducted a risk analysis (Table 2).

The analysis confirms the need to implement a sales strategy and a system of sales standards and improve customer service. A regulated and executed sales process minimizes the fall in the plans set for the sales department and standardizes the execution of tasks.

The system of social and governance standards for ESG investment corresponds to the normal cycle of company management. It consists of developing a goal (tree of goals), planning the achievement of the set results, monitoring and control of the

Table 2 Analysis of the problems in the social standard for ESG investing

The problem	The description of the problem	The solution to the problem
The employees of the construction industry are not prepared for new standards	1. Lack of understanding and motivation among employees in achieving new tasks 2. Unpreparedness and lack of knowledge among employees to work according to new standards	1. Involving employees in setting goals and objectives when developing a strategy. Implementation of a transparent system of remuneration and bonuses to employees for the implementation of the plan 2. Education, instruction, and advanced training of employees on an ongoing basis, organization of quarterly training
Low level of planning and control	1. Lack of leadership planning culture 2. Irrelevant and inconsistent sales plans	1. Approval of monthly sales plans with the team of the construction industry enterprise 2. Permanent and continuous monitoring of construction
Lack of monitoring	Lack of implemented tools for analysis and monitoring of the external environment	Analysis of information on the political and economic situation in the country on changes in demand in the construction services market. Constant adjustment of the strategic objectives set

Source Compiled by the authors

results achieved, constant analysis of external and internal factors, and adjustment of decisions and tasks in accordance with the changing reality.

With the effective and long-term achievement of the goal, customer satisfaction with all services provided and received is expected.

The recommended system of sales standards is represented by three indicators:

1. Communication with clients:

- Increase in the number of clients;
- Introduction of cold calls—20 calls per day with new clients;
- Having three meetings a day with clients.

The following tasks are performed:

- Filling the database of clients to determine the status of negotiations;
- Implementation of a single database for new and existing customers;
- Fixing customer needs and reasons for refusal.

2. Assessment of customer needs:

- Having 2–3 scheduled meetings with clients.

The following tasks are performed:

- Preparation and presentation of services according to the clients' needs;
- Negotiation of contract terms.

3. Contract implementation:

- Provision of services in time and in the specified volume according to the concluded contract.

The following tasks are performed:

- Monitoring the quality of services provided;
- Filling in the customer satisfaction checklist.

5 Conclusion

Based on the monitoring of strategic criteria in the form of ESG investment in the accounting system, it can be concluded that the market development strategy is the most appropriate. It satisfies the goals and vision of the construction enterprise. As part of this strategy, the company will be able to enter the B2B market and serve new market segments. The analysis of the external environment consisted of the continuous collection of information through employees within the construction industry and through the media. Enterprises in the construction industry are encouraged to respond to changes in the external environment on weak signals.

According to forecasts, the consequences of the COVID-19 pandemic will be felt for another 2–3 years. Enterprises in the construction industry need to maintain the stability of their performance, pay special attention to monitoring serious consequences, and plan and work out possible cases of development of the situation. Special attention should be paid to minimizing the costs and expenses of firms, centralizing the security of management [12], and introducing new methods of assessment and statistical indicators of its activities into management analysis.

References

1. Kyrgyz Republic (1994) The law "On urban planning and architecture of the Kyrgyz Republic" (11 Jan 1994, No. 1372-XII, as amended 6 Jan 2021, No. 3). Bishkek, Kyrgyzstan. Retrieved from http://cbd.minjust.gov.kg/act/view/ru-ru/112164?cl=ru-ru. Accessed 1 Nov 2022
2. Kyrgyz Republic (2011) The law "On competition of the Kyrgyz Republic" (22 July 2011, No. 116, as amended 5 Apr 2013, No. 47). Bishkek, Kyrgyzstan. Retrieved from http://cbd.minjust.gov.kg/act/view/ru-ru/203356. Accessed 1 Nov 2022

3. National Statistical Committee of the Kyrgyz Republic (2018) National development strategy of the Kyrgyz Republic for 2018–2040 (approved by the Decree of the President of the Kyrgyz Republic on 31 Oct 2018, No. 221). Bishkek, Kyrgyzstan. Retrieved from mfa.gov.kg/uploads/content/1036/3ccf962c-a0fc-3e32-b2f0-5580bfc79401.pdf. Accessed 1 Nov 2022

4. Ansoff I (2009) Strategic management: textbook. Peter Publishing House, St. Petersburg, Russia

5. Attokurova NS, Mun VG (2020) Problems of development of innovative technologies and innovative activity at the present stage. J "Vestnik KRSU" 20(3):9–13

6. Demina ID, Polulekh MV, Gordova M, Sorokina VV (2022) Management (tactical and strategic) accounting: textbook. Vuzovskoe obrazovanie Publishing House, Saratov, Russia

7. Economist (2021, Mar 19) Inflation will be 7% this year. Retrieved from https://econom ist.kg/novosti/2021/03/19/inflyaciya-v-etom-godu-sostavit-7-prognoz-nacbanka/. Accessed 9 Nov 2022

8. Kametdinovm N (2021, Apr 28) A 15-minute grocery delivery service in California with Russian roots has raised $2 million. Forbes. Retrieved from www.forbes.ru/newsroom/karera-i-svoy-biznes/428121-servis-dostavki-produktov-za-15-minut-v-kalifornii-s. Accessed 7 Oct 2022

9. Kochurova TS (2021) Strategy for managing innovation processes: textbook. Litres Publishing House, Moscow, Russia

10. Online Encyclopedia of Marketing and Advertising "PowerBranding" (n.d.) Five competitive forces analysis model by Michael Porter. Retrieved from http://powerbranding.ru/biznes-ana liz/porter-model/. Accessed 10 Nov 2022

11. Popkova EG, Zavyalova E (eds) (2021) New institutions for socio-economic development: the change of paradigm from rationality and stability to responsibility and dynamism. De Gruyter, Berlin, Germany; Boston, MA. https://doi.org/10.1515/9783110699869

12. Portal on Strategic Management and Planning "STPLAN.RU" (2020, Nov 19) Why uncertainty is an ideal environment for strategic planning. Retrieved from http://www.stplan.ru/articles/practice/pochemu-neopredelennost-idealnaya-sreda-dlya-strategicheskogo-planirova niya.htm. Accessed 11 Nov 2022

13. Starikova EA, Shamanina EA (2021) Corporate practice of implementing measures to combat climate change in the Russian oil and gas companies. In: Zavyalova EB, Popkova EG (eds) Industry 4.0: exploring the consequences of climate change. Palgrave Macmillan, Cham, Switzerland, pp 221–233. https://doi.org/10.1007/978-3-030-75405-1_20

14. World Bank in the Kyrgyz Republic (2021) Overview of the Kyrgyz Republic: economy. Retrieved from https://www.vsemirnyjbank.org/ru/country/kyrgyzrepublic/overvi ew#3. Accessed 19 Nov 2022

15. Zub AT (2018) Strategic management: textbook and workshop for academic undergraduate students. Urayt Publishing House, Moscow, Russia

Current Trends in the ESG Bond Market

Andrey N. Liventsev⬤

Abstract The relevance of the research topic is confirmed by the statistics of increasing volumes of responsible investment, the annual improvement of the regulatory framework within the concept of sustainable development, and the reports of the World Economic Forum on critical global risks. In this paper, the author puts forward hypotheses on the interrelation of inflation rates, military expenditures, and dynamics of COVID-19 pandemic diseases with the volume of ESG bonds issue, examining their attractiveness at present, when the leaders of most countries give preference to short-term stability. The research aims to analyze the dynamics of the ESG bond market. The scientific novelty of this research is that the author reveals the correlation of dependence and statistical significance between indicators of green investments and inflationary pressure, military expenditures of countries, and COVID-19 incidence. Moreover, global trends of ESG bonds market development are determined. The research methods include conceptual and empirical research methods, a set of methods of economic and statistical analysis, and the analysis of literary and electronic sources. The research shows a direct rather than inverse correlation between green bonds and military spending. This means that governments are indeed concerned about the environment and the consequences of military action, which offers hope for further positive trends in ESG bond issuance. Additionally, it is found that COVID-19 disease rates have no influence on the dynamics of ESG bond issuance in 2022.

Keywords ESG bonds · Responsible investing · Sustainable development concept · Investments · Correlation

JEL Classification C10 · G10 · O44 · O51 · O52 · O53 · O57 · Q01 · Q54 · Q56 · Q58

A. N. Liventsev (✉)
MGIMO University, Moscow, Russia
e-mail: liventsevandrey98@gmail.com

E. G. Popkova (ed.), *Smart Green Innovations in Industry 4.0 for Climate Change Risk Management*, Environmental Footprints and Eco-design of Products and Processes, https://doi.org/10.1007/978-3-031-28457-1_65

1   Introduction

According to the World Economic Forum's (WEF) Global Risk Report 2022, environmental and social sector issues rank in the top 8 in terms of scale and likelihood of realization. The whole world is concerned about combating climate change, lack of livelihoods, and other global risks, which can be realized either on a time horizon of 0–2 years or 5–10 years.

The UN climate summit kicked off in Sharm el-Sheikh on November 6, 2022. While discussing the climate agenda, calls were heard to force Russia to negotiate and stop its aggression. The organization believes that fewer promises and commitments are likely to be made at this summit than in previous years. The new British Prime Minister has advocated the development of energy independence. Ahead of the summit, the UN circulated a billboard stating that the Russia-Ukraine conflict has exacerbated the global inflation and energy and food crises. A German representative said the country's climate targets would be reduced in the short term. The US-China climate working group announced in 2021 suspended its activities. Poland's president has said that Polish and EU residents are more concerned about their standard of living than the climate agenda. The World Bank has decided to change its policy on climate finance (RBCF). The bank now plans to provide money not before but after achieving its climate change outcome and meeting national greenhouse gas reduction targets.

Analysts and experts of the Analytical Center "Forum" [1], "Bloomberg" [2], "Cbonds" [3], the World Bank [6], the UN [9], the World Economic Forum [10], and many others contributed to the development and study of the concept. Based on statements by world leaders, the author poses the following questions:

- Will multinational corporations (MNCs) maintain their commitment to sustainability?
- Will external factors (e.g., inflation, military spending, and the impact of the COVID-19 pandemic) affect the demand for ESG bonds;
- What is the current appeal of ESG bonds?

The term ESG bonds refer to a type of debt security that is a loan for projects that meet the principles of sustainable development. There are green bonds (GBP), transition bonds, social bonds (SBP), sustainability bonds (SBG), and sustainability performance bonds (SLBs). The advantage of these securities in times of crisis is clear: bonds are a more reliable fixed-income financial instrument. ESG bond issuers have better disclosure practices. The disadvantages are that there is no standardized report on the implementation of the ESG strategy. Responsible finance meets the needs and values of the Millennial and Zoomer generations. With recovery from the COVID-19 pandemic, geopolitical tensions, and undermining international cooperation, many governments opt for short-term stability. Various economic institutions estimate that $120 trillion to $290 trillion in the capital will be needed to successfully achieve the UN goals, 20% of which will come from financing through debt markets.

Over time, the ESG agenda will become increasingly demanding: regulatory aspects will evolve, and the number of theoretical frameworks will increase.

The world does not have a common understanding of the terms green and social agenda. However, everyone understands the importance of exploring the issue in depth. There is still a lack of commonly agreed corporate information, metrics, and methodologies that could provide a basis for comparing different products or strategies. The author suggests that the role of responsible finance, Russia's high profile on the climate and social agenda, and the increasing cooperation with Asian countries are huge opportunities. Thus, the role of transition finance tools should not be underestimated. Nowadays, many companies have questioned the need to invest in green projects for fear of diminishing returns. Nevertheless, the fact that ESG bonds have the undisputed leading growth rate is undeniable.

2   Methodology

As the global community became aware of the need for a common approach to accounting for responsible investment, including bond instruments, this responsibility was assigned to the International Capital Markets Association (ICMA) and the Climate Bonds Initiative (CBI). According to the standards developed by these international organizations, an important aspect of classifying bonds as green or social is their consistency with national or supranational taxonomies of green, social, or transition projects. In analyzing the dynamics of ESG bond issuance, the author referred to data from the Climate Bonds Initiative, according to which the aggregate volume of green, social, transition, and SLBs exceeded $3.5 trillion by September 30, 2022. The USA, France, and China are the countries with the largest share of the ESG bond market. Debt securities were issued in 63 currencies. Some 82% of the market for sustainability bonds is in euros, US dollars, and Chinese yuan. The popularity of the climate agenda in Europe has encouraged the euro to finance 42% of the market. Total ESG bonds in Q3 2022 reached $152.3 billion, 35% lower than in Q2 2022 and 45% lower than in Q3 2021. Geopolitical and macroeconomic factors have left their mark on the volume of debt issuance in all areas of responsible investment. By the end of Q3, total debt securities linked to the ESG sector amounted to $635.7 billion. Green bonds accounted for more than half of the total (52.3%). Sustainability bonds accounted for 22.4%, social bonds for 14.8%, SLBs for 10%, and transition bonds for the smallest share at 0.5%. The green bond market expanded rapidly; by the end of Q3, it had surpassed $2 trillion. However, to make a significant contribution to addressing global climate risks, annual issuance needs to be at least $5 trillion from 2025 onwards [5]. Let us look at the dynamics and key trends of the ESG bond market in the world.

2.1 Asian Countries

The Monetary Authority of Singapore (MAS) announced the debut offering of sovereign green bonds. The issue amounted to 2.4 billion Singapore dollars (S$) (about $1.75 billion) with a yield of 3.04%. The 50-year S$2.35-billion bond was offered to institutional investors, with the remaining S$50 million bond to be offered to individual investors. The Aug-72 bond is the first 50-year bond issued by the Singapore government and the longest green bond issued by a sovereign to date. The Singapore authorities have previously announced that they plan to issue up to S$35 billion worth of green bonds by 2030. Singapore's Green Bond Concept [7], released on the eve of the offering, outlines the government's intended use of green bond proceeds, a governance structure for selecting eligible projects, an operational approach to managing green bond proceeds, and annual distribution and impact reporting obligations. The bond proceeds will be used to finance costs in support of Singapore's 2030 Green Plan, including the Jurong Region High-Speed Rail Line and Cross Island Line.

The China Securities Regulatory Commission (CSRC) has instructed the Shanghai and Shenzhen stock exchanges to revise the rules to translate green bond issuance in line with the China Green Bond Principles, published in July 2022 [4]. According to the principles developed by the Chinese regulator, 100% of the proceeds from green bond issues must be used for green purposes or projects. Previously, only issuers of bank green bonds were required to use 100% of the proceeds for green projects. The requirement of 70% is applied to listed corporate bonds (regulated by the CSRC), and a minimum of 50% is applied to green bonds of state corporations (regulated by the National Development and Reform Commission, NDRC). Thus, Chinese supervisory authorities have ensured uniform requirements regarding the share of proceeds allocated for green purposes.

2.2 Australia

Australian supervisory authorities are introducing requirements for investment managers to combat greenwashing. The new rules were developed in response to increased investor interest in ESG and climate risk, which led to a proliferation of investment products and services positioned as green, sustainable, or contributing to zero emissions without clear rules to inform investors of the methodology and criteria used related to sustainability. The regulator required emission reduction targets to be set for investment portfolios with a particular focus on emissions assessment and climate risk reporting in line with the TCFD Recommendations.

2.3 Russia

In Russia, the ESG bond market has begun to develop since 2017. The starting point can be considered the development of the Moscow Exchange sector responsible for the environment, environmental protection, and socially significant projects. The Russian government has also been actively involved in developing the market. At the end of 2019, the Bank of Russia approved the Standards for issuing green and social bonds. Since November 2020, VEB.RF has been the methodological center for financial instruments in responsible investment. In 2021, a taxonomy was developed that approved the eligibility criteria for green projects. In the author's view, the ESG bond market cannot develop solely on the initiative of issuers and investors. The securities market in the sustainable development sector needs government support measures. For example, banks demand the abolition of mandatory provisioning in the Central Bank of the Russian Federation for green and social bonds whose issuers they are, the abolition of taxation on income from the sale of such securities, and compensation of costs associated with the issue of ESG bonds. Despite external factors, Russia recognizes the importance of the climate agenda and remains committed to the UN's sustainable development goals: the plans to achieve carbon neutrality by 2060 are unchanged. The rationale behind the transition bond format is to promote early decarburization of the economy, which would then lead to lower anthropogenic impacts on the climate. In Russia, this issue is particularly important because the country's carbon-intensive industries play a key role. By the end of 2022, the area of responsible investing should be enriched with standards for ESG climate transition bonds and other sustainability-related bonds. The development of a regulatory framework is essential for developing transition bonds to understand real sector ownership of the UN vision and goals and minimize the risk of a geopolitical slowdown in the growth of ESG bonds.

2.4 France

In May 2022, France floated inflation-linked green bonds, becoming the first sovereign issuer to do so. The 4.3-billion-euro deal attracted a 27.5-billion-euro order book thanks to growing investor concerns about the transition to a green economy and rising inflation. The bond is indexed by the European Consumer Price Index. Thus, coupon payments provide investors with protection against rising inflation. The bond price was set slightly outside the inflation-linked FRTR curve, but it moved inside after a month. The order book included 230 names; more than half of the bonds were placed with green investors. At the end of June 2022, France was the largest issuer of sovereign green bonds, with 49.4 billion euro ($56 billion) maturing between two bullet bonds and a new inflation-linked deal. The French state agency Caisse d'Amortissement de la Dette Sociale (CADES) is the largest issuer of 2022 social bonds, having issued a total of $35.9 billion in euros, US dollars, and Swedish krona, overtaking the second largest issuer of the year, the Asian Development Bank (ADB), which had placed bonds worth a total of $4.0 billion by the end of September. In the third quarter alone, CADES placed two bonds: a 5-year 3-billion-euro ($2.9 billion) bond and a 10-year 5-billion-euro ($4.9 billion) bond. The proceeds of these bonds will be used for health, employment, and equality spending. The UoP is in line with Sustainable Development Goals (SDGs): SDG 1 (No poverty), SDG 3 (Good health and well-being), SDG 10 (Reduced inequalities), and SDG 11 (Sustainable cities and communities).

2.5 Germany

German development bank KfW was the largest non-sovereign issuer of green bonds in Q3 2022, raising a total of $4.3 billion. Most of this amount came from a 7-year 4-billion-euro ($4.1 billion) deal. Smaller amounts were issued in forints and Hong Kong dollars; a Chinese renminbi bond was reopened. By the end of September, Climate Bonds had registered $8.3 billion in green bonds for 2022 from KfW, making it the third largest issuer of green bonds for 2022 after the European Union ($17.3 billion) and the European Investment Bank (EIB) ($9.6 billion).

2.6 USA

Wells Fargo placed its debut sustainability bond in mid-August. The $2-billion 4-year deal accounted for more than half of the $3.9 billion sustainability bond volume in the third quarter from 16 issuers from the corporate financial sector. Other issuers included Berkshire Hills Bancorp ($100 million), Gunma Bank ($70 million), and Raiffeisen Bank Romania ($104 million).

2.7 Brazil

At the end of August 2022, Brazilian meat processor JBS closed its fifth SLB deal worth $968.5 million. The deal was tied to a 19.1% emissions reduction target. Climate Bonds analysts note that this deal is a revocable deal, with the first revocation date falling on January 15, 2027.

2.8 Results

In answering the main questions of the research, the author examines the relationship between green bond issuance and global military expenditure over the time horizon of 2015–2021 and determines the statistical significance of the resulting relationship.

In the calculation presented in Table 1, the author obtained the following empirical regression coefficients:

$$b = (\overline{yx} - \overline{y} * \underline{x})/\sigma_x^2 = (405,630.652 - 209.771 * 1880.67)/7971.06 = 1.3951$$

$$a = \overline{y} - b * \underline{x} = 209.771 - 1.3951 * 1880.67 = -2413.9441$$

The empirical regression equation is as follows:

$$\hat{Y} = -2413.94 + 1.3951x$$

The author calculated a linear pairwise correlation coefficient: the relationship between trait Y and factor X is high and direct:

$$r_{xy} = b * \frac{\sigma_x}{\sigma_y} = 0.914$$

The coefficient of determination confirms that 83.55% of global green bond issuance depends on global military expenditures and 16.45% on other factors. Next, the author checks the significance of the correlation coefficient. The hypotheses are as follows:

- H_0: $r_{xy} = 0$, the random nature of the revealed dependence;
- H_1: $r_{xy} \neq 0$, statistical significance.

$$t_{nab} = r_{xy} * \frac{\sqrt{n-2}}{\sqrt{1 - r^2(xy)}} = 5.039$$

Using Student's table with significance level $\alpha = 0.05$ and degrees of freedom k = 5, the author found t_{crit}:

$$t_{crit}(n - m - 1; \ \alpha/2) = t_{crit}(5; \ 0.025) = 3.163$$

where m = 1—the number of explanatory variables.

Since $|t_{nab}| > t_{crit}$, the correlation coefficient is statistically significant.

On average, the estimated values deviate from the actual values by 22.07%. Since the error is greater than 7%, this equation is not desirable as a regression. These statistics seem quite interesting to the author because the study showed a direct rather than inverse correlation of green bond issuance and military spending, which means

Table 1 Construction of regression equation

t	Date	Issue of green bonds in the world, in billions of U.S. dollars (Y)	Military expenditures in the world, in billions of dollars (X)	XY	X^2	Y^2	$\hat{Y}$	$|(Y - \hat{Y})/Y|$
1	2015	42	1778	74,661	3,160,040	1764	66	0.57
2	2016	81	1787	144,736	3,192,869	6561	79	0.03
3	2017	156	1810	282,390	3,276,788	24,336	111	0.29
4	2018	171	1859	318,638	3,455,993	29,378	180	0.05
5	2019	267	1932	514,910	3,733,088	71,022	282	0.06
6	2020	270	1992	536,909	3,969,020	72,630	365	0.36
7	2021	482	2007	967,172	4,026,363	232,324	385	0.20
SUMM	28	1468	13,165	2,839,415	24,814,160	438,015	1468	1.54
Average value	4	210	1881	405,631	3,544,880	62,574	210	0.22
σ		136	89					
σ^2		18,570	7971					
a		−2413.9	$\hat{Y} = -2413.94 + 1.3951x$					
b		1.4						
r_{xy}		0.9						
R_{xy}		83.5%						
A		22.1%						

Source Developed and compiled by the author based on data from the Climate Bonds Initiative [5] and Stockholm International Peace Research Institute (SIPRI) [8] that governments are concerned about the environment and military consequences, which gives hope for a further positive trend in ESG bonds.

Consider the relationship between ESG bond issuance and the incidence of COVID-19.

In the calculation presented in Table 2, the author obtained the following empirical regression coefficients:

$$b = (yx - y * x)/\sigma_x^2 = 0.000004$$

$$a = y - b * x = 209.771 - 1.3951 * 1880.67 = 31.9449$$

The empirical regression equation is as follows:

$$\hat{Y} = 31.9449 + 0.000004x$$

The author calculated a linear pairwise correlation coefficient: the relationship between trait Y and factor X is weak and direct:

$$r_{xy} = b * \frac{\sigma_x}{\sigma_y} = 0.2216$$

The coefficient of determination confirms that 4.91% of global ESG output depends on COVID-19 incidence rates and 95.09% on other factors.

Next, the author tests the significance of the correlation coefficient. The hypotheses are as follows:

- H_0: $r_{xy} = 0$, the random nature of the relationship detected;
- H_1: $r_{xy} \neq 0$, statistical significance.

$$t_{nab} = r_{xy} * \frac{\sqrt{n-2}}{\sqrt{1 - r^2(xy)}} = 0.682$$

Using Student's table with significance level $\alpha = 0.05$ and degrees of freedom k $= 9$, the author found t_{crit}:

$$t_{crit}(n - m - 1; \; \alpha/2) = t_{crit}(9; \; 0.025) = 2.685$$

where m = 1—the number of explanatory variables.

Since $|t_{nab}| < t_{crit}$, the correlation coefficient is of the random nature of the relationship identified and is not statistically significant.

On average, the estimated values deviate from the actual values by 43.73%. As the error is greater than 7%, it is not desirable to use this equation as a regression.

The research result shows that COVID-19 indicators do not affect the dynamics of ESG bond issuance in 2022.

Table 2 Construction of the regression equation

t	Date	Global issuance of ESG bonds (Y)	COVID-19 incidence rate (X)	XY	X²	Y²	Ŷ	|(Y − Ŷ)/Y|
1	Jan. 22	41.3	1,229,430	50,779,147.3	1,511,498,124,900	1705.9	36.8	0.1
2	Feb. 22	44.2	3,124,549	138,067,571.2	9,762,806,453,401	1952.6	44.3	0.0
3	Mar. 22	33.9	1,657,586	56,149,068.2	2,747,591,347,396	1147.4	38.5	0.1
4	April 22	44.0	1,067,949	46,970,532.9	1,140,515,066,601	1934.4	36.2	0.2
5	May 22	59.8	457,853	27,402,044.2	209,629,369,609	3581.9	33.7	0.4
6	June 22	47.3	524,681	24,822,658.1	275,290,151,761	2238.2	34.0	0.3
7	July 22	30.8	936,463	28,860,853.2	876,962,950,369	949.8	35.6	0.2
8	Aug. 22	18.7	909,315	16,963,271.3	826,853,769,225	348.0	35.5	0.9
9	Sept. 22	44.0	652,293	28,683,932.4	425,486,157,849	1933.7	34.5	0.2
10	Oct. 22	18.9	239,344	4,517,378.7	57,285,550,336	356.2	32.9	0.7
11	Nov. 22	12.6	382,157	4,824,350.0	146,043,972,649	159.4	33.5	1.6
SUMM	66	395.5	11,181,620	428,040,807.4	17,979,962,914,096	16,307.6	395.5	4.8
Average value	6	36.0	1,016,510.9	38,912,800.7	1,634,542,083099.6	1482.5	36.0	0.4
σ		13.8	775,401.6					
σ²		190.1	601,247,654798.8					

(continued)

Table 2 (continued)

		$\hat{Y} = 31.9449 + 0.000004x$
A	31.9449	
B	0.000004	
r(xy)	0.2216	
R(xy)	4.91%	
A	43.73%	

Source Developed and compiled by the author based on data from Climate Bonds Initiative [5]

3   Conclusions

The USA, France, and China are the countries with the largest ESG bond market share. Debt securities were issued in 63 currencies. Some 82% of the market for sustainability bonds is in euros, US dollars, and Chinese yuan. The popularity of the climate agenda in Europe has encouraged the euro to finance 42% of the market. The green bond market expanded rapidly; by the end of Q3, it had surpassed $2 trillion. However, to make a significant contribution to addressing global climate risks, annual issuance needs to be at least $5 trillion from 2025 onwards [5].

The research results show the following:

- There is a direct rather than inverse correlation between green bond issuance rates and military spending, meaning that governments are genuinely concerned about the environment and the impact of military action, which gives hope for a further positive trend in ESG bonds;
- COVID-19 incidence rates have no impact on the dynamics of ESG bond issuance in 2022.

Despite external factors, Russia recognizes the importance of the climate agenda and remains committed to the UN Sustainable Development Goals: the plans to achieve carbon neutrality by 2060 are unchanged. The formation of a regulatory framework is essential for developing "transition" bonds to recognize real sector ownership of the UN vision and goals and minimize the risk of ESG bond growth decline due to the geopolitical situation.

References

1. Analytical Centre "Forum" (2022) Responsible investing and ESG standards: events and trends. Moscow, Russia. Retrieved from https://mfc-moscow.com/assets/files/analytics/ESG%20MONITORING/monitoring_ESG_October_2022.pdf. Accessed 15 Nov 2022
2. Bloomberg (n.d.) Environmental, social & governance (ESG). Retrieved from https://www.bloomberg.com/professional/solution/sustainable-finance/. Accessed 3 Nov 2022
3. Cbonds (n.d.) Cbonds estimation. Retrieved from https://cbonds.ru/cbonds_estimation/. Accessed 21 Nov 2022
4. China Securities Regulatory Commission (2022, Jan 27) CSRC's annual work conference sets priorities for 2022. Retrieved from http://www.csrc.gov.cn/csrc_en/c102030/c1805625/content.shtml. Accessed 21 Nov 2022
5. Climate Bonds Initiative (2022) Q3 2022 market summary. Retrieved from https://www.climatebonds.net/files/reports/cbi_susdebtsum_highlq32022_final.pdf. Accessed 21 Nov 2022
6. Inderst G, Stewart F (2018) Incorporating environmental, social and governance (ESG) factors into fixed income investment. World Bank Group Publication, Washington, DC. Retrieved from https://documents1.worldbank.org/curated/en/913961524150628959/pdf/Incorporating-environmental-social-and-governance-factors-into-fixed-income-investment.pdf. Accessed 10 Oct 2022

7. Monetary Authority of Singapore. Public offer now open for individual investors. Retrieved from https://www.mas.gov.sg/news/media-releases/2022/singapore-prices-2-4-billion-50-year-inaugural-sovereign-green-bond-public-offer-now-open-for-individual-investors. Accessed 21 Nov 2022
8. Stockholm International Peace Research Institute (SIPRI) (n.d.) SIPRI military expenditure database. Retrieved from https://milex.sipri.org/sipri. Accessed 21 Nov 2022
9. United Nations (2022) The sustainable development goals report 2022. Retrieved from https://unstats.un.org/sdgs/report/2022/The-Sustainable-Development-Goals-Report-2022.pdf. Accessed 21 Nov 2022
10. World Economic Forum (2022) Global risks report 2022, 17th edn. Retrieved from https://www3.weforum.org/docs/WEF_The_Global_Risks_Report_2022.pdf. Accessed 21 Nov 2022

Perspectives on Combating Climate Change in the Economy and Business 5.0 in the Fifth Industrial Revolution (Conclusion)

The environmental footprints of Industry 4.0 in the Decade of Action are largely driven by the use of climate-smart technology and climate-responsible innovation in the economy and business. Corporate environmental responsibility is currently one of the important components of the digital competitiveness of business. Therefore, climate responsibility must be reflected in the eco-design of products and processes. The Fourth Industrial Revolution laid a solid foundation for the green economy and climate-resilient business in the Decade of Action. However, the launched processes will get a new impetus to development in the long term (up to 2050 and beyond).

The prospects for combating climate change in the Economy and Business 5.0 in the context of the Fifth Industrial Revolution are even more impressive. By socializing AI into sustainable communities, the climate agenda will gain even more mass support. Given that the first signs of the Fifth Industrial Revolution are already visible and its manifestations will intensify in the Decade of Action, this raises new research questions.

One of these questions is the potential to involve artificial intelligence in the creation of climate-smart innovations and the development of eco-design of products and processes. The potential of R&D automation is still unknown. Some scholars point to the limitations of artificial intelligence and machine learning, while others highlight the growing creativity of artificial intelligence that creates art objects and writes music.

The question is whether an unconventional view from artificial intelligence can bring something new and useful to the task of combating climate change. So far, smart technology has made it possible to automate environmental monitoring. It is also used in the implementation of green innovations. This provides ever greater results, suggesting that the circular practices of the future after the Decade of Action will be saturated with even more intense climate innovations continuously generated by the automation of the R&D process.

E. G. Popkova (ed.), *Smart Green Innovations in Industry 4.0 for Climate Change Risk Management*, Environmental Footprints and Eco-design of Products and Processes, https://doi.org/10.1007/978-3-031-28457-1

Another question concerns how the digital economy will change under Industry 5.0. The high energy intensity of automation is still a serious disadvantage of artificial intelligence from the perspective of its use in the fight against climate change. Nevertheless, this shortcoming can be overcome if new clean energy solutions can be found that make automation more affordable and efficient to use.

This book is a platform for the search for answers to the research questions raised. It is suggested that future research be devoted to working on these research questions.